UNIVERSITY OF ROCHESTER LIBRARIE
3 9087 01015077 1
T5-CFR-486

Fundamentals of Medical BACTERIOLOGY and MYCOLOGY

Fundamentals of Medical BACTERIOLOGY and MYCOLOGY

QUENTIN N. MYRVIK, PH.D.
Professor
Department of Microbiology and Immunology
Bowman Gray School of Medicine
Wake Forest University
Winston-Salem, North Carolina

RUSSELL S. WEISER, PH.D.
Emeritus Professor of Immunology
Department of Microbiology
University of Washington School of Medicine
Seattle, Washington

Second Edition

Lea & Febiger

Philadelphia 1988

Lea & Febiger
600 Washington Square
Philadelphia, PA 19106-4198
U.S.A.
(215) 922-1330

First Edition, 1974
Reprinted, 1976
Reprinted, 1978
Second Edition, 1988

Library of Congress Cataloging in Publication Data

Myrvik, Quentin N., 1921–
Fundamentals of medical bacteriology and mycology.

Includes bibliographies and index.
1. Bacteriology, Medical. 2. Medical mycology.
I. Weiser, Russell S., 1906– . II. Title.
[DNLM: 1. Bacteriology. 2. Mycology. QW 4 M998f]
QR46.M97 1987 616′.014 87–2866
ISBN 0–8121–1087–0

PRINTED IN THE UNITED STATES OF AMERICA

Print number: 5 4 3 2 1

Preface

The concerns about medical education expressed in the first edition of this text in 1974 are still fully applicable to the second edition. The paragraph on medical perspectives continues to deserve emphasis, especially because of the ceaseless emergence of new pathogens and the profound societal impact exerted by new advances in the control of infectious diseases, as depicted in Chapter 1.

In the second edition we have renewed our view that students of medicine profit most from texts in medical microbiology in which due emphasis is placed on the fundamentals that determine the natural history of disease. Special attention has been placed on mechanisms of pathogenesis and host resistance, particularly at the cellular and molecular levels. We continue to believe that fundamental knowledge provides the most valuable resource to predict the outcome of disease as well as the most effective base for diagnosis, therapy and research of infectious disease.

All of the chapters in the second edition have been revised and in many cases totally rewritten. We have included new information pertaining to the new technologies that are rapidly being applied to the field of microbiology.

New information on virulence factors and sites of tissue injury is described and discussed in the context of devising measures to prevent infections and to arrest the development and progress of disease.

Knowledge of the complex interactions of the components of the immune system has been included, particularly as it applies to opportunistic pathogens. In this regard, the devastating outcome of infection with human immunodeficiency virus, which leads to AIDS and the attending severe immunodeficiency with unrelenting microbial infections, represents the ultimate model of immunodeficiency and fatal opportunistic infections.

New advances for controlling infectious diseases continue apace, largely as a result of the application of modern DNA techniques concerned with the identification, isolation and production of protective epitopes and and monoclonal Abs for respective use as vaccines and specific antisera; the burgeoning field of molecular biology has provided techniques for the cloning of DNA segments and the synthesis of pure immunogens or single protective epitopes (Ag determinants) using recombinant DNA technology. Furthermore, amino acid sequencing of purified epitopes has facilitated the synthesis of immunogenic (protective) peptides which may prove useful as specific vaccines.

We sincerely hope that the second edition will be of value to students of medicine and related biologic sciences.

Winston-Salem, North Carolina — Quentin N. Myrvik
Seattle, Washington — Russell S. Weiser

Contributors

Chapter 3 Bacterial Viruses, Henry Drexler, Ph.D., Professor, Department of Microbiology and Immunology, Bowman Gray School of Medicine

Chapter 4 Bacterial Genetics, Henry Drexler, Ph.D.

Chapter 31 Rickettsia and Coxiella, Henry Drexler, Ph.D

Chapter 32 Chlamydia, Henry Drexler, Ph.D.

Chapter 6 Antimicrobial Chemotherapy, Joseph E. Johnson, III, M.D., Dean, University of Michigan School of Medicine, Ann Arbor, MI

Chapter 10 Streptococcus, Ivo van de Rijn, Ph.D., Associate Professor, Department of Microbiology and Immunology, Bowman Gray School of Medicine

Chapter 9 Staphylococcus, Arnold S. Kreger, Ph.D., Professor, Department of Microbiology and Immunology, Bowman Gray School of Medicine

Chapter 35 Miscellaneous Pathogenic Bacteria, Arnold S. Kreger

Contents

1

THE IMPACT OF MICROBIOLOGY ON THE HEALTH OF MAN

A. MILESTONES IN THE DEVELOPMENT OF MEDICAL MICROBIOLOGY

Among the major afflictions of man—famine, pestilence and war—the greatest progress has been made in controlling pestilence; in terms of lives saved, measures for controlling infectious diseases stand as the most important achievements in the history of medicine. Even more remarkable is the fact that essentially all of these achievements have been accomplished within the last century. No longer do pandemics of infectious diseases sweep continents with great destruction of life, such as occurred when the black plague crossed Europe in the early 14th century killing one-fourth or more of the population; no longer do chronic diseases such as syphilis and tuberculosis ravage populations on a worldwide basis, as in Koch's time in the 1880s when one of every three persons died of tuberculosis by middle age; and no longer are certain acute infections, such as streptococcal meningitis, invariably fatal.

The great progress made in the control of infectious diseases stemmed from the discovery of bacteria by Anton van Leeuwenhoek in 1683—a clever man of great talent but modest means, whose hobby was microscopy. Leeuwenhoek presented the first reliable evidence of the existence of bacteria. He was a self-trained expert at lens grinding, and among the hundreds of single-lens microscopes he built many were superior to any others of his time, having effective resolution powers up to 300×. Although secretive about his techniques, which probably included some form of darkfield illumination, his scientific reports were voluminous and comprised some 200 letters to the Royal Society of London written over a period of many years. He examined essentially every conceivable object that struck his fancy from the hairs on a louse to the "beasties" or "animalcules" (small animals) in scrapings from his teeth. His discourses on the shapes and movements of various bacteria of the mouth were remarkably accurate; he described them as varying greatly in their movements, as being a million times smaller than the eye of a louse, and as being more numerous in a man's mouth than "all of the men in the kingdom." The only reward of his hobby was the reward of discovery; but what a reward it must have been, for he spent essentially every spare moment at his microscopes during many years extending from 1660 until nearly the time of his death at the age of 91 years in 1723. He was elected to membership in the Royal Society of London, to which he bequeathed many of his microscopes, the remainder, some 247, being sold at auction for 61 pounds.

The germ theory of disease was proposed centuries before Leeuwenhoek's time (Varro, 116–26 B.C.; Columella, 60 B.C.) and was most clearly conceived by the illustrious poet-physician Gerolamo Fracastoro in 1546, who expressed the view that infection is transmitted by the tiniest of particles *(seminaria)* that we cannot perceive with our senses, *particulae minimae et insensibilis.* He envisioned infectious agents to be specific living organisms capable of multiplying and spreading within the body. Despite the fact that the germ theory was well documented in Fracastoro's book *De Contagione et Contagiosis Morbis* in 1547, the discovery of bacteria by Leeuwenhoek did not promptly spark the idea that these "animalcules" could be the causative agents of infectious disease. Instead microbiology languished for almost two centuries as a purely descriptive and abstract science literally awaiting the perfection of the compound microscope and the development of pure culture techniques. Other investigators had only limited success in confirming Leeuwenhoek's observations. It is unfortunate that he was so secretive about his "darkfield technique," for had he made it known to others, progress in bacteriology might have been greatly advanced. It is quite possible that with his technique the anthrax bacillus could have been detected in the blood of infected animals, a feat which was not accomplished until some two centuries later by Davaine with the aid of the compound microscope. Like all sciences, the history of microbiology is largely the history of breakthroughs in methodology and instrumentation; the perfection of the compound microscope was one such breakthrough.

For more than a century after Leeuwenhoek's time the attention of bacteriologists was focused principally on the theory of spontaneous generation of life. The world of "animalcules" remained the last stronghold for the proponents of spontaneous generation despite strong opposing evidence provided by the classical experiments of Spallanzani (1776) and Schwann (1836) on the preservative effects of heat treatment on organic materials. The much-investigated but unsettled question of spontaneous generation absorbed the early attentions of Louis Pasteur, whose experiments culminated in the famous concluding remarks of his Sorbonne lecture of 1864 in which he stated that his sterile medium did not become contaminated "because I have kept it sheltered from the only living thing that man does not know how to produce; from the germs which float in the air, from Life, for Life is a germ and a germ is Life. Never will the doctrine of spontaneous generation recover from the mortal blow of this simple experiment."

Although Pasteur is regarded as the father of experimental and theoretical microbiology, he was also a practical man intent on science serving mankind. He was an unusually vigorous, perceptive, and imaginative investigator, a chemist by training and a superb experimentalist. By well-designed and painstaking experiments he avoided the pitfalls which plagued most investigators of his day and thus was able to solve problem after problem rapidly. He settled the dispute about the role of yeast in alcoholic fermentation and established that different fermentations were carried out by different and specific microbes, each demanding particular cultural conditions for its growth and activities. He introduced autoclaving to destroy heat-resistant forms (spores) and devised liquid media for growing cultures. He was also the first to employ synthetic media and to discover microbes that could live without air (anaerobes). Pasteur's interests were unusually wide and ranged from "diseases of beer and wine" to diseases of plants, animals, and man. Through his work on pebrine of silkworms, anthrax of sheep, and spoilage of wine he came to be regarded as the savior of the silk, sheep, and wine industries of France. In view of his training, it is not surprising that his interests remained focused throughout his life on the biochemical activities of microbes and that he regarded animal diseases literally to represent fermentations in the animal body.

About 1880 Pasteur directed his attentions to antimicrobial immunity. Among his many contributions one of the most important was his discovery of the enduring principle that mutant strains of pathogens of attenuated virulence can be isolated and that they commonly make the best vaccines. In recognition of his successful development of the rabies vaccine in 1885, many research institutes named in his honor were established throughout the world, the largest of which is the Institut Pasteur in Paris.

Another giant in the early history of microbiology was Robert Koch, a country doctor who abandoned the practice of medicine for a research career in an effort to gain new insight into the causes and nature of infectious diseases. His motivation for changing to a research career stemmed from his feeling of helplessness in dealing with human infections and is reflected in his statement, "mothers come to me crying—asking me to save their babies, but what can I do?—grope, fumble, reassure them, when I know there is no hope. How can I cure diphtheria when I do not even know what causes it, when even the wisest doctor in Germany does not know." Koch's wife bought her husband a microscope so that he could test the dictates enunciated by Henle, his former professor, for determining the causative agents of infectious diseases. In 1876 Koch grew the anthrax bacillus in pure culture, thus erasing any doubt that the organism was the sole cause of anthrax. Six years and some 271 experiments later Koch reported his epoch-making discovery of the tubercle bacillus, the cause of tuberculosis.

Koch was a careful and tireless investigator who planned experiments with great ingenuity and carried them out with precision and skill. His staining and pure culture methods permitted precise studies on the cultural characteristics of bacteria and conclusive establishment of these organisms as the specific causative agents of disease according to his famous postulates (Chap. 8). His agar plating technique for isolating pure cultures is routine to this day. As the result of epidemiologic studies on waterborne disease, he proposed the use of sand filtration for purifying water supplies, a practice still in common use. Koch's pure culture technique was so successful that the causative agents of most of the common bacterial diseases of man were discovered within the next two decades.

Since both Pasteur and Koch were men of great compassion, it is no surprise that they were attracted to the study of infectious disease, the greatest medical problem of their day.

With the recent development of techniques for purifying and handling macromolecules a new and promising area of research in medical microbiology has opened, the molecular aspects of the pathogenesis of infectious diseases.

Bacteria have provided the best of all models for genetic engineering; recent advances have made it possible to *splice together DNA from different sources* and to transplant the resulting "hybrid DNA" into bacteria to enable them to produce all manner of biologic substances, such as insulin and human interferon. Hybrid DNA-derived organisms with altered pathogenicity and virulence can be produced.

Bacterial plasmids, discovered by Watanabe and Fukasawa (1960), which are intracytoplasmic, double-stranded DNA structures that can replicate independently of chromosomal DNA, provide another tool for the genetic engineer; additionally they serve to explain many of the natural genetic changes that occur in bacteria involving pathogenesis, virulence and drug resistance (Chap. 4). Earlier discoveries of similar usefulness were those relating to transformation, sex-factor conjugation, and transduction by bacteriophages.

Increasing knowledge relative to the mechanisms by which bacteria adhere selectively to the epithelial cells of various mucosae and, in consequence, colonize and often induce their own phagocytosis by target cells has clarified many questions about host-parasite relationships. In particular, it has provided new insight on the manner by which many bacteria selectively invade, parasitize, and destroy host cells. Since the chemical components (adhesins) responsible for adherence commonly serve as virulence factors, advances in this area have opened new approaches for combatting infections, including the preparation and administration of more effective vaccines.

Recent discoveries relating to the capacity of some intracellular pathogens to destroy phagosomal membranes or to so alter them that phagosome-lysosome fusion fails have greatly advanced knowledge on the mechanisms of virulence and pathogenicity.

The development of the hybridoma technique for producing monoclonal Abs by Kohler and Milstein in 1975 promises to be the most valuable contribution made to biology and medicine in the 20th century. The technique is being used in essentially every area of biology and medicine ranging from basic research to practical applications of every sort, including all aspects of clinical medicine; its greatest impact has been in the areas of immunology and medical microbiology, especially as they relate to host-parasite relationships and control of infectious diseases. The work of Kohler and Milstein was based on the 1960 discovery of natural hybrids formed spontaneously in mammalian mixed-cell cultures as the result of the fusion of 2 cells of different types. Such hybrids were useful for short-term genetic studies but could not be cultured permanently. The value of the hybridoma technique was the discovery that fusion of a normal lymphocyte with a tumor cell of the lymphocyte series (myeloma) conveyed the properties of *immortality* and Ab production to the hybrid such that it could be cultured indefinitely as an Ab-producing cell. The original technique of fusing activated (immune) B cells to myeloma cells to form B cell hybridomas can also be used to produce T cell hybridomas.

Monoclonal Abs, the product of cloned B cell hybridomas, are precise reagents for identifying and purifying Ags at the molecular and submolecular (epitope) levels. They provide a tool for dissecting the immune response and for identifying and purifying pathogenetic agents, virulence factors and protective immunogens. The use of monoclonal Abs will undoubtedly lead to major advances in the diagnosis, treatment, prognosis and prevention of infectious diseases as well as in other areas of medicine, such as cancer and organ transplantation. Hybridomas produced with T cells also promise to be useful for studying the immune response and host-parasite relationships. Lymphokines produced by T cell hybridomas may prove to be of practical use in the area of infectious diseases.

Advances in other areas of immunology have contributed to a better understanding of host-parasite relationships. For example, studies on the T suppressor cell-contrasuppressor cell circuit, which began with the discovery of T suppressor cells by Gershon and Konda in 1970, is of singular importance; indeed it now appears that some pathogens can suppress the immune response by overstimulating the production of T suppressor cells.

Recent discoveries relating to lymphokines and monokines, (interleukins) that regulate the activities of various leukocytes, are also having an important impact on medical microbiology. Being among the regulators of the specific immune response (SIR), they present possibilities for its manipulation to advantage, either by enhancing it or by blocking mechanisms by which pathogens subvert the SIR.

Late discoveries concerning various mechanisms by which complement (C) contributes to antimicrobial defense have revealed that they are more complex and far-reaching than had been previously thought. For example, certain cell wall components of bacteria can trigger C activation in the absence of specific Ab with resulting deposition of C3b on the cell wall surface where it can serve as an opsonin by reacting with C3b receptors on the phagocyte surface; C3b-9 complexes are also formed which are bactericidal for Gram-negative but not Gram-positive bacteria because the latter are protected against the cidal action of C by a thick peptidoglycan layer in their cell walls. Whereas Gram-negative bacteria can trigger C activation by both the alternative pathway (AP) and the classical pathway (CP), Gram-positive bacteria can only activate C *via* the AP.

Among the most significant of recent findings on C is the observation that capsules and certain other surface substances that serve as virulence factors, act, in part at least, by masking underlying outer-cell wall components that activate C. For example, the thickness of the capsule of the pneumococcus and its ability to trigger C activation by the AP are major determinants of virulence. In accord with expectation, major reduction or loss of masking components leads to avirulence by permitting C activation, opsonization and phagocytosis to occur.

Another observation of interest is that C-reactive protein (CRP) binds to the phosphorylcholine moiety of teichoic acid in the cell walls of Gram-positive bacteria, such as the pneumococcus, and in so doing triggers the CP of C activation with resulting C3b deposition on the outer cell wall. The capsular polysaccharides of certain pneumococcal types can also bind CRP. Although the injection of CRP into mice can protect against challenge with capsulated pneumococci, it remains to be determined whether protection results from C activation and C3b-induced phagocytosis or from some other mechanism that might involve special CRP receptors on phagocytes.

Still another significant observation on C activities is that the engagement of C3b and/or Fc receptors on monocytes by Abs and components of C facilitates the intracellular killing of bacteria they engulf. The blockade or removal of these receptors blocks the killing of ingested bacteria.

Another developing area of great promise concerns the synthesis of immunogenic peptides specific for microbial Ags and their use for producing protective vaccines and Abs. The identification and amino acid sequences of immunogenic peptides that serve as epitopes on native microbial protein molecules is made possible by the combined use of monoclonal Abs and recombinant DNA technology.

Space does not permit a detailed recounting of the many contributions that have played important roles in the development of medical microbiology. However, a limited list of milestones is presented in Table 1–1.

B. The Role of Infectious Diseases in Population Dynamics

Many a man stands living here
That your father, in the nick of time,
Snatched from the fever's burning rage
When he put limits to the plague.
Goethe

Table 1–1. Some Important Milestones in the Development of Medical Microbiology

Approximate Date	*Principal Contributor(s)*	*Contribution(s)*
1546	Fracastoro	Clearly enunciated the germ theory of disease, including modes of transmission and the development of specific acquired immunity.
1683	Leeuwenhoek	Gave the first accurate account of the visualization of yeasts and bacteria.
1798	Jenner	Introduced vaccination against smallpox with cowpox virus.
1835	Bassi	Showed that "animalcules" cause a transmissible fungal disease (muscardine) of silk worms.
1836–1837	Cagniard-Latour; Schwann	Clearly established that microbes (yeasts) are biochemically active (alcoholic fermentation).
1839	Schönlein	Showed that certain skin diseases are due to fungi.
1843–1847	Holmes; Semmelweis	Showed that physicians and attendants often transmit infectious disease from one patient to another and demonstrated the effectiveness of control procedures.
1849	Snow	Proved that cholera is waterborne and set forth the epidemiologic principles of infectious diseases.
1850–1872	Davaine	First to observe bacteria clearly in tissues (blood) during infection (anthrax) and demonstrated increased virulence with animal passage.
1861–1885	Pasteur	Demonstrated that different kinds of bacteria have distinct biologic activities and cultural needs. Demonstrated that some microbes can grow without oxygen. Disproved the theory of spontaneous generation. Introduced sterilization by steam under pressure. Proved the value of attenuated vaccines.
1870	Hansen	Observed *Mycobacterium leprae* in the tissues of lepers; this was the first time that bacteria were seen in human lesions.
1871	Weigert	Introduced successful methods for staining bacteria with aniline dyes.
1872	Cohn	Established the basis of systematic classification of bacteria.
1876–1883	Koch*	Observed the sporulation cycle of *Bacillus anthracis* and conclusively proved it to be the specific cause of anthrax. Introduced pure culture methods using solid media and perfected methods for staining bacteria. Discovered the tubercle bacillus and described delayed hypersensitivity to tuberculin. Set forth criteria for ascertaining the causative agents of infectious diseases (Koch's postulates).

Table 1–1. *Continued*

Approximate Date	*Principal Contributor(s)*	*Contribution(s)*
1878	Tyndall	Proved conclusively that microbes are associated with dust particles in air and that they consist of thermolabile (vegetative) and thermostable (spore) forms.
1878	Lister	Introduced the practice of antiseptic surgery; obtained the first pure cultures of bacteria by an in vitro method, extinction dilution.
1880–1883	Ogston	Established that wound suppuration is commonly caused by cocci of two types, cocci in clusters and cocci in chains; introduced the concept that systemic infection can arise by spread from localized infection.
1884	Chamberland	Introduced the bacterial filter; perfected the autoclave.
1884	Gram	Developed the Gram stain.
1884	Metchnikoff*	Introduced the concept that antimicrobial cellular immunity is effected by phagocytes.
1884–1886	Salmon and Theobald Smith	Introduced the use of killed vaccines.
1889	Roux and Yersin	Proved that bacterial toxin can account for the pathogenicity of an organism; e.g., *Corynebacterium diphtheriae.*
1890	H. Buchner	First to clearly enunciate the theory of humoral immunity.
1890	von Behring* and Kitasato	Discovered Abs (tetanus and diphtheria antitoxins).
1894	Pfeiffer	Described specific bacteriolysis by immune serum in vitro and in vivo (Pfeiffer phenomenon).
1894	Denys and Havet	Established that immune serum and phagocytes can act conjointly in antibacterial defense.
1895–1901	Bordet*	Conclusively established that specific Ab and "alexine" (later termed "complement") can lyse certain bacteria and mammalian cells and with Gengou, developed the complement-fixation test.
1896	Durham and Gruber	Provided the first clear description of specific bacterial agglutination.
1896	Widal	Introduced specific serodiagnosis of infectious diseases with his "Widal test" for typhoid fever.
1896–1909	Ehrlich*	Introduced methods for standardizing toxins and antitoxins. Introduced the procedure of acid-fast staining. Set forth a theory of Ab production. Introduced chemotherapy (arsenicals for syphilis).
1897	E. Buchner*	Discovered the role of yeast zymase in alcoholic fermentation.

Table 1–1. ***Continued***

Approximate Date	*Principal Contributor(s)*	*Contribution(s)*
1897	Kraus	Developed the precipitin test for measuring the reaction of Abs with soluble Ags.
1898	Nocard, Roux, and associates	Discovered *Mycoplasma mycoides* (the cause of pleuropneumonia of cattle) the first of the Mycoplasma described.
1898	Beijerinck	Discovered filterable viruses (tobacco mosaic virus).
1903	Wright and Douglas	Discovered specific opsonization.
1905–1908	von Pirquet	Advanced the concept that allergy plays a major role in determining the lesions, symptoms and course of infectious diseases.
1909	Ricketts	Discovered *Rickettsia rickettsii* (the cause of Rocky Mountain spotted fever) the first of the Rickettsiaceae described.
1913	Schick	Introduced the diphtheria toxin skin test (Schick test) for measuring immunity to diphtheria.
1915–1917	Twort; d'Herelle	Discovered bacteriophage.
1921–1925	Zinnser and associates	Provided evidence supporting the concept advanced earlier by von Behring that antimicrobial immunity and delayed hypersensitivity "are parallel and perhaps causally related phenomena."
1923	Ramon	Introduced the principle of converting toxin to a nontoxic immunizing agent, toxoid.
1922–1928	Fleming*	Discovered the bacteriolytic agent lysozyme. Discovered penicillin, the first antibiotic later used successfully for antibacterial chemotherapy.
1930–1937	Dienes and Schoenheit; Dienes and Edsall	Described the capacity of tubercle bacilli to direct the immune response of associated Ags toward the development of delayed hypersensitivity. First to recognize wall-less bacteria (formerly called L forms by Klieneberger-Nobel, 1935).
1935	Boivin and Mesrobeanu	Discovered the endotoxins of Gram-negative bacteria.
1935	Domagk*	Discovered Prontosil, the precursor of sulfanilamide, and ushered in the modern era of chemotherapy.
1934–1941	Ruska; Marton	Developed the electron microscope.
1937	Freund	Established the principles of adjuvanticity. Developed Freund's adjuvant mixture, a tubercle bacillus-water-in-oil mixture that enhances Ab production and "directs" the immune response toward the development of delayed hypersensitivity and cellular immunity.

Table 1–1. *Continued*

Approximate Date	*Principal Contributor(s)*	*Contribution(s)*
1940	Florey,* Chain,* and colleagues	Purified penicillin and developed procedures for its production and use as a chemotherapeutic agent.
1942–1945	Landsteiner* and Chase; Chase	Accomplished passive transfer of delayed type hypersensitivity with lymphoid cells.
1944	S.A. Waksman* and associates	Discovered streptomycin, the first antibiotic effective against tuberculosis.
1944	Avery, MacLeod, and McCarty	Discovered genetic transformation of bacteria by DNA.
1952	Bruton	Presented the first example of an hereditary immunologic deficiency disease, sex-linked agammaglobulinemia.
1951–1953	Freeman; Groman	Discovered lysogenic conversion of bacteria by bacteriophage (toxin production by *C. diphtheriae*).
1952	Westphal and associates	Purified and determined the structure of endotoxin.
1957	Germuth and associates; Dixon and associates	Demonstrated that soluble Ag-Ab complexes can cause allergic injury.
1959	Watanabe	Discovered episomal transfer of multiple antibiotic resistance in bacteria.
1960	Shepard	Produced the first experimental infection with *M. leprae,* using the mouse footpad.
1962	Mackaness and associates	Demonstrated that lymphocytes are initiator cells and that macrophages are effector cells in specific antimicrobial cellular immunity.
1969	H. Muller-Eberhard	Made major contributions toward an understanding of the complexity and functions of the complement system.
1970	Gershon and Konda	Discovered T suppressor cells.
1975	Kohler* and Milstein*	Developed the hybridoma technique for producing monoclonal Abs.

*Recipients of Nobel Prize.

Factors concerned in the ecology of pathogens and their hosts as they bear on the dynamics of host populations are many and highly complex. Nevertheless, studies on the interrelationships of these factors is crucial for designing measures for the control and potential eradication of infectious diseases. In any such study it should be kept firmly in mind that high population densities favor the transmission of many diseases to the point where epidemics with high mortalities often occur. Consequently consideration must be given to circumstances that influence population densities, such as standards of living, fertility rates, hygiene, sanitation, mass immunization and medical practices.

As illustrated below, the basic principles relating to epidemics and the impact of infectious diseases on host populations are best studied in isolated or closed populations uninfluenced by human activities that restrict deaths due to infectious diseases, such as vaccination and practices of sanitation.

The predisposition to epidemics exerted by high population densities is favored by high density of both the *susceptibility group* and the *infectious group* (shedder group) within a population, as well as by high transmissibility of the infectious agent. Shedders may include carriers as well as patients with active disease. Whereas highly infectious agents can initiate epidemics in populations in which the densities of susceptibles and shedders are relatively low, other less-infectious agents initiate epidemics only when the densities of susceptibles and/or shedders are relatively high.

An epidemic will tend to die out when the numbers of susceptibles and/or shedders in the population drops to a level where the probability of encounter between a shedder and a susceptible individual approaches zero.

Epidemics due to a single agent that engenders long-lasting immunity tend to occur intermittently, the interval between epidemics representing the time required for the buildup of a new generation of young susceptibles to an epidemic level. For example, epidemics of measles in unvaccinated populations occur every 2 years.

By contrast, in diseases in which shedders include carriers as well as patients with active disease, such as diphtheria, an epidemic will continue until essentially all susceptibles become infected, after which the number of carriers rapidly declines and a new buildup of susceptibles occurs. The interval between successive epidemics of diphtheria is about 1 year.

Since each infectious agent in ecologic equilibrium with its host population exacts a certain death toll due to the infections it produces, the total death toll within a population and, in consequence, the achievable population density in the area will be influenced by the number of pathogens indigenous within the population and the respective death tolls they exact.

Although the profound depopulating effects of war, famine and pestilence are well-known, their interrelationships and relative roles in depopulation are not generally appreciated. War commonly exerts secondary effects that predispose to infectious diseases, such as famine, overcrowding, poor sanitation and other dislocations. Famine can also exert profound secondary effects including, crowding, poor sanitation and malnutrition.

In the course of history infectious diseases have taken a far greater toll of human life than weapons of war, starvation and all other causes of death combined. During past centuries, infant mortality, which was due primarily to infectious diseases, was often so high that fewer than 3 out of 10 children lived beyond the age of 3 years.

It is an immutable truth that in the absence of human intervention involving measures that substantially reduce deaths due to infectious diseases, this cause of death has always been the major regulatory force restricting overpopulation.

1. Historic Aspects

Because of accessibility of data, populations of Western Europe during past centuries provide the best models for historic studies on population dynamics. During the *agricultural era* (preindustrial era) in Europe covering many centuries, cities with high density populations sprang up. This resulted in frequent and often widespread epidemics of diseases, such as smallpox, plague, typhus, typhoid fever, and cholera. Smallpox alone took an average toll of some 8% of each generation and the death tolls from plague epidemics are alleged to have ranged as high as 25% or more. Malaria and tuberculosis were also major killers. Following epidemics, underpopulation often became a problem. During the intervals between epidemics populations tended to be restored within a decade or two because of high fertility rates, only to be decimated later when population densities rose to epidemic levels. Since epidemics were largely confined to cities and villages the post-epidemic inflow of immigrants from rural areas contributed substantially to the repopulation of the depopulated areas. Population growth rates during the *agricultural era* were slow and seldom exceeded 0.1% per annum despite coexisting high fertility rates; they were reversed during the *industrial era.*

This reversal, which began about 1750, resulted almost entirely from reductions in deaths due to infectious diseases, reductions that stemmed from improvements in standards of living, nutrition, hygiene, sanitation, and, most recently, from the increasing use of antimicrobics and mass immunization. In developed countries the sharp drop in mortality due to infectious diseases, which began with the birth of microbiology in the late decades of the last century, has accelerated to the present low level where it represents only 0.5% of deaths from all causes. The decrease in deaths due to infectious diseases during the industrial era is largely responsible for a simultaneous increase in life expectancy which has risen from 35 years in 1750 to 76 years in 1986.

It is of singular interest that population increases during the industrial era did not keep pace with decreases in mortality, primarily because fertility rates tend to fall with rises in standards of living. Indeed, during the past century there has been a good correlation between decreases in fertility, decreases in mortality due to infectious diseases, increases in life expectancy and increases in standards of living.

2. Future Outlook

Although the *future impact of infectious diseases on human health in developed countries* presents a few uncertainties, it can be confidently anticipated that important advances will soon be made that will prove beneficial in almost every area and aspect of infectious diseases, including vaccines, antisera, antimicrobics, diagnosis, prognosis, therapy and carrier detection etc. The use of monoclonal Abs is but one avenue for making such advances.

A few disadvantages of controlling infectious diseases, either present or future, can be envisioned, among which are the possibilities of generating unmanageable antibiotic-resistant strains of bacteria or highly susceptible host populations lacking in either innate immunity, acquired immunity or the genetically determined ability to mount a specific immune response.

The future impact of infectious diseases on the health of man in underdeveloped countries is difficult to predict. Although at first glance the outlook would appear to be favorable, this is not assured because of many conditions that commonly exist in such countries including lack of education, poor hygiene and sanitation, low standards of living, high fertility rates, overpopulation, poverty, malnutrition, epidemic disease and frequent droughts and famine. Since most of these conditions are inextricably related, the measures to combat infectious diseases may be only temporarily effective or even counter-productive. For example, in the face of high fertility rates any marked reductions in deaths due to infectious diseases will lead to overpopulation which, in turn, predisposes to epidemics. Other effects of overpopulation, such as poverty, malnutrition, poor hygiene and sanitation also predispose to epidemics. Even the complete control of one disease, such as smallpox, may contribute little to reduce deaths from infectious diseases because of overpopulation-induced epidemics involving other pathogens.

The frustrating nature of the problem of permanently reducing deaths due to infectious diseases in underdeveloped countries is emphasized by remarks of a tribal chieftain in an overpopulated famine area of Africa who begged that diphtheria antitoxin not be sent to their camp to save children dying of diphtheria because, he maintained, "it is easier to die of diphtheria than by starvation."

The levels of mortality from infectious diseases in underdeveloped countries today are equivalent to those existing in now-developed countries some 150 years ago. For example in one rural area of Africa measles accounts for about a half million deaths annually and world-wide deaths in underdeveloped countries due to enteric infections and pneumonias alone total some 18 million; infant mortality in some areas exceeds about 50%.

C. ERADICATION OF INFECTIOUS DISEASES FROM HUMAN POPULATIONS

Since most of the important infectious diseases of man are those in which man serves as the sole natural host, the present discussion will be limited to this category.

A number of human pathogens and the diseases they produce possess features that favor the feasibility of eradicating them from human populations. Some of these features are: (a) the disease should be easy to recognize and diagnose. (b) man should be the only natural host and reservoir of infection. (c) infection should be limited to person-to-person transmission. (d) the organism should lack the ability to produce either latent disease with recrudescence or chronic carriage. (e) a vaccine should be available that produces solid and long-lasting immunity.

1. Regional Populations

Regional eradication or near-eradication of some infectious diseases has been accomplished in developed countries. Unfortunately regional eradication is temporary or is stopped short of completion because new sources of infection arrive continuously from foreign countries, including travelers and immigrants who carry causative agents.

Among various infectious diseases that have been reduced to zero or near zero in the USA are: poliomyelitis, measles, pertussis, diphtheria, typhoid fever, and tuberculosis.

2. World Populations

If all countries in the World were to become "developed," an unlikely event in the foreseeable future, it would be reasonable to expect that global eradication of a number of infectious diseases would be well within the realm of possibility. However, under existing circumstances the global eradication of any infectious disease is extremely difficult because of various conditions that obstruct the effective application of eradication measures, notably the lack of education, sanitation and communication together with geographic isolation, poverty and political, religious and social prejudices. To ferret out the last remote foci of a smoldering disease often proves to be an interminable task.

Global eradication of infectious diseases of man, a long-sought goal of the WHO, now appears to have been realized in the case of one of the most deadly scourges of mankind, smallpox. This stands among the greatest achievements in the history of microbiology and hopefully presages further successes. The final change in strategy that led to the eradication of smallpox was to promptly identify and deal with each new case and its possible contacts using measures to block disease transmission, including blanket vaccination of all persons in the locales involved.

Unfortunately only a few pathogens have features that would make their global eradication conceptually feasible. Programs of the WHO have not been limited to those infectious diseases that hold high promise for global elimination but instead have included others that promise to effect major, albeit not necessarily complete eradication. These programs have made little progress, including large programs on malaria and yaws. However, with proper strategies poliomyelitis and measles hold reasonable promise for global eradication.

REFERENCES

Anderson, R.M. and May, R.M. (eds.).: Population Biology of Infectious Diseases. Report of the Dahlem conference held at Berlin March 14–19, Berlin, Springer-Verlag. 1982.

Avery, O.T., MacLeod, C.M., and McCarty, M.: Induction of transformation by a deoxyribonucleic acid fraction isolated from pneumococcus type III. J. Exp. Med. *79*:137, 1944.

Bullock, W.: The History of Bacteriology. London, Oxford University Press, 1938.

Bruton, O.C.: Agammaglobulinemia. Pediatrics *9*:722, 1952.

Chase, M.W.: The cellular transfer of cutaneous hypersensitivity to tuberculin. Proc. Soc. Exp. Biol. Med. *59*:134, 1945.

Dienes, L., and Schöenheit, E.W.: Certain characteristics of the infectious processes in connection with the influence exerted on the immunity response. J. Immunol. *19*:41, 1930.

Dixon, F.J.: Characterization of the antibody response. J. Cellular Comp. Physiol. *50*:27, 1957.

Dixon, F.J.: The role of antigen-antibody complexes in disease. The Harvey Lectures Series 58, 1962–63.

Engleman, E.G., Foung, S.K.H., Larrick, J.L., and Raubitschek, A.: Human Hybridomas and Monoclonal Antibodies. New York, Plenum Press, 1985.

Fleming, A.: On a remarkable bacteriolytic element found in tissues and secretions. Proc. Roy. Soc. Ser. B *93*:306, 1922.

Fleming, A.: On the antibacterial action of cultures of a penicillium, with special reference to their use in the isolation of *B. influenzae.* Br. J. Exp. Pathol. *10*:226, 1929.

Freeman, V.J.: Studies on the virulence of bacteriophage-infected strains of *Corynebacterium diphtheriae.* J. Bacteriol. *61*:675, 1951.

Freund, J., and McDermott, K.: Sensitization to horse serum by means of adjuvants. Proc. Soc. Exp. Biol. Med. *49*:548, 1942.

Florey, H.W., et al.: Antibiotics. Vol. I and II. London, Oxford University Press, 1949.

Germuth, F.G., and McKinnon, G.E.: Studies on the biological properties of antigen-antibody complexes. Bull. Johns Hopkins Hosp. *101*:13, 1957.

Groman, N.B.: Evidence for the induced nature of the change from nontoxigenicity to toxigenicity in *Corynebacterium diphtheriae* as a result of exposure to specific bacteriophage. J. Bacteriol. *66*:184, 1953.

Joiner, K.A., and Frank, M.M.: Complement and Bacteria: Chemistry and Biology of Host Defense. Annu. Rev. Immunol. *2*:461, 1984.

Landsteiner, K., and Chase, M.W.: Experiments on transfer of cutaneous sensitivity to simple compounds. Proc. Soc. Exp. Biol. Med. *49*:688, 1942.

Lüderitz, O., and Westphal, O.: Uber die chromatographie auf rundfiltern. Zeit. fur Natursforsch. G. *76*:136, 1952.

Mackaness, G.B.: The immunology of antituberculous immunity. Am. Rev. Respir. Dis. *97*:337, 1968.

Marton, L.: La microscopie electronique des objets biologiques. Bull. Acad. Roy. Belg. Cl. Sci. *20*:439, 1934.

Marton, L.: The electron microscope. J. Bacteriol. *41*:397, 1941.

McMichael, A.J., and Fabre, J.W.: Monoclonal Antibodies in Clinical Medicine. London, Academic Press, 1982.

Muller-Eberhard, H.J.: Complement. Annu. Rev. Biochem. *38*:389, 1969.

Nisonoff, A. and Gwish, M.F.: From Network Theory Toward Vaccination by Monoclonal Antibodies. New Horizons in Microbiology. Sanna, A., and Morace, G. (eds.): Amsterdam, Elsevier Science Publishers, 1984.

Nocard, E., Roux, E.R., Borrel, M., Salimbeni and Dujardin-Beaumetz: Le microbe de la peripneumonie. Ann. Inst. Pasteur *12*:240, 1898.

Von Pirquet, C.: Allergy. München Med. Wschr. *30*:1457, 1906.

Ricketts, H.T.: A microorganism which apparently has a specific relationship to Rocky Mountain spotted fever. JAMA *52*:379, 1909.

Ruska, E.: Uber fortschritte im bau und in der leistung des magnetischen elektronenmikroskops. (Aus dem Hochspannunginstitut Neubabelsberg der Technischen Hochschule Berlin). Z. Physik. *87*:580, 1934.

Schoolnik, G.K., Lark, D., and O'Hanley, P.: Bacterial Adherence and Anticolonization Vaccines. Current Clinical Topics in Infectious Disease, Vol. 6, Remington, J.S., and Schwartz, M.N. (eds.), New York, McGraw-Hill Book Co., 1985.

Shepard, C.C.: The experimental disease that follows the injection of human leprosy bacilli into footpads of mice. J. Exp. Med. *112*:445, 1960.

Shinnick, T.M., et al.: Synthetic peptide immunogens as vaccines. Annu. Rev. Microbiol. *37*:425, 1983.

Stuart-Harris, C.: Prospects for the eradication of infectious diseases. Rev. Infect. Dis. *6*:405, 1984.

Watanabe, T.: Infective heredity of multiple drug resistance in bacteria. Bacteriol. Rev. *27*:87, 1963.

Watanabe, T., and Fukasawa, T.: "Resistance transfer factor," an episome in *Enterobacteriaceae.* Biochem. Biophys. Res. Comm. *3*:660, 1960.

Zinnser, H., and Mueller, J.H.: On the nature of bacterial allergies. J. Exp. Med. *41*:159, 1925.

2

BIOLOGY OF BACTERIA

The purpose of this chapter is to give a brief overview of the biologic and physiologic properties of bacteria with particular reference to the medically important bacteria. The biology of fungi is discussed in later chapters. Microorganisms play major roles in man's environment by participating in the nitrogen, carbon and sulfur cycles. They also are an integral component in the food chain upon which all living biologic entities depend. In this regard, bacteria and fungi are employed extensively in food production and food processing as well as in manufacturing. It is somewhat ironic that microbes are used to produce antibiotics that are used against them to control infectious diseases.

Bacteria have a wide range of sizes as well as shapes. Most bacteria are 1 to 6 μm in length by 0.2 to 1.5 μm in diameter. Although living bacteria can be seen in suspension with the light microscope, phase-contrast microscopy, which depends on changes in the phase pattern rather than changes in amplitude of light waves, reveals more detailed morphology. However, bacteria are most readily seen with the light microscope on glass slides after staining with basic dyes.

In 1968 the bacteria and blue-green algae were removed from the Plant Kingdom and placed in a new Kingdom, the *Procaryotae* (cells with a primitive nucleus), because their DNA is not enclosed within a membrane, like eucaryotic cells. In addition, eucaryotic cells, unlike procaryotes, engulf or endocytose food particles. Collectively, several observations have prompted the concept that eucaryotic cells evolved from procaryotic cells by a process referred to as endosymbiosis. For example, prevailing data support the idea that mitochondria of eucaryotes originated as the result of endosymbiotic events that took place millions of years ago.

Bacteria are single-celled organisms that reproduce by simple binary fission. Unlike eucaryotes,[1] they do not contain 80S ribosomes composed of 60S and 40S subunits or membrane-bound organelles such as nuclei, lysosomes, mitochondria, endoplasmic reticulum, or golgi. Instead, bacteria contain 70S ribosomes which are composed of 50S and 30S subunits and naked single circular chromosomes composed of double-stranded deoxyribonucleic acid (DNA). The bacterial chromosome, commonly referred to as a "nucleoid," replicates amitotically. The cell membrane of bacteria is the site of transport and all energy-producing and specialized biosynthetic functions. Some bacteria produce external microfibrils (fimbriae or pili), which are used for adhering to surfaces. Bacteria that exhibit motility possess one or more single-filament flagellar structures. Almost all bacteria contain a rigid cell wall consisting of peptidoglycan. Only a small number of species of bacteria, such as the mycoplasma, lack classical rigid cell walls. The basic shapes of cell-wall producing bacteria include spheres (cocci), rods (bacilli) and curved or spiral shaped cells.

A. Classification of Bacteria (Table 2–1)

Traditionally all terms of the Latinized scheme for classifying bacteria are either italicized or underlined and standardized endings are used for various groups.

Bacteria are classified by Kingdom, Class, Order *(-ales)*, Family *-aceae)*, Tribe

---

[1]Yeasts and fungi have a membrane-bound nucleus and therefore are eucaryotes.

Table 2–1. Medically Important Bacteria

Family-(aceae) *Genus* *Species*	*Characteristic Properties*
GRAM-POSITIVE BACTERIA	
COCCI	
Micrococcaceae	Occur in irregular clusters singly and in pairs; facultatively anaerobic, metabolism respiratory and fermentative, nonmotile, catalase (+)
Staphylococcus	
S. aureus	
S. epidermidis	
S. saprophyticus	
Streptococcaceae	Occur in pairs or chains; nonmotile, facultatively anaerobic, metabolism fermentative, complex nutritional requirements,
Streptococcus	Catalase (−), produce lactic acid only (homofermentative)
S. pyogenes	
S. agalactiae	
S. pneumoniae	
SPORE-FORMING RODS	
Bacillaceae	Often peritrichously flagellated
Bacillus	Strict aerobes or facultative anaerobes
B. anthracis	
Clostridium	Spores usually distend the organism, most species strictly anaerobic
C. botulinum	
C. perfringens	
C. tetani	
NON-SPOREFORMING RODS	
Lactobacillaceae	Occur singly or in chains; complex nutritional requirements
Lactobacillus	Nonmotile, catalase (−), anaerobic or facultative, homolactic fermentation
L. acidophilus	
Listeria	Motile, aerobic, catalase (+)

Erysipelothrix	Nonmotile, often filamentous, catalase (−)
E. rhusiopathiae	
Propionibacteriaceae	Pleomorphic rods or filaments
Propionibacterium	Nonmotile, anaerobic to aerotolerant, propionic and acetic acids are major fermentation products
P. acnes	
Corynebacteriaceae	Straight to slightly curved rods, irregular staining, generally aerobic
Corynebacterium	Nonmotile, catalase (+)
C. diphtheriae	
Mycobacteriaceae	Nonmotile, acid-fast, aerobic, many slow growers
Mycobacterium	
M. tuberculosis	
M. leprae	Has not been grown in vitro
Nocardiaceae	Branching filamentous forms produced, nonmotile, aerobic
Nocardia	Some species acid-fast
N. asteroides	
Actinomycetaceae	Nonmotile, nonacid-fast, predominantly diphtheroid in shape, tend to produce branching filaments
Actinomyces	Anaerobic to facultatively anaerobic, catalase (±), major fermentation products are formic, acetic, lactic, and succinic acids, no gas
A. israelii	
A. naeslundii	
Arachnia	Facultatively anaerobic, catalase (−), major fermentation products are propionic and acetic acids and CO_2
A. propionica	
Streptomycetaceae	Aerobic soil organisms, true mycelia, aerial spores, source of antibiotics, mostly nonpathogens
Streptomyces	
S. griseus	
Micromonosporaceae	Primarily saprophytic soil forms. Spores on aerial and/or substrate mycelium
Micropolyspora	Some facultative thermophiles
M. faeni	

Table 2–1. *Continued*

Family-(aceae) *Genus* *Species*	*Characteristic Properties*
GRAM-NEGATIVE BACTERIA	
COCCI AND COCCOBACILLI	
Neisseriaceae	Aerobic, coccal to rod-shaped, nonmotile
Neisseria	Coffee bean-shaped diplococci, catalase (+), oxidase (+), a few sugars fermented
N. meningitidis	
N. gonorrhoeae	
Branhamella	Diplococci, catalase (+), oxidase (+), nitrates reduced, no acid from carbohydrates
B. catarrhalis	
Moraxella	Plump rods in pairs or short chains, strict aerobes, catalase (+), oxidase (+)
M. lacunata	
Acinetobacter	Rod-shaped to spherical, strict aerobes, catalase (−), oxidase (+)
Veillonellaceae	Anaerobic cocci, complex growth requirements, cytochrome oxidase (−), catalase (−), nonmotile, may produce volatile acids
Veillonella	Pyruvate utilized, carbohydrates not fermented
V. parvula	
AEROBIC BACILLI AND COCCOBACILLI	
Pseudomonadaceae	Straight or curved rods, motile, aerobic
Pseudomonas	Growth factors not required
P. aeruginosa	
P. mallei	
P. pseudomallei	
P. cepacia	
P. maltophila	

Organism	Characteristics
Brucellaceae	
Brucella	Coccobacilli, nonmotile, facultative intracellular parasites, fastidious growth requirements, strict aerobe, catalase (+)
B. abortus	
B. melitensis	
B. suis	
Bordetella	Small coccobacilli, polytrichous if motile, strict aerobes
B. pertussis	
Francisella	Pleomorphic rods require cysteine for growth
F. tularensis	
Legionellaceae	
Legionella	Stains Gram-negative poorly, catalase (+), rods
L. pneumophila	
FACULTATIVELY ANAEROBIC BACILLI	
Enterobacteriaceae	Small rods, facultative anaerobes, metabolism respiratory and fermentative, catalase (+), oxidase (−), nitrates reduced to nitrites
Escherichia	Motile, mixed acid fermentation, CO_2 and H_2 produced, lactose fermented by most strains, H_2S (−), citrate not utilized
E. coli	
Edwardsiella	Usually motile, mixed acid fermentation, lactose not fermented, indole (+), H_2S (+)
E. tarda	
Citrobacter	Usually motile, citrate utilized, lysine not decarboxylated
C. freundii	
Salmonella	Usually motile, lactose not fermented, indole (−), H_2S (+), lysine and ornithine decarboxylated
S. typhi	Acid but no gas from glucose
S. cholera-suis	Acid but no gas from glucose
S. enteritidis	
Shigella	Nonmotile, lactose not fermented, no gas from glucose, H_2S (−)
S. dysenteriae	
S. flexneri	
S. boydii	
S. sonnei	

Table 2–1. *Continued*

Family-(aceae) / *Genus* / *Species*	*Characteristic Properties*
Klebsiella	Nonmotile, encapsulated, 2,3-butanediol fermentation, citrate utilized
K. pneumoniae	Lactose fermented (acid and gas)
Enterobacter	Motile, 2,3-butanediol fermentation, lactose fermented with acid and gas, citrate utilized
E. aerogenes	
Serratia	Lactose not fermented, motile, H_2S (−), lysine decarboxylated, citrate utilized, pink to red pigment at room temperature
S. marcescens	
Proteus	Lactose not fermented, motile, urea hydrolyzed
P. mirabilis	
P. vulgaris	
Yersinia	Lactose not fermented, cells ovoid to rod-shaped, some species motile
Y. enterocolitica	
Y. pestis	
Vibrionaceae	Straight or curved rods, polar flagella, oxidase (+), catalase (+), metabolism respiratory and fermentative
Vibrio	
V. cholerae	
V. parahaemolyticus	Halophilic
V. alginolyticus	Halophilic
V. fluvialis	Halophilic
GENERA OF UNCERTAIN TAXONOMIC RELATIONSHIPS	
Spirillum	
S. minus	Polar flagella, aerobic to obligately microaerophilic
Flavobacterium	Coccobacilli to slender rods, motile or nonmotile, pigmented, aerobic
F. meningosepticum	
Haemophilus	Coccobacillary to rod-shaped, strict parasite, requires growth blood factors

H. influenzae	
Pasteurella	Ovoid or rod-shaped, bipolar staining, catalase (+), fermentative
P. multocida	
Actinobacillus	Bacilli with coccal elements, nonmotile, fermentative
A. lignieresii	
Streptobacillus	Rods form chains or filaments, nonmotile, may convert to L forms, fermentative
S. moniliformis	
Calymmatobacterium	Pleomorphic rods, usually encapsulated, nonmotile
C. granulomatis	
ANAEROBIC BACILLI	
Bacteroidaceae	Obligate anaerobic rods, non-sporeforming
Bacteroides	Fermentation products include lactic succinic, acetic, formic, lactic, and propionic acids
B. fragilis	
Fusobacterium	Butyric acid a major fermentation product
F. nucleatum	
Leptotrichia	Straight or slightly curved, ends rounded or pointed, nonmotile, lactic acid is major fermentation product
L. buccalis	
HELICAL CELLS	
Spirochaetaceae	Slender, flexible, coiled, produce axial filaments, aerobic to anaerobic
Treponema	Anaerobic, motile, tight regular or irregular spirals
T. pallidum	Has not been grown in vitro
Borrelia	Anaerobic, 3–10 coarse, uneven coils, motile
B. recurrentis	
B. burgdorferi	
Leptospira	Aerobic, tightly coiled, one or both ends bent or hooked, motile
L. interrogans	
S. minor	Polar flagella, growth enhanced by increased CO_2, has not been grown in vitro

Table 2–1. *Continued*

Family-(aceae) / *Genus* / *Species*	*Characteristic Properties*
Campylobacter	Single polar flagellum, microaerophilic to anaerobic
C. fetus	
C. jejuni	
C. coli	
C. laridis	
RICKETTSIAS AND CHLAMYDIAS	
Rickettsiaceae	Small rods to coccoid, usually intracellular parasites associated with arthropods
Rickettsia	Grows in cytoplasm, and sometimes in nuclei
R. rickettsii	
Rochalimaea	Grows extracellularly, can be grown on bacteriologic media
R. quintana	
Coxiella	Grows in vacuoles, resistant to extracellular drying or heat
C. burnetti	
Bartonellaceae	Rod to coccoid, ring- or disc-shaped, parasites of red blood cells, grow on nonliving media
Bartonella	Characteristically in chains of segmenting organisms, occur in human and arthropod vectors
B. bacilliformis	
Chlamydiaceae	Coccoid, obligate intracellular parasites, grows in yolk sac of chick embryos or in tissue culture
Chlamydia	Nonmotile, restricted metabolic activity
C. trachomatis	
CELL WALL-LESS FORMS	
Mycoplasmataceae	Cells pleomorphic, lack true cell wall, stain Gram-negative, colonies have fried egg appearance, penicillin-resistant, require sterols for growth
Mycoplasma	
M. pneumoniae	

Modified from Krieg and Holt (eds): *Bergey's Manual of Systematic Bacteriology*, 9th ed. Baltimore, Williams & Wilkins, 1984.

(-ieae) and a specific binomial term for genus and species. The generic name is capitalized and the species name is kept in lower case. The classification scheme used in the USA is based on *Bergey's Manual of Systematic Bacteriology;* the last edition was published in 1984. It is especially noteworthy that the medically important bacteria represent only a small fraction of all bacterial species. A classification scheme of the medically important bacteria based on the 9th Edition of *Bergey's Manual* is presented in Table 2–1. Official names of organisms and late revisions of names can be found in the *International Journal of Systematic Bacteriology* which began publication about 1980.

When writing the names of microbes it is common practice to designate the genus by using the first letter as an abbreviation; for example, *Staphylococcus aureus* = *S. aureus.* Experts in the field accept this practice because their familiarity with various species usually enables them to readily recall the genus name by its association with the species name. However, beginning students find this practice confusing because in many instances the same capital letter is used to designate the genus names of a half dozen or more organisms. Consequently, to limit confusion, many of the genus abbreviations used in this text comprise more than one letter, such as *Sal.* rather than *S.* for the genus, *Salmonella;* these abbreviations are contained in the glossary.

Several important criteria used for bacterial classification are: size, shape, staining characteristics, motility, physiologic requirements (i.e. oxygen, temperature, pH), metabolic reactions, metabolic endproducts, ecologic traits, genetic traits and antigenic composition. Because of the difficulties in using a few arbitrary weighted characteristics for classification, a system of numerical taxonomy has been attempted. This system uses a large number of traits in a numerical expression as a similarity (S) coefficient. The basic idea is that if a large number of traits are used in a computerized system it removes subjectivity from taxonomy. The S coefficient is a percentage of total characteristics that are held in common between two organisms or groups of bacteria.

Other approaches to classification have involved gene-controlled metabolic patterns that exhibit stability such as cell structural macromolecules and special organelles. In addition, attempts to define nucleic acid homology based on guanine plus cytosine (GC) as the percentage of total purines and pyrimidines has proven useful. The percent of GC for microorganisms usually ranges from 25 to 75%. However, similar percentages of GC suggest but do not assure homology. The most useful technique presently available is to determine the percent molecular hybridization that occurs when two DNA strands, each from different organisms, are mixed and allowed to interact and hybridize. By using one DNA strand intact and by adding radiolabeled fragments of DNA from a different organism the percent of homology can be determined based on the amount of radiolabel that is DNAse resistant. Only single stranded DNA is sensitive to DNAse. For example, using this approach *Escherichia coli* shows an 89% homology to *Shigella dysenteriae* but only 1% homology to *Bacillus subtilis.*

B. METHODS USED FOR IDENTIFYING BACTERIA

It is of utmost importance for a clinical microbiology laboratory to be able to culture, isolate and identify the causative agents of infectious diseases. These techniques are based on a set of standard procedures that involve variations of or selected combinations of the following: (1) isolation of the agent in pure culture by streak plating techniques, (2) determination of microscopic morphology and staining reactions using bacterial smears, (3) evaluation of colonial morphology, (4) determination of biochemical characteristics, (5) evaluation of serologic reactions (agglutination tests, enzyme-linked assays), (6) determination of susceptibility to lysis by specific bacteriophages (bacterial viruses), (7) animal pathogenicity and (8) antibiotic sensitivity. Animal pathogenicity is used only in special instances and not on a routine basis. Phage susceptibility is used principally for epidemiologic studies in which strains (Phagevars) are identified.

The term *biotype* is commonly used to indicate a strain of organisms within a species that best represents or is the prototype strain of the species. Sometimes it is useful to have sub-species classification based on serotypes (Serovars) or phage susceptibility (Phagevars).

C. MORPHOLOGY OF THE BACTERIAL CELL

Medically important bacteria exhibit 3 major morphologic forms, spheres (cocci), rods (bacilli) and curved or spiral shaped forms. Cocci can occur as pairs of cocci (diplococci), chains of cocci (streptococci) or irregular clusters of cocci (staphylococci). Bacteria can be observed using light microscopy of stained smears; suspensions of bacteria can also be viewed by phase contrast microscopy, darkfield microscopy and interference microscopy. Electron microscopy is the most powerful tool for morphologic studies although the disadvantage is that killed cells must be used. In particular, freeze etching has been highly useful for determining surface structures as well as internal structures when coupled with freeze fracture.

Special structures that can be readily seen with the electron microscope are flagella, axial filaments of the spirochetes, fimbriae, sex pili, cell envelopes, capsules, plasma membranes, nucleoids, ribosomes, cytoplasmic granules and endospores. A composite diagram of bacterial structures is presented in Figure 2–1.

1. Capsule (Glycocalyx)

Medically important bacteria commonly synthesize a layer on their surfaces which can be demonstrated as a capsule when special stains are used; capsules play an important role in virulence because they usually interfere with normal phagocytosis and killing. It is important to note that specific Abs can neutralize this antiphagocytic virulence property and promote phagocytosis. In the case of *Streptococcus pneumoniae* there are more than 80 serotypes based on the serologic specificity of the capsules. Since specific Ab can cause a swelling (Quellung) reaction of the capsules, this provides a simple technique for serotyping the different isolates of *strep. pneumoniae.*

Whereas most bacterial capsules are composed of polysaccharide, the capsule of *Bacillus anthracis,* the causative agent of anthrax, is composed largely of the unnatural amino acid, D-glutamic acid, in polymerized form. The chemical composition of capsules of representative medically important bacteria is found in Table 2–2.

Capsules also appear to play an important role in bacterial adherence. For example, *Streptococcus mutans* produces a levan polymer in the presence of

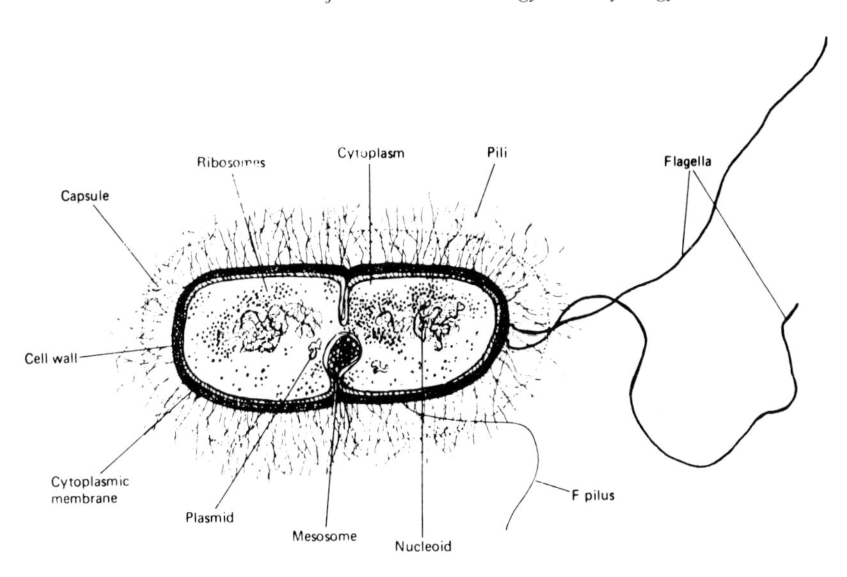

Fig. 2–1. Schematic diagram of a dividing bacterium. Pili refers to adhesion pili. F pilus is a sex pilus. Plasmid is a sub-nucleoid DNA fragment. (From: Medical Microbiology. J.C. Sherris, (ed.). Chapter 2. Bacterial Structure, Physiology and Growth. Spizizen, J., New York, Elsevier, 1984.)

sucrose. There is good evidence that this capsule facilitates attachment of this organism to the enamel on teeth and subsequently initiates caries through production of lactic acid. An unexpected, but not surprising, example of the marked tendency of many microbes to adhere selectively to foreign surfaces came to light with the discovery that infections sometimes occur on prosthetic materials including high density polyethylene, methyl methacrylate and stainless steel. In the case of artificial hips, certain strains of *Staphylococcus epidermidis* (a normal skin inhabitant) can colonize on the prosthetic material

Table 2–2. Chemical Composition of the Capsules of Representative Medically Important Bacteria

Organism	*Nature of Capsule*	*Chemical Subunits*
Bacillus anthracis	Polypeptide	Polymer of D-glutamic acid
Neisseria meningitidis	Polysaccharide	Sialic acid, hexosamines
Haemophilus influenzae	Polysaccharide	Polysugarphosphates
Streptococcus pneumoniae	Polysaccharide	
	Type II	Rhamnose, glucose, glucuronic acid
	Type XIV	Galactose, glucose, N-acetylglucosamine
Streptococcus spp	Hyaluronic acid	N-acetylglucosamine, glucuronic acid
	M-protein	
Salmonella typhi	Polysaccharide (Vi antigens)	N-acetyl-D-galactosaminuronic acid
Escherichia coli	Acidic polysaccharide	Sialic acid, uronic acids

surface where they produce polysaccharide capsules as well as polysaccharide slime (excess capsular material) that protects the organisms from host defenses. Apparently, the surfaces of the prosthetic devices used can somehow promote adherence and slime production and thus provide the means for this normal flora organism, and probably others as well, to become potentially serious pathogens.

2. Cell Wall

Bacterial cell walls are unique in that they are rigid and act as a "corset" to keep the protoplast from bursting in the usual hypotonic environment in which bacteria must survive and grow. Bacterial cell walls have no counterpart in mammalian cells which makes them a unique target for attack by antibiotics.

Bacteria exist in two large groups based on their response to the Gram stain. When a bacterial smear made from a mixture of organisms is stained with crystal violet followed by an iodine solution and brief destaining with ethanol, the purple dye-iodine complex is retained by one group, which is termed *Gram positive* and is lost by the other group which is called *Gram negative.* The red counterstain, safranin, is then applied as a final step so that any destained Gram-negative bacteria will counterstain a contrasting red color.

Since the cell walls of Gram-negative bacteria have a high lipid/peptidoglycan ratio, the alcohol extracts the lipid moieties which allows the dye-iodine complexes to solubilize and rapidly leach out the crystal violet from Gram-negative cells. In contrast, the cell walls in Gram-positive bacteria are composed, for the most part, of peptidoglycan (murein) with little or no lipid; as a consequence, alcohol penetrates the cells poorly because the murein is alcohol insoluble. Accordingly, the solubilization of the dye-complex is retarded and the intact Gram-positive cells retain most of the crystal-iodine complex and appear purple on the smear. Because of the especially high lipid content of their cell walls, certain bacteria, such as *Mycobacterium tuberculosis,* stain poorly with the Gram stain. However, they are readily differentiated from other organisms by the acid-fast stain.

The *Gram-positive cell wall* appears as being much thicker than the Gram-negative cell wall (Fig. 2–2). The major polymer in the Gram-positive cell wall is referred to as a *mucopeptide* or *peptidoglycan.* This polymer is comprised of a chain of alternating N-acetylglucosamine and N-acetylmuramic acid in a 1:4 linkage to which a tetrapeptide of alternating L- and D-amino acids is attached to the O-lactyl group of the muramic acid (Fig. 2–3). The amino acids found are D-glutamic acid, D- and L-alanine, L-lysine, glycine and a unique diamino acid, diaminopimelic acid. In addition, adjacent chains are cross-linked by peptide linkages between the 3rd amino acid and the terminal D-alanine on the other chain. About 30 to 40% of these peptides are linked to the identical peptides on adjacent sugar backbones by tetra- or pentapeptides, glycine or by another peptide having the same sequence as that linked to muramic acid.

A second polymer of Gram-positive cell walls called teichoic acid is composed of glycerol phosphate or ribitol phosphate; it can be linked to muramic acid. A schematic representation of the Gram-positive cell wall is presented in Figure 2–4.

The *Gram-negative cell wall* is more complex. A much thinner and less cross-

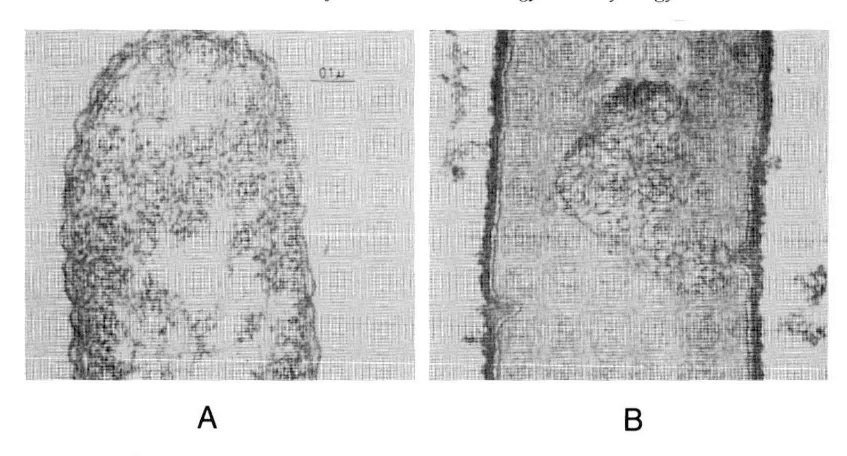

Fig. 2–2. Thin-section electron micrographs of representative bacteria. (A) *Escherichia coli* (Gram-negative); note the multilayered cell envelope. (B) *Bacillus subtilis* (Gram-positive); the thick amorphous cell wall is adjacent to the cytoplasmic membrane. Note the vesicular invagination of the membrane. (Reprinted with permission from Bacteriol. Rev. *29*:299, 1965.)

linked *peptidoglycan* is found on the inner surface adjacent to the cytoplasmic membrane. With Gram-negative cell walls the links between adjacent chains are directly through the peptides of muramic acid. The mucopeptide of Gram-negative bacteria is only about one molecule thick and is only about 15 to 30% cross-linked. On the outer surface of the peptidoglycan layer is a layer referred to as the *outer membrane*. It consists of a bilayer of phospholipids comprising *lipopolysaccharide* on the outer surface with embedded lipoprotein and proteins. The *lipoprotein* can constitute up to 40% of the cell wall dry weight. The lipopolysaccharide is comprised of lipid A and a core polysaccharide that is common to all Gram-negative bacteria. An additional polysaccharide com-

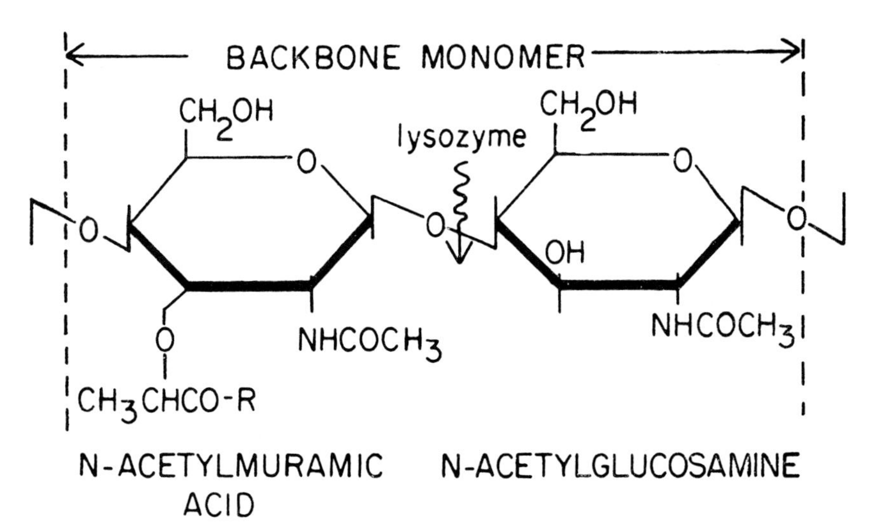

Fig. 2–3. The monomer which forms the basic polymer of peptidoglycan.

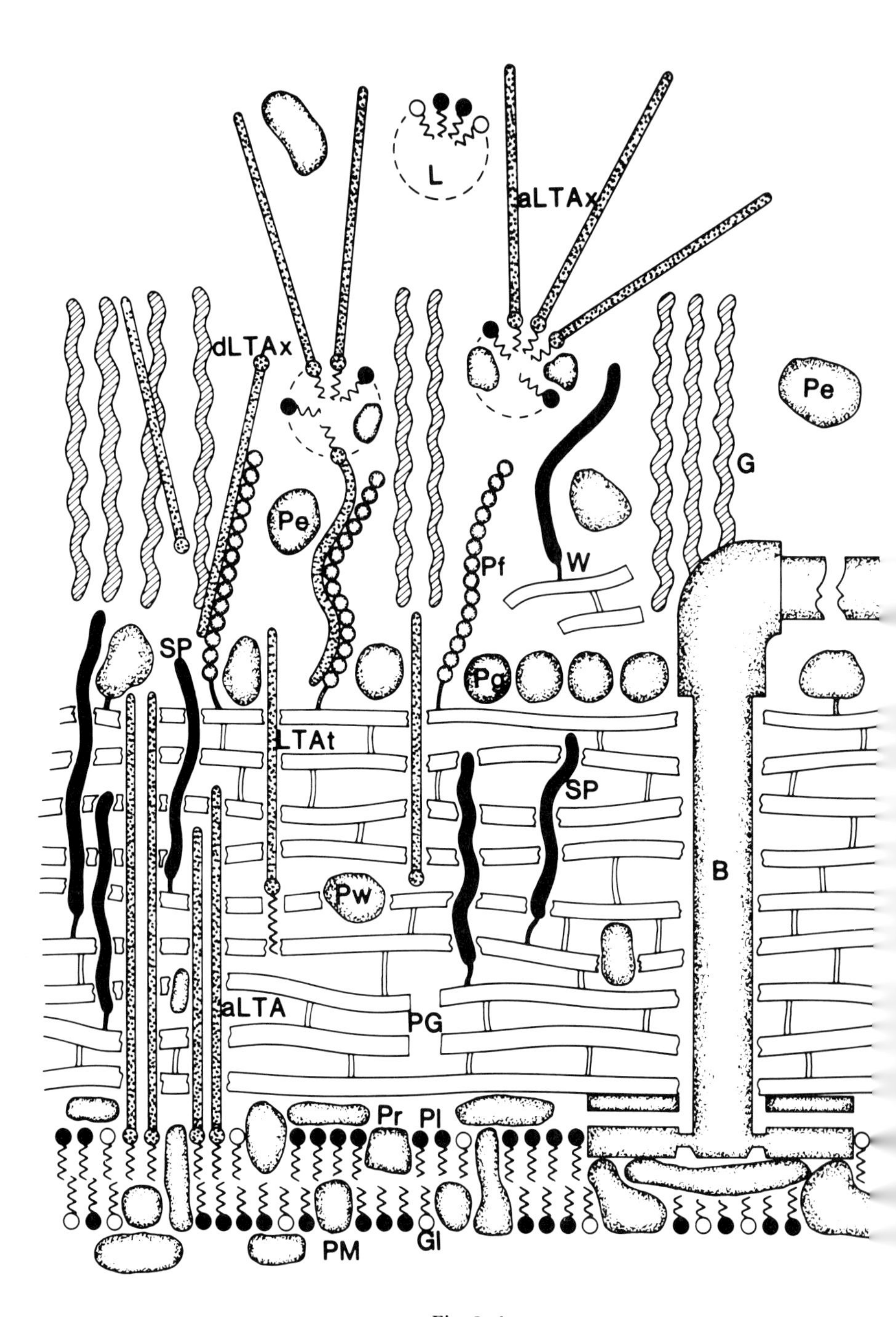

Fig. 2–4

Fig. 2–4. Schematic representation of the organization of the cell wall of gram-positive bacteria. Cross-linked peptidoglycan (PG) is shown overlaying the plasma cell membrane (PM) composed of protein (Pr), phospholipid (Pl), and glycolipid (Gl). Secondary cell wall polymers (SP) such as teichoic acids, teichuronic acids, and polysaccharides are covalently linked to peptidoglycan and extend through the peptidoglycan network with exposure, in some cases, at the surface of the rigid portion of the cell wall. Protein components of the wall include the basal bodies of flagella (B), noncovalently associated proteins (Pw) both within the peptidoglycan matrix and as part of the glycocalyx region. Covalently associated protein on the surfaces of the peptidoglycan is shown in two configurations, globular (Pg) and fibrillar (Pf). Lipoteichoic acid, in its acylated form (aLTA), extends from the upper half of the bilayer of the plasma membrane, to varying degrees depending on chain length, through the peptidoglycan net. Excreted LTA (aLTAx) may exist as mixed micellar aggregates with protein and lipid. Deacylated extracellular LTA (dLTAx) exists as monomer, and the polyglycerophosphate chains of both forms may interact ironically with surface fibrillar proteins. LTA in the process of excretion has a transient association in either the fully acylated or the deacylated form, with the peptidoglycan net (LTAt). Similar orientations are assumed for other amphiphiles in those gram-positive bacteria that do not contain LTA. Also contributing to the glycocalyx region are excreted lipids (L), proteins (Pe), and wall turnover products (W) as well as extracellular polymeric material peculiar to the capsular or glycocalyx region of the cell (G). (The diagram is intended to be illustrative of the concepts described in the text and should not be regarded as a literal depiction or necessarily drawn to scale.) (From *Bacterial Adhesions,* D.C. Savage and M. Fletcher (Eds.). Chapter 2: Wicken, A.J. New York, Plenum Press, 1985, p. 46.)

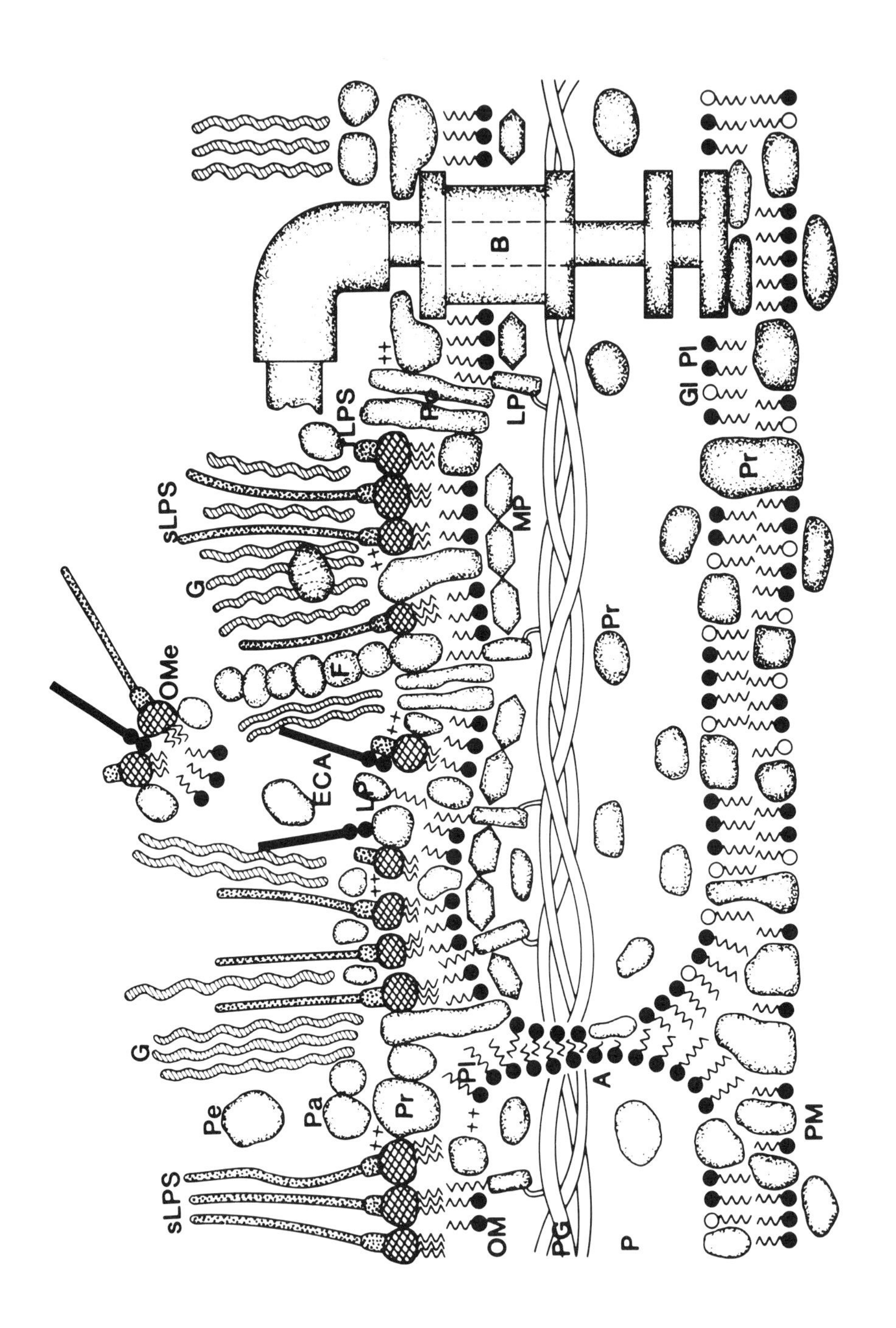
sLPS
Pe
Pa
Pr
G
OM
Pl
PG
P
A
PM
OMe
ECA
LP
F
G
sLPS
MP
Pr
LPS
LP
B
Gl
Pl
Pr

Fig. 2–5. Schematic representation of the organization of the cell wall of gram-negative bacteria. The cell wall is shown as having three regions, namely, outer membrane (OM), peptidoglycan monolayer (PG), and periplasm (P) overlaying the inner plasma membrane (PM). The two membranes show continuity at various attachment points (A) and may also be bridged by the basal bodies (B) of flagella (pili may have a similar orientation). The plasma membrane is shown as composed of phospholipid (Pl), glycolipid (Gl), and protein (Pr); some protein is also found in the periplasm. The asymmetric outer membrane is composed of phospholipid (Pl), protein (Pr), lipopolysaccharide in smooth (sLPS) and rough (rLPS) forms, and related polysaccharide polymers such as the enterobacterial common antigens (ECA) as well as divalent metal cations ($^{++}$). Peptidoglycan is closely associated with the inner face of the outer membrane, covalently bound through lipoprotein (LP) and noncovalently through a hexagonal array of matrix protein (MP). Other integral proteins of the outer membrane are shown as either restricted to one face of the membrane or spanning it and in the case of the porins (Po) forming narrow ionic channels across the membrane. Fimbriae (F) are suggested to arise from proteins in the outer face of the outer membrane and may extend for considerable distances beyond the outer membrane. Outer membrane-associated protein (Pa) and extracellular protein (Pe), excreted outer membrane fragments (OMe), the extended O-polysaccharide side chains of sLPS, as well as extracellular capsular polysaccharides (G) are shown as contributing to the heterogeneity of the glycocalyx region of the cell. (The diagram is intended as being illustrative of the concepts described in the text and should not be regarded as a literal depiction or one that is necessarily drawn to scale.) (From *Bacterial Adhesions.* D.C. Savage and M. Fletcher (Eds.). Chapter 2: Wicken, A.J. Plenum Press, New York, 1985, p. 56.)

ponent, which is bound to the core polysaccharide to complete the lipopolysaccharide complex (LPS complex or LPS), varies considerably among various strains of organisms and is referred to as the O Ag of Gram-negative bacteria. A schematic representation of the cell wall of Gram-negative bacteria is presented in Figure 2–5.

Electron microscopy and chemical analysis of mutants that are unable to synthesize all or part of their LPS complex show that the LPS complex has the general structure illustrated in Figure 2–6. Buried in the outermost lipoprotein layer of the cell wall is a complex lipid structure designated *Lipid A.* Attached to this is a *core polysaccharide* containing 5 sugars including 2-keto-3-deoxyoctonic acid (KDO), which is found only in the core region. Linked to the polysaccharide moiety of the core region are long strands of polysaccharide made up of repeating sequences of tetra- or pentasaccharides.

Biosynthesis of LPS involves individual sugar components that are attached to nucleotides, transferred to lipid carriers, and then added to the Lipid A-core "backbone" as a single unit. Each individual sugar in the repeating sequence of the specific polysaccharide chain is activated by a different nucleotide, which greatly increases the specificity of the enzyme pathways involved in the synthesis of the chain.

The LPS complex, also referred to as *endotoxin,* is the cause of fever, shock and hemorrhage that commonly develops in severe Gram negative infections. Lipid A plus the core polysaccharide represents the endotoxic component of the LPS complex.

Biosynthesis of Peptidoglycan. Although there is considerable variation in amino acid composition, it appears that the peptidoglycan layers of all bacteria have the same basic structure. The biosynthetic pathways employed by bacteria for the synthesis of peptidoglycan involve (1) nucleotide activation and synthesis of peptides on the muramyl residue; (2) transfer to the lipid carrier, modification of the amino acid side chains, and transport through the membrane; (3) attachment to the growing peptidoglycan mesh followed by cross-linking.

Each of these synthetic phases is unique biochemically and is not duplicated elsewhere in nature. Peptidoglycan synthesis has provided a unique biochemical target that has proven vulnerable to antibacterial substances that can specifically inhibit cell-wall synthesis without any harmful effect on host cells. In addition, lysozyme, an enzyme found in tears, saliva, and body fluids, specifically disrupts the Beta 1–4 linkages in the sugar backbone of the wall and ultimately causes susceptible bacteria to lyse. Lysozyme (muramidase) is particularly effective against Gram-positive bacteria possessing cell walls containing as much as 50% peptidoglycan. Gram-negative cells, whose murein layer lies beneath other cell-wall components, require special treatment with chelating agents or at an alkaline pH (which exposes the peptidoglycan layer) before lysozyme has an effect. When susceptible Gram-positive cells are treated with lysozyme, their entire peptidoglycan layer eventually dissolves leaving naked *protoplasts** bounded only by the cytoplasmic membrane. Under normal

*By convention Gram-positive cells yield *protoplasts,* whereas the analogous structures derived from Gram-negative cells are called *"spheroplasts."* This is because the latter contain residues of cell-wall material, and some are capable of reversion to rod forms when the inducing agent is removed.

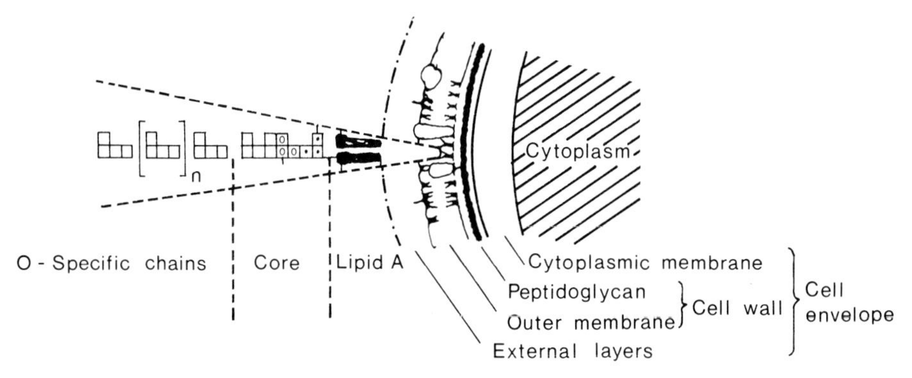

Fig. 2–6. The organization of the cell envelope of Gram-negative bacteria and the location and structure of the lipopolysaccharide of *Salmonella,* ●, Phosphate; ᴧ, phospholipid; ⬯, protein; ⋎, lipopolysaccharide; ■, D-glucosamine; ≡, long-chain fatty acids; ⊡, 2-keto-3-deoxyoctonate; ⊚, L-glycero-D-mannoheptose; □, other sugar units. (From Weckesser, J. and Drews, G., Annu. Rev. Microbiol. *33*:216, 1979.)

osmotic conditions the high osmotic pressure inside the protoplast (5 to 20 atmospheres) causes the naked protoplast to burst. If the osmotic pressure of the medium is increased to equal intracellular pressure by adding sucrose (an osmotic stabilizer), the spherical protoplasts can survive and carry out most of their normal cellular functions except division. Protoplasts are unable to divide, presumably because they have no residual template or peptidoglycan to which the newly synthesized monomeric units can attach themselves to form the cell wall, a structure that is evidently necessary for membrane septation.

Porins. The outer membranes of Gram-negative bacteria contain proteins called porins that form transmembrane pores that permit passage of low MW hydrophilic molecules. In addition, porins can serve as attachment sites for bacteriophage and some nutrients. The porins of members of the family *Enterobacteriaceae* exclude molecules with a MW greater than 600. Other Gram-negative organisms, such as *Pseudomonas* and *Neisseria,* exclude molecules in the 3000 to 9000 MW range.

Acid-Fast Bacteria. Envelopes and cell walls of acid-fast and related bacteria are of singular interest. Members of the genus *Mycobacterium* as well as a few species of the genus *Nocardia* stain poorly with Gram stain. However, they are readily stained with the red dye carbolfuchsin when a detergent has been added or heat has been applied. Once stained with this method they resist decolorization with acid-alcohol (3% HCl in ethanol) which seems to depend on mycolic acids in the cell wall. No other bacteria resist acid-alcohol decolorization. Accordingly, the species that resist acid-alcohol decolorization are referred to as acid-fast bacteria.

The cell wall of *M. tuberculosis* is comprised of equal amounts of peptidoglycan, arabinogalactan and lipid. More than 5% of the lipid comprises esterified mycolic acids. In addition, the cell wall contains peptidoglycan-linked poly-L-glutamic acid. Another class of lipids, namely, sulfolipids, are apparently located in a peripheral location within the envelope and appear to correlate with the virulence of *M. tuberculosis.*

Other components of mycobacteria include mycosides, phospholipids and lipopolysaccharides.

L-Forms and Mycoplasma. Occasionally when bacteria are repeatedly subcultured in the presence of a spheroplast inducer and an osmotic stabilizer, mutants arise that are unable to revert to the parental, osmotically stable form because of a defect in cell wall synthesis. This same phenomenon may occur during the course of an active infection which can result in the emergence of permanently fragile organisms designated as *L-forms.* These L-forms have been derived from a wide variety of pathogens. Since they lack a complete cell wall, they are resistant to penicillin and its derivatives. These organisms have been alleged to be important in a number of disease states, such as rheumatoid arthritis and nongonococcal urethritis, but definite proof of their etiologic role in these diseases is lacking.

A closely related group of microorganisms of the genus *Mycoplasma* has also received much attention in recent years. Organisms of this group, which includes the human pathogen, *M. pneumoniae* and some members of the normal flora, are nearly identical to L-forms in size and pleomorphism. Mycoplasmas have never been observed to revert to a bacterial form, a "mutation" which is a relatively common event among L-forms. In addition, recent studies employing DNA analysis and hybridization have not shown that mycoplasmas are related to other known bacteria.

3. Cytoplasmic Membrane

The cytoplasmic membrane is composed of a classic phospholipid bilayer in which proteins (many are enzymes) are interspersed. The membrane contains enzymes that mediate oxidative phosphorylation and electron transport. In addition, the membrane contains enzymes that synthesize the cell wall polymers and capsular components. Secretion of bacterial components (including enzymes) commonly occurs. Many of these enzymes are hydrolases, which break down polymers and make subunits available to the bacteria.

The periplasmic space that is found between the cytoplasmic membrane and the cell wall contains many different enzymes that are involved in active transport and degradation of large molecules.

4. Cytoplasm

Several structural entities are seen in the cytoplasmic compartment which contains most of the chemical moieties responsible for metabolism and replication (Fig. 2–1).

The *mesosome* is seen as an invagination of the cytoplasmic membrane which is involved somehow in cell division.

The *nucleus* or *nucleoid* consists of a single chromosome ($MW = 3 \times 10^9$) without a nuclear membrane. It is a double-stranded circular DNA molecule associated with basic proteins and polyamines. Some bacteria also contain extrachromosomal DNA called *plasmids.* Plasmids commonly carry genes coding for antibiotic resistance, fertility factors and factors that aid in survival in adverse environments.

Ribosomes are 70S and are commonly found in the cytoplasm. The distinction between bacterial and mammalian ribosomes provides another useful target for antibiotics that have selective action on bacterial ribosomes.

Cytoplasmic inclusions are usually represented by storage granules (glycogen or starch). Volutin granules (probably polyphosphate) found in *Corynebacte-*

rium diphtheriae stain red and help characterize this organism in stained smears.

5. External Structures

a. ***Flagella.*** Several species of bacteria express motility mediated by special locomotor organs called *flagella;* flagella, which range from 3 to 10 μm in length and 10 to 20 nm in diameter, are hollow structures that are composed of subunits arranged in a triple helix. They are attached to the cell wall and both inner and outer membranes by a basal body through a structure referred to as a hook (Figs. 2–4, 2–5). It appears that the entire flagellum can rotate from the basal body. The flagellar protein Ags, referred to as flagellin, can be used as serologic markers and are referred to as H or flagellar Ags. Flagella may occur over the entire cell (peritrichous) or at one or both ends of the cell (polar). Polar flagella exist either as a tuft of flagella (lophotrichous) or a single flagellum (monotrichous). Support for a role for flagella in virulence is limited. However, there is evidence that motility may facilitate the spread of an ascending urinary tract infection, and that flagella augment adherence to mucosal surfaces. It is of particular interest that flagellar motion and motility are linked to the chemotactic responses expressed by flagellated bacteria. Chemotaxis may be either negative or positive.

b. ***Axial filaments.*** In the case of spirochaetes the structures for locomotion are called axial filaments. These structures are similar to flagella except that they are enclosed in a sheath and are attached to and wound around the long axis of the cell. The rotation of the filament produces motility (Chap. 30).

c. ***Fimbriae or Pili.*** Filamentous protein appendages that occur on a number of Gram-negative bacteria but only on a few Gram-positive bacteria are referred to as fimbriae or pili. These are stiff, spike-like structures 0.2 to 20 μm in length and 5 to 30 nm in diameter. One class of pili, referred to as attachment pili, are involved in the initiation of infection by mediating attachment to the epithelium (Fig. 2–7). A classic example is *Neisseria gonorrhoeae* which attaches to the epithelium of the urethra. A loss of pili is usually associated with a loss of virulence. A second class is represented by pili which function as a "sex" attachment pili on "male" strains of *Escherichia coli* (Fig. 2–7). Sex pili are involved in conjugation and the transfer of DNA from the piliated "male" strain to a recipient "female" strain. It was once thought that sex pili serve as a conduit for DNA. This idea is now uncertain and there is evidence that sex pili only provide attachment and that DNA is transferred through some other mechanism.

6. Endospores

A small number of genera produce endospores which are highly resistant survival forms of the organism. Endospores are dormant structures that are resistant to high temperature, have a low water content and contain an unusual compound, *dipicolinic acid* complexed with calcium (Fig. 2–8).

In general, vegetative cells of spore-forming bacteria differentiate by forming endospores when exposed to starvation or other adverse conditions; one vegetative cell develops one spore. The spore wall consists of a thick layer of peptidoglycan covered by layers of an insoluble structural protein that protects the spore from toxic environmental components. The spore coat is encased in

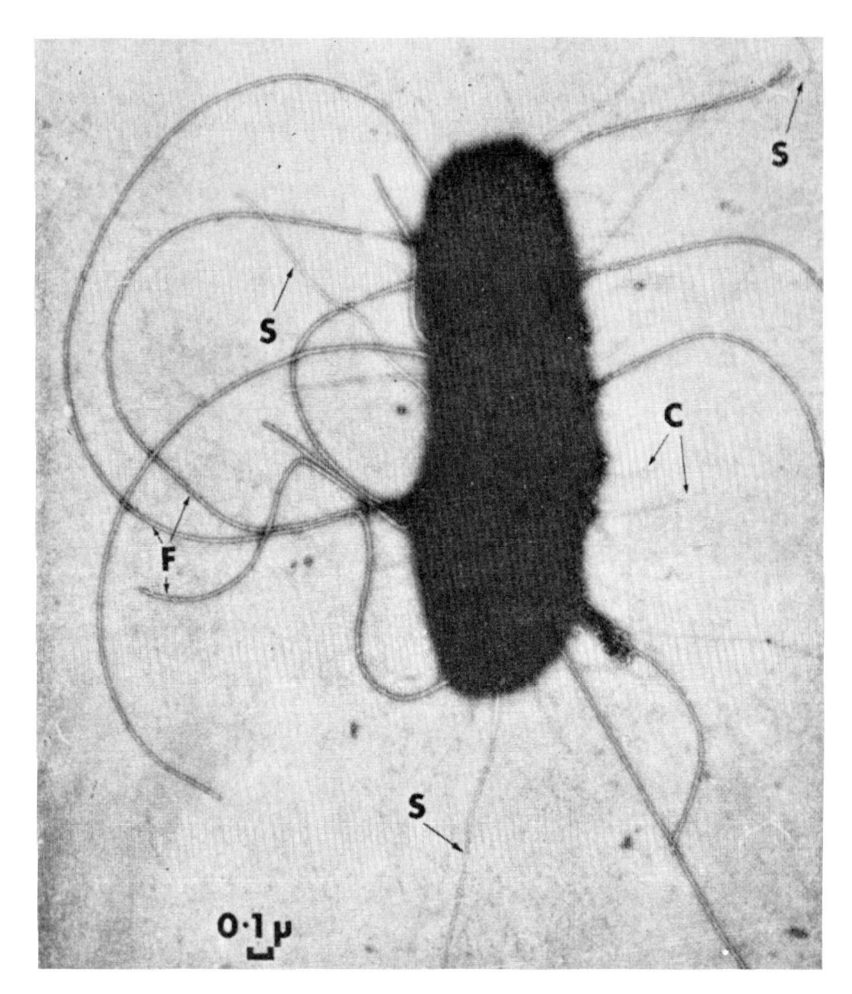

Fig. 2–7. *Escherichia coli* showing sexual pili (S), common pili (C), and flagella (F). (Reprinted with permission from Bacteriol. Rev. *32*:55, 1968.)

a lipoprotein and polysaccharide layer called the *exosporium.* When the spore is placed in a favorable environment, it gives rise to one vegetative cell.

D. METABOLISM

In order to survive and multiply bacteria must have an efficient system for generating energy. The catabolic (degradative) reactions provide some essential subunits for metabolic reactions and also generate energy for the anabolic (synthetic) processes. Similar to eucaryotic cells, energy is released in the form of high energy phosphate and stored as adenosine triphosphate (ATP).

Those bacteria that grow only in the presence of oxygen are called *obligate aerobes,* whereas those that can grow either in the presence or absence of oxygen are referred to as *facultative anaerobes.* Organisms that can only grow in the absence of oxygen are classified as *obligate anaerobes.*

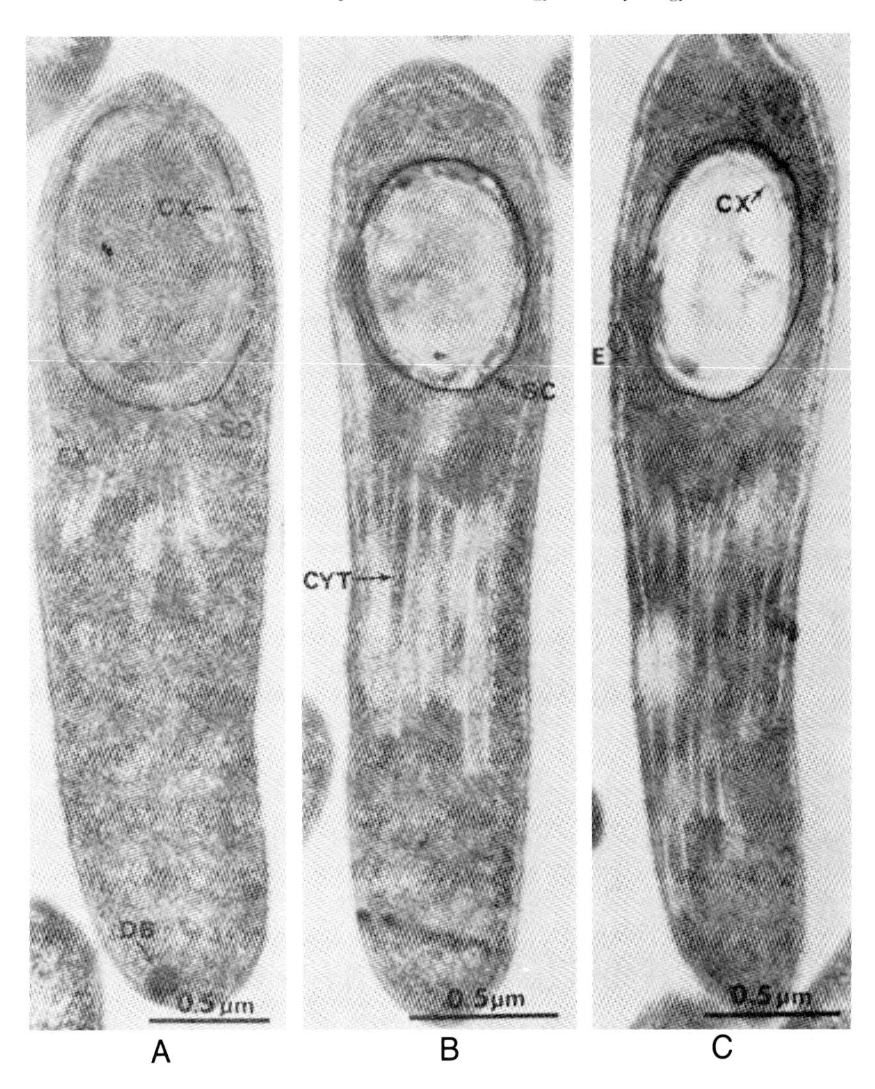

Fig. 2–8. (A) Section through an immature spore showing the deposition of coat material (SC) between the appendages and their site of origin. Note the presence of an electron-dense body (DB), increased cortical development (CX, between the arrows), and the presence of an immature exosporium (EX). (B) Section through a developing spore with a completed spore coat (SC); the appendages are no longer contiguous with their origin. Note the presence of cytoplasmic material (CYT) within the appendage tubules. (C) Longitudinal section through a mature spore within the sporangium illustrating that some of the appendages reach almost the entire length of the sporangium. The spore possesses a well-developed cortex (CX), between the arrows, and a laminated exosporium (EX). (Reproduced with permission from J. Bacteriol. *106*:269, 1971.)

Bacteria can generate energy by oxidizing carbohydrate molecules and capturing the energy generated in the form of readily utilizable ATP. The medically important bacteria do this through respiration or fermentation. Respiration is an aerobic process in which molecular oxygen acts as the final electron acceptor and the end products are H_2O and CO_2. In the case of fermentation the carbohydrate substrate molecules are usually broken down into two fragments; one fragment ends up oxidizing the second fragment, a reaction which also yields energy in the form of ATP.

Bacteria are classified as *chemoorganotrophs* if they derive their energy from oxidation-reduction reactions of organic compounds; in this case organic compounds like glucose also serve as electron donors. This is in contrast to *photoorganotrophs* that obtain energy from light but use organic compounds as electron donors. Two additional types of bacteria are found in nature; *photolithotrophs* that use light as energy, CO_2 as a carbon source, and inorganic compounds (S, H_2S) as electron donors, and *chemolithotrophs* that use CO_2 as a carbon source, oxidation-reduction reactions for energy and inorganic compounds as electron donors (H_2, S, H_2S, Fe, NH_3).

All of the medically important bacteria derive their energy from oxidation-reduction reactions of organic molecules like glucose and are referred to as chemoorganotrophs or heterotrophs. Since the bacterial cell requires building blocks for cell material, an organism like *Escherichia coli* growing aerobically will oxidize about 50% of the glucose to CO_2. This results in sufficient ATP to convert the remaining glucose substrate into building blocks to be used in cell growth and division.

1. Glycolytic Pathway

The major route of glucose catabolism in bacterial cells is referred to as glycolysis or the glycolytic pathway. This pathway is outlined in Figure 2–9 as the Embden-Meyerhof-Parnas glycolytic scheme. Phase I of glycolysis involves the phosphorylation of glucose by either ATP or phosphoenolpyruvate followed by cleavage to form glyceraldehyde 3-PO_4. During Phase II this 3-carbon intermediate is converted to lactic acid in a set of reactions coupled to the phosphorylation of adenosine diphosphate (ADP). It can be noted that 2 moles of ATP are consumed but 4 moles of ATP are generated thus giving a net energy yield of 2 moles of ATP. This pathway is relatively inefficient because the reaction products of glucose contain carbon that is still reduced and still available for further oxidation. The fate of pyruvate is dependent on the ability of the cell to generate oxidized nicotinamide adenosine dinucleotide (NAD+) from the reduced NADH.

2. Pentose Phosphate Pathway (Hexose monophosphate shunt)

This pathway is used in the fermentation of hexoses, pentoses and other carbohydrates. Although this pathway is a major energy-yielding system for some organisms, the majority use it to generate reduced NADH for synthetic reactions, to provide pentoses for nucleotide synthesis and to oxidize pentoses by the glycolytic pathway. The net yield of ATP is one-half (one mole of ATP/mole of glucose) of the yield of the Embden-Meyerhof-Parnas pathway.

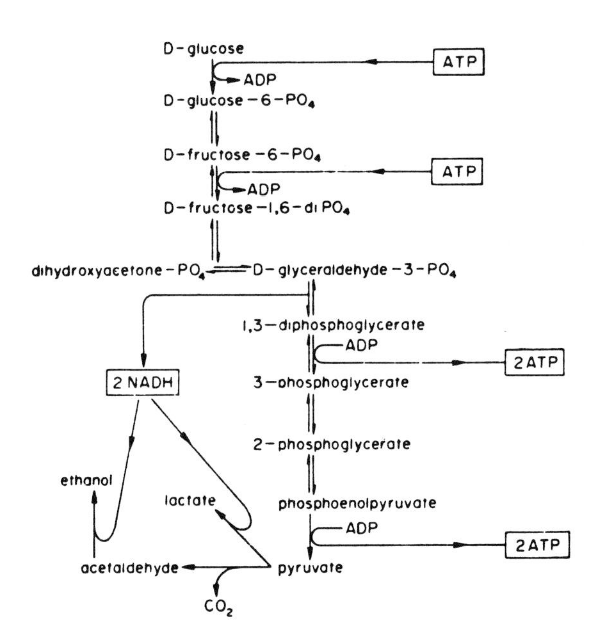

Fig. 2–9. The Embden-Meyerhof-Parnas glycolytic scheme. The energy yield of glucose through this pathway is 56,000 Cal/mole. (From Joklik et al., *Zinsser Microbiology,* 18th Ed. Courtesy of Appleton-Century-Crofts, 1984.)

3. Alternate Pathways of Glucose Fermentation

The fermentation of glucose always requires the expenditure of one ATP molecule to form glucose 6-PO_4. When this molecule is converted to pyruvate, NAD is reduced and therefore, must be reoxidized.

Incomplete oxidation of carbohydrates is the common outcome of bacterial fermentation because of limitations of available O_2 and the preferential anaerobic metabolism of some organisms. Whereas the end product produced by some organisms is either ethanol or lactic acid, other organisms metabolize pyruvate to either acetic acid, butyric acid, butanol, propionic acid, acetone, isopropanol or 2,3-butanediol.

A composite diagram illustrating how some of the fermentative end products are formed is illustrated in Figure 2–10. Obviously, bacterial fermentations not only have great commercial importance, but, in addition, the end products are useful in the identification of pathogens.

4. Aerobic Respiration

Bacteria that are obligate aerobes obtain their energy by the complete oxidation of the substrate thus leaving CO_2 and H_2O as end products. The most important pathway for aerobic oxidation is the Krebs tricarboxylic acid cycle (TCA) (Fig. 2–11). This cycle provides energy as well as carbon building blocks for synthesis of cell components.

In aerobic respiration pyruvate, which is formed from the glycolytic pathway,

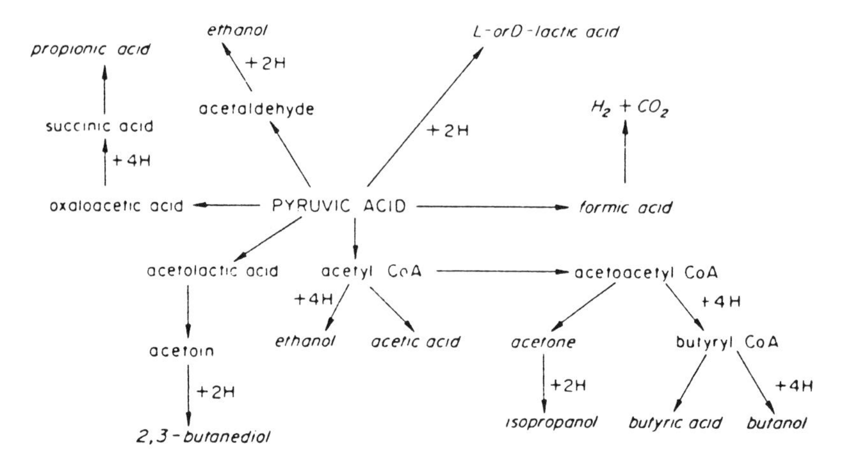

Fig. 2–10. Fate of pyruvate in major fermentations by microorganisms. (From Joklik et al., *Zinsser Microbiology*, 18th Ed. Courtesy of Appleton-Century-Crofts, 1984.)

is oxidized enzymatically to acetyl-CoA. At this point the electrons provided by NAD and accepted from pyruvate are carried by NADH to the respiratory chain. Acetyl-CoA is oxidized to CO_2 with concomitant reduction of NAD, NADP and flavine adenosine dinucleotide (FAD). Acetyl-CoA also enters the Krebs TCA cycle through the citrate synthase reaction. This allows oxaloacetate and acetyl-CoA to condense to form citric acid (Fig. 2–11). In turn this molecule is decarboxylated and oxidized to regenerate oxaloacetate with the 2 carbon atoms forming CO_2. As a consequence, 4 pairs of electrons are enzymatically removed from intermediates in the cycle. Accordingly, anything capable of forming acetyl-CoA can be oxidized by this cycle.

If bacteria utilize fatty acids or acetate as the only carbon source, acetyl-CoA is formed with the participation of pyruvate. In this case, growth on acetate induces the enzymes, malate synthase and isocitric lyase, which mediate a modification of the TCA cycle referred to as the glyoxylate cycle (Fig. 2–12). The CO_2 evolving steps of the TCA cycle are bypassed. It can be noted that succinate formed by the glyoxylate cycle is converted to oxaloacetate in the TCA cycle, which in turn can condense with acetyl-CoA to enter another turn of the cycle. Four dehydrogenations occur for each turn of the TCA cycle. Nicotinamide adenine dinucleotide serves as the electron acceptor in three of these dehydrogenations, whereas FAD serves as the fourth electron acceptor.

The reduced coenzymes are reoxidized by a passage of electrons involving a series of freely reactive compounds (quinones, cytochromes) and finally combine with oxygen in a reaction catalyzed by a terminal oxidase. The decline in free energy when a pair of electrons passes from NADH to molecular oxygen is sufficient to generate the synthesis of one molecule of ATP from ADP and inorganic phosphate.

5. Electron Transport System

The redox carriers and enzymes that participate in electron transport are located in the bacterial plasma membrane. Pyridine and flavin-linked dehydrogenases, coenzyme Q, cytochromes and iron-sulfur proteins represent the

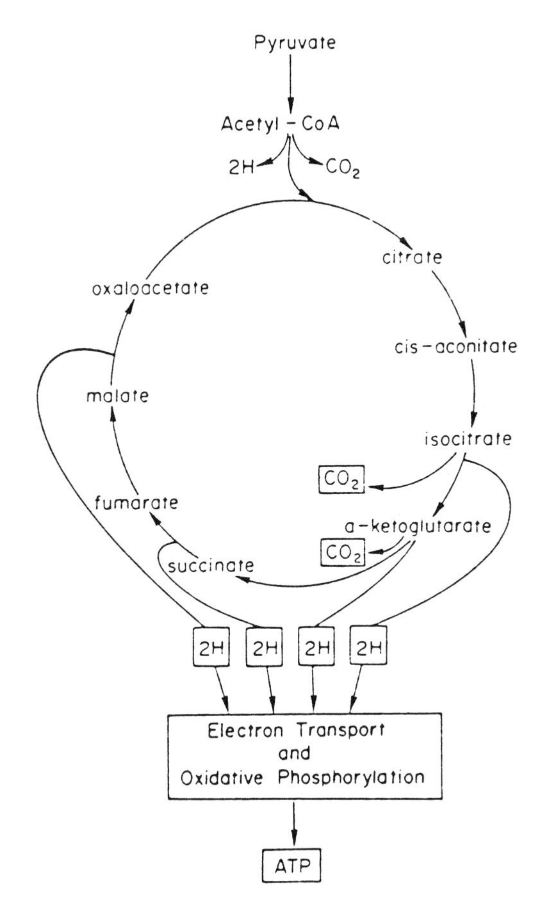

Fig. 2–11. The tricarboxylic acid cycle. The four pairs of H atoms liberated are fed into the respiratory chain. A net total of 38 moles of ATP are produced in the oxidation of 1 mole of glucose to CO_2 and H_2O. Approximately 266,000 calories are captured since one high energy bond of ATP yields about 7,000 calories. This represents 42 percent efficiency in terms of energy yield from glucose. (From Joklik et al., *Zinsser Microbiology,* 18th Ed. Courtesy of Appleton-Century-Crofts, 1984.)

major constituents involved in the transport of electrons from an organic compound to oxygen.

Pyridine-Linked Dehydrogenases. The coenzyme most frequently employed by bacteria as an acceptor of electrons from substrate is NAD+. These NADP-linked dehydrogenases serve to transfer electrons from catabolic intermediates to anabolic intermediates.

Flavin-Linked Dehydrogenases. These riboflavin-containing enzymes occur as flavin mononucleotide (FMN) or FAD. A highly important flavin-linked dehydrogenase is NADPH dehydrogenase which mediates the transfer of electrons from NADH to the next moiety in the electron transport chain.

Coenzyme Q. Coenzyme Q occurs in membranes and serves as an electron acceptor for one set of enzymes and subsequently as an electron donor to the next constituent in the electron transport chain.

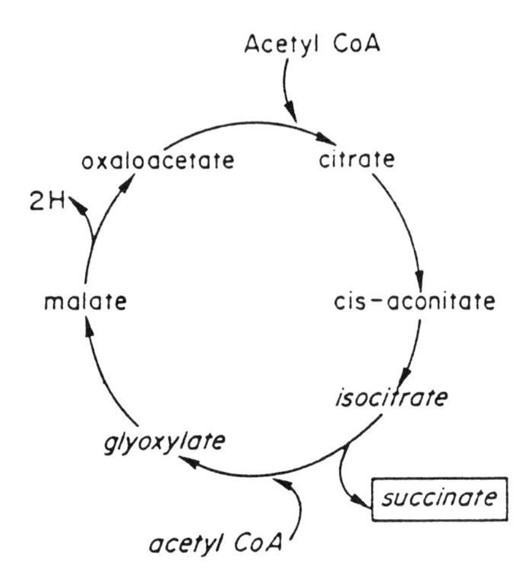

Fig. 2–12. The glyoxylate cycle. This cycle provides both energy and 4-carbon intermediates for biosynthetic purposes. In each turn of the cycle, two molecules of acetyl-CoA enter, and one molecule of succinate is formed. The reactions in the pathway between isocitrate and malate are catalyzed by auxiliary enzymes. All of the others are reactions of the TCA cycle. (From Joklik et al., *Zinsser Microbiology,* 18th Ed. Courtesy of Appleton-Century-Crofts, 1984.)

Iron-Sulfur Proteins. These proteins can function as electron carriers through reversible oxidation-reduction of iron. The ferredoxins are examples found in some nitrogen-fixing clostridia.

Cytochromes. Cytochromes are the iron-porphyrins of the electron transport chain. They sequentially transport electrons toward molecular oxygen utilizing reversible oxidation-reduction reactions of iron. There is considerable diversity among bacterial cytochromes. Furthermore, some bacteria can utilize the cytochromes to transport electrons to non-oxygen acceptors.

In summary, oxygen is the only terminal electron acceptor of obligate aerobes, cytochromes a, d and o functioning as terminal oxidases. On the other hand, bacteria growing anaerobically utilize non-oxygen electron transport systems. In particular, reductive processes coupled with phosphorylation provide mechanisms to transport electrons anaerobically and generate energy.

Bacteria can secrete hydrolytic enzymes (proteases, collagenases, lipases, nucleases, amylases and lecithinases) which provide a means for utilizing large MW compounds in their environment.

E. NUTRITION AND GROWTH REQUIREMENTS

The basic requirements of medically important bacteria are: a carbon source, a nitrogen source, phosphorus, sulfur and trace mineral elements. As a rule glucose or a similar organic compound serves as the metabolizable carbon source that provides energy for growth. Some medically important bacteria require increased CO_2 for growth because it is a precursor in many biosynthetic pathways. For example, *Neisseria gonorrhoeae* requires increased concentrations of CO_2 for isolation and the maintenance of growth in vitro.

The nitrogen requirements of most bacteria is met by ammonia or ammonium ion. This requirement can be satisfied by enzymatic deamination of amino acids or transamination of keto acids. Ammonia can be utilized directly or by a transamination reaction of alpha-ketoglutaric acid to form glutamic acid, a key amino acid in protein anabolism.

Phosphorus is essential because of its critical role in activities related to nucleic acid, ATP and certain coenzymes. Iron is a constituent of cytochromes and sulfur is required for certain amino acids.

Some species of bacteria require extra preformed growth factors including coenzymes, vitamins, hematin, certain amino acids, purines and pyrimidines. It is thought that in the course of evolution obligate human pathogens, such as certain group A streptococci, progressively shed some of their biosynthetic capabilities because needs provided by these capabilities could be met by utilizing preformed host components. The apparent consequence of the metabolic adaptation of a pathogen to a single natural host is that it greatly limits its capacity to infect other hosts.

As previously discussed, bacteria, for the most part, carry out the same biosynthetic processes that occur in other cells. All bacteria possess a glycolytic pathway, some form of a tricarboxylic acid cycle, and, with the exception of obligate anaerobes, an electron transport mechanism geared to generate ATP via oxidative phosphorylation. *The major difference between bacteria and eucaryotic cells is that bacteria have a greater capacity to adapt to new environments by undergoing phenotypic changes.* For example, *E. coli* in its normal habitat in the colon proliferates anaerobically by fermenting amino acids and sugars but can grow equally well aerobically in a simple solution of salts with glucose as the only organic constituent. The genetic potential and regulatory changes which allow this marked diversity of life styles will be discussed in Chapter 4.

Many bacteria, including most pathogens, can be cultivated in vitro on media composed of meat or plant digests plus an extract of yeast. For fastidious pathogens, such as *Streptococcus pneumoniae,* human or animal blood may be added. In general the more parasitic an organism becomes, the more fastidious it becomes with respect to its nutritional requirements. Presumably as organisms have adapted to a parasitic existence, they have gradually lost much of their biosynthetic capacity through mutation and selection. Under conditions of obligate parasitism many of their complex nutritional requirements are supplied by the host; consequently the enzymes and genes concerned with such nutritional requirements are no longer needed by the bacterial cell and are not retained in the course of their evolution.

Knowledge of the nutritional peculiarities of microorganisms can be used for designing culture media that will prevent or inhibit the growth of unwanted bacteria and/or permit differentiation between pathogens and nonpathogens. For instance, some *selective media* used to isolate enteric pathogens contain relatively high concentrations of citrate which inhibits the growth of *E. coli* but is harmless for species of *Salmonella* and *Shigella.* This allows the clinical microbiologist to isolate etiologic agents which may be present in relatively low numbers, from fecal specimens containing huge numbers of normal flora coliforms. The addition of lactose plus certain indicators provides a mechanism for monitoring changes in the color of colonies that ferment lactose; thus lactose-fermenting pathogens can be differentiated from nonpathogens that do not ferment lactose *(differential medium).* Judicious employment of differential and selective media can markedly simplify the isolation and identification of pathogens from clinical specimens.

F. KINETICS OF GROWTH

Bacteria divide by transverse binary fission and, therefore, express a $\log_2$ function during the period when the maximum rate of growth occurs (log phase) (Fig. 2–13). When bacteria are inoculated into fresh liquid medium there is a characteristic *lag phase* in the growth cycle during which a biosynthetic "tooling-up" occurs and the cells grow in size. During the lag phase the cells adjust their phenotype to the new environment by turning on or turning off certain functions, such as utilization of a specific sugar, amino acids, proteins or fats as sources of nutrients. The lag period is usually influenced by the growth history of the inoculum. For example when a log-phase inoculum is transferred to a new batch of medium the resulting lag period tends to be short or absent; on the other hand, organisms transferred during the stationary or death phase of the growth cycle exhibit a long lag phase.

After an early adjustment period the inoculated organisms in aging inocula begin to enlarge during the latter part of the lag phase and then begin to divide exponentially; this represents the initiation of the phase of most rapid multiplication, the *exponential phase.* Later, as the environment becomes unfavorable the multiplication rate gradually declines until the *stationary phase* is reached, during which the multiplication rate is equal to the death rate; subsequent to this the *decline phase* occurs, during which death exceeds multiplication. The maximum rates of multiplication attainable by different bacteria vary markedly. For example, *Vibrio cholerae* under ideal circumstances has a generation time of approximately 20 minutes, whereas *M. tuberculosis* can divide only once every 14 hours under optimal conditions. However, the generation times in vivo are usually longer than optimal in vitro environments due primarily to the forces of host defense and nutritional limitations. Compared with the maximum growth rate of mammalian cells (generation time

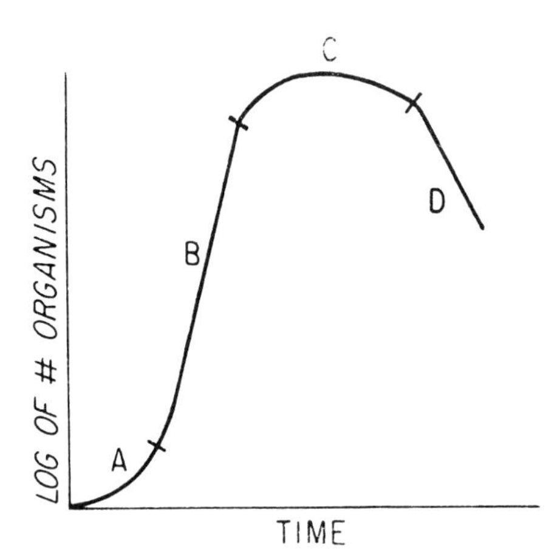

Fig. 2–13. A typical growth curve for a bacterial population. Because of the large numbers of cells produced the counts are generally expressed as Log of the actual number. A, lag phase; B, exponential phase; C, stationary phase; D, decline phase.

about 8 hours), the growth rate of most bacteria is nothing short of phenomenal. The great speed of bacterial growth, which is largely a function of the exponential nature of their multiplication, is a difficult but important concept for the beginning student to grasp. Theoretically, for example, if a culture of *E. coli* could be provided with continuing optimal conditions for growth, a single cell would give rise to a cell mass 4000 times that of the earth within 2 days!

Bacteria seldom sustain exponential growth for more than short periods because of increasing nutritional limitations and/or the accumulation of toxic metabolites in their environment. Even under ideal conditions the maximum population density attainable seldom exceeds 10^{10} per ml.

The basic aspects of cell division can be readily demonstrated using Ab, specific for cell walls, which has been conjugated with a fluorescent dye to permit visualization of the growing cell wall. After the cells have been treated with the labeled Ab, they are allowed to grow; the points of growth show up as dark areas of increasing size. As shown in Figure 2–14, cocci and rods exhibit different patterns of growth. The cocci enlarge and develop a narrow equatorial band which produces a constriction resulting in separation of the two cells; rod-shaped organisms elongate as the result of random or dispersed growth until septation and division take place.

Following division the cells may or may not separate; consequently either single cells or aggregates of cells result, the nature of which varies and is a characteristic of the species. For example, most of the enteric Gram-negative rods occur singly, whereas the Gram-positive cocci tend to remain attached to one another. Among the Gram-positive cocci the pattern of cell aggregation

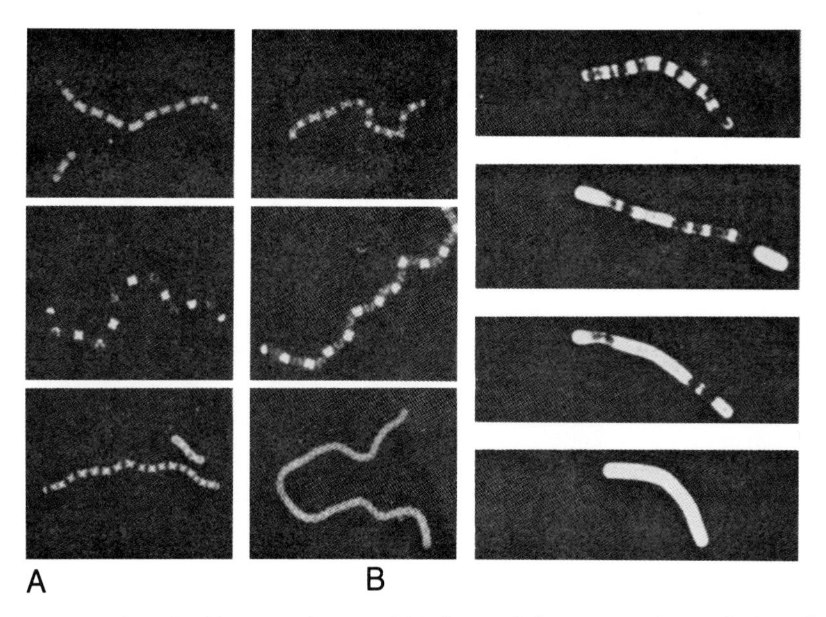

Fig. 2–14. Growth of bacteria after initial labeling with fluorescent Ab specific for cell-wall components. The dark areas represent new wall growth, the brightly stained portions old wall. (A) *Bacillus cereus;* (B) *Streptococcus pyogenes.* (Reprinted with permission from Bacteriol. Rev. *29*:326, 1965.)

varies, depending on the order of division in different planes. Whereas staphylococci divide randomly in 3 planes to form grape-like clusters, pneumococci and streptococci divide in a single plane which results in aggregates of 2 or more cells arranged linearly. In the case of *Corynebacterium diphtheriae* a "snapping motion" commonly follows septation causing the cells to line up in parallel fashion or to form patterns resembling Chinese letters. These species' characteristics are often reflected in the morphology of colonies growing on artificial media. For example, organisms that lyse when they reach the stationary phase of growth yield colonies with depressed centers composed of dead cells and elevated margins composed of actively growing cells. Organisms that display "snapping" post-fission movements tend to form wrinkled, "heaped-up" colonies.

G. PURE CULTURES

To identify a bacterial pathogen with certainty, it is usually necessary to isolate the organism in pure culture. In most instances this is readily accomplished by depositing a small amount of a specimen at the margin of a Petri plate containing an appropriate solid agar growth medium. The specimen is then mechanically dispersed over the surface of the medium by "streaking" with a sterile wire loop. By dragging bacteria away from the initial inoculation site they are in effect diluted out until some of the cells are deposited singly along the line of streaking. Since most motile bacteria are unable to move about on the relatively dry medium surface, the dividing bacteria accumulate and form visible colonies, many of which are the progeny of but a single organism. Colonies that arise from a single cell represent a pure culture; other colonies may also represent pure cultures providing they arise from organisms of a single kind even if more than one cell is initially present. The latter is referred to as a *colony-forming unit.* Since it is not always possible to be certain that a colony arose from a single cell or more than one cell of the same type, it is common practice to select isolated colonies and repeat the streaking procedure to ensure that the culture isolate is pure.

Once a pure culture has been obtained a panel of tests is applied to identify the unknown isolate. Because only a limited number of tests may be required to identify a given species of microbe, standard keys are usually consulted since they include outlines of tests needed for identification as well as systematic procedures. Once a pure culture has been obtained the identification tests commonly employed by the diagnostic medical microbiologist include (1) the Gram-stain reaction, (2) cell morphology and the presence or absence of a capsule, (3) cell grouping, (4) motility and arrangement of flagella, (5) ability to ferment certain sugars, (6) utilization of certain amino acids, (7) identification of unique indicator enzymes, (8) susceptibility to environmental conditions including oxygen tension and specific inhibitors of growth, (9) serologic tests for specific Ags, (10) animal pathogenicity, and (11) phage typing. These procedures can be found in manuals or books on clinical microbiology. The clinical microbiologist is well-advised to keep abreast of the latest advances in identification procedures; many of which are published periodically by the Centers for Disease Control, Atlanta, GA.

There are a few species of medically important bacteria that have not been grown on artificial media. These include *Mycobacterium leprae, Treponema pallidum* and species of *Chlamydia* and *Rickettsia*. The latter two genera can be grown inside host cells in tissue culture. However, this has not been accomplished as yet with the first two organisms.

H. BACTERIAL TAXONOMY

Taxonomic schema are important communication systems which biologists continually struggle to improve. Microbiologists have developed taxonomic schema similar to those employed for multicellular organisms by which bacteria and fungi are placed into orders, classes, families, genera, and species. The principle behind the system is that the more closely related two organisms are, the greater the number of features they should have in common (Sect. A). Since all of the traits used are phenotypic and since not all traits are given equal weight, the system is artificial. However, this means of identifying bacteria is of great practical value to clinical microbiologists because they can label an isolate possessing a given set of characteristics by name and thus communicate their findings to physicians and other microbiologists by this simple expedient.

In Table 2–1, most of the medically important genera of bacteria are grouped according to some of their primary characteristics such as the Gram-stain reaction, shape, cell arrangement, morphologic features, and oxygen requirements. In many instances it is possible to assign an isolate to a genus with no additional information. However, differentiation of a species within a genus (speciation) requires additional tests.

I. QUANTIFICATION OF BACTERIA

It is common practice to determine the number of viable bacteria present in clinical specimens, such as a sample of urine from a patient suspected of having a urinary tract infection, because if it exceeds 10^5 per ml the patient is probably infected. The principle commonly used for determining the number of viable bacteria in a specimen is based on diluting the specimen and plating aliquots on the surface of a suitable agar medium. The specimen must be diluted to a point where the colonies that grow on the surface are dispersed sufficiently so that they can be counted. The number of colonies is a reflection of the number of live bacteria (per ml) in the original urine specimen or, more accurately, the number of colony forming units per ml since clumping often occurs.

Other applications in which bacterial enumeration techniques are used include analyses of water, milk, ice cream, and other food products.

J. BACTERIAL ATTRIBUTES OF SPECIAL MEDICAL IMPORTANCE

The pathogenicity of bacteria rests on their ability to enter the host and to multiply and injure tissues by whatever virulence factors they are genetically capable of expressing. These include secretory products as well as structural components. For example, encapsulated organisms and smooth colonial forms of Gram-negative organisms possess an enhanced resistance to phagocytosis. In the case of the pathogens that can colonize the intestinal and urogenital tracts, hold-fast pili assist in establishing an infection and sexual pili permit the rapid transfer of DNA including DNA that mediates resistance to antimicrobial drugs (Sect. C). The chemical composition of the cell surfaces of all pathogens prevents, to a certain extent, their destruction by host defense mechanisms. Indeed some bacteria can even multiply within phagocytes. In the host, as well as in nature, bacteria are remarkably well suited to adapt to constantly changing environments which accounts for the constant challenge microbes present to the human host and to the microbiologist.

REFERENCES

Christensen, G.D., et al.: The molecular basis for the localization of bacterial infections. Adv. Intern. Med. *30*:79, 1984.
Costerton, J.W.: The role of electron microscopy in the elucidation of bacterial structure and function. Annu. Rev. Microbiol. *33*:459, 1979.
Freifelder, D.: Molecular Biology, 2nd Ed., Boston, Jones and Bartlett, 1987.
Gerhardt, P., et al.: Manual of Methods for General Bacteriology. Washington, D.C., American Society for Microbiology, 1981.
Gottschalk, G.: Bacterial Metabolism, 2nd Ed., New York, Springer-Verlag, 1986.
Gray, M.W. and Doolittle, W.F.: Has the endosymbiont hypothesis been proven? Microbiol. Rev. *46*:1, 1982.
Holt, J.G. and Krieg, N.R.: Bergey's Manual of Systematic Bacteriology. Baltimore, Williams & Wilkins, 1984.
Lennette, E.H.: Manual of Clinical Microbiology, 4th Ed., Washington, D.C., American Society for Microbiology, 1985.
Old, D.C.: Bacterial adherence. Med. Lab. Sci. *42*:78, 1985.
Rogers, H.J.: Bacterial Cell Structure, Washington, D.C., American Society for Microbiology, 1984.
Shockman, G.D. and Barrett, J.F.: Structure, function and assembly of cell walls of Gram-positive bacteria. Annu. Rev. Microbiol. *37*:501, 1983.
Smith, H.: Microbial surfaces in relation to pathogenicity. Bacteriol. Rev. *41*:475, 1977.

3

BACTERIAL VIRUSES

Virtually all types of living cells are subject to invasion by viruses, and bacteria are no exception. The viruses that infect bacteria are called *bacteriophages,* or simply *phages.* Because bacteria can be easily cultured and grow rapidly, studies on phage replication and infection are more readily accomplished than similar studies using animal viruses. Studies on bacteriophages have been of great importance in microbiology, genetics, and molecular biology. Medical microbiology, for example, has profited from phage studies in the following ways: first, the mechanism of replication of phages and viruses that infect animals are similar, second, phages (like all viruses) are highly specific with regard to the cells they can infect; therefore it is possible to use the infection pattern with a panel of known phages to identify bacterial strains. This procedure, known as *phage typing,* has been useful in epidemiologic studies concerned with tracing the spread of infection due to specific strains of bacteria. Finally, certain kinds of phage can infect bacterial cells and direct the production of potent bacterial exotoxins that are responsible for diseases such as diphtheria, scarlet fever, and botulism.

Many phages have been discovered. The best studied phage is called *lambda* (λ) which attacks *Escherichia coli.* The entire complement of 48,514 base pairs which make up the double stranded chromosome of this phage has been sequenced and all protein producing genes identified. Therefore λ provides an excellent model for understanding and illustrating phage infection.

The morphology of phage λ is shown in Figure 3–1. Although the morphology of λ and many other phages seems to be more complex than the morphology of most animal viruses, there are certain similarities between viruses of the two groups. Like all viruses, the λ particle consists essentially of a piece of nucleic acid surrounded by a protein coat *(capsid).* The complete virus particle is called a *virion.* When outside its host cells, the mature λ virion is inert. However, when the phage nucleic acid reaches the cytoplasm, it is capable of diverting cellular metabolism toward production of virus particles.

When phage λ encounters an appropriate host cell, the base plate of the phage's "tail" attaches to a specific site *(receptor)* on the bacterial surface. Once

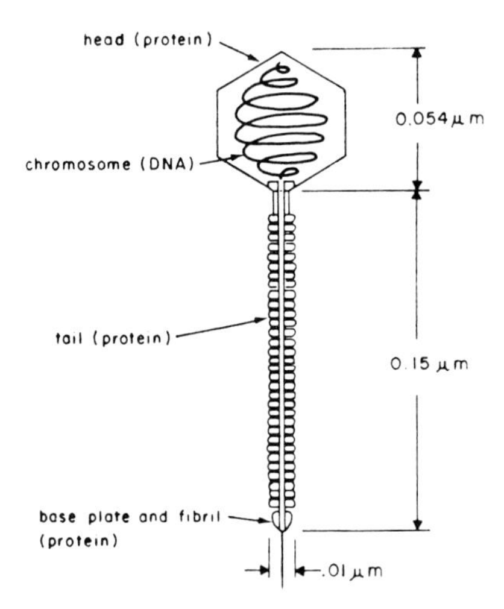

Fig. 3–1. Bacteriophage λ. (Adapted from The Bacteriophage Lambda. Ed. by A.D. Hershey. Cold Spring Harbor, NY: Cold Spring Harbor Laboratory, 1971.)

specific attachment has occurred, the viral DNA is injected into the cell, a process called *penetration,* leaving the protein capsid outside. Thus, the phage tail functions as an organelle to facilitate specific adsorption to cells and as a tube to deliver the phage DNA through the thick bacterial cell wall and into the cytoplasm.

There are two distinct types of life cycles which phage λ can undergo (Fig. 3–2).

A. THE VEGETATIVE (LYTIC) LIFE CYCLE

About 45 minutes after a λ chromosome has entered a host cell, the cell usually lyses and releases approximately 200 mature virus particles. The in-

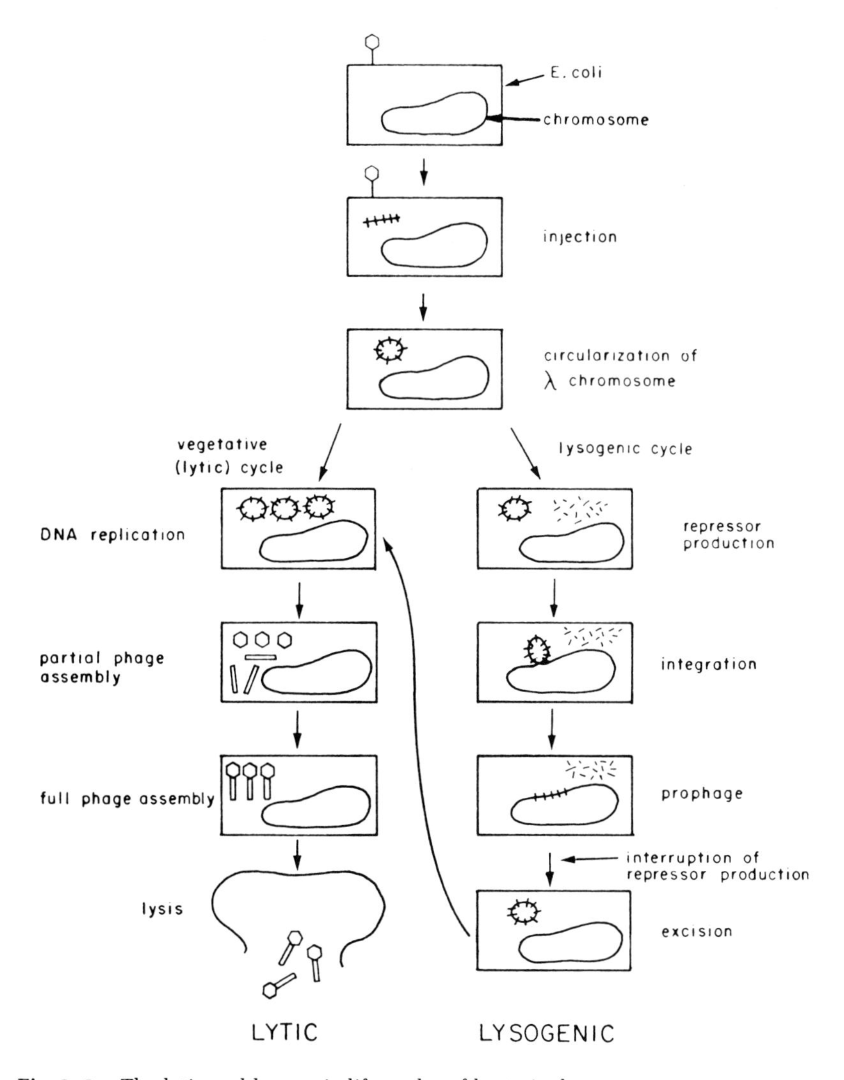

Fig. 3–2. The lytic and lysogenic life cycles of bacteriophage λ.

terval between penetration and host cell lysis is called the *latent period.* Much of what the virus needs to complete its vegetative life cycle is coded for by the phage chromosome. Therefore, the latent period has provided an excellent opportunity for studying gene expression. The extensive genetic mapping and study of the functions of λ have demonstrated that its chromosome is highly organized, the genes determining related functions being grouped into adjacent clusters (Fig. 3–3).

In mature, infectious particles, the chromosome of phage λ is a linear molecule of double-stranded DNA about 1% as long as the host *E. coli* chromosome. Soon after infection, the mature ends of the λ chromosome join to form a circle which serves as a template for the synthesis of messenger ribonucleic acid (mRNA). The early control functions shown in the map in Figure 3–3 are expressed first. Then the functions controlling recombination, DNA replication, and late control are expressed. Finally, the functions that control lysis and head and tail functions are expressed. Although synthesis of both phage nucleic acid and protein is substantial throughout the latent period, no mature phage

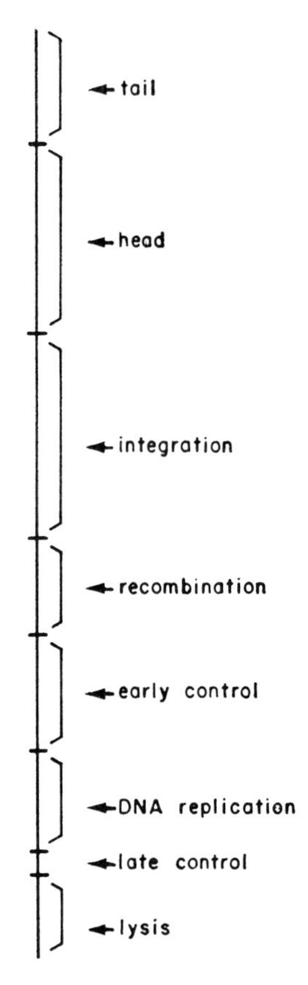

Fig. 3–3. The genetic map of bacteriophage λ. Genes that control related functions are grouped together on the chromosome of λ. Therefore each of the bracketed regions represents a cluster of genes.

particles appear during the first half of this period. This early part of the latent period, during which no mature virions are present, is called the *eclipse period.* During the second half of the latent period which is designated the *maturation period,* mature virus particles are assembled and finally released by cell lysis.

The mechanisms by which the expression of λ genes are controlled, as well as the act of gene expression itself, have been well characterized. Although some areas are still not well understood, it is appropriate to recount certain important events which occur during the eclipse phase. The steps involved are:

1. Genes specifying the proteins which are involved in the control of transcription are expressed first (i.e., early control).
2. Following this, products needed for recombination and DNA replication are synthesized; copies of the virus chromosome are then synthesized.
3. A product is synthesized which is needed for efficient expression of the genes involved in lysis as well as in head and tail construction (i.e., late control); the head, tail and lysis proteins then begin to accumulate.
4. Heads and tails begin to form which, together with DNA, are subsequently assembled into mature virus particles. Assembly of the first mature virion signals the end of the eclipse period and the beginning of the maturation period.

During the maturation period, synthesis of many virus components continues, while at the same time final assembly of components into mature virus particles leads to the accumulation of infectious virions in the cytoplasm of the cell. The formation of mature virus takes place in an assembly-line fashion and is notably different from the multiplication of bacteria, which occurs as the result of binary fission. As a terminal event, the accumulation of a phage-induced endolysin causes lysis of the cell wall and release of the phage particles.

B. THE LYSOGENIC LIFE CYCLE

For reasons which are not entirely clear, some λ chromosomes are only partially expressed and then become integrated into the bacterial chromosome, a process known as *lysogenization.* It has been found that the probability of lysis or lysogeny is significantly influenced by the level of cyclic AMP in *E. coli* at the time of infection. Expression of phage genes themselves is not directly affected by cyclic AMP; however, the *efficient* expression of the phage genes whose protein products stimulate lysogeny (see below) is dependent on a cell protein which is efficiently produced only when cyclic AMP is present in relatively large amounts. Therefore, the scale tips toward lysis of *E. coli* by λ when the cyclic AMP concentration is relatively low and toward lysogeny when cyclic AMP is abundant. What this means is that λ will tend to lyse *E.* coli when the bacteria are obtaining their energy from glucose and lysogenize when glucose is not present. Preliminary to integration another protein is synthesized, namely, a repressor, that eventually represses the production of all mRNA of the phage except that which specifies the repressor molecule. The integrated λ chromosome is called a *prophage.* The DNA of the prophage is replicated along with the bacterial chromosome; however, except for the production of repressor, the prophage is a silent partner in the phage-bacterium association.

A notable finding is that the presence of prophage repressor in the lysogenic cell renders the cell "immune" to superinfection with infectious phage λ because the repressor (often referred to as an "immunity repressor") blocks the production of mRNA by any newly introduced λ chromosome.

The stability of prophage in a culture is illustrated by the observation that certain lysogenic strains of bacteria have been maintained by serial transfer for over 20 years. Nevertheless, the prophage is potentially lethal to the cell that carries it. In about one of every 10^5 cells the synthesis of repressor is interrupted and, as a result, λ genes are expressed; the prophage is excised from the chromosome of the organism, and λ genes initiate the lytic (vegetative) cycle leading to the production of mature λ particles and lysis of the host cell. Release of λ phage particles into the culture medium of a lysogenic strain is of no consequence to the remaining lysogenic cells in the culture since they are immune to superinfection with infectious λ particles.

The probability of a switch from the prophage state to the vegetative state can be increased by treating lysogenic cells with certain chemical and physical agents which mediate a process known as *induction*. For example, appropriate exposure of *E. coli* (λ) (i.e., *E. coli* lysogenic for bacteriophage λ) to ultraviolet (UV) irradiation leads to induction of the prophage in nearly every cell in the population and, after a period of time, there is a massive release of phage from the induced cells.

Other interesting phenomena associated with bacterial viruses (not necessarily λ) will be discussed since some of them also occur among the animal viruses.

C. PHENOTYPIC MIXING

The assembly-line manner of production of mature virions can lead to unusual types of virus particles in which components of similar, but genetically distinct, viruses are combined. One example involves the so-called *host range mutant* (Fig. 3–4). The cell-wall sites which serve as specific attachment receptors for the complementary sites on the tip of the phage tail sometimes

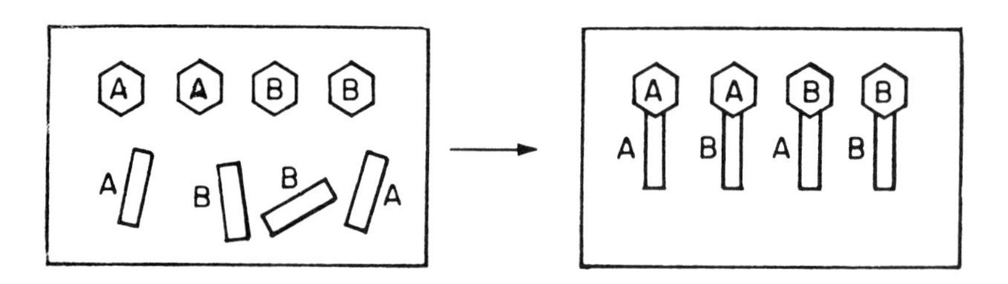

Fig. 3–4. Phenotypic mixing. The assembly-line production of complete bacteriophages from component parts can lead to phenotypically mixed particles. For example, tails specified by the genes of phage A can be connected to a head containing a B-type chromosome and vice versa. Phages A and B must be related for phenotypic mixing to occur.

change due to mutation, so that they can no longer permit the adsorption of the particular phage. The phage, in turn, can sometimes undergo mutation involving one of the subunits of its base plate concerned with cell attachment, so that the phage can attach to both mutant and wild-type (nonmutant) host cells. For example, *phage A* might be capable of adsorbing to and growing in bacterial *strain a* but not *strain b.* In contrast, *phage B* (a mutant of phage A) might be capable of attaching to and injecting its DNA into both *strain a* and *strain b.*

Now suppose that a cell of strain a simultaneously is infected with phage A and phage B and that, except for one mutation (presumably caused by the alteration of a single base pair), these two phage strains are identical. As shown in Figure 3–4, during the assembly of progeny phage particles, a tail made by phage A could become attached to a head containing the DNA of phage B or vice versa. This means that a phage can be formed possessing the genotype of phage A but the phenotype of phage B, or vice versa. Because of these possibilities, it is possible to infect cells of type B with the DNA of phage type A. As would be expected, the chromosome of phage type A will direct the production of only type A phage regardless of what kind of tail was associated with the original chromosome.

The mutation of phage A to phage B also illustrates a mechanism by which phages can vary so as to attack a wide range of hosts. In the particular model cited, phage B would be referred to as a *host range mutant* of phage A.

D. VIRAL EXCLUSION OR INTERFERENCE

There are very few individual bacterial cells which can simultaneously support the replication of two unrelated phages. The reasons for this undoubtedly differ widely from one situation to another. Competition for essential constituents such as ribosomes, enzymes, and nucleotides is probably an important contributing factor in many cases of viral exclusion. Some bacteriophages, such as the coliphage T4, can exclude the growth of almost all unrelated viruses because T4 makes nucleases that destroy most heterologous DNA molecules.

E. MULTIPLICITY REACTIVATION

Exposure of phage particles to harmful agents such as low doses of UV irradiation, often causes a precipitous drop in viral infectivity. However, the members of a population of noninfectious irradiated phages often can cooperate to produce an infection. It is presumed that the cooperating phage particles have been damaged in different parts of their chromosomes; however, as the result of genetic recombination of undamaged genes, the cooperating chromosomes give rise to undamaged phages.

F. CROSS REACTIVATION

A phenomenon similar to multiplicity reactivation is cross reactivation, a circumstance in which an undamaged virus serves to repair a damaged phage. Cross reactivation can occur between heterologous as well as between homologous phages; evidently it results from complementation of the damaged genes by complementary genes of the undamaged phage as well as through genetic recombination.

REFERENCES

Freifelder, D.: Molecular Biology. 2nd Ed. Boston, Jones and Bartlett, 1987.

Hendrix, R.W., et al.: Lambda II. Cold Spring Harbor, New York, Cold Spring Harbor Laboratory, 1983.

Hershey, A.D., (ed): The Bacteriophage Lambda. Cold Spring Harbor, NY: Cold Spring Harbor Laboratory, 1971.

Kucera, L.S. and Myrvik, Q.N.: Fundamentals of Medical Virology. 2nd Ed. Philadelphia: Lea & Febiger, 1985.

Mathews, C.K. et al.: Bacteriophage T4. Washington, D.C., The American Society for Microbiology, 1983.

Stent, G.S. and Calendar, R.: Molecular Genetics, 2nd Ed. San Francisco, W.H. Freeman and Company, 1978.

Stewart, F.M. and Levin, B.R.: The population biology of bacterial viruses: why be temperate? Theor. Popul. Biol. *26*:93, 1984.

Yanagida, M. et al.: Molecular organization of the head of bacteriophage T seven: underlying design principles. Adv. Biophys. *17*:97, 1984.

4

BACTERIAL GENETICS

Microbes are notorious for their potential variability. In fact, genetic changes in infecting organisms often occur during the course of infection. Rapid genetic changes in infecting organisms do not occur because the mechanisms of genetic change among bacteria are basically different from those that occur in higher organisms; instead, the unusually short generation time of bacteria simply leads to a more rapid selection of those organisms in a population which are best equipped to survive a given environment.

It is assumed that the student has at least a cursory knowledge of the structure of deoxyribonucleic acid (DNA), the role of DNA in heredity, and the way in which mutations occur. The purpose of this chapter is to enable a student to build a framework of knowledge which will be of practical value in clinical medicine. Four aspects of microbial genetics are considered: (1) bacteria as genetic entities; (2) the exchange of genetic information between bacteria; (3) control of genetic expression, and (4) microbial genetics in medicine.

A. BACTERIA AS GENETIC ENTITIES

Like all forms of life, bacteria are able to transmit their characteristics to their progeny, and the progeny in turn are able to translate this information (genetic potential or *genotype*) into products (genetic expression or *phenotype*). With respect to microbial genetics, the most studied species is *Escherichia coli.* From corollary studies with organisms related to *E. coli,* such as *Salmonella* species as well as with members of the unrelated genus *Bacillus,* it has become apparent that much of the information obtained with *E. coli* has general application.

The chromosome *E. coli* is about 1400 nm long and contains approximately 4×10^6 base pairs. Since an average gene contains 1000 base pairs, *E. coli* has approximately 4000 genes; over 500 of these genes have been characterized and located on the chromosome. This has permitted detailed studies on mutation, gene expression, and heredity.

Mutation is defined as a heritable change in the genome which does not result from the incorporation of genetic material from another organism. It is well recognized that mutations occur at a relatively constant rate in all dividing cells. Certain agents *(mutagens)* are able to accelerate mutation by mechanisms known to cause natural or spontaneous mutation. The way in which a cell responds to a given mutagen is not always simple or predictable. Mutagens commonly cause mutations by inducing pairing errors of various sorts between the nitrogenous bases of DNA during its replication. Such errors may result from either chemical or physical damage to DNA by the mutagen or a variety of other causes. It is notable that many of the cells present in a population of 10^8 *E. coli* will have experienced a mutagenic event. In fact it is probable that a large population of *E. coli* contains at least one mutant for every gene possessed by the cells. As a consequence of evolution, microbes have undergone mutation and selection for perhaps millions of years; this has resulted in a large number of species that specifically fit their own particular ecologic niches. Therefore, most mutants derived from wild-type *E. coli* have a reduced ability to survive in nature! Accordingly, most mutants are usually at a selective disadvantage if they must compete in the natural habitat of the wild-type species. When the natural habitat is altered, mutants may have greater ability to survive than the wild-type strain.

For example, if a large population of *E. coli* is exposed to appropriate concentrations of the antibiotic, streptomycin, the chances are good that before adding the drug at least one organism in the initial population would have experienced a random mutation yielding streptomycin resistance. Consequently, all of the wild-type streptomycin-sensitive cells should soon be killed, and further incubation of the streptomycin-containing culture should lead to the growth of streptomycin-resistant mutants. This illustrates the important evolutionary principle that the ability to undergo random mutation endows each microbial species with a remarkable degree of flexibility for meeting unfavorable alterations in the environment.

Once DNA has undergone a mutational change, there is a certain time lapse before the new mutant phenotype is expressed *(phenotypic lag).* This time lapse may occur because the product of the formerly unmutated gene must be depleted before the new phenotypic change is expressed. A lapse may also occur because many bacterial cells have two copies of the chromosome, only one of which is mutated. When the cell divides, only one of the daughter cells obtains the mutant copy.

It is important to remember that mutation occurs spontaneously and randomly; because of the short generation time of bacteria, a bacterial mutant can often be selected in an altered environment preferentially favorable to it and dominate a population within a day or two.

It is also noteworthy that phenotypic changes occur in bacteria which are not the direct result of a mutation of genetic exchange but instead involve a complex system of gene control. At any given moment, a large number of the genes of an organism are not expressed, and moreover, it is only in the face of specific stimuli *(inducers)* that certain genes are expressed. The altered response of an organism by specific stimuli is called *adaption* or *induction* and is discussed in Section E_2 of this chapter. It occurs much more rapidly than selection of a mutant; in general, all members of a population may become fully induced within a few hours. In this regard, pathogenic organisms can undergo adaptive change in response to host factors during the course of infection. Such change is not the result of mutation but nevertheless represents an advantage provided by evolution to enable a pathogen to adjust rapidly to the hostile environment normally presented by its natural host.

B. TRANSPOSONS

Transposons or transposable elements (the terms are used interchangeably) are specific sequences of DNA which can "transpose" or "hop" from one DNA site to another. A large variety of transposons have been discovered and they range in length from 0.8 to 40kb (kilobases). Some of the smaller transposons seem to code for nothing except the ability to transpose and these are often referred to by the special term "insertion sequence" or IS-module. Large transposons usually carry a readily identifiable gene such as antibiotic resistance and some carry transcriptional promotors which can cause constitutive transcription of adjacent bacterial DNA.

A wide variety of transposons have been identified, each with its own individual characteristics. The purpose of the discussion in this section is to define the terminology used with transposons, the features common to transposons, the effect of the transposition of transposons on microbial genetics, and the role played by transposons in medical microbiology.

As mentioned above, the IS-modules (IS-1, IS-2, etc) code only for products essential for transposition. IS modules are usually 0.8 to 2kb in length. Larger transposons which code for other products, in addition to those needed for transposition, are usually designated Tn3, Tn5, Tn9, etc. However, other types of terminology do exist (e.g. Tnγ). Identical IS-modules can sometimes cooperate to form a composite Tn. For example, there is a 9.3kb transposon designated as Tn10 which consists of two identical IS-1 modules at each end of a 7.3kb length of DNA which contains a tetracycline resistance gene. Many transposons which carry antibiotic resistance genes are not composites of IS-modules but rather unique entities.

All transposons have the following features in common: 1) a fixed length and sequence (these characteristics vary from transposon to transposon but for any particular transposon the length and sequence are constant), 2) identical (or nearly identical) inverted repeats at their termini, 3) the formation of short (5 to 12kb) direct repeats of the target DNA located immediately adjacent to the transposon, 4) the ability to code for one or more proteins essential for transposition, 5) the ability to move from one site to another through self-replication. This characteristic means a transposon which hops from one site to another remains at the original site, 6) the frequent causation of cointegration or joining of independent replicons (discussed more fully in Section C below), and 7) the formation of genetic rearrangements or alterations near the site of their insertion.

Transposition is not a common event and usually occurs only once in 10^5 cells per generation. Experimental data suggest that a specific transposon-coded protein (either alone or in cooperation with cellular proteins) recognizes a transposon terminus and a target site (most targets appear to be random sequences of DNA but certain transposons select specific targets). One strand of the transposons is probably cut and joined to the cleaved target site; each strand of the transposon is then replicated, and finally the other terminus is then cut and joined to the target. The exact mechanism by which transposons accomplish transposition to a new site remains unknown.

The transposition of a transposon, which leads to a duplication of the transposon, naturally leads to the duplication of a stretch of homologous DNA. When a transposon hops to a new site in the same chromosome, homologous recombination between the two stretches of homologous DNA can have far reaching effects. For example, if the two copies of the transposon are in the same orientation, a single reciprocal recombination event will lead to a deletion of all the DNA located between the transposons; if the copies are in the opposite orientation, recombination between them will cause the inversion of all the intervening DNA. In addition, insertion of a transposon into a gene will nearly always result in a loss of gene function and, furthermore, all operon genes separated from their promoter by transposon insertion will usually be inactivated. As mentioned above, insertion of certain transposons can sometimes lead to constitutive transcription of adjacent bacterial genes. Therefore, transposons are potent mutagens.

In addition to causing mutations, transposons can affect the genetics of microorganisms in another important way. If a transposon hops from one replicon to another, a single reciprocal recombination event between the resulting duplication of homologous DNA can lead to "co-integration" or joining of the replicons. For example, if a 10kb transposon hopped from a 50kb plasmid to a 400kb bacterial chromosome, a recombination event between the duplicated homologous, 10kb DNA would lead to a phenomenon generally referred to as integration of the plasmid into the bacterial chromosome but what in fact is co-integration or joining of the two replicons. In the example cited a new 470-kb chromosome would be formed.

Transposons are important in research as either a simple method of causing mutations or a mechanism to "tag' plasmids with antibiotic resistance genes. Likewise, transposons are often used as a method of joining different replicons. In medical microbiology transposons provide the explanation of how antibiotic resistance can first become part of a plasmid (some of which are able to catalyze conjugation between bacteria) and then move to new cells, eventually ending up in pathogens. This mechanism was undoubtedly one of those which has been responsible for the movement of antibiotic resistance genes from relatively harmless, free-living organisms to human pathogens; furthermore, it demonstrates how antibiotic resistance genes brought into a pathogen by a plasmid can hop to the bacterial chromosome and become a permanent part of the pathogen.

C. PLASMIDS

Plasmids are extrachromosomal pieces of DNA which are able to replicate without physically joining the bacterial chromosome. Analogous to the word "operon," which is a coordinate unit of transcription, the word "replicon" is used to describe a unit of replication. The bacterial chromosome is a replicon and so is the chromosome of a vegetative bacteriophage. In its prophage form the chromosome of a temperate phage becomes part of the bacterium's replicon (see Chap. 3). Therefore, plasmids which replicate independently of the bacterial chromosome are also replicons. The minimum content of any replicon is a site at which the DNA polymerase can bind and begin replication with DNA. Most replicons also carry information specifying proteins which control DNA replication.

Plasmids are circles of double stranded DNA which can be thought of as parasites of the bacterial cell. However, the intimate association of plasmids with their host cell precludes most of the harmful effects which parasites often display toward their host. On the contrary, in exchange for providing a home for a plasmid, a bacterium may acquire a plasmid function which gives a cell selective advantages. "Basic" plasmids, with no known function other than the ability to replicate, are the exception rather than the rule. Most plasmids have genes which extend their activities beyond mere replication. There are three activities of plasmids which concern us here: (1) promotion of conjugation and gene transfer to recipient cells; (2) ability of a plasmid to be inserted into a bacterial chromosome, and (3) expression of genes which make cells antibiotic resistant.

1. Conjugative Plasmids

Certain plasmid-containing cells are able to initiate mating with compatible recipients. Mating leads to the transfer of DNA from the plasmid-containing cell (called a "donor" or "male") to the recipient (called a "female"). Recipients are usually closely related cells which do not contain the plasmid. Genes which control mating are numerous and the process is complex. A specific conjugative plasmid (the F factor of *E. coli*) is discussed below. Therefore, only a brief and general description is given here. A plasmid becomes a conjugative plasmid when it acquires a group of genes which code for the proteins which form sex pili. Compared to common pili, sex pili are much longer and have a more complex structure. The sex pilus serves to make contact with a female cell, thus joining the mating pair. Furthermore, a conjugative plasmid must be able to initiate a special type of DNA replication that permits the plasmid to bypass normal replication-control and transfer a single strand of plasmid DNA to the recipient. Finally, by unknown means, plasmids often promote transfer of the DNA of other plasmids and even of the donor bacterium to recipient cells.

2. Drug Resistant Plasmids

Bacteria may carry plasmids which cause them to be resistant to an antibiotic. In fact, a cell may carry a single plasmid which specifies resistance to more than one antibiotic or several different types of plasmids each of which specifies resistance to only one antibiotic. An interesting feature of plasmid-mediated antibiotic resistance is that the genes which specify resistance do not seem to have originated in the pathogen in which the plasmid is found. By the use of DNA:DNA hybridization it is possible to determine the degree of relatedness of different pieces of DNA. There is little complementarity between the DNA of the resistance genes and the host chromosome. The origin of the genes which code for antibiotic resistance in medically important organisms must, therefore, remain a mystery at present. The mechanisms by which antibiotic resistance is effected by plasmid genes vary from antibiotic to antibiotic and plasmid to plasmid; however, a common mechanism is the enzymatic inactivation of the antibiotic.

Antibiotic resistance plasmids can be transferred from cell to cell by a variety of methods. For example, in bacterial genera with no known conjugation mechanisms, plasmids can be transferred via transduction. Most antibiotic resistance genes in plasmids are associated with transposons (Sect. B).

3. Complex Plasmids

As mentioned above, a "basic" plasmid may acquire groups of genes which extend the activity of the plasmid. Some plasmids acquire several groups of genes and, for example, are both conjugative and drug resistant. An example of a complex plasmid which carries antibiotic resistance genes and is conjugative is the R factor of *E. coli* or *Salmonella*. A conjugative-integrative plasmid (the "F" factor of *E. coli*) is discussed below.

4. Integration of Plasmids

As mentioned in part B of this chapter, plasmids and bacterial chromosomes can co-integrate or join. The best data available suggest that co-integration is nearly always brought about through the duplication of homology which results from transposon duplication. Plasmids and bacterial chromosomes may, by chance, carry identical transposons or the transposon of one replicon may hop to the other and thus "create" homology. Some transposons are known to form co-integrates as an intermediate in transposition and these intermediates may sometimes become fixed or "trapped" in the co-integrate form. It will be shown in part D of this chapter that co-integration of conjugative plasmids and bacterial chromosomes has a profound effect on the exchange of bacterial genes between different types of bacteria. The co-integration of a plasmid and bacterial chromosome is frequently referred to as the integration of the plasmid into the bacterial chromosome.

D. GENETIC EXCHANGE IN BACTERIA

The exchange of genetic material between bacterial cells is known to occur by transformation, conjugation, and transduction. Although bacteria of any given species usually do not exchange genetic information by all three mechanisms, cells of all species exchange genetic material by at least one of the three mechanisms. Studies of gene transfers between bacteria have made possible the mapping of some bacterial chromosomes.

1. Transformation

The exchange of genetic information by the transfer of naked DNA from one cell to another is called *transformation.* This phenomenon was first described in an in vivo experiment in mice, with pneumococci. Subsequently, the results of in vitro experiments established that bacteria of other genera can also exchange genetic information by this means. Segments of donor DNA are taken into recipient cells and combined with recipient chromosomes. Because the piece of DNA taken up by the recipient cell is usually small, only a few genes are usually transferred, a circumstance which has greatly aided in mapping the chromosomes of some bacteria.

Transformation provides a means for assaying the biological activity of DNA and consequently of relating the specific activity of mutagens which directly alter DNA. In addition, it provides a useful tool for determining the effects of common chemicals on DNA function. Transformation is also a promising means for assessing the functional attributes of DNA that has been synthesized in vitro.

2. Conjugation

Bacteria of some species exchange genetic information by a form of sexual recombination, in which "male" cells conjugate with "female" cells. During conjugation a part of, or sometimes even the entire, chromosome of the donor male bacterium is transferred to the female recipient. Subsequent genetic recombination of the DNA results in incorporation of the transferred genes.

Elucidation of the genetic elements that control bacterial conjugation has led to a better understanding of reactions between chromosomes and extrachromosomal elements. Obviously, bacterial conjugation is a useful tool with which to study genetics in general and bacteral genetics in particular. The direct transfer of hereditary characteristics from one bacterium to another is of great importance in medicine as well, e.g., the transfer of antibiotic resistance.

It is appropriate that a discussion of conjugation begin with the *E. coli* F factor (fertility or sex factor) since no other fertility factor model is so well understood. The F factor of *E. coli* is an episome consisting of a piece of DNA about the same size as bacteriophage λ (i.e., about 1% of the size of the bacterial chromosome). The F factor is both a conjugative and an integrative plasmid.

a. ***Autonomous F Factor.*** In *E. coli* the autonomous form of the F factor evidently replicates in approximate synchrony with the bacterial chromosome. Cells with an autonomous F factor are called F^+ ("male") and those without are termed F^- ("female").

The functions controlled by F factors are not well understood. In addition to controlling its own autonomous replication, an F gene directs the synthesis of a product called *F protein,* which is a constituent of *F pili.* Although a bacterium may have many pili of various kinds, a given F^+ cell usually does not possess more than a few F pili and these are longer than hold-fast pili. Mating between F^+ and F^- cells occurs by attachment of the pilus of the F^+ cell to the F^- cell. The pilus then retracts to draw the cells close together thus permitting bridge formation.

The DNA component which is transferred most efficiently from F^+ and F^- cells is the F factor itself. Upon acquiring the F factor, the F^- cell becomes an F^+ cell. It has been demonstrated that a brief exposure of an F^- population to a few F^+ cells causes a rapid conversion of essentially the entire population to F^+ cells. This is feasible because newly formed F^+ cells can, in turn, transfer the F factor to F^- cells. It is believed that the F factor itself is replicated at the time of transfer and that possibly the energy associated with replication is the driving force for DNA transfer.

The weight of experimental evidence supports the concept that an autonomous F factor can also "mobilize" the transfer of other bacterial DNA from F^+ to F^- cells. However, the transfer of other bacterial DNA through the mediation of an autonomous F factor is inefficient compared to the transfer of the F factor itself. The primary means by which bacterial DNA is transferred when male and female cells are mixed will be discussed below.

Autonomous F factors are easily deleted from F^+ cells by a variety of agents such as UV irradiation and acridine dyes. A cell which loses its F factor is said to have been "cured," and such cells become typical F^- cells capable of acting as recipients during conjugation.

b. ***Integrated F Factor.*** In a population of F^+ cells there are always a few cells in which the F factor has become integrated into the bacterial chromosome. Cells with integrated F factor are called *Hfr (high frequency of recombination) cells.* The integrated form of F is stable, and such cells can be obtained in pure culture. Like F^+ cells, the Hfr cells produce F pili.

The most striking feature of F factor integration is that the transfer of chromosomal DNA from Hfr to F^- cells is much more efficient than the transfer from F^+ to F^- cells. In fact, it is generally believed that most of the transfer of bacterial DNA which occurs when F^+ and F^- cells are mixed is mediated by the few Hfr cells found in most F^+ populations.

The way in which bacterial DNA is transferred from an Hfr to an F^- cell depends on where and how the sex factor is integrated into the bacterial chromosome. Unlike phage λ, which can integrate at only one site, the F factor can integrate at any one of a number of different locations. Furthermore, depending on the particular site at which the sex factor is integrated, F can be oriented in either direction. In Figure 4–1, the F factor is depicted as an arrow, which can be pointed (or oriented) in either a clockwise (Fig. 4–1b) or a counterclockwise (Fig. 4–1a) direction. This is important because it determines the direction in which the bacterial chromosome is transferred to recipient cells.

It is believed that during conjugation the chromosome breaks at the site (the origin) of integration of the F factor, and that the gene originally next to it enters the recipient cell first. Replication of the bacterial chromosome begins at the origin and the DNA is transferred as a single, continuous strand. In Figure 4–1a, the order of transfer of the genes would be A, B, C, D and in Figure 4–1b, the order would be C, B, A, D. The probability that a recipient gene will acquire the leading gene may, under ideal conditions, be as high as 25%. The farther a gene is from the leading end of the chromosome (i.e., from the origin) the less its chances of transfer. The time required for transfer of an entire chromosome of *E. coli* is about 90 minutes under ideal conditions. Because the link between conjugating male and female bacteria is fragile and easily disrupted, transfer of the entire bacterial chromosome is uncommon. Therefore, the transfer of the integrated F factor at the very end rarely occurs, and most recipient cells remain F^-.

c. ***Intermediate F Factor.*** On occasion, the integrated F factor leaves the bacterial chromosome and carries with it a piece of adjacent bacterial DNA. A departing episome which acquires bacterial DNA often leaves behind some episomal DNA; hence the terms *intermediate* or *substituted* F factor. A cell with an intermediate F factor is called an F′ ("F prime") cell. The intermediate

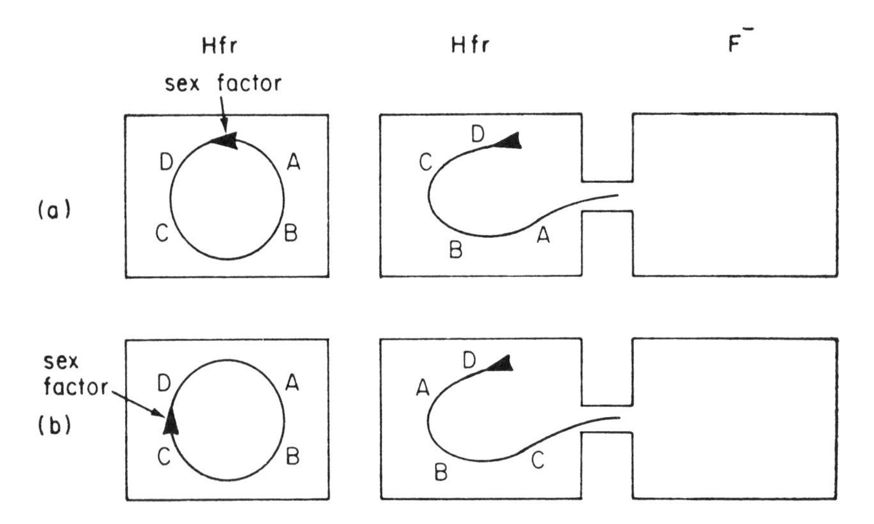

Fig. 4–1. Conjugation between Hfr and F^- strains of *Escherichia coli.* The two parts (a and b) of the figure demonstrate the effect of location of the integrated F factor on the origin and direction of transfer.

sex factor, including the bacterial genes it carries can be transferred to a female with an efficiency which resembles the transfer of an autonomous F episome. Except for the genes attached to the sex factor itself, F′ cells do not mediate the transfer of bacterial genes with a high efficiency.

The cell which acquires an intermediate F factor becomes diploid for the region of the chromosome carried by the intermediate F. The bacterial genes carried by the intermediate F, in rare instances, can undergo genetic recombination with the homologous genes in the bacterial chromosome but more often the cell survives as a partial diploid. The intermediate F is somewhat unstable and F′ cells are easily cured of their fertility factor. It is of singular importance that the F′ particle bears a close resemblance to certain episomes which mediate the rapid transfer of multiple drug resistance among enteric pathogens (Chap. 18).

3. Transduction

Certain bacteriophages are able to carry bacterial genes from one bacterium to another by a process known as *transduction.* There are two types of transduction, specialized and generalized.

Specialized transduction will be mentioned only briefly. Occasionally when phage DNA, such as that of phage λ, is excised from the bacterial chromosome an error is made. Excision of phage λ may result in a piece of bacterial DNA remaining attached to the excised λ DNA while a piece of λ DNA is left behind. These hybrid particles are defective and are called λ*dg;* they are defective because they lack the λ functions which are left behind. They also carry the adjacent bacterial genes which control the ability to ferment the sugar, galactose. Phage λ*dg* bears a certain resemblance to the intermediate sex factors just discussed. A number of other bacteriophages have been found which are able to transduce only those genes which are located near the site of phage attachment.

Of broader interest is the observation that certain bacteriophages are able to transfer any gene from one host cell to another by a process known as *generalized transduction.* The simplest and most widely accepted explanation of the mechanism of generalized transduction is that during the assembly-line production of mature phage particles, a phage head can occasionally be filled with a piece of bacterial DNA rather than phage DNA. Thus, the production of generalized transducing particles is caused by a "phenotypic mixing phenomenon" in which a piece of bacterial DNA has replaced phage DNA in an otherwise normal phage particle. The transducing particles have the same host range as the phage. The probability that a given gene will be accidentally incorporated in a phage head and eventually transferred to a recipient cell is approximately equal for most genes. However, there is a wide variety of factors which can lead to a greater probability of a particular gene being picked up and transferred.

Transduction is known to occur among bacteria of many genera and therefore may be the most important mechanism of genetic exchange among bacteria in nature. Transduction is known to be an important mechanism of transfer of drug resistance between certain types of pathogens. For example, plasmid-mediated penicillin resistance, which is of vital concern in infections with *Staphylococcus aureus,* can be transferred from one cell to another by means of transduction. Thus, transduction is not only a means by which bacterial-chromosomal genes can be transferred but may also serve as a means of transfer of an entire plasmid.

E. CONTROL OF GENETIC EXPRESSION

As previously mentioned, not all bacterial genes function simultaneously. Rather, it has been observed that genes which control the individual steps in a particular metabolic pathway are often transcribed or alternatively, repressed en bloc. This form of gene regulation is important to the energy economy in a bacterium since it enables the cell to "switch on" blocks of genes when they are needed. Two metabolic pathways, whose control mechanisms have been well studied, should serve to explain some of the principles of gene regulation.

Wild-type *E. coli* can utilize the disaccharide lactose as a source of energy because it is able to manufacture two enzymes. One of these enzymes, called β-galactosidase, splits lactose into glucose and galactose, and the other, called lactose permease, promotes the entry of lactose into the cell. The regulatory mechanism whereby *E. coli* makes copious amounts of β-galactosidase and lactose permease in the presence of lactose and almost none in the absence of lactose depends on a regulation unit called the *operon.*

Figure 4–2A illustrates the genes in the lactose operon as well as the mechanism by which the operon is repressed in the absence of lactose. A regulator gene codes for a repressor which attaches to the operator gene, thereby preventing the initiation of transcription of the operon's genes by the cell's RNA polymerase.

Figure 4–2B illustrates what happens when lactose is present. Lactose attaches to and, as a consequence, inactivates the repressor. Therefore, transcription of the β-galactosidase and lactose permease genes is coordinately repressed in the absence of lactose and coordinately derepressed in the presence of lactose.

In the lactose system, lactose is referred to as an *inducer* since it induces the

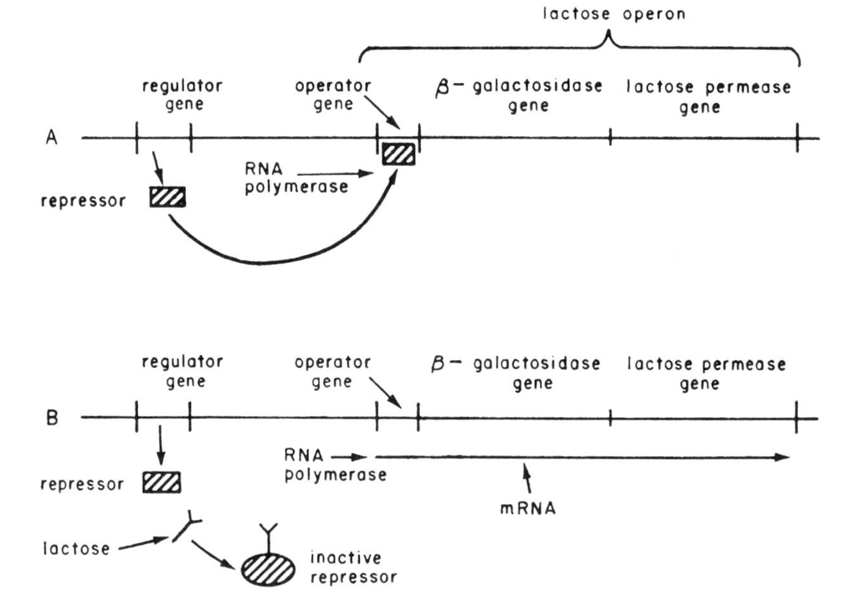

Fig. 4–2. The lactose operon. (A) The repression of messenger RNA synthesis by repressor substances. (B) The induction (derepression) of the lactose operon caused by the inactivation of repressor by an inducing agent (lactose).

production of the enzymes required for the fermentation of lactose. The overall process of enzyme production by adding lactose is called *induction* or *induced enzyme formation.* Similar terminology has already been introduced in Chapter 3 to describe the induction of a prophage by an inducing agent.

The basic function of the operon is to coordinate repression or derepression of transcription of genes involved in a metabolic pathway. This principle is not only used by cells for degradative (catabolic) pathways but also for biosynthetic (anabolic) pathways.

When wild-type *E. coli* cells are supplied with a surplus of exogenous tryptophan the cells cease making tryptophan. Figure 4–3A illustrates the fact that a regulatory gene makes an inactive product or *aporepressor,* which is unable to prevent the transcription of the genes in the tryptophan operon. Figure 4–3B illustrates that a surplus molecule of tryptophan which is either supplied exogenously or made endogenously can act as a *corepressor* (or effector). Acting together, the aporepressor and tryptophan form a full-fledged repressor capable of switching off the transcription of the genes in the tryptophan operon. This phenomenon is usually referred to as *repression* in contrast to the phenomenon of induction which has just been discussed in reference to lactose fermentation.

Finally, it has been found in tryptophan biosynthesis that the end product (i.e., tryptophan) can react with and inactivate the first enzyme in the biosynthetic pathway (enzyme E1 in Figure 4–3A). This phenomenon is called *end* product (or *feed back*) inhibition. Thus, *E. coli* has two mechanisms to prevent the production of tryptophan when tryptophan is present in excess: end product inhibition and repression.

There are many other examples of gene regulation. In most of the systems studied, the operon or some variation of the operon controls the system. The

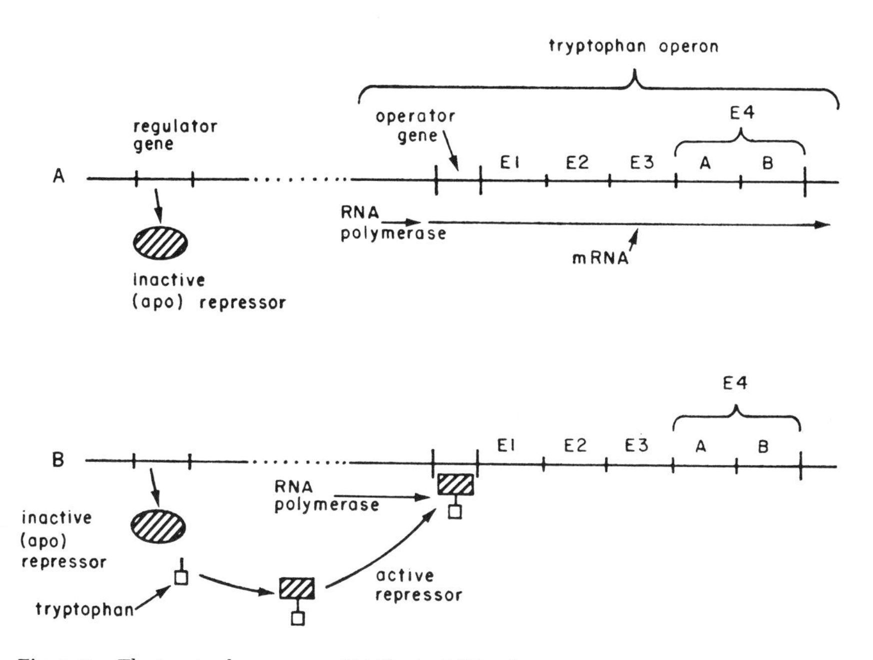

Fig. 4–3. The tryptophan operon. (A) The inability of aporepressor to prevent messenger RNA synthesis. (B) The activation of aporepressor by tryptophan leading to the repression of the synthesis of messenger RNA.

ability of a cell to control coordinately the transcription of the genes which direct the elements of a given metabolic pathway represents an important means by which the cells can rapidly adapt to a hostile environment. For example, certain cells carry the genetic potential to produce penicillinase, an enzyme which inactivates the antibiotic, penicillin. The price the cells have to pay to possess the potential to produce penicillinase is merely the energy and raw materials needed to replicate the specific genes necessary for the enzyme's production. When placed in the presence of penicillin, the potentially resistant cells are able to reorder their energy priorities and produce penicillinase, thus permitting survival of the cells.

F. SOME EXAMPLES OF THE APPLICATION OF MICROBIAL GENETICS TO PROBLEMS IN MEDICAL MICROBIOLOGY

Many aspects of microbial genetics are important to medical microbiology from both a practical and theoretical standpoint. Several interesting applications in microbial genetics to medicine are presented below.

1. Mutation

Mutations among microbes of medical importance can lead to confusion in diagnosis as well as problems with chemotherapy. A notable mutation among pathogens is from antibiotic sensitivity to resistance. Mutation to drug resistance is apparently a natural evolutionary event which will constantly complicate therapy and require a continuing search for new antimicrobial drugs. However, it is the responsibility of every physician to be fully aware that the overuse and misuse of antibiotics can increase the chances that drug-resistant strains will be selected.

Since mutations are relatively rare, the possibility that an organism will mutate simultaneously to resistance to more than one antibiotic is extremely rare. Hence, within the limits set by drug antagonism, it is possible in special instances to avoid selection of drug-resistant mutants of bacteria by treating the patient simultaneously with more than one antibacterial agent (Chap. 6). Alternatively, since resistance to high levels of certain antibiotics is sometime acquired in a step-wide fashion and furthermore, because of the phenomenon of phenotypic lag, it is often possible to avoid the selection of drug-resistant mutants by utilizing a high concentration of a given antibiotic from the first moment of treatment.

2. Genetic Exchange

A complete understanding of the attributes of microbes which convey pathogenicity must include knowledge of the genetic characteristics of the organism. Thus far, only one species has been intensively studied, *E. coli,* and even in this instance only about 15% of the genes have been characterized and mapped! Extensive "genetic engineering" has been accomplished with *E. coli.* It is conceivable that the knowledge gained with this organism will make it possible to reduce the pathogenicity of various microbes. It may also be possible through the techniques of genetic engineering to "create" new and more effective live vaccines.

3. Genetic Engineering

Modern techniques have enabled scientists to isolate genes from many different forms of life and incorporate or "clone" them into plasmids. In turn, simple, easy to grow cells such as *E. coli* can be induced to take up (i.e., be transformed by) plasmids. Some plasmids have a high copy number of *E. coli.* For example, the 4.4kb plasmid pBR322 (a popular cloning vector) has a copy number of about 50 per *E. coli* cell. Therefore, the transformation of *E. coli* with pBR322 containing a specific cloned gene can lead to the production of at least 5×10^9 copies of the gene per milliliter of overnight culture; if the cloned gene is expressed (i.e., transcribed and translated) the cell will manufacture very large amounts of the gene product.

There are great advantages in obtaining large amounts of a particular gene or its product. With sufficient quantities of a gene it is possible to sequence it and, as a result, better understand its protein product. In addition, the DNA of a gene can be labeled with radioactive isotopes and used as probes to look for similar genes or for specific mRNA. Large quantities of a gene product can be used to study the structure and function of the protein. Researchers are now acquiring the ability to produce relatively large amounts of proteins which are potential therapeutic or immunizing agents. At one time research with the rare brain somatostatin required extraction of the hormone from thousands of sheep brains, whereas today workable amounts of somatostatin can be prepared from one liter of an overnight culture of an *E. coli* strain which contains a cloned somatostatin gene. Today cattle are being immunized with surface antigens of the Aphthovirus virus which causes foot and mouth disease.

In the future, the ability of researchers to obtain large amounts of proteins by manufacturing the proteins in bacteria will undoubtedly see increasing use. There is even thought that it may be possible to engineer DNA vectors which can be taken up by human cells and perpetuated in such a way as to permanently cure patients with genetically caused, chemical deficiencies such as diabetes.

REFERENCES

Bachman, B.: Linkage map of *Escherichia coli* K12. Microbiol. Rev. *47*:180, 1983.

Dressler, D. and Potter H.: Molecular mechanisms in genetic recombination. Annu. Rev. Biochem. *51*:727, 1982.

Freifelder, D.: Molecular Biology. 2nd Ed., Boston, Jones and Bartlett, 1987.

Holmes, W.M., Platt, T., and Rosenberg, M.: Termination of transcription in E. coli. Cell *32*:1029, 1983.

Kleckner, N.: Transposable elements in prokaryotes. Annu. Rev. Genet. *15*:341, 1981.

Lake, J.A.: The Ribosome. Sci. Am. *245*:84, 1981.

Levy, S.B., et al.: Molecular Biology. Pathogenicity and Ecology of Bacterial Plasmids. New York, Plenum Press, 1981.

Lewin, B.: Genes, 2nd Ed., New York, John Wiley and Sons, 1985.

Pettijohn, D.E.: Structure and properties of the bacterial nucleoid. Cell *30*:667, 1982.

Platt, t.: Transcription termination and regulation of gene expression. Annu. Rev. Biochem. *55*:339, 1986.

Reznikoff, W.S., et al.: The regulation of transcription initiation in bacteria. Annu. Rev. Genet. *19*:355, 1985.

Smith, H.O. and Danner, D.B.: Genetic transformation. Annu. Rev. Biochem. *50*:41, 1981.

Stent, G.S. and Calendar, R.: Molecular Genetics. 2nd Ed. San Francisco, W.H. Freeman-and Company, 1978.

Watson, J.D.: Molecular Biology of the Gene. 3rd Ed. New York, W.A. Benjamin, 1976.

Willets, N. and Skurray, R.: The conjugation system of F-like plasmids. Annu. Rev. Genet. *14*:41, 1980.

5

STERILIZATION AND DISINFECTION

In view of the exceedingly large numbers of diverse microorganisms that populate our environment, the need for controlling them under many circumstances is obvious. Not only must disease-producing microbes be controlled, but also those responsible for food spoilage and deterioration of various other materials. Common techniques that are employed for controlling microorganisms include sterilization of surgical instruments with heat (autoclaving), application of disinfectants on skin before surgery, canning techniques for preserving foods, use of fungicides to prevent deterioration of non-food materials, and the use of chemotherapeutic agents to eradicate disease-producing microbes.

Sterilization procedures must destroy all forms of microbial life because the term *sterile* literally means the absence of all life. In this regard, microorganisms vary markedly in their resistance to various adverse environmental conditions. Bacterial spores are the most resistant forms of life, and techniques employed for sterilization must be designed to destroy these unique resting-stage microorganisms which characterize the genera *Bacillus* and *Clostridium.* Spores survive drying indefinitely, whereas some vegetative bacterial cells die quickly and others survive drying for varying, but limited, periods. Techniques used by the canning industry are specifically designed to kill spores with minimum alteration of the food being processed.

Most of our basic concepts concerning sterilization emerged from the innovative experiments of Pasteur which ultimately disproved the dogma of spontaneous generation in 1864. However, as late as 1876, Bastian published experimental results which opposed Pasteur's earlier claim that contaminated urine "sterilized" by boiling remained free of growth on subsequent incubation. Bastian declared that if the urine was made alkaline prior to boiling, growth often took place. Pasteur repeated the experiment and was forced to admit the truth of Bastian's claims. Two possible explanations could have accounted for the results: (1) *the urine was sufficiently acid so that it did not support the growth of the bacteria present,* or (2) *boiling was more effective in the acid than alkaline range of pH employed, particularly wth respect to bacterial spores.*

As a consequence of this experience, Pasteur adopted the practice of heating fluids to 120°C by using steam under pressure and thus introduced the autoclave into the laboratory. In addition, he initiated the practice of sterilizing glassware by dry heat at 170°C. Tyndall, who observed that vegetative bacteria were easily destroyed by boiling, developed a sterilization procedure consisting of repeated heatings with appropriate incubation intervals between them. Such intervals allowed the heat-resistant spores to germinate into vegetative cells which were susceptible to killing at 100°C. This method, which was first described by Tyndall in a letter to Huxley in 1877, is still known as *tyndallization.* Present sterilization methods are based on the above observations, together with a more exact knowledge of the heat resistance of bacterial spores. For example, it is well known that spores are more susceptible to heat in an acid medium than in a neutral or slightly alkaline medium.

Compared to moist heat (steam), hot air has low penetrability and heat transfer is relatively slow. In particular, dry heat is relatively ineffective if articles are wrapped with heat-insulating material.

In choosing a sterilization procedure, the type of item to be sterilized and convenience are considered as well as the cost and efficiency of the method.

A. DEATH CURVE OF POPULATIONS OF MICROORGANISMS

When populations of microorganisms are subjected to the common sterilization procedures, they are killed exponentially (Fig. 5–1). Accordingly, the number of bacterial cells dying in each time interval is a function of the number of survivors present. The exponential death process can be expressed by the following formula:

$$S = S_0e^{-kt}$$

S_0 is the number of bacteria surviving at 0 time, and S represents the number of survivors at a later time t. The rate of exponential death is expressed by the term $-k$ when ln (S/S_0) is plotted vs time.

It can be seen in that it took 30 minutes to reduce the population of bacteria from 10^3/ml to 10^0/ml. Therefore only one organism/ml was viable after 30 minutes of sterilization. Based on the exponential death rate function, there would be 10^{-1} organisms/ml at 40 minutes and 10^{-2} organisms/ml at 50 minutes of sterilization. This means that after 40 minutes one viable organism would be present in 10 ml and at 50 minutes only one organism would be viable in 100 ml. After 120 minutes of sterilization only one organism should be viable in 1000 liters. As a general rule a reduction of this magnitude can be considered a safe and practical sterilization procedure in most instances. It should be apparent that killing microorganisms is basically a problem of statistics and a sizable margin of safety must be built into any sterilization procedure.

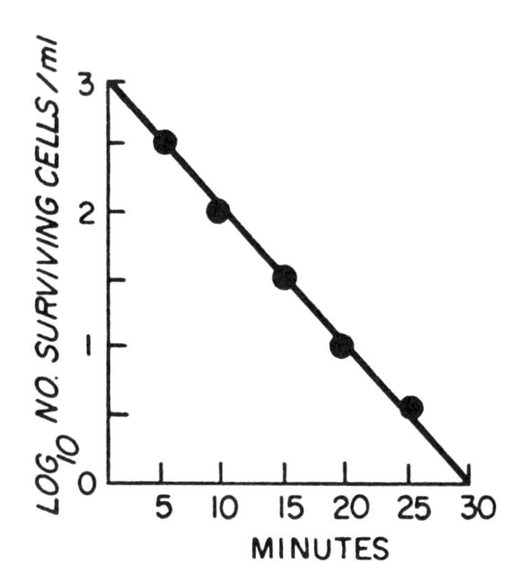

Fig. 5–1. Exponential death curve of bacteria subjected to common sterilization procedures.

B. STERILIZATION BY HEAT

Heat may be employed in several ways for sterilizing objects. (1) *Incineration* can be used to sterilize transfer loops, coverslips, etc. (2) *Hot air.* Objects are usually heated at 160°C for 2 hours or 170°C for 1 hour. This method is commonly used for glass Petri dishes, flasks, test tubes, pipettes, and glass syringes. Obviously, most liquids cannot be sterilized by this method. (3) *Steam under pressure (autoclaving)* is the most effective practical method for heat sterilization. Exposure to steam in an autoclave at 15 pounds pressure (121°C) for 20 minutes will kill all forms of life. Steam under pressure has high penetrability and is particularly effective since it condenses on a cool object to release 596 cal per g (heat of condensation). This condensation allows more steam to make contact with the objects being sterilized and as a result, the process of heat transfer continues until temperature equilibrium is reached. (4) *Flowing live steam* under ambient atmospheric pressure is highly effective against vegetative bacteria but is not a reliable method for killing bacterial spores unless the time-consuming tyndallization technique is used. At sea level, the boiling temperature attained approaches 100°C, but at high elevations the boiling point of water is so reduced that this technique is relatively ineffective. (5) *Boiling water* has been used to prepare objects, such as syringes and instruments, for minor surgery. As one would expect, this technique is not reliable for killing spores and should not be used when an autoclave is available. (6) *Pasteurization* is a process whereby a substance is heated to accomplish the killing of nonsporing pathogens or certain organisms that cause spoilage of milk or food products. In the case of milk, the holding process consists of heating at 62.5°C for 30 minutes and then cooling quickly. Milk may also be "flash pasteurized" by heating it in a thin layer for 3 to 5 seconds at 74°C. Pasteurization is useful for killing important pathogenic microorganisms that may be shed into milk by infected cows or may gain entrance into milk during its processing and distribution. However, pasteurized milk is *not sterile* since spores and certain nonsporing microbes, such as *Streptococcus lactis,* usually survive this treatment.

Virus particles are much less resistant to thermal inactivation than bacterial spores. However, it appears that serum hepatitis virus is protected in an organic medium like serum. It has been reported that single-stranded virus nucleic acids also are inactivated by a first-order process. For example, when a single phosphodiester bond is cleaved the RNA of tobacco mosaic virus is inactivated. On the other hand, a single depurination inactivates the DNA of phage ϕ X 174 nucleic acid.

C. STERILIZATION BY FILTRATION

There are many types of filters on the market which remove bacteria from liquids and accordingly are useful for "sterilizing" solutions containing heat-sensitive materials, such as serum proteins. Since viruses usually pass through such filters, the materials filtered are only "sterile" with respect to nonviral agents. The early filters were prepared from special purified clays. Subsequently, filters composed of asbestos pads were devised. Still later, fritted glass filters composed of masses of porous Pyrex glass were developed. All of these filters have pore sizes which exceed the size of bacteria. Accordingly, the effectiveness of these filters does not result from sieve action but by absorption of the bacteria due to differences between the respective electrical charges of the bacteria and filter material.

Membrane filters, composed of biologically-inert cellulose esters having a uniform porosity, are now readily available. These filters are available with porosities ranging from 0.005 to 1.0 μm and depend on sieve action almost exclusively for their effectiveness. A 0.22 μm filter effectively removes all bacteria, yeast and molds from biologic and pharmaceutical fluids, whereas the smallest pore size filters available retain most viruses. The filter is mounted in a special apparatus and sterilized prior to use. A vacuum is usually applied to the collecting flask or air pressure can be applied to the filter.

Membrane filters are especially useful for isolating and identifying organisms from water and air. The organisms retained by the filter are transferred with the filter to appropriate growth media.

D. GAS STERILIZATION

Under special circumstances, gases have been employed for sterilizing mattresses, bed clothing, heat-labile objects such as plastics, and even entire rooms or buildings.

1. Ethylene Oxide

Ethylene oxide (CH_2OCH_2) has a boiling point of about 11°C. It is usually mixed with 80 to 90% CO_2 to minimize its explosiveness, or it can be used as an 11% concentration in halogenated petroleum. The mixture is fed into an enclosed chamber resembling an autoclave in order to control humidity, temperature, and pressure. Ethylene oxide is a highly effective microbicidal agent and functions by alkylating protoplasmic nitrogenous components. However, it does not alter most organic materials that would be damaged by heat or moisture. It has good penetrability and dissipates readily from the sterilized material.

Ethylene oxide readily penetrates polyethylene wrapping material used for disposable items, but cellulose acetate, polyvinylidene chloride, and polyester wrapping materials are less permeable and should not be used.

The primary antimicrobiologic activity of ethylene dioxide results from the alkylation of the guanine and adenine components of DNA. It should be noted that bacterial spores are 5 to 10 times more resistant to ethylene oxide than vegetative cells. Accordingly, *Bacillus subtilis* spores are used as the biologic indicator for monitoring ethylene oxide sterilization cycles. Care must be taken in removing the gas from the sterilization chamber because it is highly toxic for humans. The 1958 Capital Food Additive and Amendment to the Food Drug and Cosmetic Act prohibits the use of ethylene dioxide to sterilize food products.

2. Propylene Oxide

Gaseous propylene oxide is used to sterilize food products. The mode of action depends on the alkylation of guanine of DNA which results in single-stranded breaks of DNA. The main reason it can be used with foodstuffs is that propylene glycol is formed in the presence of water. In contrast, ethylene oxide forms the toxic compound ethylene chlorohydrin when it undergoes hydrolysis in the presence of water.

3. Betapropiolactone

Betapropiolactone $(CH_2)_2CO$ is a liquid at 20°C which may be stored at 4°C. It has been used to sterilize vaccines and tissue grafts but it must be used in a closed chamber where temperature and humidity can be controlled. Exposure for 2 to 3 hours at concentrations of 2 to 5 mg/L of air is a highly effective sterilizing procedure. Betapropiolactone is 4,000 times more active than ethylene oxide; it is noncorrosive and noninflammatory but has weak penetrability. This chemical is generally recommended as a disinfectant for rooms. Its antimicrobial activity is primarily due to alkylation of DNA.

4. Formaldehyde

Formaldehyde (CH_2O) has been used to a limited extent as a gaseous sterilant. Its sterilizing activity is due to its reactivity for amino groups in protein molecules. For example, in the disinfection of rooms, it has been recommended that 1 ml of formalin (37.5% formaldehyde) be vaporized for each cubic foot of space with an exposure time of 10 hours at 70°F and 50% relative humidity. A major limitation of the procedure is that the room must be aired for several days to remove the unpleasant odor.

Melamine formaldehyde can be used as a source of formaldehyde gas in special applications for sterilizing items after packaging. A strip of paper impregnated with melamine formaldehyde is incorporated during the packaging process of needles, blades, and plastic ware. The packages are heated to 50 to 60°C to release formaldehyde within the package from the melamine polymer. Effectiveness of this type of sterilization is monitored by exposing paper strips containing 10^5 to 10^6 bacterial spores to an identical packaging and heating process.

E. DISINFECTANTS

In some textbooks the term *disinfection* is defined as "the destruction of potential pathogens on inanimate objects," whereas the term *antisepsis* is defined as "the destruction of such microbes in surface wounds or on body surfaces". In this chapter *disinfection* will be used in a general sense, namely to denote the destruction of any microorganism that is capable of producing an infection regardless of its location. If a disinfectant is lethal for bacteria or fungi, it is referred to as being *bactericidal* or *fungicidal.* If it prevents multiplication, it is commonly referred to as being *bacteriostatic* or *fungistatic.* The most useful disinfectants have broad-spectrum bactericidal or fungicidal activity.

There are two general groups of disinfectants. The toxicities of disinfectants of the first group are sufficiently low so that they can be used topically on skin and open wounds. In this application they are commonly referred to as *antiseptics.* However, topical disinfectants in this category are too toxic for systemic (parenteral) use and should not be confused with chemotherapeutic agents which can be administered systemically as well as topically. Disinfectants of the second group are generally too toxic for topical application on skin but are useful for disinfecting liquids and inanimate objects.

Standardization of disinfectants with respect to their effectiveness is a difficult problem. The official method to determine effectiveness is the test specified by the Food and Drug Administration. Strains of *Salmonella typhi, Staphylococcus aureus* and *Pseudomonas aeruginosa* with known susceptibility to phenol are tested and compared with the available disinfectants, this provides the basis for calculating the phenol coefficient.* Accordingly, the use of the *phenol coefficient* for standardizing disinfectants represents an attempt to rate them by comparing their activities with those of phenol under standard conditions. For example, if a disinfectant has a phenol coefficient of 3, it is 3 times as effective as phenol when compared in a standard bactericidal test. Unfortunately the activities of disinfectants vary greatly depending on a number of environmental conditions. Consequently, the phenol coefficient is only a rough estimate of the effectiveness of an agent; under field conditions the estimate may be completely erroneous. For example, the presence of excessive organic material can markedly alter the efficiency of many disinfectants and, as a consequence, the effectiveness of a disinfectant must be certified under actual field conditions.

1. Some Compounds That Are Used to Disinfect Skin and External Wounds

a. ***Iodine.*** A 2% tincture of iodine in 70% alcohol is useful for small cuts and abrasions as well as for reducing the number of bacteria on skin prior to venipuncture or surgery. Maximum bactericidal action of iodine occurs at pH values below 6. At pH values above 7.5 the form I_2 becomes I^- as a consequence of hydrolysis. This form as well as I_3^- have little microbicidal activity.

b. ***Iodophors.*** An iodophor is a loose complex of elemental iodine or triiodide with a carrier that not only increases solubility of the iodine, but also provides a sustained release of the iodine. In aqueous solution the iodophor forms micellar aggregates which become dispersed in an aqueous environment. Iodine in this form has little odor and is relatively nonirritating. The best known iodophor is povidone iodine (USP XX, Betadine, Isodine) which is a 1-vinyl-2-pyrrolidinone polymer with no less than 9% and no more than 12% iodine. A commercially-available iodophor, Betadine, is gaining popularity as a preoperative antiseptic. It has the advantage of being highly bactericidal, nonirritating, and nonstaining.

c. ***Mercurials.*** Organic mercurials such as Metaphen, Merthiolate and Mercurochrome have been widely used as antiseptic agents. Their bacteriostatic properties reside in the ability of mercury to react with the thio groups of proteins, thus inactivating bacterial enzymes. Thio groups of the host can reverse bacteriostatic activity.

d. ***Silver Nitrate.*** Silver nitrate is commonly used in a 1% aqueous solution for application to an infant's eyes at birth to prevent infection and guard against blindness of the newborn due to *Neisseria gonorrhoeae.* Silver ions have the property of combining with and inactivating enzymes.

e. ***Hexachlorophene.*** This compound is chlorinated biphenol with surface-active properties which disrupt the cellular membrane. Hexachlorophene appears to have low toxicity and is bactericidal in low concentrations. This com-

---

*The phenol coefficient of a compound is the ratio of the minimum sterilizing concentration of phenol to that of the test compound.

pound is commonly used in concentrations ranging from 1:1000 to 1:10,000 to irrigate surgical wounds and is usually incorporated into surgical soaps. The residual hexachlorophene helps suppress the growth of skin bacteria during and following surgery.

During the period 1961 to 1971 it was common practice to bathe newborn infants in water containing hexachlorophene to reduce colonization of the umbilical stump by *Staphylococcus aureus.* However, the U.S. Food and Drug Administration placed a strict ban on this practice in 1971 because of evidence that dermal absorption resulted in neurotoxicity.

f. ***Quaternary Ammonium Compounds.*** Chemically these compounds have 4 carbon atoms linked to the nitrogen atom through covalent bonds. The anion in the original alkylating agents becomes linked to the nitrogen by an electrovalent bond. Examples of commercially-available monoalkyltrimethyl ammonium salts include cetyltrimethyl ammonium chloride (CTAB) and alkyltrimethyl ammonium chloride (Arquad 16).

Examples of monoalkyldimethylbenzyl ammonium salts include alkyldimethylbenzyl ammonium chlorides as BTC 824 and Hyamine 3500.

Dialkyldimethyl ammonium salts available in this group are didecyldimethyl ammonium halides (Deciquam 522 and Bardac).

In addition, there are heteroaromatic salts, polysubstituted quaternary ammonium salts, bis-quaternary ammonium salts and polymeric quaternary ammonium salts. In this latter group Mirapol A-15, which is poly [N-[3-(dimethylammonio) propyl]-N^1-[3-(ethyleneoxyethylene-dimethylammonio) propyl]urea dichloride] is of special interest because of its bactericidal action against *Pitrosporum ovale, Staphylococcus aureus* and *Escherichia coli* (100 ppm). These polymeric quaternaries are unique in that they have remarkably low toxicity, do not foam and possess good bactericidal properties as compared with other quaternary compounds.

g. ***Ethyl Alcohol and Isopropyl Alcohol.*** Both ethyl and isopropyl alcohols are commonly used in a 70% concentration; higher concentrations are less effective. Their bactericidal activity is inactivated by mucus or pus, and they are not effective against bacterial spores. However, these alcohols are moderately effective against vegetative bacteria on skin, especially in combination with green soap. The action of alcohols involves their capacity to precipitate cellular proteins.

2. Some Chemical Agents for Disinfecting Liquids and Inanimate Objects

Since toxicity is usually not a problem, many of the disinfectants commonly used to kill microorganisms on inanimate objects are either too toxic for human use or are used in concentrations exceeding those that can be safely used on skin or minor wounds. Many of the disinfectants listed above can also be used on inanimate objects.

a. ***Wescodyne.*** Wescodyne (a trade name) is a commercially available iodophor that combines the cleansing action of a nonionic detergent with the bactericidal action of iodine. This compound is recommended over alcohol for sterilizing and storing thermometers.

b. ***Phenol.*** Phenol (C_6H_5OH) in a 1% solution kills essentially all vegetative forms of bacteria in 20 minutes. A 5% solution, which is commonly used to disinfect surgical instruments and excreta, will kill bacterial spores in a few hours; its effectiveness is altered only slightly by organic matter. Phenol is no longer used as a major disinfectant because there are phenol derivatives that are much less caustic.

c. ***Cresols.*** The most common alkyl phenols are the cresols. A mixture of cresols emulsified in green soap is sold under the trade name of Lysol and Creolin. Lysol is about 4 times as effective as phenol. Ortho-, meta-, and paracresols are employed in a mixture known as tricresol, which is also more effective than phenol.

d. ***Mercuric Chloride.*** When used under optimal conditions at a concentration of 1:1000, mercuric chloride is a powerful germicide. Mercuric ions form mercaptides with sulfhydryl groups which inactivate bacterial enzymes. However, the activity of mercuric chloride is markedly reduced by the presence of organic matter.

e. ***Chlorine.*** Chlorine is commonly used to disinfect drinking water and water in swimming pools. It is important to monitor the free chlorine content and maintain effective concentrations. Chlorine gas added to water at a pH above 2.0 forms hypochlorous acid and hydrochloric acid. The hypochlorous acid (HOCl) is the bactericidal agent and acts through its oxidizing capacity. Another source of chlorine is NaOCl (sodium hypochlorite); it is available in grocery stores under the trade name Clorox.

f. ***Cationic Detergents.*** These compounds are highly effective as general disinfectants. Cationic detergents are extensively employed to cleanse and disinfect food utensils.

g. ***Formalin.*** Formalin (formaldehyde in solution) is a potent antimicrobial agent due to its high reactivity with proteins. It can block free amino groups at pH 9.5; it also reacts with the nitrogen of the imidazole ring of histidine at pH 6.0, and is fixed to amide and guanidyl groups at pH 5.0 or below.

h. ***Glutaraldehyde.*** This agent has been used extensively recently as a cold sterilant for surgical instruments. At the present time it is the only disinfectant recommended by the Centers For Disease Control for respiratory therapy equipment. Although glutaraldehyde is much less toxic for humans than formaldehyde it is about 10 times more potent as an antimicrobial agent.

i. ***Hydrogen peroxide.*** At a concentration of 3% this agent is a useful disinfectant for inanimate materials. It has been used to a great extent to disinfect medical-surgical devices. More recently it has been found to be useful for sterilizing soft plastic contact lenses.

j. ***Dyes.*** Derivatives of triphenylmethane, which represent the *aniline dyes,* such as brilliant green, crystal violet and malachite green, have marked antimicrobial activity.

The acridine dyes are also bactericidal, largely because they interfere with the synthesis of nucleic acids and proteins; they are also toxic for eucaryotic cells.

F. RADIATION

1. Ultraviolet Light

Ultraviolet light has moderate microbicidal activity when applied properly. Its effectiveness is limited by its extremely low penetrability; it can penetrate quartz but not glass. The wave length most commonly used is about 260 nm.

Ultraviolet light is useful for *reducing* the population of airborne microorganisms in operating rooms, tissue culture rooms, and "sterile" rooms. It is also used widely in meat-packing houses and bakeries for controlling molds. The bactericidal and fungicidal action of ultraviolet light is due directly to the formation of pyrimidine dimers from adjacent monomers on the same DNA strand and indirectly to the production of peroxides in the medium which act as oxidizing agents. The dimers may be mutagenic or lethal for bacteria. Resistance to pyrimidine dimers depends on the effectiveness of enzyme repair systems.

2. Ionizing Radiations

Ionizing radiations fall into two general categories; radiations of the first category have mass and may be charged or uncharged. Those of the second category only possess energy. Ionizing radiations can be products of radioactive decay (α-, β-, γ- rays) produced by an x-ray machine; they can also result from particle bombardment or nuclear reactors. The most practical ionizing radiations for sterilization are generated by electromagnetic x-rays, γ-rays and particulate cathode rays (generated by accelerating electrons). These cathode rays are also referred to as *high-energy electrons.* Because of their high energy cathode rays can penetrate more effectively than other rays.

Gamma radiation generated by cobalt 60 is effective for sterilizing packaged plastic wares such as plastic Petri dishes. It has been used for sterilizing dressings, catheters, and syringes and could find a special application in preserving foods except for the disagreeable flavors resulting from this procedure. About 2.5 Mrads are required to reduce microbial populations by a factor of 10^7.

3. High-Energy Electrons

High-energy electrons have been used to sterilize plastic and rubber articles. Linear accelerations in the range of 1 to 5 MeV are used for this purpose.

G. GENERAL CONSIDERATIONS

Personal biases influence the selection of a general antiseptic for use on skin surfaces. Preparations containing hexachlorophene, organic iodine solutions, or cationic detergents are most often selected for preoperative skin preparation. On the other hand, minor skin abrasions are commonly cared for by cleansing thoroughly with soap and water and protecting the area from subsequent contamination.

Cationic detergents may be chosen to disinfect areas such as tabletops because they do not have the unpleasant odor of phenol or cresol. Regardless of the agent used it is important to allow sufficient time for the agent to act in order to be effective. *Merely wiping a clinical thermometer with an iodophor like Wescodyne just prior to its use for each of a series of patients is not an acceptable practice.* Adequate exposure of an object to a disinfectant must always be allowed.

If the types of organisms contaminating a solid object or liquid were known, it would be possible to select the least drastic of the sterilization methods that would be effective. In practice, however, one must assume that spores are present; accordingly the method chosen is not directed toward a particular organism but against all organisms or at least all pathogenic organisms that may be present on the item. In many instances, surgical instruments could be sterilized with either the autoclave, the hot air oven, the gas sterilizer, or by gamma radiation without harming the instruments. However, autoclaving is the most practical means for ensuring the sterility of such objects.

REFERENCES

Ackland, N.R., et al.: Controlled formaldehyde fumigation system. Appl. Environ. Microbiol. *39*:480, 1980.

Block, S.S.: Disinfection, Sterilization and Preservation, 3rd Ed. Philadelphia, Lea & Febiger, 1983.

Caputo, R.A., et al.: Recovery of biological indicator organisms after sublethal sterilization treatment. J. Parent. Drug Assoc. *34*:394, 1980.

Chipley, J.R.: Effects of microwave irradiation on microorganisms. Adv. Appl. Microbiol. *26*:129, 1980.

Davis, K.W., et al.: DS gamma radiation dose setting and auditing strategies for sterilizing medical devices. In Sterilization of Medical Products. Vol. II. Ed. E.R.L. Gaughran and R.F. Morrissey. Montreal, Multiscience Publ. Ltd., 1981.

Evans, C.A., and Mattern, K.L.: The bacterial flora of the antecubital fossa: The efficacy of alcohol disinfection of this site, the palm and the forehead. J. Invest. Dermatol. *75*:140, 1980.

Favero, M.S.: Iodine-champagne in a tin cup. Infect. Cont. *3*:30, 1982.

Favero, M.S.: Sterilization, disinfection, and antisepsis in the hospital. In Manual of Clinical Microbiology. 3rd Ed. Washington, D.C., American Society for Microbiology, 1980, pp. 952.

Gorman, S.P., et al.: A review. Antimicrobial activity, uses and mechanism of action of glutaraldehyde. J. Appl. Bacteriol., *48*:161, 1980.

Harm, W.: Biological Effects of Ultraviolet Radiation. New York, Cambridge University Press, 1980.

Hollis, C.G., and Smalley, D.L.: Resistance of *Legionella-pneumophila* to microbicides. Dev. Indust. Microbiol., *21*:265, 1980.

Levine, W.L., et al.: Disinfection of hydrophilic contact lenses with commercial preparations of 3% and 6% hydrogen peroxide. Dev. Indust. Microbiol. *22*:813, 1981.

Mallison, G.F.: Hospital disinfectants for housekeeping: floors and tables. Infect. Control *5*:537, 1984.

Ojajärvi, J., et al.: Failure of hand disinfection with frequent handwashing: a need for prolonged field studies. J. Hyg. Camb. *79*:107, 1977.

Perkins, J.J.: Principles and Methods of Sterilization in Health Sciences. 2nd Ed. Springfield, Charles C Thomas, 1982.

U.S. Department of HEW, FDA: OTC topical antimicrobial products: over-the-counter drugs generally recognized as safe, effective, and not misbranded. Federal Register, *43*:1210, 1978.

Zamora, J.L.: Chemical and microbiologic characteristics and toxicity of povidone-iodine solutions. Am. J. Surg. *151*:400, 1986.

6

ANTIMICROBIAL THERAPY

*(See Kucera, L.S. & Myrvik, Q.N.: Fundamentals of Medical Virology, 2nd Ed. Lea & Febiger, 1985.)

A. MEDICAL PERSPECTIVES

The empiric treatment of infectious disease dates back many centuries and is rooted in ancient folk-medicine. For example, an early Chinese remedy for boils was the application of moldy soybean curd, and foot infections were treated by wearing mold-covered sandals. However, not until the achievements of Pasteur, a little over a century ago, in laying the ground work for understanding the etiology of infectious diseases was a rational scientific approach to chemotherapy possible.

At the beginning of the twentieth century, the German chemist Paul Ehrlich used his knowledge that the salts of heavy metals were helpful in treating the surface lesions of syphilis; thus began the first systematic search for a chemical agent to kill the etiologic agent of syphilis, *Treponema pallidum.* His concept of a "magic bullet," an agent which would be lethal to the pathogen without harming the host, remains the ideal of modern chemotherapy. This ideal is still not fully realized, because all presently known antimicrobial agents have the potential for causing harm to the patient. Ehrlich's relentless search was crowned with success in 1910 when salvarsan, the 606th arsenic compound tried, proved effective although not without toxicity.

A systematic search for naturally occurring antibiotics, originally defined as substances produced by living cells which are antagonistic to other cells, was carried out in the early 1920s by Gratia and Doth, but it was the serendipitous observations of Alexander Fleming in London in 1928 which led to the discovery of the first true "wonder drug," penicillin. Fleming noted that the mold, *Pencillium notatum,* which had accidentally contaminated an agar culture of staphylococci, inhibited the growth of the bacteria. The application of this observation awaited the isolation of penicillin by Florey and Chain at Oxford in 1940 and the subsequent mass production of penicillin by the U.S. pharmaceutical industry during World War II.

During the interval between Fleming's observation and the eventual production of penicillin, another group of agents, the sulfonamides, was described in 1936 by Domagk. These are properly called antimicrobial compounds and are not technically antibiotics, since they are not produced by living organisms. The term *antimicrobial* is a more general term which encompasses all types of chemotherapeutic agents used against infectious diseases. However, since many antibiotics have been synthesized, and the distinction is less important the term *antibiotic* is often used synonymously with antimicrobial agent.

The discovery of new classes of antibiotics was continued in 1944 by Waksman with the development of streptomycin, an aminoglycoside produced by *Streptomyces.* Subsequently, many "broad-spectrum" agents have been introduced. The principal sources of antibiotics in nature have been microorganisms found in soil, including members of the genera *Bacillus, Penicillium, Streptomyces* and, more recently, *Micromonospora.* The initial source of the cephalosporin antibiotics was a mold, *Cephalosporium acrimonium* isolated from Sardinian sewage. Thousands of antimicrobial agents have been identified and tested, including chemical modifications of biologically produced agents, as well as de novo synthesis of others. Well over a hundred of these have been used at one time or another in patients. As new agents are developed, other more toxic or less useful agents are discarded.

An active search for new and more useful antibiotics continues at a rapid pace for several reasons. Widespread clinical use of antibiotics has led to the emergence of microorganisms resistant to the effects of commonly used agents. In addition, the use of broad spectrum antibiotics or the use of combinations of several antibiotics has led to significant problems with *superinfection* by organisms of the normal or indigenous flora of humans and the environment. Finally, virtually all known antibiotics have significant potential for producing adverse effects in patients, including toxic side effects and hypersensitivity reactions. Accordingly, the search for Ehrlich's ideal "magic bullet" continues.

B. DETERMINANTS OF RESPONSE TO ANTIMICROBIAL THERAPY

Before selecting an antimicrobial agent for the treatment of disease, the physician should consider a number of factors which determine the response to therapy. The key concept of the "magic bullet" is that of *selective toxicity;* the antimicrobial agent must be more toxic to the infecting microorganism than to the cells of the host. Even agents which show little or no direct toxicity for the host cells at a given dose, nevertheless, have the potential for sensitizing the host (as in penicillin allergy) or eradicating certain host flora that restrict the overgrowth of potentially pathogenic, drug-resistant microorganisms. Thus, there is no truly innocuous antibiotic.

A related concept, which takes into account the dose levels of antibiotic achieved in host tissues, is that of the *therapeutic index,* which is the ratio of the toxic dose to the effective dose of a drug. In the case of the penicillins, many-fold higher concentrations of drug than are needed for killing susceptible bacteria can be reached before significant toxicity is observed. Thus, penicillin has a high therapeutic index. Other agents such as the aminoglycoside antibiotics (streptomycin, gentamicin, etc.) and the polyene antibiotic (amphotericin B) have a low therapeutic index and produce significant toxicity at levels not much higher than are needed for effective killing of microorganisms.

Antibiotics such as the penicillins, cephalosporins and aminoglycosides possess another desirable attribute in that they directly kill susceptible microorganisms, either by inhibition of cell-wall synthesis (penicillins and cephalosporins) or by a lethal inhibition of protein synthesis (aminoglycosides). These agents are thus *bactericidal* since they directly produce death of the microbe, and thus do not depend to a great extent upon host defense mechanisms. Other antibiotics (chloramphenicol, erythromycin and the tetracyclines) inhibit bacterial cell replication but do not kill the microorganisms. They are classified, therefore, as *bacteriostatic* (producing inhibition of cell growth but not death), and rely upon normal functioning of the body's defense mechanisms for optimal effectiveness. Since the body's defense mechanisms normally are quite effective, the selection of a "cidal" rather than a "static" drug may not be critical in many infections. *Bactericidal* drugs are most advantageous in severe infections and especially in patients whose antimicrobial defenses have been markedly compromised. Examples of "compromised hosts" are those whose immune system is deficient, as in certain congenital immunodeficiency diseases, as well as in patients with lymphoreticular malignancies or diabetic keto-acidosis. Also, increasing numbers of seriously ill patients are receiving immunosuppressive drug therapy or radiotherapy which depresses host defenses, such as phagocytic cell function, a key "first line" defense mechanism.

Additionally, it should be noted that the distinction between "cidal" and "static" drugs depends on the drug concentration and the properties of the infecting microorganism. Bactericidal drugs at low concentrations may sometimes have a bacteriostatic effect, and some bacteriostatic drugs at appropriate concentrations may ultimately produce killing of the microorganism.

For certain types of infections in which the organisms survive and grow predominantly in an intracellular environment within the host (tuberculosis, brucellosis, typhoid fever, rocky-mountain spotted fever, etc.), the time required for an effective response to antimicrobial therapy is often more prolonged than for predominantly extracellular infections, since antibiotics do not penetrate well into host cells. For example, the penetration of streptomycin into the phagosomes of macrophages infected with tubercle bacilli is much inferior to that of isoniazid, which is therefore a more effective therapeutic agent than streptomycin for tuberculosis.

During the selection of an appropriate antibiotic for treatment of an infectious disease, the physician must consider both bacterial and host determinants. The most critical *bacterial determinant* is the sensitivity of the organism to the drug. The physiologic state of the organism may also play a role. Antibiotics such as penicillin, which act by inhibiting cell-wall synthesis, require that the organism be in an active growth phase in order to achieve optimum antibacterial effects. Organisms in a vegetative phase and not actively dividing, as may be the case with meningococci present in the pharynx of an asymptomatic "carrier," are typically indifferent to the effects of penicillin despite being quite sensitive to penicillin under conditions of active growth.

Host determinants of response to antibiotic therapy include the status of host defenses: adequacy of phagocytic function, immunosuppression, serious underlying diseases, and malnutrition. The concentrations of antibiotic attained in particular tissues are also critical. For example, although penicillins ordinarily do not penetrate the normal blood-brain barrier effectively, the presence of meningeal inflammation enhances transport of penicillin into the cerebrospinal fluid, facilitating the therapy of meningitis. Other drugs such as chloramphenicol readily cross even the uninflamed blood-brain barrier. In contrast, the penetration of a drug into a diseased organ may be markedly impaired. For example, in the urinary tract the antibacterial, nitrofurantoin, normally reaches high concentrations in the urine, but in patients with impaired renal function levels of nitrofurantoin in the urine are markedly decreased.

Genetic differences in the metabolism of certain drugs may also be important. Some patients are "rapid-acetylators" of the antituberculous drug isoniazid, while others metabolize the drug more slowly. This difference in the rate of inactivation may be important when designing intermittent drug dosage schedules for tuberculosis.

Other potentially important host determinants of response include the presence of foreign bodies in infected tissues which may interfere with defense mechanisms, the degree of vascularization and viability of the infected tissues, and the conditions of oxygen tension and acidity of the tissue. Antibiotic effectiveness is usually decreased in abscesses or obstructed viscera. In the case of abscesses, drug penetration may be adequate, but the organisms may be in a vegetative phase and relatively indifferent to many antibiotics. Furthermore, phagocytic function may be suboptimal in the acidic environment within the abscess. Certain agents, such as the sulfonamides, are less effective in the presence of purulent material because of the presence of cell-breakdown products (thymidine, purines, etc.) which tend to reverse the inhibitory effect of the drugs on bacterial cell folate synthesis. All of these considerations reinforce the doctrine that optimum therapy of abscesses includes surgical drainage, with or without antibiotics, depending on the circumstances.

C. MECHANISMS OF ACTION OF ANTIMICROBIAL AGENTS AND ANTIMICROBIAL RESISTANCE

Sensitivity of a microorganism to an antibiotic requires that the organism possess an appropriate target for the drug (e.g., a receptor for the drug). Five major mechanisms have been identified by which antimicrobial agents may produce killing or inhibition of bacterial cell growth. These include: (1) inhibition of cell wall synthesis (penicillin and the cephalosporins); (2) inhibition of protein synthesis (including the bactericidal aminoglycosides, streptomycin and gentamicin, and the bacteriostatic agents, tetracyclines, chloramphenicol, and erythromycin); (3) irreversible damage to bacterial cell membranes (polymyxins); (4) inhibition of the synthesis of a critical metabolite (sulfonamides), and (5) inhibition of nucleic acid metabolism (nalidixic acid, griseofulvin) (Table 6–1).

Organisms which lack an appropriate target for antibiotic action are said to possess intrinsic resistance. Resistance may develop in previously susceptible organisms in one of two basic ways: (1) mutation—a spontaneous, random, permanent alteration in the DNA base sequences of the organism and (2) *transfer of genetic information* from one organism to another, either through chromosomal recombination (transformation, transduction, conjugation) or by extrachromosomal transfer (plasmids, episomes). Successful mutation usually produces a single chromosomal change at one time and this results in chance increased resistance to a single antibiotic. Under the selection pressure of antibiotic usage, drug resistant organisms may emerge. Typically, organisms develop resistance in a stepwise fashion involving multiple mutations over a long period of time, as has been the case with the gradually increasing penicillin-resistance of gonococci.

A potentially devastating mechanism of antibiotic resistance is the transfer of genetic information from one organism to another. Thus, multiple drug resistance may be transferred in a single step by a plasmid containing appropriate genetic information. Such extrachromosomal elements are designated resistance factors (R-factors) and have two components: resistance transfer factor (RTF) which carries the genes coding for replication and a factor which allows sexual transmission of the plasmid.

The mechanisms by which organisms resist antibiotics include the following: (1) production of enzymes which inactivate the drug. Examples include the beta-lactamases, produced by staphylococci and certain Gram-negative bacilli, which inactivate penicillin (penicillinase) or cephalosporins (cephalosporinases). In addition, a variety of enzymes which inactivate aminoglycosides have been identified. These include acetylating, phosphorylating and adenylating enzymes; (2) alteration of bacterial cell permeability to the drug. This is an important mechanism in tetracycline resistance since the drug must gain access to the interior of the cell to exert its action; (3) development of alternate

Table 6–1. Mechanisms of Action of Antimicrobial Drugs

Nature of Injury	*Antimicrobial Drug*	*Mode of Action*
Defective wall cell mucopeptide	Penicillins and cephalosporins	Prevent final peptide bond between D-alanine and glycine
	Cycloserine	As structural analogue of D-alanine, it inhibits enzymes responsible for synthesis of D-alanyl-D-alanine, an essential component of mucopeptide
	Bacitracin and vancomycin	Block transfer to cell membrane of sugar pentapeptide from site of synthesis in cytoplasm
Damaged cytoplasmic membrane	Polymyxins	Disorganize lipoproteins by inserting lipophilic moiety into membrane lipid
	Polyenes	React with steroids in fungal membranes so that permeability is altered
Impaired function of ribosomes	Aminoglycosides	Bind to 30S ribosomal unit, causing ribosomes to leave mRNA prematurely; also interfere with attachment of tRNA and distort triplet codons so that message is misread
	Tetracyclines	Bind to 30S unit and block binding of tRNA so that new amino acids cannot be introduced into peptide chain
	Chloramphenicol	Attaches to 50S subunit of ribosomes and prevents peptide-bond formation by inhibiting enzyme peptidyltransferase
	Erythromycin and lincomycin	Same as chloramphenicol
	Thiosemicarbazones	Disrupt polyribosomes
Impaired nucleic acid synthesis	Rifampicin	Blocks bacterial RNA formation by inhibiting DNA-dependent RNA polymerase
	Sulfonamides and diaminopyrimidines	By preventing synthesis of folic acid, they block formation of thymidine and purines needed for nucleic acid synthesis
	Metronidazole	After partial reduction of the nitro group, the activated drug causes strand breaks in DNA
	Nalidixic acid	Inhibits DNA gyrase

Adapted from: Braude, A.I. (ed): Medical Microbiology and Infectious Diseases. Philadelphia, W.B. Saunders Co., 1981.

metabolic pathways, bypassing the reaction which the drug inhibits. This occurs in sulfonamide resistance; and (4) development of altered structural targets for the drug. Erythromycin resistance occurs when the usual binding site on the 50S ribosome is altered; alterations on the 30S ribosomes of certain bacteria also are associated with aminoglycoside resistance.

The antimicrobial sensitivity of an organism may effectively be altered when the organism is exposed simultaneously to two antibiotics. An *indifferent* (or additive response) occurs when the effect is roughly equivalent to the sum of the individual effects of the two drugs. A *synergistic* response occurs when the effect is greater than the sum of the two agents acting alone. Thus, enterococci which are moderately sensitive to penicillin and relatively resistant to aminoglycosides such as streptomycin show markedly enhanced susceptibility to killing when exposed to appropriate concentrations of both drugs together. In this instance, penicillin, through its effect on the bacterial cell wall, enhances the penetration of streptomycin into the interior of the cell where its action is exerted. A less desirable consequence of using a combination of antibiotics is *antagonism,* in which one agent actually interferes with the action of another. Typically, this may occur when a bactericidal agent such as penicillin is used together with a bacteriostatic agent, such as tetracycline. The mechanism in this case is that inhibition of active growth of the organism by the bacteriostatic agent renders the organism relatively insensitive to penicillin, which requires cell wall synthesis to exert its bactericidal action. Antagonism is probably relatively unimportant in clinical situations except when minimal doses of the cidal agent are used along with high doses of the static agent in diseases, such as meningitis, in which host defenses are initially at a disadvantage. This was the case in studies of bacterial meningitis in which the mortality rate was significantly higher when a cidal and static drug were used together than with the cidal drug alone.

Emergence of antibiotic-resistant microorganisms poses a continuing problem for the physician and has important implications for selection of appropriate antibiotics, dosage and duration of therapy, and avoidance of inappropriate usage, including inadequate clinical indications or overuse of prophylactic antibiotics, since all of these factors influence the development of antibiotic resistance.

D. COMMONLY USED ANTIBACTERIAL AGENTS

1. Inhibition of Cell Wall Synthesis

The Penicillins. Following Fleming's discovery of penicillin in 1929, and its isolation a decade later, a large number of modified penicillins have been produced and tested. The basic mechanism of action of the penicillins is interference with synthesis of the peptidoglycan structure of the bacterial cell wall by preventing cross-linkage between peptide chains. An unstable cell wall results, usually leading to lysis of the cell. Most Gram-positive cocci, except for enterococci and penicillinase-producing staphylococci, are susceptible to penicillin, as are gonococci, meningococci, *Treponema pallidum,* some actinomycetes and certain Gram-negative enteric bacilli. The major "natural" penicillins are penicillin G, the prototype, and penicillin V, which, because of greater acid-stability, is preferred for oral administration. All penicillins have a common nucleus of 6-aminopenicillanic acid (6-APA) (Fig. 6–1) which contains a β-lactam ring. Variations are achieved by alteration of the side chain. The β-lactam ring, also found in the cephalosporins, is the site of attack of inactivating enzymes, the penicillinases. By controlled biosynthesis of the 6-APA nucleus and subsequent in vitro chemical attachment of side chains, a variety of semisynthetic penicillins with differing characteristics (antibiotic spectrum, acid-stability, penicillinase-susceptibility, etc.) can be produced.

Penicillin

Acyl Group — 6-AMINOPENICILLANIC ACID

β-Lactam Ring — Thiazolidine Ring

R — $-\overset{O}{\overset{\|}{C}}-\overset{H}{N}-\overset{H}{C}-\overset{H}{C}$ — S — $C(CH_3)_2$ — $\underset{\|}{C}=O$ — N — $\underset{H}{C}-COO^- Na^+$

Fig. 6–1. Basic structure of penicillin indicating the relationship of the acyl group to the 6-aminopenicillanic acid molecule. (Adapted from Adv. Intern. Med. *16*:373, 1970).

Thus, several major classes of penicillins are now in use. In addition to penicillin G, the *oral penicillins* (acid-stable), penicillin V (phenoxymethyl penicillin) and phenoxyethyl penicillin (phenethicillin) are available (Fig. 6–2). The principal classes of semisynthetic penicillins include: the *antistaphylococcal penicillins* (methicillin, oxacillin, nafcillin) (Fig. 6–3) whose major advantage is resistance to the action of penicillinases; the *aminopenicillins* (ampicillin and amoxicillin) (Fig. 6–4) which have a broader antibacterial spectrum against certain Gram-negative bacilli, including *Haemophilus influenzae, Escherichia coli, Salmonella,* and *Shigella;* and the *antipseudomonas penicillins* (carbenicillin, ticarcillin) (Fig. 6–5), whose spectrum extends to *Pseudomonas aeruginosa,* indole-positive *Proteus,* and *Bacteroides fragilis.*

The *newer penicillins* (4th generation) include piperacillin, mezlocillin, and azlocillin (Fig. 6–6). They are more active against Gram-negative organisms

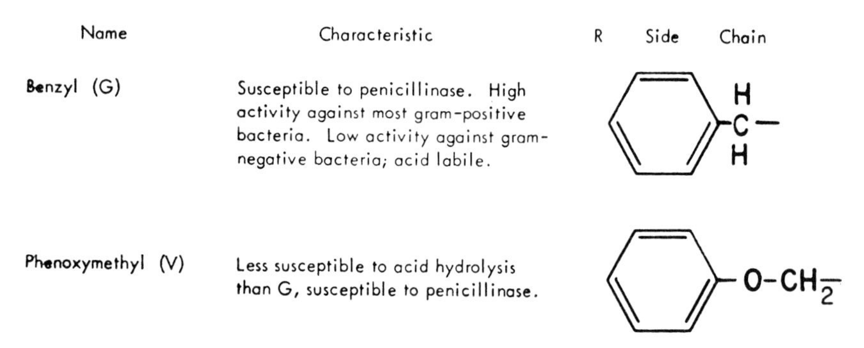

Name	Characteristic	R Side Chain
Benzyl (G)	Susceptible to penicillinase. High activity against most gram-positive bacteria. Low activity against gram-negative bacteria; acid labile.	$C_6H_5-CH_2-$
Phenoxymethyl (V)	Less susceptible to acid hydrolysis than G, susceptible to penicillinase.	$C_6H_5-O-CH_2-$

Fig. 6–2. Characteristics and structure of the acyl groups of two natural penicillins. If corn steepliquor is supplied with β-phenylethylamine, penicillin G is produced, whereas, if phenoxyacetic acid is added, penicillin V is synthesized by *Penicillium chrysogenum.* Note site of attachment of acyl group to the 6-aminopenicillanic acid molecule in Figure 6–1. (Adapted from Adv. Intern. Med. *16*:373, 1970).

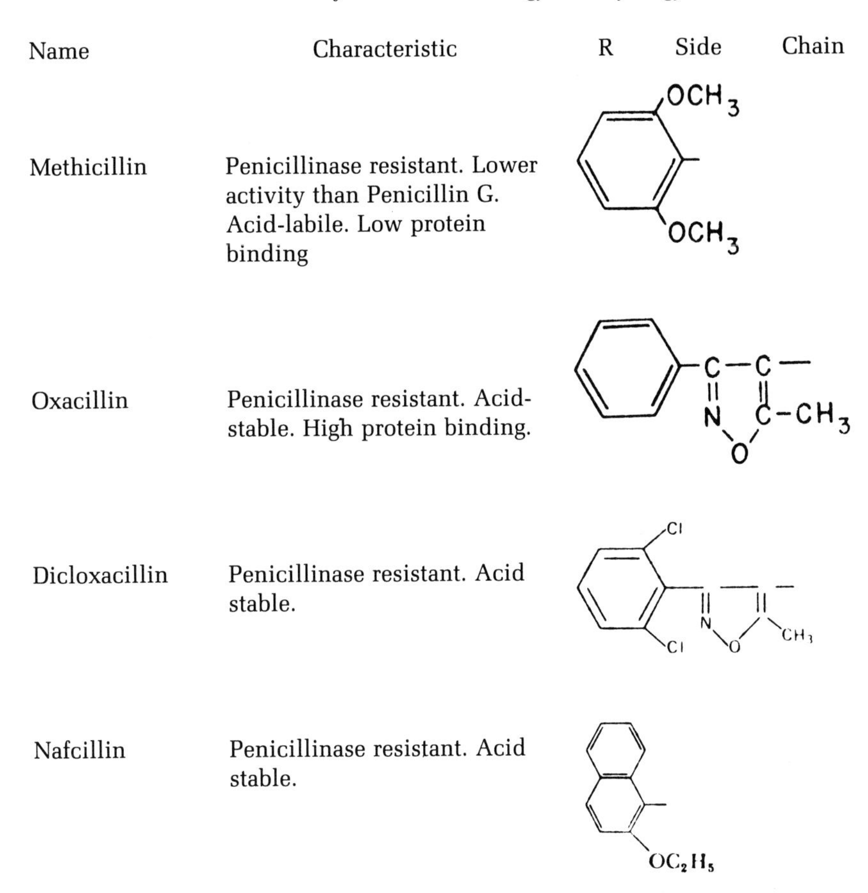

Name	Characteristic	R Side Chain
Methicillin	Penicillinase resistant. Lower activity than Penicillin G. Acid-labile. Low protein binding	OCH_3, OCH_3
Oxacillin	Penicillinase resistant. Acid-stable. High protein binding.	C—C—, N, C—CH_3, O
Dicloxacillin	Penicillinase resistant. Acid stable.	Cl, Cl, N, O, CH_3
Nafcillin	Penicillinase resistant. Acid stable.	OC_2H_5

Fig. 6–3. Antistaphylococcal penicillins. Methicillin is resistant to penicillinase but is acid-labile, whereas oxacillin, dicloxacillin, and nafcillin are resistant to both.

Name	Characteristic	R Side Chain
Ampicillin	Penicillinase sensitive. Wider spectrum. Acid stable.	CH—, NH_2
Amoxicillin	Penicillinase sensitive. Wider spectrum. Acid stable.	HO—, C — CH — C, O, NH_2

Fig. 6–4. Aminopenicillins. These penicillins have a wide spectrum of activity, are acid stable but are penicillinase sensitive.

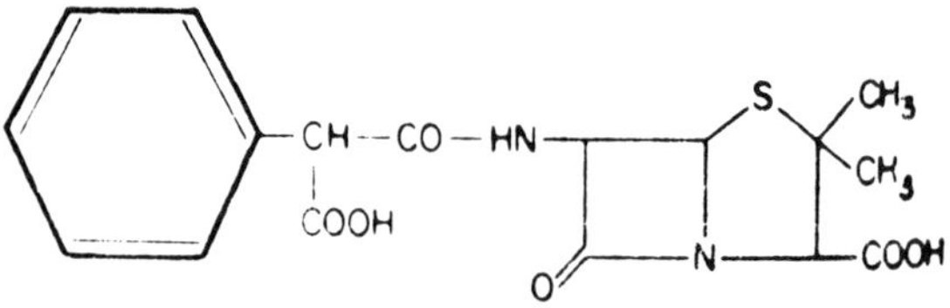

Carbenicillin

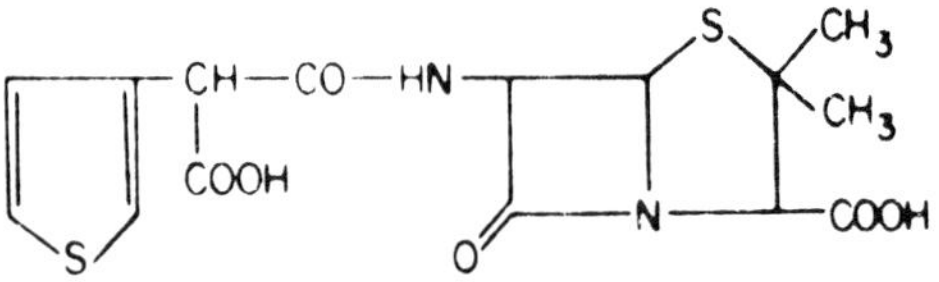

Ticarcillin

Fig. 6–5. Antipseudomonas penicillins. They are penicillinase sensitive, have a wide antibacterial spectrum but are acid labile. These penicillins are active against indole positive Proteus species and *Bacteroides fragilis.* The acid lability of carbenicillin in the stomach has been overcome by converting it to 5-indanyl ester.

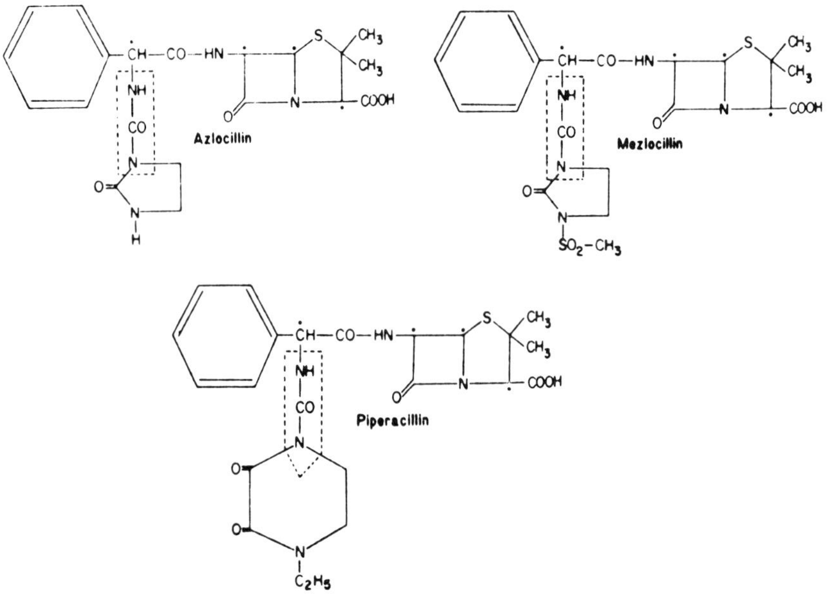

Azlocillin

Mezlocillin

Piperacillin

Fig. 6–6. Structural formulas of the newer penicillins. They inhibit streptococci at concentrations similar to those of penicillin G or ampicillin. They have a greater affinity for penicillin binding proteins but they are less stable to β-lactamases. However, they are relatively resistant to the cephalosporinases. These compounds are derivatives of ampicillin with conversion of the α-amino group to a ureido group.

than ticarillin and carbenicillin. For example, piperacillin and azlocillin are 4 times more active than ticarcillin and mezlocillin against *Pseudomonas aeruginosa.* Unlike the cephalosporins, all the new penicillins are active against the enterococci.

Temocillin, another new penicillin under investigation, is the sodium salt of 6-β-(2-carboxy-2-thien-3-yl acetamido)-6-α-methoxy-penicillin acid (Fig. 6–7). It differs from ticarcillin by the introduction of a methoxy group at the 6-α position of the β-lactam ring. *Temocillin* is resistant to the β-lactamases but is not active against *Pseudomonas* sp. and Gram-positive organisms. It is particularly active against *Haemophilus influenzae* and pathogenic *Neisseria* species. One of the most striking characteristics of temocillin is its high β-lactamase stability.

In patients with serious infections *penicillins* are usually administered parenterally rather than orally, because of limitations of absorption and problems of acid-stability by the oral route. However, the oral route is acceptable for certain non-life-threatening infections. Modified forms of penicillin G (procaine penicillin and benzathine [dibenzyl] penicillin G), when given intramuscularly, lead to the slow release of penicillin over extended periods. For example, a single injection of benzathine penicillin each month is effective prophylaxis against streptococcal infections which may lead to rheumatic fever. In general, the penicillins have a high therapeutic index and exert almost no toxicity at the usual doses. The major limitations to their use are the lack of susceptibility of certain bacterial strains and the occurrence of penicillin hypersensitivity (allergy) in some patients (see Myrvik & Weiser: Fundamentals of Immunology, 2nd ed. Lea & Febiger, 1984).

The antibacterial spectrum of certain penicillins may be extended by combining them with the penicillinase inhibitor *clavulanic acid.* Such combinations have been effective both orally (amoxicillin plus clavulanic acid) and intravenously (ticarcillin plus clavulanic acid) in treating penicillinase-producing staphylococci and some Gram-negative bacilli.

The Cephalosporins. The cephalosporins are similar to the penicillins in structure (Fig. 6–8); however, they possess a 6-membered rather than a 5-membered thiozolidine ring fused to a β-lactam ring. This difference helps render the cephalosporins moderately resistant to penicillinases but, in turn, susceptible to the cephalosporinases; it is also the basis for the fortunately limited cross-hypersensitivity to cephalosporins seen in some penicillin-allergic patients. In general, the cephalosporins have a broader antibacterial spectrum than the penicillins, including activity against most pathogenic Gram-positive cocci plus a variety of Gram-negative bacilli. Many cephalosporins have become clinically available; they differ principally in their pharmacologic characteristics such as absorption, stability, etc., and in their antimicrobial spectra. The most widely used parenteral agents are cephalothin and cefazolin

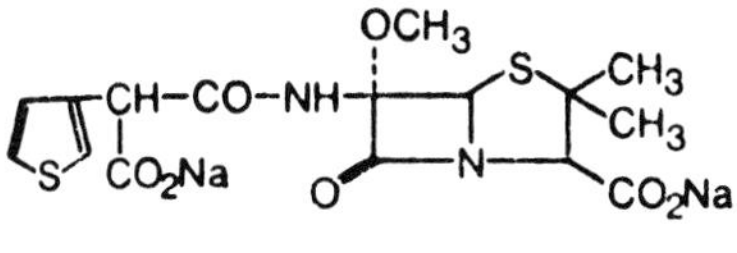

Fig. 6–7. Structural formula for temocillin, a new penicillin that differs from ticarcillin by the introduction of a methoxy group at the 6-α-position of the β-lactam ring.

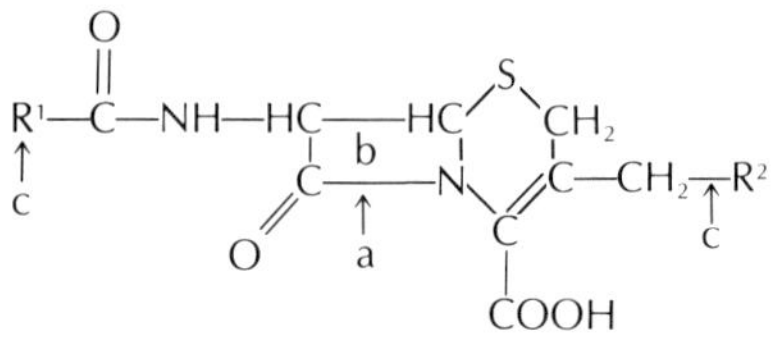

7-Aminocephalosporanic acid

Fig. 6–8. Basic structure of cephalosporins. (a) Site of action of β-lactamases; (b) β-lactam ring; (c) Side chains control spectrum and resistance to β-lactamases.

(Fig. 6–9). Among the principal oral agents is cephalexin (Fig. 6–9). Cefoxitin, which is technically a cephamycin due to possession of a methoxy group on the β-lactam ring, is more active against the important opportunistic pathogenic anaerobe, *Bacteroides fragilis,* than other cephalosporins. Cefamandole and Cefuroxime (parenteral) and cefaclor (oral) have high activity against *Haemophilus influenzae* (Fig. 6–9). The third generation of cephalosporins includes *cefotaxime, moxalactam, cefoperazone* and *ceftizoxime* which are restricted to parenteral use (Fig. 6–10). These new cephalosporins have an expanded antibacterial spectrum with high activity against *Neisseria gonorrhoeae* and *Haemophilus influenzae.* They are particularly effective against β-lactamase-producing strains and have antibacterial activity against *Pseudomonas aeruginosa* and several species of Gram-negative anaerobes.

Other new cephalosporins include *Ceftriaxone, Ceftazidime, Cefmenoxime,* and *Cefsulodin*; the latter two are still under investigation.

The broader spectrum of the cephalosporins, along with their relative penicillinase-resistance offers potential clinical advantages in treating certain

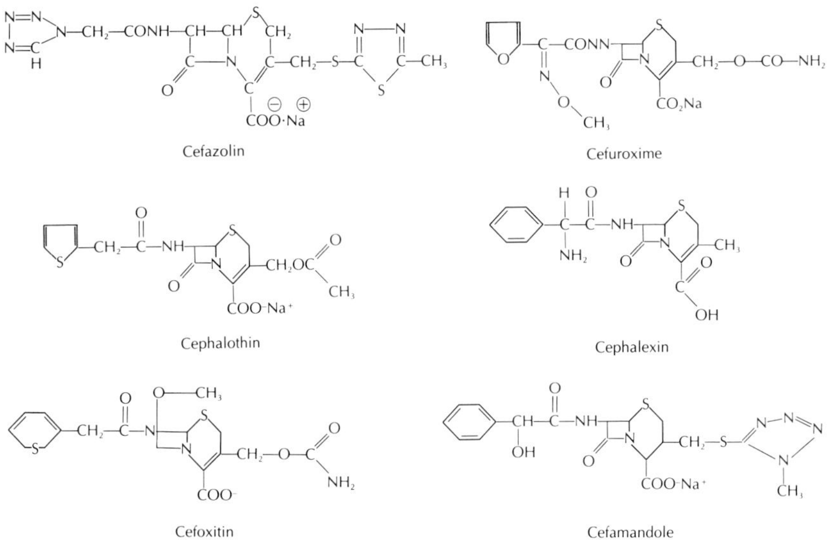

Cefazolin

Cefuroxime

Cephalothin

Cephalexin

Cefoxitin

Cefamandole

Fig. 6–9. Structural formulas of the most commonly used cephalosporins. Note the differences in side chains at acylation and deacetylation sites.

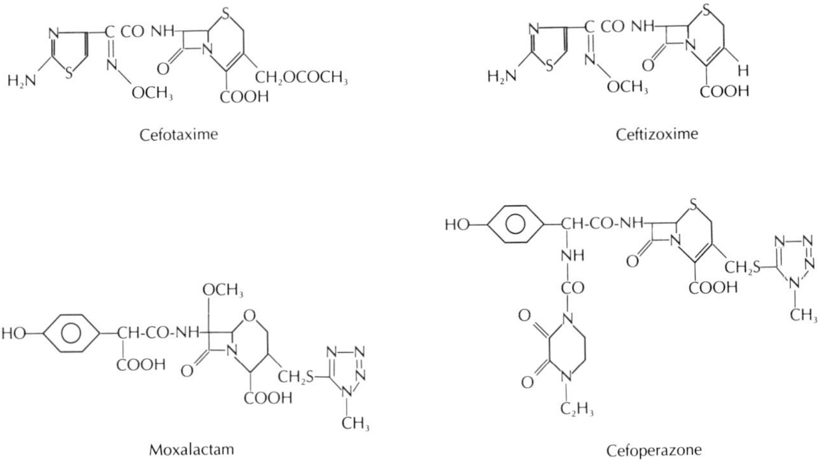

Cefotaxime

Ceftizoxime

Moxalactam

Cefoperazone

Fig. 6–10. Structural formulas of the newer cephalosporins (3rd generation).

mixed infections and serious infections of unknown etiology; however, they have the disadvantage of favoring superinfection and overgrowth of drug resistant opportunistic pathogens.

The *carbapenems* are a new class of antibiotics resembling the penicillins and cephalosporins. The principal available agent, imipenem (derived from thienamycin), has the broadest spectrum and is active against most Gram-positive cocci and Gram-negative rods. It acts by preferential binding to penicillin-binding protein (PBP-2) within the bacterial cell wall. Because imipenem is rapidly metabolized by renal dehydropeptidases, a second agent, the dehydropeptidase inhibitor cilastatin, has been added to produce a combination with enhanced effectiveness.

A further variation on the penicillin molecule has resulted in another category of agents, the *monobactams,* which contain only a β-lactam ring plus a variable side chain. These agents bind to PBP-3 in the bacterial cell wall, producing changes that ultimately destroy the organism. The first commercially available agent in this class is aztreonam.

Vancomycin is a relatively toxic drug that is poorly absorbed from the GI tract and must be used i.v. for systemic infections. Intravenously administered vancomycin may cause thrombophlebitis; nevertheless, it is a highly effective bactericidal agent for the treatment of serious staphylococcal infections, especially in penicillin-allergic patients. It is also effective against other serious infections; e.g., endocarditis due to enterococci and α-hemolytic streptococci. Vancomycin is useful in the oral therapy of antibiotic-associated enterocolitis caused by the overgrowth of certain broad spectrum antibiotic resistant bacteria in the bowel.

Bacitracin. This highly toxic drug is now restricted to topical use for skin infections due to common pathogens, such as staphylococci and streptococci.

2. Inhibition of Protein Synthesis

Bactericidal Agents. The aminoglycoside antibiotics, which include a number of related polycationic compounds (amino sugars connected by glycosidic linkages), are produced by *Micromonospora*) (Fig. 6–11). They exert bactericidal effects against a variety of Gram-negative bacilli (and some Gram-positive organisms) by binding to the specific receptor sites on the 20S ribosome and lethally inhibiting protein synthesis. They all have a low therapeutic index with potential toxicity for the kidneys and the 8th cranial nerve. As a group, they are often poorly absorbed when administered orally and penetrate minimally into the CNS. Aminoglycosides, when used with penicillins, show synergistic effects against certain organisms (particularly enterococci, α-hemolytic

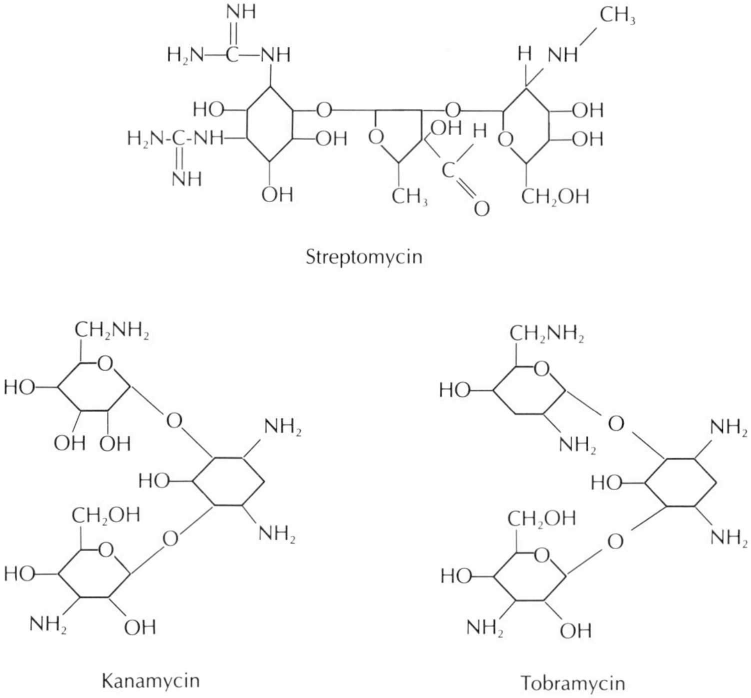

Streptomycin

Kanamycin

Tobramycin

Fig. 6–11. The basic structure of the aminoglycosides, streptomycin, kanamycin and tobramycin. The inositol residue is common to both streptomycin and kanamycin. Amikacin and tobramycin are structurally related to kanamycin. In amikacin, the lower amino group is replaced with a side chain consisting of:

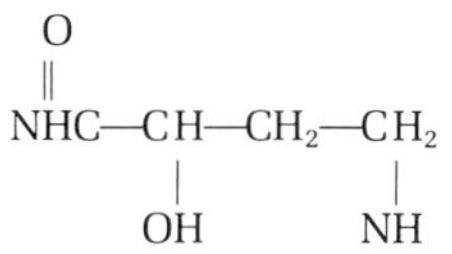

$$NHC(=O)—CH(OH)—CH_2—CH_2(NH)$$

which increases its resistance to bacterial inactivating enzymes. The main differences between tobramycin and kanamycin are the absence of the first two hydroxyl groups in the upper ring and the presence of an amino group.

streptococci, *Streptococcus bovis, Staphylococcus aureus* and a few Gram-negative bacilli).

The first effective aminoglycoside, *Streptomycin,* was discovered by Waksman in 1944; it remains as an important antituberculous agent and is also useful in bacterial endocarditis caused by certain streptococci, as well as in infections due to Gram-negative bacilli. *Kanamycin* has been superseded by a related derivative, *amikacin. Gentamicin* has also been widely used in serious infections caused by Gram-negative bacilli. *Tobramycin* has an antibacterial spectrum similar to gentamicin but with greater activity against *Pseudomonas aeruginosa* and somewhat less toxicity. *Neomycin* is too toxic to be used systemically and is restricted to topical use. It is poorly absorbed orally and has been used by this route to suppress gastrointestinal flora. Additional aminoglycosides are currently under development . They are an important group of agents for the treatment of serious Gram-negative bacillary infections.

A new aminoglycoside, *netilmicin,* is now available for clinical use (Fig. 6–12). Netilmicin appears to be as effective as tobramycin. Animal studies indicate that nephrotoxicity and ototoxicity appear to be less with netilmicin than with gentamicin or tobramycin.

Bacteriostatic Agents

Tetracyclines. The tetracyclines are a group of closely related polycyclic compounds which exert primarily a bacteriostatic effect (Fig. 6–13) by binding to the 30S ribosomal subunits and inhibiting bacterial protein synthesis. The first of these compounds, *chlortetracycline* was followed by *oxytetracycline* and *tetracycline.* Later modifications of the tetracycline structure have produced compounds with specific advantages but also new adverse effects.

Although effective against a broad spectrum of both Gram-negative and Gram-positive organisms, widespread use of the tetracyclines has been associated with increasing microbial resistance to these agents. However, they remain particularly useful in rickettsial, chlamydial and mycoplasmal infections, as well as in many of the common sexually-transmitted diseases, uncomplicated urinary tract infections, and chronic bronchitis. They appear to be useful both topically or systemically in long-term, low dose therapy for severe pustular acne. *Doxycycline,* a longer acting compound, does not accumulate in patients with renal insufficiency and accordingly is the tetracycline of choice in such patients. *Minocycline* has high activity against some anaerobic organisms; it also has been found useful for eradicating the meningococcal carrier state.

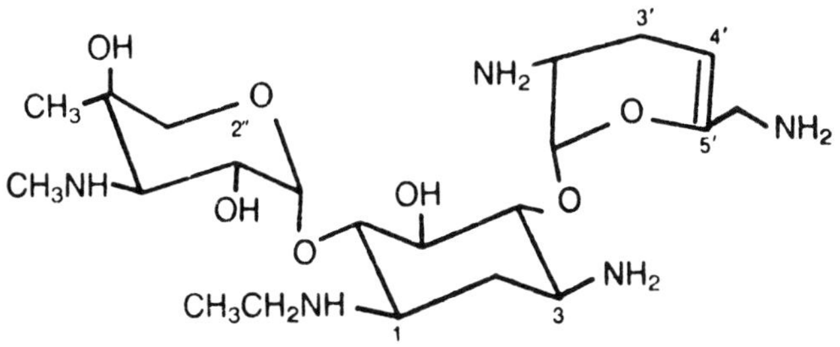

Fig. 6–12. Structural formula for netilmicin.

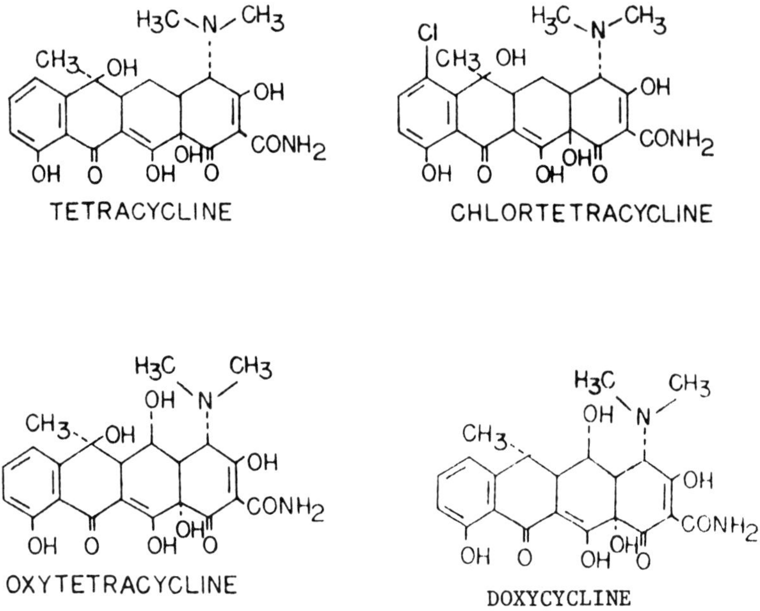

TETRACYCLINE

CHLORTETRACYCLINE

OXYTETRACYCLINE

DOXYCYCLINE

Fig. 6–13. Structures of the common tetracyclines.

Important adverse effects of the tetracyclines include staining of teeth in children under 8 years and when used in pregnancy, interference with fetal dental development, and fetal bone damage. These drugs, which are given primarily by the oral route, also may cause gastrointestinal side effects. When used i.v. hepatotoxicity is a potential risk. Photosensitivity skin reactions are most common with the formulations containing chloride, particularly *demeclocycline.*

Chloramphenicol, which was originally isolated from a soil actinomycete, is now produced by chemical synthesis (Fig. 6–14). Along with erythromycin, lincomycin and clindamycin, chloramphenicol inhibits bacterial protein synthesis by binding reversibly to the bacterial 50S ribosomal subunit. In fact, these drugs should not be used concomitantly because of potential competition for the common binding site. The antibacterial spectrum of chloramphenicol is similar although somewhat broader than that of the tetracyclines. An advantage of chloramphenicol is its excellent penetration into the CNS, the eye, and most other tissues. The principal toxic effect of chloramphenicol is a reversible, dose-dependent suppression of the bone marrow. Much less common is an irreversible, fatal aplastic anemia. Chloramphenicol must be administered in low doses to infants, because their detoxification system involving hepatic glucuronide conjugation is not fully developed, and the "gray syndrome" results from high serum levels of unconjugated drug. Major indications for chloramphenicol comprise severe *Salmonella* infections, including typhoid fever, rickettsial infections, CNS infections, (including *Haemophilus influenzae* meningitis) and certain anaerobic infections.

Erythromycin, the prototype of the macrolide antibiotics is the only one of

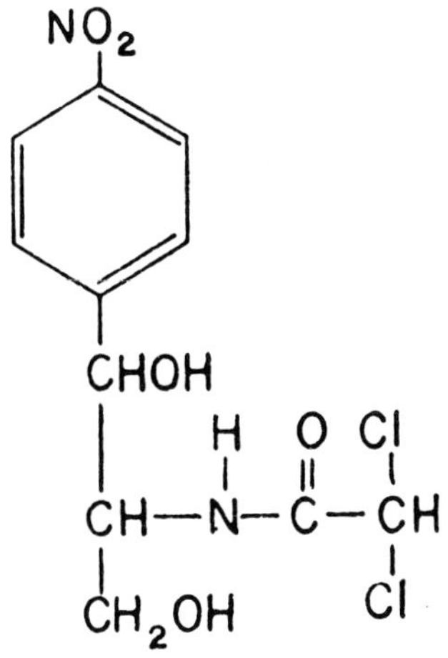

Fig. 6–14. Structure of chloramphenicol.

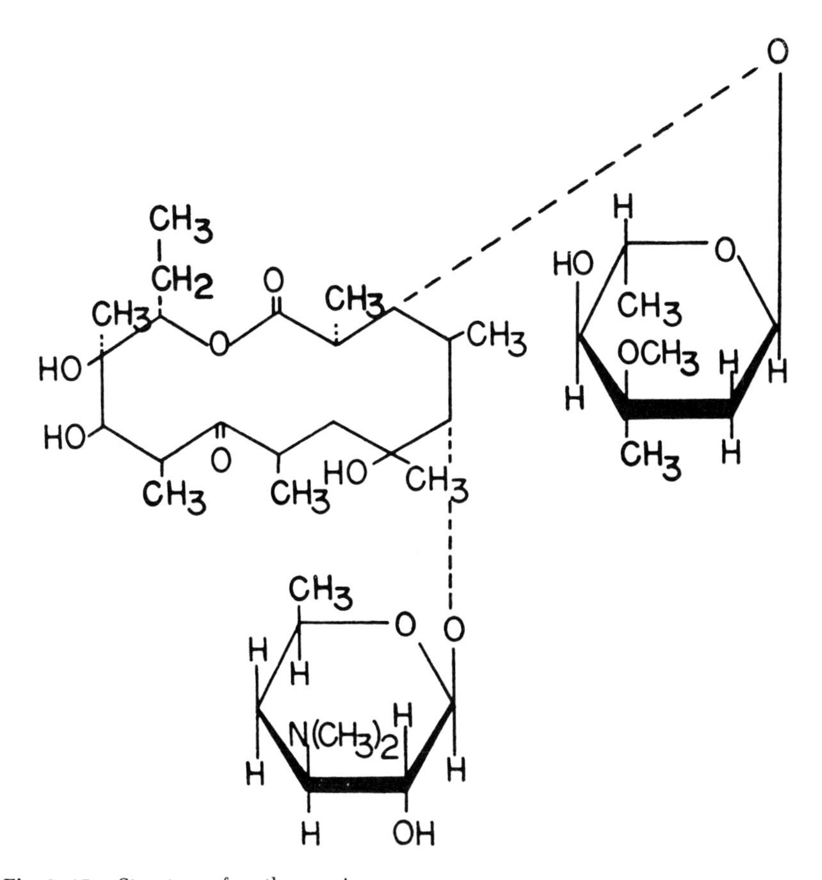

Fig. 6–15. Structure of erythromycin.

this group still in common use (Fig. 6–15). Its antibacterial spectrum against Gram-positive organisms is similar to that of penicillin; it has been especially useful in penicillin-allergic patients. Erythromycin is an alternative to tetracycline in mycoplasmal and chlamydial infections. It has also proven to be helpful in the therapy of *Legionella pneumophila* infections (Legionnaire's disease).

Lincomycin and *clindamycin* (7-chlorolincomycin) are bacteriostatic agents with antibacterial spectra and mechanisms of action similar to erythromycin (Fig. 6–16). They are useful in the treatment of severe Gram-positive infections, especially in penicillin-allergic patients. Clindamycin is more potent in vitro than lincomycin against most Gram-positive organisms and has been especially effective against anaerobic bacteria, particularly *Bacteroides fragilis.* A potential problem with these two drugs which is occasionally seen with most other antibiotics, is the occurrence of antibiotic-associated pseudomembranous colitis, a potentially debilitating and even fatal diarrheal condition apparently resulting from the overgrowth of toxin-producing *Clostridium difficile,* a member of the bowel flora.

3. Agents Affecting Membrane Permeability

The polymyxin antibiotics, polymyxin B and colistin (polymyxin E) are polypeptides which act directly on the bacterial cell membrane to produce irreversible damage leading to a fatal alteration in membrane permeability. In addition, the antifungal polyene agents discussed below have a similar locus of action. Therapeutic indications for the polymyxins are highly limited be-

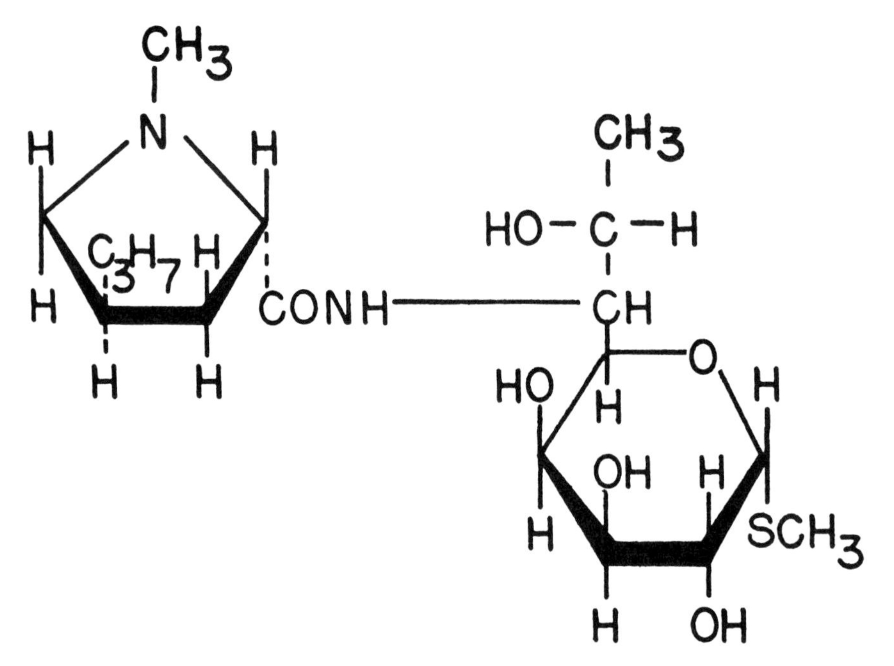

Fig. 6–16. Structure of lincomycin.

cause of their significant systemic nephrotoxicity and neurotoxicity. Consequently, they are used principally for the topical treatment of bacterial diseases of the skin, often in combination with neomycin and/or bacitracin.

4. Antimetabolites

Sulfonamides. In 1935, Domagk initiated the era of widespread antibacterial chemotherapy by demonstrating that a sulfonamide-containing dye, prontosil, was effective for treating mice infected with streptococci. It was later found that prontosil was metabolized to yield the active moiety sulfanilamide. Subsequently, thousands of related compounds were synthesized and tested. The sulfonamides are structural analogs of para-aminobenzoic acid (PABA) which is a key precursor in the synthesis of folic acid (Fig. 6–17). Sulfonamides act as competitive inhibitors of PABA, thus blocking synthesis of folic acid, which is not transported into bacterial cells (in contrast to mammalian cells) and must be synthesized intracellularly. This difference between the metabolism of bacterial and mammalian cells is the basis of the selective toxicity of sulfonamides for bacteria.

Although effective against both Gram-positive and Gram-negative organisms, increasing bacterial resistance plus a moderately high incidence of toxic and other adverse reactions has limited the use of the sulfonamides. Principal indications at present include uncomplicated urinary tract infections (sulfisoxazole and sulfamentoxazole), chancroid, lymphogranuloma venereum, nocardiosis, and some meningococcal and *Haemophilus* infections. Topical sulfonamides are used in burns to limit infection and locally in the eye, vagina, etc. Poorly absorbed oral agents are used to suppress bowel flora (sulfathiazole) and to treat ulcerative colitis (sulfasalazine) (Fig. 6–18).

Another antimetabolite, trimethoprim, a competitive inhibitor of dihydrofolate reductase, is now used in combination with sulfamethoxazole; the combination (trimethoprim-sulfamethoxazole, TMP-SMX, or co-trimoxazole) exerts a synergistic antibacterial effect (Fig. 6–19). Indications for the TMP-SMX combination include recurrent or resistant urinary tract infections, some respiratory

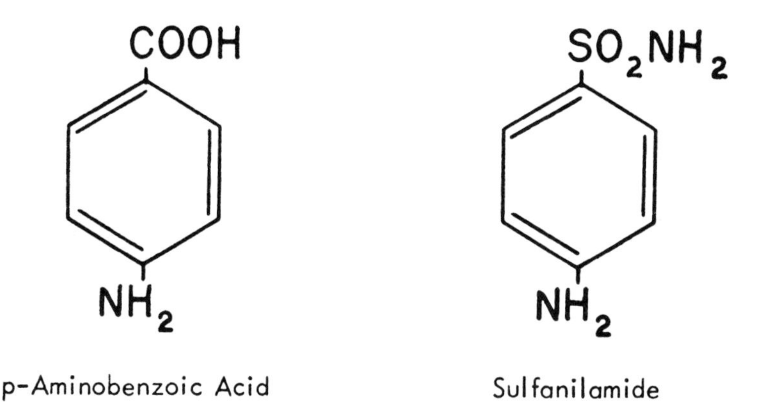

Fig. 6–17. The structural similarity of PABA and sulfanilamide explains the basis of competitive inhibition of folic acid synthesis.

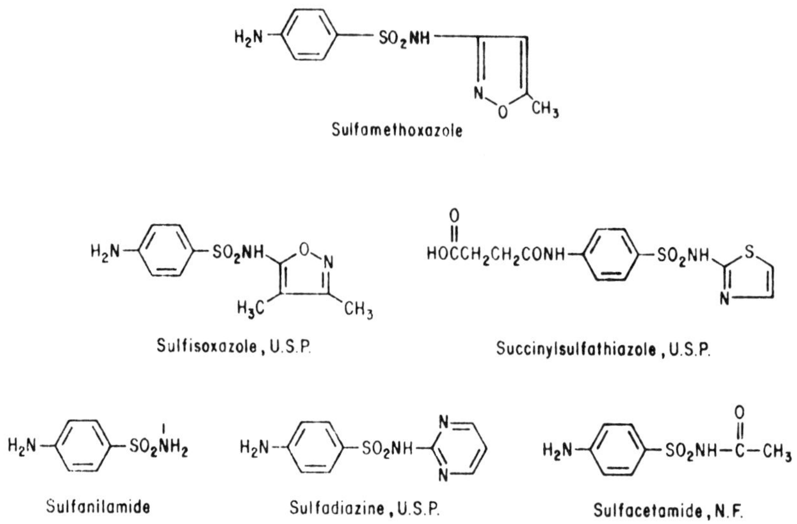

Sulfamethoxazole

Sulfisoxazole, U.S.P.

Succinylsulfathiazole, U.S.P.

Sulfanilamide

Sulfadiazine, U.S.P.

Sulfacetamide, N.F.

Fig. 6–18. Structure of common sulfanilamide compounds.

tract infections (including the opportunistic protozoan infection with *Pneumocystis carinii*), otitis media, and some enteric infections.

Another group of antimetabolites are the sulfones, including an important drug for the therapy of leprosy, dapsone (diaminodiphenyl sulfone or DDS) (Fig. 16–20).

5. Urinary Tract Antiseptics and Related Agents

Important agents for the treatment of urinary tract infections previously discussed include the soluble sulfonamides, ampicillin, the tetracyclines and cotrimoxazole. In addition, certain other agents are concentrated principally in the urine, achieve low tissue levels elsewhere and, thus, are indicated only in the treatment of urinary tract infections. These include *nitrofurantoin, nalidixic acid, oxoclinic acid,* and *methenamine.* These agents are used mainly to prevent recurrent urinary tract infections. All may produce untoward or adverse effects.

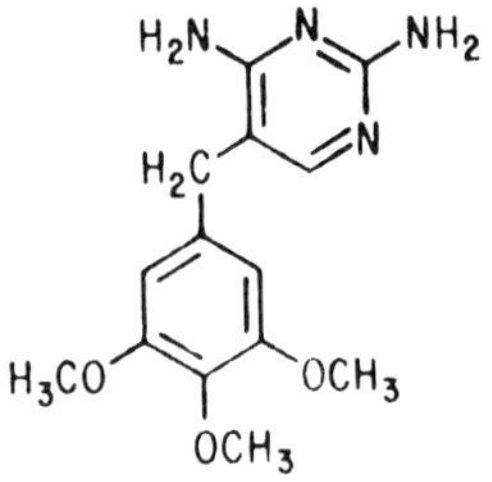

Trimethoprim

Fig. 6–19. Structure of the diaminopyrimidine trimethoprim. Trimethoprim, a competitive inhibitor of dihydrofolate reductase, is used in combination with sulfamethoxazole.

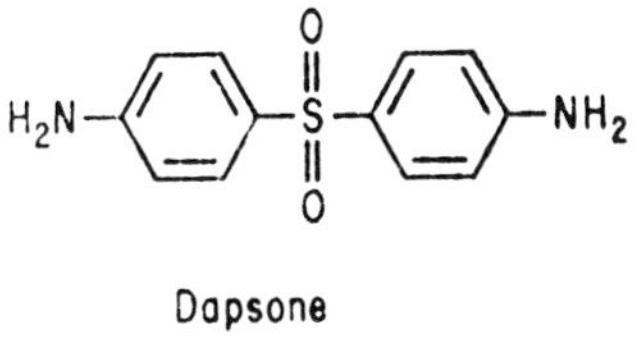

Dapsone

Fig. 6–20. Note the structural similarities between dapsone and sulfonamides.

Nitrofurantoin, for example, in addition to nausea and vomiting, may occasionally produce polyneuritis, pulmonary fibrosis, hemolytic anemia, or hepatitis. Methenamine, is itself not bactericidal, but at acid pH is hydrolyzed to ammonia and formaldehyde, the active bactericidal agents. Accordingly, when methenamine is used, an acidifying agent (such as ascorbic acid) should also be used to keep the urine pH at or below 5.5.

An important new class of synthetic orally administered antibacterial agents related to naladixic acid are the *fluoroquinolones.* These agents, including norfloxacin and ciprofloxacin, achieve significant extrarenal tissue levels and have a broad antibacterial spectrum. Therefore they have significant potential for the treatment of systemic infections.

E. ANTIMYCOBACTERIAL AGENTS

The treatment of tuberculosis requires long term therapy (typically 1 to 2 years) and the simultaneous use of two or more drugs to prevent emergence of drug-resistant mycobacteria during therapy. At the present time, the major (or "first-line") agents are isoniazid, ethambutol, streptomycin and rifampin. In addition, a second group of more toxic agents is employed for drug-resistant infections (including certain of the atypical mycobacterial infections). These include paraminosalicylic acid (PAS), cycloserine, viomycin, pyrazinamide, capreomycin, ethionamide and kanamycin.

Isoniazid (INH) is a particularly important agent since it is bactericidal for tubercle bacilli and penetrates well into most organs, including the CNS, as well as cells such as macrophages in which tubercle bacilli find sanctuary (Fig. 6–21). During the treatment of active tuberculosis it is particularly important to employ a second agent to prevent development of isoniazid resistance. Isoniazid may be used alone for chemoprophylaxis of exposed or recently infected individuals ("tuberculin converters") as well as other high risk groups with asymptomatic or latent infection. In contrast to the high "infectious bacillary load" in patients with active disease, the low "infectious bacillary load" in the

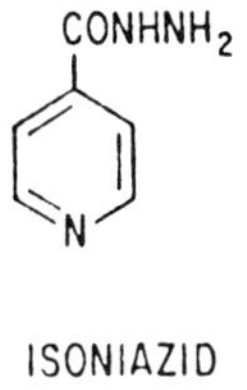

ISONIAZID

Fig. 6–21. Isoniazid is a derivative of isonicotinic acid.

above individuals reduces the likelihood that isoniazid-resistant mutants will emerge during chemoprophylaxis.

A second major agent to treat tuberculosis is *rifampin*, an orally active semisynthetic member of the rifamycin group of antibiotics (Fig. 6–22). Unlike isoniazid, rifampin has a wide spectrum of activity against Gram-positive bacteria, certain Gram-negative bacteria, including *Neisseria*, various species of *mycobacteria* and even viruses. Rifampin specifically inhibits DNA-dependent RNA polymerase by blocking steps in RNA chain initiation. The two major clinical uses of rifampin at present are in the therapy of mycobacterial infections and in the short term treatment of meningococcal carriers.

F. ANTIFUNGAL AGENTS

Systemic mycoses, including histoplasmosis, blastomycosis, coccidioidomycosis and sporotrichosis, as well as the "opportunistic" mycoses (candidiasis, aspergillosis, phycomycosis and cryptococcosis) may require therapy with *amphotericin B, flucytosine* or *miconazole*. Other experimental drugs are under investigation. For cutaneous forms of sporotrichosis, saturated potassium iodine solution (Lugol's solution) is effective, and for some forms of blastomycosis, *hydroxystilbamidine* may be used. Although not true fungi, *Nocardia* and certain actinomycetes produce clinical diseases resembling systemic mycoses. Nocardiosis may be treated effectively with sulfonamides, and actinomycosis with penicillin.

The common superficial mycoses (epidermophytosis—"ringworm", "athletes foot" and candida skin infections) usually respond to a variety of topical agents. For particularly difficult dermatophytoses, the oral agent, *griseofulvin* is used.

Amphotericin B, and *nystatin* are antibiotics which bind selectively to sterols in the cell membranes of fungi, producing lethal alterations in membrane permeability. Since they may also bind to sterols in mammalian cell membranes, these agents are also potentially toxic to the host. *Amphotericin B*, accordingly,

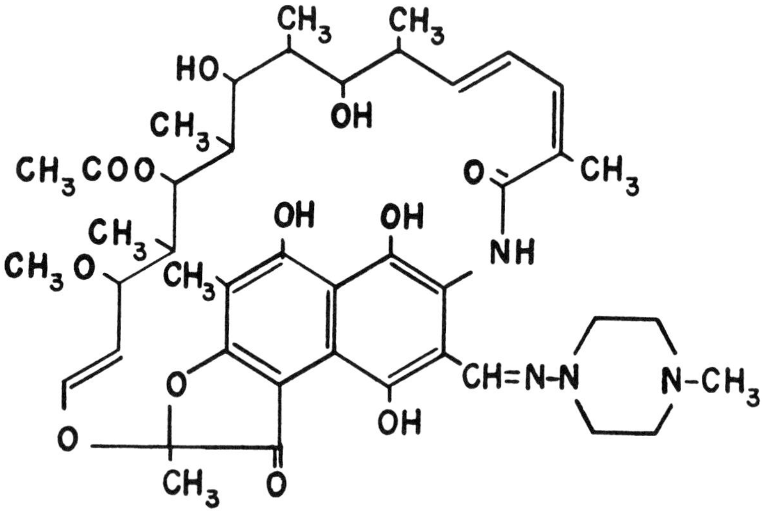

Fig. 6–22. Structure of rifampin.

is a relatively toxic drug and its systemic use is confined to serious infections; it also has limited topical use. *Nystatin* is used topically for candida skin infections and orally for intestinal candidiasis.

5-fluorocytosine *(flucytosine)* is transported into fungal cells where it is hydrolyzed to 5-fluorouracil (5-FU); 5-FU is incorporated into fungal RNA and results in "faulty" RNA production by the fungus. Flucytosine has been used successfully in the therapy of *candida* and *cryptococcal* infections. Since flucytosine-resistant strains may emerge, it is used in combination with amphotericin B, especially in meningitis due to *Cryptococcus neoformans.*

Miconazole and *ketoconazole*, are synthetic broad-spectrum antifungal agents used in treating infections caused by dermatophytes, dimorphic fungi and yeasts. *Miconazole* can be used topically and is also administered i.v. Ketoconazole is effective by the oral route against superficial and deep mycotic infections. These drugs alter the permeability of the cell membrane causing the leakage of phosphorus-containing intermediates as well as potassium ions. Cell membrane changes result from inhibition of ergosterol synthesis.

Griseofulvin, which is used orally against the superficial dermatophytes, *Microsporum* and *Trichophyton,* accumulates in the keratin layer, where it acts selectively on dermatophytes by inhibition of nucleic acid synthesis (Fig. 6–23). Dermatophyte infections require prolonged therapy, and in the case of infections of the nails, months of therapy are required.

G. ANTIVIRAL AGENTS*

H. SELECTION OF APPROPRIATE ANTIBIOTICS

The choice of appropriate antibiotic therapy must take into account a variety of factors including the bacterial and host determinants previously discussed. It is important that a correct identification of the etiologic agent be made or alternatively, that the *most likely* diagnosis be made so that antibiotics will not be used for inappropriate indications. For example, antibacterial agents are of no benefit in viral infections and may lead to adverse drug reactions, and/or superinfection with other organisms possessing either natural or acquired drug-resistance.

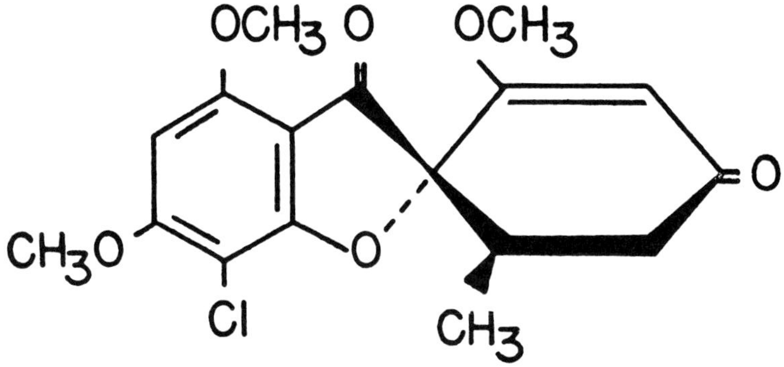

Fig. 6–23. Structure of griseofulvin.

*See Kucera, L.S. and Myrvik, Q.N.: Fundamentals of Medical Virology, 2nd Ed. Lea & Febiger, 1985.

The prophylactic use of antibiotics to prevent infection may be indicated in specific situations but is frequently excessive and inappropriate. In general, chemoprophylaxis is most effective when aimed at a single organism, as in the use of penicillin to prevent β-hemolytic streptococcal infections and its sequela of recurrent rheumatic fever, or when used for a short duration, as in the use of a brief pulse of antibiotics just prior and for one or two doses following dental procedures in individuals with damaged heart valves (Chap. 10). Such individuals are at high risk of developing endocarditis on their damaged valves when experiencing a transient bacteremia that often follows dental procedures. Attempts to erect an antibiotic "umbrella" over the infected patient usually fail and are fraught with hazard. The route and dose of an antibiotic appropriate to reach the infected tissue, as well as its potential toxic effects must be balanced; cost factors must also be considered. Bactericidal agents are usually indicated in situations in which host defenses are impaired.

The *antibiotic sensitivity of the etiologic agent is clearly a key factor in selection.* Ideally, the antibiotic sensitivity of the pathogen(s) isolated from the patient should be determined. In rapidly progressing, life-threatening diseases, and in circumstances in which appropriate cultures cannot be obtained, therapy must be instituted without this information. Adjustments in therapy should be made later if subsequent sensitivity data so indicate. Thus, in serious staphylococcal infections, the presumption of penicillin-resistance is made pending sensitivity-testing of the isolate. At times, our knowledge about the sensitivity of hemolytic streptococci to penicillin makes routine sensitivity determinations unnecessary. Laboratory methods for antibiotic sensitivity testing of microbial isolates include the paper disc method and the tube-dilution technique.

The disc method entails the use of standardized commercial paper discs, impregnated with known concentrations of particular antimicrobial agents, being tested; the discs are placed on agar culture plates seeded with the isolate to be tested. After incubation, a clear zone surrounds the disc if the test organisms are sensitive to the antimicrobial agent which has diffused into the agar. The size of the zone of inhibition generally correlates with the sensitivity of the organism, but must be standardized for each organism and each antibiotic. For example, the diffusability of some agents, such as the polypeptide antibiotics (polymyxin) is less than that of other antibiotics which results in small zones of inhibition. The presence of PABA in the culture media will markedly affect the bacteriostatic activity of sulfonamides. Special cultural conditions are necessary for the correct determination of methicillin-resistance of staphylococci.

In the tube-dilution technique, the antibiotic is serially diluted in liquid culture media and each tube is seeded with a standard inoculum of the test organism. The lowest concentration of the antimicrobial agent that will inhibit the growth of the test organism (minimal inhibitory concentration or MIC) can be determined by this method, and also the lowest concentration needed to kill the organism in vitro, (minimal bactericidal concentration or MBC). This is determined by subculturing the tubes and noting which dilutions result in no visible growth. The MBC is the lowest concentration in which the subculture is sterile.

In special circumstances, such as following the treatment of a patient with endocarditis, it is useful to determine the antimicrobial effectiveness of the patient's serum as a reflection of the level of antibiotic present. This can be done by a modification of the tube dilution technique in which the patient's serum is diluted and tested against a standard organism of known sensitivity, or against the etiologic agent isolated from the patient. This technique may be particularly helpful in documenting synergism or antagonism when two or more antibiotics are being used. In the case of renal failure, which markedly influences the serum levels of some antibiotics, monitoring of antibiotic blood levels may be indicated. This may be carried out by a biologic assay using a standard test organism of known sensitivity, or, in some instances, by chemical determination (e.g., gentamicin).

I. ADVERSE DRUG REACTIONS

The major limiting factor in antimicrobial chemotherapy is the potential occurrence of a variety of untoward or adverse effects of therapy. During treatment with drugs which have a low therapeutic index, such as the aminoglycosides, direct *toxicity* to the *kidneys* or the eighth cranial nerve is a major concern and is particularly likely to be a problem in patients with a reduced renal function in whom toxic levels of the drug may accumulate in the serum unless the dosage is appropriately modified. Other adverse effects, such as nausea, vomiting and diarrhea sometimes associated with erythromycin represent annoying *side effects* which may be alleviated by changing the manner of drug administration, for example giving the drug along with food. A particularly dangerous side effect is the occurrence of *superinfection* or overgrowth of drug resistant organisms producing, for example, antibiotic-associated pseudomembranous enterocolitis. True *hypersensitivity* or allergic reactions are less common than pharmacologic toxic or side effects, but may present major problems as in the case of the variety of reactions ranging from skin rashes to fever, hemolytic anemia, and anaphylaxis occurring, in penicillin-allergic patients. It is important that the physician be aware of potential adverse effects, so that risks and benefits of therapy may be weighed and early detection of untoward effects assured.

J. REASONS FOR FAILURE OF ANTIBIOTIC THERAPY

There are many possible reasons for the failure of antibiotic therapy to suppress or eradicate the etiologic agent. The most obvious reason is that the antibiotic is being given for an inappropriate indication for which no benefit can be expected (antibacterial agent used to treat a disease of viral etiology) or that the bacteria being treated are resistant to the particular antibiotic given. Organisms in a vegetative state, e.g., pharyngeal meningococci in an asymptomatic carrier, may be temporarily "phenotypically resistant" to penicillin, because susceptibility requires active cell-wall synthesis.

Other considerations relate to the pharmacology of the drug. Dosage must be adequate to achieve therapeutic levels in the infected tissues and the drug must be capable of penetrating into the infectious area. For example, in the treatment of meningitis due to *Haemophilus influenzae,* a relatively high dose of ampicillin is recommended (400 mg per Kg of body weight per day) to insure that adequate levels are achieved in the cerebrospinal fluid. Aminoglycosides, in general, penetrate poorly into the meninges, even when inflamed.

Drugs such as nitrofurantoin which become concentrated in the urine achieve low levels in the blood and non-renal tissues. Thus, although an organism may be reported by the laboratory to be sensitive to nitrofurantoin, the drug will have virtually no effect if the infection is in an extra-renal locus. The route of administration may be critical. Penicillins given orally may be substantially inactivated by gastric acidity, especially when given on a "full stomach". Oral tetracyclines may be bound by calcium in antacids. Some agents are incompatible when mixed in intravenous fluids (e.g., carbenicillin and gentamicin). Pharmacologic interactions, competition, or antagonism may also occur in the patient.

Physiologic conditions may affect the action of the antibiotic. Certain aminoglycosides are most effective at an alkaline pH, and alkalinization of the urine (e.g., with oral sodium bicarbonate) may be required in treatment of urinary tract infections. On the other hand, urinary antiseptics such as mandelamine and nitrofurantoin optimally require an acid pH in the urine.

Potentially successful therapy may be obscured by the occurrence of adverse drug reactions, including superinfection with a drug-resistant pathogen, or the appearance of a drug hypersensitivity reaction such as "drug fever."

Finally, even when adequate doses of the appropriate antibiotic are given, a variety of host factors may prevent a satisfactory response: the presence of foreign bodies, devitalized tissues, abscess formation, obstructed viscera or tubular structures, and impaired host defense mechanisms.

REFERENCES

Bauer, A.W., et al.: Antibiotic susceptibility testing by a standardized single disk method. Am. J. Clin. Pathol. *45*:493, 1966.

Cluff, L.E., et al.: Clinical Problems with Drugs. Philadelphia, W.B. Saunders Co., 1975.

Cluff, L.E., and Johnson, J.E., III: Clinical Concepts of Infectious Diseases. 3rd Ed. Baltimore, Williams & Wilkins, 1982.

del Busto, Ramon: An overview of the newer antibiotics. Henry Ford Hospital Med. J. *32*:90, 1984.

Dums, F., et al.: Aminoglycosides and penicillins can be inactivated by one enzyme, penicillinase. Naturwissenschaften *72*:484, 1985.

Finland, M., et al.: Summary of the Symposium: Trimethoprim-sulfamethoxazole. Rev. Infect. Dis. *4*:185,1982.

Gardner, P., and Provine, H.T.: Manual of Acute Bacterial Infections: Early Diagnosis and Treatment. 2nd Ed. Boston, Little, Brown and Co., 1984.

Gilman, A.G., et al.: The Pharmacological Basis of Therapeutics. 7th Ed. New York, Macmillan, 1985.

Kagan, B.M.: Antimicrobial Therapy. 3rd Ed. Philadelphia, W.B. Saunders Co., 1980.

Mattie, H.: Antibiotics. Contemp. Issues Clin. Biochem. *3*:264, 1985.

Menninger, J.R.: Functional consequences of binding macrolides to ribosomes. J. Antimicrob. Chemother. *16*:A23, 1985.

Murray, B.E., and Moellering, R.C., Jr.: Cephalosporins. Annu. Rev. Med. *32*:559, 1981.

Neu, H.C.: Structure-activity relations of new β-lactam compounds and in vitro activity against common bacteria. Rev. Infect. Dis. *5*:S319, 1983.

Omura, S.: Philosophy of New Drug Discovery. Microbiol Rev. *50*:259, 1986.

Pestka, S.: Inhibitors of ribosome functions. Annu. Rev. Microbiol. *25*:487, 1971.

Pratt, W.B.: Chemotherapy of Infection. New York, Oxford University Press. 1977.

Thompkins, L.S., et al.: Molecular analysis of R-factors from multiresistant nosocomial isolates. J. Infect. Dis. *141*:625, 1980.

Van Landuyt, H.W., et al.: In vitro activity of temocillin against clinical isolates. Drugs *29* (Suppl. 5):1, 1985.

Young, L.S.: Current needs in chemotherapy for bacterial and fungal infections. Rev. Infect. Dis. *7*:S380, 1985.

7

THE NORMAL MICROBIAL FLORA

One of the most remarkable aspects of host-parasite relationships involves the wide array of microbes that colonize normal human beings from the moment of birth. In the course of evolution these microorganisms, which are termed the normal microbial flora, have adapted specifically to various parts of the body and represent persistent parasites of essentially every normal human being. It is of singular ecologic interest that each animal species has its own characteristic natural microbial flora which is remarkably constant among individuals and that such flora becomes rapidly established in the newborn by contact with adults of the species. It is highly important that students of medicine understand the relationship between man and the microbes that normally inhabit the body because this relationship is an integral part of man's total ecology including the agents of infectious disease. The following facts attest to the vital role that the normal microflora plays in body economy. Members of the normal microflora (1) can assume the role of pathogens if the host's defenses are defective, (2) can interfere with colonization and/or invasion by true pathogens, (3) can immunize the host against pathogens when related or cross-reacting Ags are shared, and (4) can be confused with the true etiologic agent of a disease because of their ubiquity on body surfaces and mucous membranes and the fact that many of them closely resemble pathogens.

The normal microflora is comprised of those organisms which universally colonize various regions of the body and in a sense represent "standard equipment" of the normal healthy host; e.g., *Staphylococcus epidermidis* on skin or *Escherichia coli* in the intestinal tract. They are distinct from the so-called "transient flora" consisting of organisms that usually colonize a small percentage of the host population only temporarily or on an intermittent basis. Members of the transient flora may include facultative parasites which normally exist as saprophytes, opportunistic pathogens, or true pathogens. A pathogen may establish a short-term or long-term carrier state in its host. The long-term carrier state of pathogens obtains because of acquired immunity of the host, together with special anatomic or microenvironmental conditions which permit limited propagation of a pathogenic microorganism in a host for extended periods without any apparent harm; e.g., carriers of *Salmonella typhi* or coagulase-positive *Staphylococcus aureus.* The carrier state is the major means whereby pathogenic microorganisms persist in their natural hosts, the carrier functioning automatically as the reservoir of the infectious agent.

A. DISTRIBUTION PATTERNS OF THE NORMAL MICROBIAL FLORA

The various microorganisms that commonly and in many instances consistently inhabit the skin, nose, pharynx, mouth, conjunctivae, lower intestine, external genitalia, and vagina are listed in Table 7–1. The flora of skin varies depending on the anatomic site. As a rule *Staph. epidermidis* and *Propionibacterium acnes* are the most abundant inhabitants of skin. The latter organism is normally found in sebaceous glands and hair follicles. *Mycobacterium smegmatis* is commonly present on the external genitalia and adjacent skin.

Within 12 hours after birth α-hemolytic streptococci are found in the upper respiratory tract and become the dominant organism of the oropharynx. Other nonpathogenic organisms that become permanent residents in the mouth and throat include coagulase-negative staphylococci, neisseriae, lactobacilli, *Branhamella, Bacteroides,* corynebacteria, and spirochetes. In addition, species of

Table 7–1. Microorganisms Found Consistently[1] on Various Integuments of the Human Body

SKIN:	
Staphylococcus epidermidis	*Corynebacterium* (Diphtheroids)
Propionibacterium acnes	*Pityrosporum ovale* (scalp and other skin areas)
CONJUNCTIVA:	
Corynebacterium species[2]	*Haemophilus* species
NOSE:	
Staphylococcus epidermidis	*Streptococcus mitis*
Corynebacterium species	*Streptococcus salivarius*
Haemophilus species	*Moraxella lacunata*
Branhamella catarrhalis	
MOUTH:	
Staphylococcus epidermidis	Anaerobic micrococci
Streptococcus salivarius	*Streptococcus mitis*
Lactobacillus acidophilus	Anaerobic streptococci
Corynebacterium species	*Neisseria* species
Actinomyces bifidus	*Bacteroides* species
Leptotrichia buccalis	*Actinomyces israelii*
Treponema dentium	*Fusobacterium* species
Mycoplasma species	*Candida albicans*
Spirillum sputigenum	*Branhamella catarrhalis*
Streptococcus mutans	*Veillonella alcalescens*
PHARYNX:	
Streptococcus salivarius	*Streptococcus mitis*
Neisseria species	Anaerobic streptococci
Corynebacterium species	*Veillonella alcalescens*
Fusobacterium species	*Bacteroides* species
Treponema dentium	*Vibrio sputorum*
Klebsiella aerogenes	*Actinomyces israelii*
Proteus species	*Haemophilus* species
Branhamella species	
LOWER INTESTINE:	
Streptococcus mitis	Anaerobic micrococci
Steptococcus faecalis	*Streptococcus salivarius*
Lactobacillus species	Anaerobic streptococci
Escherichia coli	*Clostridium* species
Pseudomonas aeruginosa	*Alcaligenes faecalis*
Bacteroides species	*Klebsiella aerogenes*
Mycoplasma species	*Fusobacterium* species
Candida albicans	*Eubacterium* species
Bifidobacterium bifidum	
EXTERNAL GENITALIA:	
Staphylococcus epidermidis	*Streptococcus* species
Streptococcus faecalis	Anaerobic streptococci
Escherichia coli	*Spirillum sputigenum*
Bacteroides species	*Treponema dentium*
Mycobacterium smegmatis	*Candida albicans*
Fusobacterium species	*Mycoplasma* species
Corynebacterium species (Diphtheroids)	
VAGINA:	
Anaerobic micrococci	*Corynebacterium* species (Diphtheroids)
Neisseria species	*Mima vaginicola*
Mima polymorpha	*Treponema dentium*
Haemophilus vaginalis	*Mycoplasma* species
Streptococcus faecalis	*Lactobacillus* species
Candida albicans	*Trichomonas vaginalis*
Escherichia coli	

[1]This list is not intended to contain all of the organisms of the normal microbial flora.
[2]More than one species.

Haemophilus, Actinomyces, and *Mycoplasma* are usually present in a large majority of healthy individuals. As a rule α-hemolytic streptococci predominate in the pharynx and coagulase-negative staphylococci predominate in the nose. It is of interest that the normal lower respiratory tract is bacteriologically sterile even though the trachea may contain numerous bacteria. This exemplifies the remarkable defense system presented by the lung, involving alveolar macrophages and mucociliary clearance.

The surface of the esophageal wall usually contains only the bacteria swallowed with saliva and food. Because of low pH the stomach is virtually sterile except during the early intervals after eating. In patients with carcinoma of the stomach or achlorhydria due to pernicious anemia, the stomach literally becomes a fermentation vat; this attests to the role of HCl in maintaining a low population of organisms in the normal stomach.

The number of bacteria increases progressively from a point below the duodenum to the colon, being comparatively low in the small intestine. The largest number of bacteria exist in the contents of the colon and can approach 10^{10}/g of feces. In breast-fed infants lactobacilli predominate in the feces, evidently as the result of environmental conditions created by ingested human milk rich in lactose. As the consequence of a more diverse diet, a relative increase in numbers of *E. coli, Bacteroides* sp, clostridia, and enterococci occurs. Members of the genus *Bacteroides* represent by far the most numerous organisms in the adult colon.

The vagina becomes colonized with lactobacilli immediately after birth, followed by staphylococci, enterococci, and diphtheroids. With the onset of puberty and the child-bearing years, lactobacilli predominate in the vagina, allegedly because of the high glycogen content of the epithelial cells. Lactobacilli are evidently responsible for the acidity of vaginal secretions during the child-bearing age. The onset of the postmenopausal period is attended by a drop in the numbers of lactobacilli and the return of the prepubertal flora.

The normal urethra is usually sterile above the urethrovesicular junction. A few bacteria from the external mucous membranes of the genitalia may contaminate the lower segment of the urethra.

Small numbers of organisms such as diphtheroids, *Haemophilus* sp, *Branhamella* sp, non-hemolytic streptococci, and coagulase-negative staphylococci can be isolated from the conjunctivae.

A discussion of the principle of selective attachment of bacteria to mammalian cells is presented in Chapter 8. Whereas the mechanisms operative in attachment are not known with certainty, it is likely that components on the bacterial surface are, in part responsible for the special distribution pattern of indigenous microbes. In this regard, it must be emphasized that the microorganisms must also find the habitat of their choice compatible with their survival.

B. POTENTIAL BENEFITS OF THE NORMAL MICROBIAL FLORA

Microorganisms of the normal flora appear to exert mutual control on each other through some type of microbial antagonism. Bacterial antagonism presented by the normal flora also appears to play a role in controlling implantation and colonization of pathogenic microorganisms. It is well recognized that if the gastrointestinal flora is reduced by treatment with streptomycin, experimental animals will exhibit a marked increase in their susceptibility to *Salmonella enteritidis* introduced by the oral route.

Several examples of the principles by which normal flora organisms exclude pathogens are available. For example, the *Bifidobacteria* in the colon of breast-fed infants block the colonization of enteric pathogens. In addition, ingestion of maternal IgA also can provide some short term immunity against intestinal pathogens. Antibiotic therapy can result in the overgrowth of *Candida albicans* in the mouth, vagina and anal region. Viridans streptococci can block or markedly reduce colonization of *C. albicans* on oral epithelium.

In addition, the normal microflora can provide an important stimulus to the development and maturation of the organ systems responsible for acquired immunity. In particular, endotoxin may influence the immunologic response of the host and enhance the activity of some of the immunologic systems.

The normal microflora of the intestinal tract can contribute vitamin K and several B vitamins which could be beneficial under some circumstances. Accordingly, the availability of bacterial-produced vitamins would be curtailed during administration of broad-spectrum antibiotics.

C. POTENTIAL LIABILITIES OF THE NORMAL MICROBIAL FLORA

Normal flora microorganisms can assume the role of serious opportunistic pathogens when host defense factors are depressed or the equilibrium of the flora is altered by prolonged antibiotic therapy. It has been suggested that endotoxins produced by the normal flora of the intestinal tract effect chronic low-grade toxicity. This is based on the observation that animals raised on diets that suppress endotoxin-producing organisms grow faster than animals on standard diets. Another potential liability is that several *Bacteroides* sp in the colon produce a mutagen which appears to contribute to the incidence of colon cancer.

It has also been observed that when penicillinase-producing staphylococci are associated with chronic gonococcal urethritis, they may interfere with the capacity of penicillin to eradicate the gonococci.

There is an accumulating body of evidence which suggests that certain oral streptococci can induce the formation of plaque, a deposit of microbial dextrans and organisms on teeth. It has been proposed that anaerobic streptococci convert sucrose to dextrans, which in turn are fermented by the cariogenic streptococci and possibly lactobacilli present in the plaque with resulting acid demineralization of the enamel. In addition, proteolytic organisms, including actinomycetes and bacilli, apparently produce damage to the dentin. Undoubtedly, genetic, hormonal, and nutritional factors are important in dental caries.

D. SHIFTS IN THE NORMAL MICROBIAL FLORA AS A CONSEQUENCE OF CHANGES IN THE ENVIRONMENT

Special manipulations that select for particular organisms can have a marked influence on the so-called normal flora. For example, penicillinase-producing *Staph. aureus* is not a member of the constant normal flora and commonly colonizes fewer than 20% of the members of a healthy population. However, in some hospital populations (employees and patients) colonization with *Staph. aureus* may range from 50 to 100%. This increased colonization rate is due to the high prevalence of the organism in the hospital environment, which in turn results from the high incidence of staphylococcal infections and carriers. These hospital strains of *Staph. aureus* are not capable of maintaining this state of high colonization when staphylococcus-infected patients are moved out of the hospital. In other words, the hospital strains of *Staph. aureus* have not evolved to the point where they are as well equipped for permanent colonization of normal individuals in the community as *Staph. epidermidis.* This principle applies to other microorganisms and has led to some confusion as to what organisms should be classified as normal flora. A similar principle obtains when army recruits experience increased colonization rates of *Neisseria meningitidis* and *Strep. pyogenes* group A as the result of barracks life coupled with sporadic infections by the above organisms.

It is of singular interest that diet markedly influences the flora of the gut. Restriction of carbohydrate intake will markedly reduce the numbers of lactobacilli and *Strep. mutans* in the mouth because synthesis of glucan from glucose is required for the adherence of *Strep. mutans* to the tooth surface.

E. MEDICAL PERSPECTIVES

Many attempts have been made to establish a more beneficial normal flora. However, since colonization commonly takes its normal course after birth, it is unlikely that much can be achieved in this direction because of the insults to the flora caused by the widespread use of antibiotics and the increased prevalence of certain pathogens in hospital environments.

The observation that animals reared in germ-free environments live longer than animals reared in natural germ-laden environments suggests that the conventional microflora in natural environments may do more harm than good; however, quite to the contrary, if the protective conventional microflora were nonexistent man would doubtlessly soon succumb to the onslaught of pathogens.

REFERENCES

Bitton, G., and Marshall, K.C. (eds.): Absorption of Microorganisms to Surfaces. New York, John Wiley and Sons, 1980.

Evans, C.A., and Strom, M.S.: Eight year persistence of individual differences in the bacterial flora of the forehead. J. Invest. Dermatol. *79*:51, 1982.

Evans, C.A., et al.: The bacterial flora of the forehead and back of Alaskan native villagers in summer and winter. J. Invest. Dermatol. *82*:294, 1984.

Feingold, D.S.: Cutaneous microbial flora. Cutis *36*:1, 1985.

Leyden, J.J., et al.: Skin microflora. J. Invest. Dermatol. *88*:665, 1987.

Mackowiak, P.A.: The normal microbial flora. N. Engl. J. Med. *307*:83, 1982.

Miller, C.P., and Bohnhoff, M.: Changes in the mouse's enteric microflora associated with enhanced susceptibility to *salmonella* infection following streptomycin treatment. J. Infect. Dis. *113*:59, 1963.

Noble, W.C.: Microbiology of Human Skin. London, Lloyd-Luke Medical Books, 1981.

Rolfe, R.D.: Interactions among microorganisms of the indigenous intestinal flora and their influence on the host. Rev. Infect. Dis. *1*:S73, 1984.

Rosebury, T.: Microorganisms Indigenous to Man. New York, McGraw-Hill Book Co., 1962.

Skinner, F.A., and Carr, J.G.: The Normal Microbial Flora of Man. New York, Academic Press, 1974.

Sutter, V.L.: Anaerobes as normal oral flora. Rev. Infect. Dis. *1*:S152, 1984.

8

PARASITISM AND THE PATHOGENESIS OF BACTERIAL AND MYCOLOGIC DISEASES

Since much of the subject matter discussed below is covered in the companion text, *Fundamentals of Immunology** (Chaps. 21 and 22), as well as in various chapters in this text, the present discussion is presented as an overview with emphasis on the pathogenesis of infectious diseases. The development of the concept of contagious disease is covered in Chapter 1 of this text.

A. THE NATURE OF PARASITISM

When an organism lives in or on the body of a larger organism the former is called a *parasite* and the latter its *host.* The broader term *symbiosis* designates the "living together" of two dissimilar organisms in close association without any implication of benefit or harm on the part of the symbionts. If a parasite cannot exist independent of a living host, it is called an *obligate parasite;* however if it normally exists as a saprophyte but can adapt to a parasitic existence, it is called a *facultative parasite.* Whereas the term, *parasitism,* implies benefit to the parasite, the host may be either benefitted, harmed or unaffected.

In order to perpetuate itself in a host species an obligate parasite must be able to pass indefinitely through a succession of individuals of the species.

1. Adherence to and Colonization on Host Cells

Since parasites must live on or in their hosts, it is usually necessary for the parasite to adhere, at least temporarily, to host cells of an external integument in order to colonize and grow. Most bacteria and fungi possess surface components (adhesins) which enable them to adhere specifically to host cells by combining with corresponding receptors on the surfaces of host cells. A few organisms can adhere to host cells because the parasite and host cell surfaces are attracted to each other by force of interfacial tension; in the case of phagocytes, the greater the hydrophilicity of the phagocyte and the greater the hydrophobicity of the microbe, the greater the forces of adhesion leading to phagocytosis. For example, tubercle bacilli are readily phagocytized without the aid of opsonins because their surfaces are highly hydrophobic.

Microbial adhesins comprise components of capsules, outer membranes and the fimbriae or fibrilliae (pili) carried by many parasitic microbes.

Being specific, adhesins account for the selective adherence and colonization of different microbial parasites at respective sites on mucosae.

2. States of Parasitism

As might be expected, different parasites establish different relations with their hosts; hence different states of parasitism are recognized. When both the host and the parasite benefit, as in the case of the permanent normal microbial flora, the state is called *mutualism.* In contrast, when the host is harmed and disease results, the state is called *pathogenism* and the parasite is called a *pathogen.* The organism benefits by being able to live and multiply in the host until such time that it may be rejected by the developing forces of immunity or the animal dies. Additionally, some of the organisms usually escape, either before or after death, to infect new hosts. In some instances infection may be

*Myrvik, Q.N. and Weiser, R.S.: Fundamentals of Immunology, 2nd Ed., Lea & Febiger, 1983.

followed by a permanent carrier state that provides a reservoir of infection. Another state of parasitism which is more often theoretical than real is *commensalism,* a state in which the parasite benefits by having a "home" and the host is neither harmed nor benefitted. Most examples of alleged commensalism are actually states of mutualism.

When a parasite becomes established permanently in a host species, the species is called a *natural host;* parasites may have one or more natural hosts. In turn, most species of animals serve as natural hosts for many species of parasites. When an organism parasitizes a host that is not its natural host, the host is called an *accidental, incidental,* or *unnatural* host. A host that transmits a parasite from one host species to another is called a "transport host". On the other hand, a parasitized individual host that fails to transmit the parasite to other individuals is called a *terminal host.* An individual who is parasitized intermittently by the same parasite is called an *intermittent host.*

B. HOST DEFENSES AGAINST PARASITES

Antimicrobial defenses can be divided into two categories, *innate* or *constitutive immunity* and *acquired* immunity. (*Fundamentals of Immunology,* Chap. 22).

1. Innate Immunity

Innate immunity is constitutive for the species and is the sole means of defense during primary infection before acquired immunity develops; it comprises external and internal defenses.

External defense is provided by the external integuments and organs comprising various mucosae, skin, ears and eyes. In addition to the mechanical barrier provided, the antibacterial agents of external defense include chemicals such as acids, mucins, and lysozyme, together with resident phagocytes of mucosae, such as pulmonary macrophages; additional mechanisms of external antimicrobial defense include the cough reflex and the mechanical action and antibacterial action of fluids, such as mucus, tears, saliva, and urine.

Internal defenses contributing to innate immunity consist of the antimicrobial activities of phagocytes and the promotion of microbicidal action by the inflammatory response and the reticuloendothelial system (RES).

Certain antimicrobial agents are present in serum, such as β-lysin and lysozyme, serum complement activated through the alternative pathway by certain bacteria can also lead to opsonization upon release of C3b.

2. Acquired Immunity

Specific acquired humoral immunity is due to specific Abs of various Ig classes, including secretory IgA, anti-adhesin Abs that block microbial adherence to host cells, IgM and IgG Abs that collaborate with C to effect the killing and sometimes lysis of Gram-negative bacteria; other Abs comprise IgM Abs that collaborate with C to serve as opsonins, IgG Abs that serve as opsonins and Abs of all of the above classes that serve to neutralize microbial enzymes and toxins. The opsonic activity of IgG Abs can be augmented by C.

Antibodies specific for a given microbe may result from parasitization by the microbe, administration of a specific vaccine or contact with organisms carrying cross-reactive Ags. Specific Abs can also be acquired passively, either artificially by injection of an antiserum (all Ig classes) or naturally from the mother by placental transfer (IgG Abs) or the ingestion of colostrum (IgA Abs). Another aspect of developing immunity during acute infection is that the serum levels of certain nonspecific antimicrobial substances rise; they include lysozyme and the so-called "acute phase substances."

Acquired antibacterial cell-mediated immunity is complex but rests primarily on circulating specifically-sensitized TH memory cells which on exposure to Ag secrete lymphokines that activate and mobilize macrophages to sites of infection where they exert their heightened capacity to ingest, inhibit and kill microbes. Marked and extended stimulation of TH memory cells incites the formation of the allergic granuloma and causes some macrophages to form epithelioid cells and giant cells. The allergic granuloma is a lesion uniquely structured to restrict and destroy infecting microbes at the site of infection (Sect. F2).

Total immunity represents the interplay of various forces, a number of which can act synergistically *(Fundamentals of Immunology).*

C. PARASITISM BY NONPATHOGENS

Nonpathogenic microbes that become associated with animal hosts include members of the normal flora and microbes from external sources, such as parasites from other animals and saprophytes from soil; associations involving microbes from external sources tend to be sporadic and intermittent (Chap. 7).

Members of the normal flora behave as perfect parasites because they can benefit healthy normal hosts in the manner mentioned above and it is only when the host's antimicrobial defenses, either local or systemic, are compromised that they may act as pathogens and harm their hosts (Sect. H). It is difficult to draw a fine line between organisms deserving of being considered part of the normal flora and other carried organisms. Whereas organisms that are constantly present on body surfaces are accepted as part of the normal flora, others of doubtful classification are carried intermittently and vary from those that rarely act as opportunists to those that frequently act as opportunists (Sect. H).

D. PARASITISM BY PATHOGENS

1. The Nature of Pathogens and Reservoirs of Infection

By definition a *pathogen* is "a parasite which is capable of producing apparent illness in a significant number of normal individuals who have not encountered the parasite previously or who have no significant acquired immunity to the parasite."

The primary requirement for perpetual propagation of a true pathogen is that it has at least one natural host that permits the parasite to multiply, escape and infect new hosts. The vast majority of pathogenic microbes that have man as their sole natural host, perpetuate themselves by serial passage, the reservoirs being either patients with clinically active infection, or more often, carriers. For the perpetuation of the pathogen in a natural host it is noteworthy that a host-parasite equilibrium must exist which preserves the chain of transmission. If a pathogen destroys all or nearly all hosts in a population, the chain of transmission will be broken. As stated in section (K) host-pathogen relations appear to continuously evolve in the direction of "perfect" parasitism; during such evolution the host species becomes more resistant and the microbe less virulent. Infections in unnatural hosts can be devastating because evolutionary adjustments between the two are lacking. In fact, the majority of highly fatal human infections involve man as an unnatural host, e.g., anthrax and bubonic plague. It is of singular interest that many pathogens that seldom, if ever, attack unnatural hosts because routes of transmission or communicability are lacking, can be highly pathogenic if introduced artificially by injection, e.g., experimental tuberculosis in the guinea pig. The need of a suitable route of transmission for the persistence of a pathogen in any host species is again emphasized by the observation that pathogens that accidentally infect an unnatural host do not long persist in the species because they either lack a suitable route of transmission or break the chain of transmission by killing a large proportion of their hosts.

Microbes, that freely reach or parasitize normal individuals with intact antimicrobial defenses but seldom or never produce disease, are not considered to be "true pathogens". When such organisms produce opportunistic infections in persons with compromised defenses, they are called *facultative pathogens.*

It has been suggested that the term *pathogenicity* be used to designate the degree of disease produced by different species of microbes and that the term *virulence* be used to designate degrees of pathogenicity of different strains of a given species of microbe. However, the terms *virulence* and *pathogenicity* are often used synonymously.

2. The Fetus and the Newborn as Special Hosts

The fetus and the newborn are discussed at this point because associations between these groups and microbial pathogens are unique *(Fundamentals of Immunology).*

The fetus is anatomically well protected from invasion by parasites normally encountered or carried by the mother during pregnancy and, in the large majority of instances, the fetal compartment remains sterile throughout pregnancy. There are only a few parasites that readily invade the fetus to cause congenital infections, e.g. *Treponema pallidum.*

During fetal development, the mother transfers her representative spectrum of IgG Abs to the fetus to provide the neonate with an important form of passive immunity until active acquired immunity is established. In general, this form of passive immunity is remarkably effective and plays a highly important role in preventing infections in the neonate.

The neonate is capable of producing IgM immediately after birth, whereas competence to synthesize IgA develops about 2 to 3 weeks later. In the case of IgG, synthesis begins when the neonate is 4 to 6 weeks old. Cell-mediated immunity can also be generated shortly after birth. It should be emphasized that immune responses are comparatively weak during the first year of life and moreover, that passively transferred maternal IgG Abs can dampen specific immune responses of the neonate. As soon as maternal IgG declines to a low level (6 to 12 months) due to normal catabolism, the infant reaches an "immunologic null period" which is gradually overcome as microbial Ags are encountered. Accordingly, the early years of life are normally plagued with numerous infections until exposure to the antigenic spectrum of the common infectious agents has been experienced. With increasing age, the immune response becomes stronger and more effective. This is largely due to accumulative and reinforcing experiences with an increasing number of protective immunogens of the parasites that are encountered. As a consequence, the incidence of various infections drops markedly during late childhood and puberty.

E. THE BASES OF PATHOGENICITY AND VIRULENCE

1. Colonization, Infection and Invasion

As stated in section A_2, many microbes can colonize on the surface of external integuments for long periods without producing clinical evidence of injury. The normal flora of the healthy host are examples of permanent colonization; they do not usually reach subepithelial tissues or the blood stream unless local defenses are depressed or a mechanical break of the external integuments occurs. Whenever a microbe grows on or in an external integument to cause injury or gains entrance to tissues below the external integument and grows, the condition is referred to as *infection.* Providing that body defenses do not arrest the growth of the infecting organisms, the initial state before evidence of injury occurs, called *silent* or *inapparent infection,* can progress to *clinical* or *apparent infection.* In some instances an organism may invade deep tissues and lie dormant for extended periods thus producing a prolonged silent infection that may only activate to become clinically evident when host defenses become lowered, e.g., tuberculosis.

With few exceptions, microbial parasites must penetrate external integuments and multiply within tissues in order to produce disease. In the case of actual wounds and abrasions, the mode of entry is obvious; however, most infections occur without readily discernible breaks in integuments. Nevertheless, microscopic breaks in certain integuments are probably frequent, such as in the mucosae of the mouth and alimentary tract, which are constantly subjected to the abrasive action of foods. In addition it appears that some bacteria may be carried through certain intact mucosae by phagocytes, whereas others adhere to and pass directly through epithelial cells. Since penetration of integuments by some microbes requires considerable time, the capacity to penetrate integuments often demands that the organisms tolerate local conditions and colonize. Many pathogens colonize selectively on certain mucosae; e.g., *Bordetella pertussis* on ciliated respiratory epithelium and *Shigella dysenteriae* on the epithelium of the colon. Such colonization permits the organisms to produce disease either with or without subsequent invasion into subepithelial tissues. *Without attachment and colonization disease production by many microbes would not occur.* One way by which the normal flora may oppose colonization by pathogens is that of preempting receptor sites on epithelia; another is by antibiosis. As stated earlier, specific adherence is effected by the binding of specific adhesins on bacterial surfaces to corresponding receptors on host cell surfaces; adherence also facilitates invasion. Certain pathogens, such as *Bord. pertussis* specifically attach to but do not invade the epithelium, whereas others attach to and invade only cells of the epithelial layer *(Sh. dysenteriae);* still others breach the epithelium and reach subepithelial tissues *(Salmonella typhi).* In the case of *Sh. dysenteriae,* which selectively attaches to and invades colonic epithelium, the organism in some unique fashion excites its own engulfment by colonic epithelial cells. An adhesin is evidently a virulence factor because avirulent strains of *Sh. dysenteriae* lack the capacity to adhere to and penetrate colonic epithelium. Moreover, in immune individuals specific IgA Abs block the adherence of microbes to host cells and their subsequent colonization.

In some circumstances organisms colonize in large numbers on integuments and, evidently as a result, produce sufficient injury to the epithelium to incite inflammation and produce "breaks" through which they pass. Once they penetrate an external integument, organisms tend either to remain localized or to be swept passively with the lymph flow to the lymph nodes; indeed some species of microbes have a marked tendency to enter lymphatic channels and pass rapidly to the draining *lymph nodes* and beyond. For example, *T. pallidum* spreads from the point of entrance with such rapidity that it reaches the blood and distant organs in a matter of minutes to hours.

A few species of pathogens that localize on but do not penetrate beyond surface integuments may nevertheless be highly fatal because of locally produced toxins; e.g., *Vibrio cholerae* and *C. diphtheriae.* Still other organisms grow within the integument and do not produce systemic injury; e.g., *Epidermophyton floccosum.*

Microorganisms synthesize iron-binding compounds, called siderophores, which are important in scavenging iron from transferrin when growing *in vivo.* Accordingly, the ability to scavenge iron from transferrin could represent an important virulence attribute. For example, pathogenic *Neisseria meningitidis* and *N. gonorrhoeae* can use partially saturated transferrin as a sole source of iron, whereas most of the non-pathogenic *Neisseria* species cannot obtain iron from transferrin and fail to grow under comparable conditions. In addition, *E. coli* strains isolated from patients with bacteremia are likely to produce colicin V which is associated with the production of the dihydroxamate siderophore which permits the organism to compete successfully for iron. It appears that these characteristics are the result of plasmids.

2. Multiplication

Multiplication of organisms within tissues depends on their abilities to satisfy their minimum basic biophysical and biochemical needs in host microenvironments and at the same time to evade, resist, or suppress host defenses, both natural and acquired. The oxygen tension in tissue is an important variable; for example *Clostridium tetani* require anaerobic conditions. Most microbes do not grow in vivo at maximum rates attainable in vitro either because of host defenses and/or because of unfavorable conditions in the microenvironments they inhabit. Host environments include not only extracellular and intracellular environments in healthy tissue, but also intralesion environments, including abscesses and areas of necrosis, such as may occur in the center of a tubercle. Since host microenvironments and defenses vary markedly among animal species and among individuals of a species, it is no surprise that patterns of disease vary widely among different hosts.

a. ***Adequacy of Nutritional Factors.*** Most pathogens can readily meet their nutritional needs within the extracellular environment of the host; however, in a few instances, specific nutritional needs are not met. For example the placentae of ungulates, which are susceptible to abortion by *Brucella abortus,* contain erythritol, an important growth factor for the organism, whereas the placentae of resistant species of animals lack erythritol. Iron shortage in vitro enhances the formation of several bacterial toxins; e.g., toxins of *C. diphtheriae, Cl. tetani, Cl. perfringens,* and *Staphylococcus aureus* enterotoxin. However, iron also influences host-parasite relations in other ways, and the role of free iron in infectious diseases is not fully understood. Growth requirements of a pathogen are not necessarily limited to basic nutritional needs but may include special precursors or cofactors needed for synthesizing virulence factors such as capsules or toxins. A few pathogens can meet their nutritional needs in the intracellular environment.

b. ***Antimicrobial Factors in Extracellular Environments.*** Specific humoral defenses, such as the antibody-complement system, or nonspecific humoral factors, such as lysozyme and beta-lysin, can be markedly antibacterial for some microorganisms. Tissue breakdown products resulting from cell death, or excesses of cell metabolites at sites of inflammation or ischemia can also exert marked antibacterial activity.

c. ***Antimicrobial Factors in Intracellular Environments.*** Certain bacteria gain entrance into cells other than professional phagocytes, by inciting endocytosis (phagocytosis) by the host cell. Environments within phagocytes are hostile to most pathogenic bacteria and only a few species, the so-called intracellular parasites, can survive in these environments for periods sufficient to produce disease.

Polymorphonuclear and mononuclear phagocytes differ markedly with respect to the microbes they can engulf and inhibit or destroy. Certain organisms can actually be protected as a result of being phagocytized by macrophages because they escape phagocytosis by neutrophils which would otherwise destroy them; e.g., *Yersinia pestis.* Other organisms, such as *Mycobacterium tuberculosis,* are more resistant to destruction by neutrophils than macrophages. However, since neutrophils are short-lived cells, the protection they afford tubercle bacilli is likewise short-lived; consequently, at any given time after infection, the majority of ingested tubercle bacilli seen in tissues are usually within macrophages.

The antimicrobial activities of phagocytes, either professional or nonprofessional, are presumed to result largely from lysosomal components discharged into the phagosome, which probably often act in concert rather than singly.

3. Evasion and Depression of Host Defenses

Substances responsible for evasion or depression of host defense activities have been called *aggressins* or *virulence factors.* Virulence factors that permit the early lesion to flourish often differ from factors that cause death of the host. Little is known about the nature of virulence factors directed at nonspecific humoral agents of defense, such as beta-lysin and lysozyme. Experimentally, however, products of virulent strains of certain organisms have been noted to protect avirulent strains against destruction by nonspecific antimicrobial components in normal serum. More is known about virulence factors that protect organisms against engulfment by phagocytes. Their action is one of either repelling or injuring the phagocyte or of shielding the organism against intracellular killing. For example, the organism may produce a substance which blocks chemotaxis or opposes engulfment, as in the case of the capsule of

Streptococcus pneumoniae; the organism may produce a substance which kills or injures phagocytes by acting extracellularly (streptolysin O) or intracellularly (streptococcal NADase) or the organism may either destroy the phagosomal membrane and escape to the cytoplasm or block the fusion of lysosomes with the phagosome. The properties of capsular materials which account for resistance to engulfment have not been defined; in some instances they appear to be surface tension effects. Whereas nonpathogenic bacteria tend to be engulfed, killed, and destroyed by phagocytes, most pathogenic bacteria either resist phagocytosis or resist destruction by phagocytes after engulfment.

Depression of host defenses as the result of microbial aggressive activity can be either local or systemic. Examples of substances that produce local aggressive effects are numerous and will not be considered in detail here; they include antiphagocytic surface components and leukocidins, etc. For example, *Staph. aureus* produces both a leukocidin, which destroys phagocytes, and another substance, "protein A," which combines with the Fc portion of IgG thus interfering with its opsonic activities. Most organisms produce more than one substance that could oppose local host defenses and since these substances often act in concert, it is exceedingly difficult to delineate their possible roles in depressing local defenses.

Systemic depression of host defense as the result of microbial aggressive activity may be specific and/or nonspecific. Fortunately, marked nonspecific systemic immunodepression seldom occurs as the result of infection; in consequence an individual suffering from one infectious disease is commonly able to mount a near normal immune response to another.

Absence or marked suppression of specific immune responses, especially CMI responses, to Ag(s) of the infecting agent is well documented. The reasons for these defects are obscure but could result from genetic incompetence, antigen mimicry, immunologic tolerance, immune deviation, or competition of Ags *(Fundamentals of Immunology).*

Whereas some pathogens escape recognition as being foreign by synthesizing surface components that are similar to host components (antigen mimicry), others produce components that lack antigenicity (hyaluronic acid) or, alternatively, adsorb host Ags and masquerade as host cells.

4. Spread of Pathogens in Tissue and Organ Systems

Dissemination of organisms from a focus of infection is opposed by various mechanisms of defense including inflammation, arrest in lymph nodes, and blood clearance involving circulating phagocytes and fixed phagocytes of the RES *(Fundamentals of Immunology).* Although most bacteria gain entrance into the circulation via the lymphatic system, *they may sometimes enter the blood directly* as, for example, when a lesion, such as caseating tubercle, ruptures into a vein. Spread within cavities and organs can likewise occur by rupture of a lesion into a duct, a tube, or a cavity.

The *lymphatic system* exerts a powerful localizing influence on the spread of microbes and only a few organisms, such as *T. pallidum,* can readily pass through the lymphatic system to the blood.

The *blood circulatory system* likewise has great powers for localizing bacteria, and it is only when highly virulent organisms are able to colonize and grow at foci in the circulatory system that a secondary progressive bacteremia occurs.

The *central nervous system* is not readily invaded by most bacteria because of a certain degree of gross and histologic isolation (the so-called "blood-brain barrier") which even restrains macromolecules, including Abs. The *placenta* also offers marked resistance to transgression by most bacteria.

Bacterial products, including enzymes and toxins, may promote spread of infecting agents. Although much has been written about such substances, including hyaluronidases, fibrinolysins, and collagenases, there is little certainty that they play important roles in the spread of infection. Instead the tendency to spread may be strongest among organisms which produce the least tissue injury early in infection and thus escape the localizing effects of inflammation; e.g., *T. pallidum.*

Forces of acquired immunity that oppose the spread of infection include hypersensitivity reactions, opsonins, antitoxins, and bacteriolysins.

Selective localization of pathogens may occur at various loci on integuments or, following blood-borne spread, at various sites in tissues and organs. Different organisms show uniquely different patterns of localization, for example, *Mycobacterium leprae* tends to invade nerves and *M. tuberculosis* rarely localizes in thyroid, pancreas, or heart. The reasons for selective localization within tissues and organs are largely obscure but probably relate to the ability of the organism to adhere to host cells, and to tolerate, suppress or escape defense forces at such sites and/or because local environments best provide the biophysical and biochemical needs of the organism. For example, *Mycobacterium balnei* cannot grow above 34°C and hence only grows and produces lesions in "cool areas" of the body. Another example of the effect of the physical environment on localization is, *Cl. tetani,* which will only grow in anaerobic areas; e.g., wounds where the blood supply is deficient.

5. Mechanisms by which Pathogens Injure Their Hosts

When infection has progressed to the point where clinical disease is evident, the host has sustained appreciable injury. However, the events leading to such injury are often exceedingly complex and usually involve both primary effects of microbes or their activities or products and secondary effects involving the liberation of harmful host mediators or tissue components. *From a theoretical standpoint, there are several major ways, both direct and indirect, by which a parasite might injure its host.* It could: produce a toxin which may cause local injury or diffuse to cause systemic injury, compete with host cells for some vital nutrient, produce injurious metabolic products, block vessels or injure cells by physical effects, release inflammatory substances which by producing clots, swelling, and scarring may block channels (vessels, ducts, airways, gut, etc.), cause excessive accumulation or proliferation of host cells or fluids to the point where pressure and crowding-out effects cause injury and finally, by synthesizing an excess of Ags or unique Ags that cross-react with host Ags it could so stimulate the immune response that harm would result from immune reactions, especially those involving delayed hypersensitivity, Ag-Ab complexes, and autoimmunity.

Since *host responses alone may sometimes totally account for the injury produced in infection,* the virulence of an organism could be independent of any toxins it might produce and result solely from its ability to evade host defenses and multiply to achieve a certain numerical threshold.

Whereas, in subsequent chapters, clear examples of harm resulting from each of the ways cited above will emerge, it is noteworthy that there is no single infectious disease in which all of the precise mechanisms responsible for host injury are known; this is especially true at the cellular and subcellular levels.

The extent to which various pathogens grow in the body seldom correlates with the extent of injury or disease produced. For example, organisms which can produce potent toxins or incite exaggerated allergic responses can cause injury all out of proportion to the extent of their growth.

Bacterial toxins are divided into two categories: the protein *exotoxins* which are, with certain exceptions, the extracellular products of Gram-positive bacteria, and *endotoxins,* which are complex lipopolysaccharide constituents of the cell walls of Gram-negative bacteria. In general, exotoxins are strongly antigenic, highly toxic, easily converted to toxoids, readily neutralized by antibodies, relatively heat-labile, and highly selective with respect to the tissues and cells they affect. In contrast, endotoxins, although antigenic, are relatively heat-stable, relatively less toxic, not convertible to toxoids, poorly neutralized by Abs and, irrespective of their source, have similar toxicities for various tissues and cells.

Most bacterial exotoxins are enzymes. Certain of them are the most potent lethal poisons known; for example, type D *Cl. botulinum* toxin is 3 million times more toxic then strychnine. Tetanus toxin is so potent that the amount that can kill a patient is far short of the amount necessary to immunize. Several of the bacterial exotoxins such as *Cl. perfringens* toxin are secreted as inactive "protoxins" and are rendered toxic by proteolytic enzymes which split off small fragments of the protoxin. A number of exotoxins, such as staphylococcal leukocidin, are composed of two or more molecular components which act synergistically.

An interesting development in recent years has been the discovery of *an increasing number of instances in which the synthesis of bacterial exotoxins depends on lysogeny.* For example, the erythrogenic toxins of *Strep. pyogenes,* and the toxin of *C. diphtheriae* are produced only by lysogenic strains of organisms. Thus the production of some bacterial exotoxins is a fortuitous result of parasitism by phage. Consequently it is reasonable to expect that the panel of toxins produced by strains of organisms of a species could change with the evolution of phage-host relationships and in some instances could account for changes in bacterial pathogenicity and disease patterns; indeed this may have occurred in the case of scarlet fever which appears to be less common and less severe than in past decades.

The profound selective toxicity of bacterial exotoxins is another one of their remarkable properties. Some are effective against one animal species and not another; one may attack epithelium, whereas another may attack only nerves. As an extreme example of selectivity, the alpha toxin of *Staph. aureus* attacks rabbit neutrophils but not human neutrophils! The mode of action of exotoxins is largely obscure, even in the case of the much-studied neurotoxins. For example, the fixation of tetanus toxin to ganglioside has given no clue to its mode of action at the molecular level.

Endotoxins are constituents of the cell walls of both pathogenic and nonpathogenic Gram-negative bacteria. They are complexes containing polysaccharide, protein and lipids; at least two lipids, referred to as *A* and *B*, are present. The toxic moiety appears to be associated with lipid A which is complexed to both the protein and the polysaccharide. Endotoxins stimulate the RES and various hormone and enzyme systems, including the enzymes responsible for fibrinolysis. Depending on dosage and timing they can produce leukocytosis or leukopenia, fever, stimulation or breakdown of defense systems, hemorrhage, and severe injury of cells of various organs, as well as systemic shock. Although complexing of endotoxin with Abs appears to speed its clearance from blood by the RES, it is uncertain whether frank neutralization of toxicity occurs. There is recent evidence that an Ab directed against the lipid structure can be produced that possesses antitoxic activity. *The contribution of endotoxins to the pathogenesis of infectious diseases remains controversial.* Nevertheless, it seems clear that they contribute substantially to pathogenesis.

Toxins of various types probably play an important role in the pathogenesis of most infectious diseases. The importance of establishing the precise roles of bacterial components, including toxins, in pathogenesis and of elucidating their mechanisms of action cannot be overemphasized, for if such goals could be accomplished, rational and new approaches to the control of bacterial diseases would be at hand.

It is commonly difficult to establish the relative role of a toxin, or indeed any factor, in the pathogenesis of an infectious disease, let alone elucidate its mode of action. The marked capacity of bacteria to undergo phenotypic shifts with changes in enivronment means that in vitro behavior is not a dependable criterion of in vivo behavior. Hence, failure to demonstrate a toxin by in vitro culture does not prove that a toxin is not produced in vivo; the truth of this statement has been well demonstrated in the case of anthrax. Also a lack of correlation between toxin production and virulence does not prove that the toxin does not contribute importantly to pathogenesis since some other factor, such as capsule production, may be the dominant factor in virulence. Likewise, failure of a toxin to produce symptoms and lesions characteristic of the disease does not prove that it is unimportant in pathogenesis since toxins can contribute to pathogenesis by acting in concert with other microbial products. It is always possible that a toxin behaves solely as an aggressin to enable the organism to invade and makes no major contribution to injury, particularly at the systemic level; such may be the case in gas gangrene in which antitoxin may prevent infection but is of little or no therapeutic value in the face of clinical disease.

Other pathophysiologic effects are evident in infectious diseases, but little is known about the mechanisms involved. For example, if pneumococcal septicemia progresses too far, the disease is fatal even if the bacteria are killed with massive doses of penicillin. *In effect the patient may die "bacteriologically cured."* Apparently, a pneumococcal neuraminidase can cause some type of irreversible physiologic derangement leading to death.

It is also well recognized that *many intracellular parasites, such as Francisella tularensis, Rickettsia prowazekii* and *Rickettsia rickettsii express marked toxicity for normal macrophages.* For example, *R. prowazekii* disrupts the phagosomal membrane by phospholipase action and grows in the cytoplasm. Macrophage death from engulfed organisms can be prevented by pretreating the rickettsiae with specific antiserum. In addition, phagosomal membrane damage does not occur in immune hosts. Such observations suggest that there are probably many microbial components, still to be identified, that can exert deleterious effects on the host.

6. Virulence Factors and Protective Immunogens

It is probable that virulence is seldom limited to a single unique microbial attribute even though the lack of but one such attribute may render the organism incapable of producing disease, as in the case of the avirulent pneumococcus which lacks a capsule. Obviously the pneumococcus possesses other "virulence" attributes that enable it to injure its host as indicated above. It is only by a comparative study of the capacity of virulent and avirulent organisms and their products to cause injury or to alter the course of infection that the various factors of virulence can be identified.

Virulence factors are not necessarily immunogenic; for example, the hyaluronic acid of the capsule of *Strep. pyogenes* is antiphagocytic and represents a virulence factor, second only to M protein, but is chemically identical with animal hyaluronic acid and is not immunogenic. Because of their chemical nature, many lipids are nonantigenic, and still other substances may resemble host Ags so closely that they fail to stimulate an immune response or do so but weakly (antigen mimicry). Consequently, in the search for virulence factors and protective immunogens, it should be kept in mind that any given organism may possess one or more virulence factors and that a virulence factor is not necessarily a protective immunogen.

F. HOST RESPONSES THAT CONTRIBUTE TO PATHOGENESIS

For the most part Ab-mediated hypersensitivity reactions with attending inflammation are probably beneficial to the host; an exception appears to be the lesions produced by soluble Ag-Ab complexes, although even in this instance it is difficult to rule out potential beneficial effects, such as limitation of spread of infection resulting from thrombosis of vessels, etc.

1. The Acute Abscess

The acute abscess is a striking example of a lesion representing in large part a host contribution to pathogenesis. An acute abscess resulting from infection represents the outcome of the local accumulation of large numbers of closely packed neutrophils attracted to the site of infection by chemotactic products of bacteria and injured tissue as well as by the split products of C resulting from Ag-Ab reactions. Although neutrophils are short-lived cells, their death tends to be hastened in the abscess by overcrowding and in many instances by bacterial toxins. The release of lysosomal hydrolases from dying neutrophils, plus damaging bacterial toxins and lack of adequate blood supply and nutrition, together with pressure effects leads to local destruction of tissue. *Pus,* which represents the partially liquified remains of dead leukocytes and necrotic tissue, is the hallmark of the acute abscess. In terms of host protection, the attraction of neutrophils to the site of infection to form the abscess is usually beneficial because of its value in localizing the infection. Although the pus formed may not provide a particularly good environment for the growth of most microbes, it is at the same time not a menstruum into which living leukocytes can migrate readily and carry on further phagocytosis. Hence, organisms often persist and even grow within pus, remote from host defense forces; it is only when a progressing abscess ruptures spontaneously or is surgically drained that rapid healing can occur. Abscesses in certain sites may favor spread rather than localization of infection (e.g., the palm of the hand) because pressure forces the pus along fascial planes and tendons with extension of the infection. The abscess also presents problems with respect to chemotherapy since it may not be readily penetrated by drugs and, in addition, the nonmultiplying organisms contained are not susceptible to those antibiotics that demand bacterial cell wall synthesis for their activity (Chap. 6).

In the case of a foreign body, such as a contaminated splinter, pus formation aids in its elimination by local sacrifice of tissue and discharge through the skin. *Thus the abscess can be an asset or a liability depending on the size and location of the lesion.*

2. Delayed Hypersensitivity and the Allergic Granuloma

Delayed hypersensitivity (DH) contributes importantly to the lesions of chronic infectious diseases and to immunity. For example, the very composition of the tubercle, an "immune structure of defense," is undoubtedly in large measure determined by DH. On the other hand, DH reactions can cause severe local injury to endothelium and destruction of many tissues, especially in chronic granulomatous diseases, such as tuberculosis. However, with respect to the *overall welfare of the host,* the inflammation and local destruction of tissue involved in DH reactions are usually beneficial.

The allergic granuloma, which is characteristically an immune reponse to intracellular parasites, is a highly effective anatomic structure of immunologic defense. Lymphokines generated by lymphocytes in the granuloma exert a marked regulatory influence on cellular events within the granuloma, an influence which continues so long as Ag persists locally to sensitize and activate lymphocytes. Lymphokine production evidently stems from the triggering of sensitive T lymphocytes of the granuloma by Ag diffusing from the infecting organisms. It can be envisioned that lymphokines attract lymphocytes and macrophages to the lesion by chemotaxis and that arriving macrophages, under their influence, become immobilized, activated, and often fused to form giant cells. Evidently macrophages are transformed into epithelioid cells under the influence of lymphokines. Epithelioid cells, by their strong tendency to adhere to one another, and the fusion of macrophages around organisms to form giant cells exert a strong localizing influence on the infecting agent. Moreover, even when necrosis of epithelioid cells occurs, growth of organisms is restricted in the resulting caseous material because of its characteristics, including low oxygen tension and lysozyme. It is only when restraining forces fail and the granuloma enlarges and ruptures that serious harm to the host results due to pressure effects, loss of vital tissue, hemorrhage, and spread of infection. *Thus on balance, delayed hypersensitivity reactions associated with granulomata may be beneficial or harmful depending on the magnitude and location of the lesions.*

Amyloid disease sometimes follows chronic infections, such as tuberculosis, leprosy, and staphylococcal osteomyelitis. Although the mechanisms concerned in amyloidosis are obscure, it is possible that in chronic infections it results from some accompanying derangement in the immune response.

3. Immune Complex Disease

Immune complex lesions can occur in man as the result of either acute or chronic infections. Ideal conditions for the development of systemic immune complex disease should be those favoring the development of soluble Ag-Ab complexes in the blood; namely, Ag-Ab reactions in the region of Ag excess. Theoretically, these conditions should be most prone to occur in situations in which Ab production is weak, and the supply of Ag is abundant and continuous. However, the pathogenesis of immune complex disease is more complicated than is generally believed and may vary with the nature of Ag and the nature and avidity of Ab as well as an excess of Ag. It is not unexpected that immune complex disease occurs as a complication of infectious diseases because it is presumed that the region of Ag excess is often attained; *the surprise is that it is not manifested or recognized more often.* A recent finding that may bear on this dilemma is that the export of serum IgA in the form of dimeric IgA-Ag complexes to saliva, bile and breast milk by ductal epithelium may constitute a noninflammatory means for eliminating excess serum Ag.

Soluble immune complexes tend to be deposited along basement membranes in blood vessels, including those of the glomerulus where they accumulate as "lumpy deposits" on the epithelial aspect of the basement membrane.

Acute post-streptococcal glomerulonephritis in man, a sequela of acute streptococcosis, is one model purported by some investigators to be the result of circulating complexes of streptococcal Ag and specific Ab. However, the nature of the causative Ag(s) and Abs remains a matter of dispute (Chap. 10). Recent evidence indicates that lesions due to immune complexes, occur in a number of infectious diseases, including leprosy, bacterial endocarditis, malaria and secondary syphilis.

Vasculitis, including polyarteritis nodosa, which sometimes occurs as the result of diseases such as streptococcosis and tuberculosis, is probably due to immune complexes *(Fundamentals of Immunology).* The frequent occurrence of polyarteritis nodosa in patients with circulating complexes of Australia Ag and Ab is an additional example of the pathogenetic role that immune complexes play in infectious diseases. In any event, the high concentration of microbial Ags at the sites of local lesions, together with microbial toxins that increase vascular permeability and thus promote penetration of vessel walls by immune complexes, should provide ideal circumstances for the development of immune complex vasculitis.

Rheumatic fever, a sequela of streptococcal pharyngitis, evidently represents still another model of immunologic injury that probably results from anti-streptococcal Abs which cross-react with related Ags of heart muscle (Chap. 10). An alternative possibility is that the lesions of rheumatic fever result from the local deposition and prolonged persistence of highly toxic and allergenic components of streptococcal cell walls; the bacterial Ags may adsorb to host cells and thus render them susceptible to cytotoxic damage by Ab and C.

4. Immediate Hypersensitivity Reactions

Immediate hypersensitivity reactions due to IgE Abs are not known to contribute importantly to the pathogenesis of infectious diseases although this is a possibility. By increasing vascular permeability, they could promote penetration of immune complexes into vessel walls and contribute to immune complex disease.

5. Fever

The term *fever* is used to designate high body temperature resulting from a disturbance in the thermoregulatory center in the brain. Such disturbances are due to direct injury of the brain or to blood-borne *endogenous pyrogens.* Although many, if not all, cells contain endogenous pyrogens, they are most abundant in neutrophils and macrophages. Fever commonly accompanies any bacterial infection of substantial nature, and the course of the fever is often characteristic of the particular infection. Apparently in fevers of infection, the humoral agents, which act on the thermoregulatory center in the brain, are exclusively the endogenous pyrogens liberated from stimulated or damaged cells, particularly leukocytes. Microbes bring about the release of endogenous pyrogens in various ways, including direct stimulation and injury to cells and tissues by microbial "toxins," commonly referred to as "bacterial pyrogens," as well as by hypersensitivity reactions.

The most-studied bacterial pyrogens, the endotoxins of Gram-negative bacteria, act principally on neutrophils and macrophages to cause them to release the "true pyrogens," endogenous pyrogens.

Fever is accompanied by numerous other symptoms and alterations, including changes in circulation and respiration. Fevers of infection seldom exceed 104 and 105°F for any extended time. In pneumococcal pneumonia temperatures of about 104°F are common and persist for several days. Temperatures of 105°F and above, due to any cause, are dangerous if they persist for any extended time and may lead to death or permanent brain damage. At 106°F and above, the thermoregulatory center probably soon breaks down and delirium, coma and death often occur; this is encountered especially in heat stroke.

Although it is tempting to speculate that fever should represent a general defense mechanism, no sound support for this hypothesis has been advanced.* In the few diseases caused by temperature-sensitive organisms, such as syphilis, temperatures high enough to arrest the organism appreciably do not develop and the only known beneficial effects of temperature in syphilis have resulted from high temperature induced by artificial means, a former therapeutic practice.

6. Septic Shock

The term septic shock *is used to designate a clinical state associated with infection;* it is characterized by circulatory collapse with hypotension, respiratory distress, and mental abnormalities. It usually develops rapidly, is commonly attended by bacteremia, and is often fatal. The condition is usually caused by Gram-negative bacteria and in these instances is generally thought to result principally from the effects of endotoxin. However, the term, *septic shock,* is used loosely to designate a condition which will probably prove to be the end result of many and different chains of pathogenetic events.

---

*Fever is reported to be beneficial in certain virus infections, but whether it ever exerts any appreciable benefit against invasive bacteria or fungi is debatable.

G. TRANSMISSION OF INFECTIOUS AGENTS

1. Modes and Routes of Transmission

As repeatedly stated, to persist in nature an infectious agent must have some means whereby it can leave one individual and infect another. This requires a suitable portal of exit from the donor host, who may be either a carrier or an individual with clinical disease, and a portal of entry into the new host. *Reservoirs* (sources) of infecting agents range from patients and carriers to biologic vectors, such as insects in which the agent can multiply, and finally to soil in the case of certain saprophytes that can adapt as facultative pathogens.

Portals of entry and exit include the (1) respiratory tract, (2) alimentary tract, (3) genitourinary tract, (4) skin, and (5) conjunctivae. Infectious agents may be transmitted by (1) aerosols, (2) contact, (3) fomites (inanimate objects), (4) ingestion, (5) mechanical vectors, (6) biologic vectors, and (7) placental passage.

It is readily apparent that the host-parasite cycle must be highly specialized in order to accommodate the exacting combinations of circumstances necessary for transmission of an infectious agent, including the ability of the agent to survive the various rigors of the external environment. For example, most agents that cause respiratory infections are transmitted almost exclusively by fresh *aerosols* (droplet nuclei) passing from a donor to a susceptible individual within the aerosol trajectory range generated by coughing, sneezing, and talking. This includes the organisms responsible for streptococcal sore throat, mycoplasma pneumonia, and pertussis. It is only in a few instances that infection commonly results from inhalation of microbes carried in airborne dust particles, such as the etiologic agents of psittacosis and smallpox. Such organisms must obviously withstand drying under atmospheric conditions.

Examples of microbes commonly transmitted by *contact* include *Francisella tularensis* (handling infected rabbits) and *Neisseria gonorrhoeae* (usually sexual contact). *Fomites* can be involved if an appropriate object becomes contaminated, and contact is made by the new host within a suitable period of time.

Ingestion of food or drink is a common means of transmission of the organisms responsible for a number of diseases, including typhoid fever, bovine tuberculosis, and cholera.

Vectors, such as certain insects, may transmit infectious agents solely by mechanical means (mechanical vectors) or in addition serve as reservoirs in which the organism multiplies *(biologic vectors).* In many diseases biologic vectors serve as an integral part of the host-parasite cycle because the vector provides an important site for propagation of the parasite. An example of a disease in which a biologic vector is important in disease transmission is Rocky Mountain spotted fever, which is transmitted to man by the bite of ticks infected with *R. rickettsii.*

In a few instances infection is passed from the mother to offspring in utero; e.g., congenital syphilis and granulomatosis infantiseptica due to *Listeria monocytogenes.*

2. The Special Role of the Carrier in Transmission*

The majority of pathogenic bacteria that have man as their natural host can establish the chronic carrier state in man. *Indeed most pathogens would fail*

*A carrier is a host *with a silent infection who sheds the carried pathogen.*

to persist in their natural hosts if they lacked the ability to establish chronic carriage or chronic disease. Two exceptions are smallpox and pertussis. Consequently, knowledge of the factors that determine inapparent infections and the chronic carrier state is important for understanding herd immunity and designing measures for controlling transmission of infectious agents. The density of the susceptible individuals within a host population bears importantly on the occurrence of an epidemic; the greater the density of susceptibles, the greater the chance that an occasional carrier in the nonsusceptible group will initiate an epidemic. *Many bacteria that are carried tend to induce specific immunity,* which may ultimately eliminate carriage of the organism in question; e.g., *Strep. pneumoniae* types. In other instances *the carried organisms tend to become less virulent* due to immune elimination of highly virulent over less virulent mutants; e.g., *Strep. pyogenes.*

Carrier studies of *C. diphtheriae* have been particularly enlightening. Since immunity to the toxin discourages carriage of the organism as well as clinical disease, long-continued mass immunization of a local population with toxoid reduces the carrier rate and the incidence of disease to extremely low levels and finally to extinction. Extinction of the organisms occurs because in a highly immune population the number of individuals who will assume carriage of the organism is so few that the chain of transmission is broken. The same breaking of the chain of transmission can take place in other diseases in which carriers and active disease are rare but in which silent infections that occasionally activate occur; e.g., tuberculosis in developed countries. *Public health measures to increase herd immunity and thus eliminate silent infections and chronic carriage can effectively accomplish virtual extinction of some infectious diseases within a population.* Unfortunately the disease can be reintroduced into the population by immigrants who carry the pathogen (Chap. 1).

The carriage of certain bacteria occurs even when the carrier has significant immunity to the microbe. Organisms in certain anatomic sites (*Sal. typhi* in the gallbladder) are often effectively sheltered from immune forces. Special genetic abnormalities of anatomic structure may also favor the persistence of pathogens and long-term carriage.

H. OPPORTUNISTIC INFECTIONS

The equilibria between normal humans and the microbes to which they are exposed are so delicate that even minor decreases in host defense or increases in microbial virulence can lead to opportunistic infections. Host alterations that can lead to lowered antimicrobial defense are many and varied and can involve innate and/or acquired defense systems *(Fundamentals of Immunology).* In accord with expectation abnormalities that lead to lowered defense can occur at most any point in the defense system and, in consequence, can predispose to infection by certain organisms or groups of organisms, depending on the defenses that normally afford protection against such organisms. For example, infants with Bruton's agammaglobinemia are especially prone to infection with *Haemophilus influenzae* and *Strep. pneumoniae.* The more complete the suppression of the specific immune response, the greater the number of potential opportunists. Complete suppression of the specific immune response opens the floodgates of opportunism to the point where lethal infections can only be prevented by isolation in a germ-free environment (neonates) or reestablishment of immunocompetence by successful bone marrow transplantation.

The burgeoning use of immunosuppressants, such as corticosteroids and the drugs used in organ transplantation and cancer therapy, has greatly increased the incidence of opportunistic infections, especially by fungi; in fact almost any one of some 100,000 species of saprophytic fungi is a potential opportunistic pathogen. Indeed, the list of fungi that are known to be capable of serving opportunistically as facultative pathogens has recently expanded to some 300. Needless to say, physicians who prescribe immunosuppressive drugs should be ever mindful of the attending risks of opportunistic infections.

I. CRITERIA FOR ESTABLISHING THE ETIOLOGIC AGENT OF AN INFECTIOUS DISEASE

Although Henle (1840), Koch's teacher, was the first to propose criteria for establishing etiologic agents of infectious disease, their validity awaited proof provided by Koch (1884) and thereafter were called "Koch's postulates." In substance they were stated essentially as follows:

1. The particular infecting microbe must be found in all cases of the disease, preferably in locations and numbers which could explain the lesions and symptoms of the disease.
2. The organism must be grown in pure culture in order to prove that it is an independent, animate particle.
3. Inoculation of the pure culture into a healthy host must reproduce the disease, including the presence of the same organism in the lesions.

Koch's postulates have proved to be highly useful for establishing the causative agents of many diseases. Even though it is not possible to satisfy all of Koch's postulates in every instance, the etiologic agent can often be identified with reasonable certainty; e.g., *M. leprae* as the cause of leprosy. Although the causative agent of leprosy has never been cultured, it appears to be one of the mycobacteria as judged by all other criteria. It is the only agent that is constantly present in abundance in the lesions of the disease and causes disease in animals resembling, in certain measure, human leprosy; hence it is fully accepted as the cause of leprosy.

J. VARIABLES THAT DETERMINE THE OUTCOME OF INFECTION

In considering some of the variables that determine the host-parasite relationship, it is important to recognize that many factors can markedly influence the outcome of infection. A brief discussion of some of these variables is presented at this point in order to introduce the concept of host-parasite relationships as they apply to specific situations.

1. Age of the Host

As a rule the neonatal animal is remarkably resistant to those infectious diseases that are controlled by immunity due to specific humoral Abs; this resistance results from passive transfer of maternal Abs through the placenta or by way of the colostrum (Sect. D 2). In granulomatous diseases like tuberculosis in which humoral Ab does not contribute to immunity, maximum susceptibilty exists during the neonatal and infant period because there is no transfer of specific CMI from the mother to the fetus and newborn.

Altered resistance to infections also results from hormonal change taking place during various stages of later life. In general, infections are handled more effectively with increasing age until senescence. This phenomenon is largely due to frequent exposure to various specific and cross-reactive infectious agents, which maintains the capacity to respond anamnestically.

2. Portal of Entry of the Parasite

The portal of entry of a parasite can markedly influence the outcome of infection. Organisms infecting by one route can be much more pathogenic than by another route; for example, *Bacillus anthracis* and *Yersinia pestis* entering by the respiratory route are nearly always fatal.

With many pathogens, it is apparent that a highly specialized set of circumstances must be met to establish an infection.

3. Physiologic State of the Host

If a host suffers from malnutrition, vitamin deficiencies, circulatory disturbances or is exposed to immunosuppressive agents, such as x ray and corticosteroids, resistance is frequently depressed so that many species of pathogens can readily infect and produce disease. For example, lobar pneumonia due to *Strep. pneumoniae* is an uncommon disease in the healthy adult but is common in the aged and in alcoholics suffering from malnutrition. Cholera can be devastating in a population of malnourished individuals but is relatively mild in a well-nourished population.

4. Anatomic Defects in the Host

The various orifices of the body have their characteristic microbial flora and, as a rule, are remarkably resistant to infection. However, if obstruction or some anatomic abnormality develops from any cause, such as trauma or some surgical procedure, susceptibility to infection can be markedly enhanced.

5. The Immunologic Potential and Previous Experience of the Host

The various exposures a host may have had to pathogens and related organisms greatly influences subsequent responses to infection. Some parasites of low virulence may immunize the host against organisms of other species, including related organisms with high levels of virulence. The middle-aged human host, in the course of years, encounters numerous types and strains of pathogens and, in consequence, possesses a broad coverage of acquired immunity because of accumulative infectious disease experience. The genetically-determined immunologic potential of the host is also important.

6. Size of the Infecting Dose of the Parasite

In some instances the infecting dose of a parasite can be as low as one organism, whereas in other instances hundreds of thousands of organisms may be needed to establish infection. The minimum infecting dose will be much larger for a host that has acquired partial immunity to the organism than for a host with no acquired immunity. Innate factors of immunity also play an important role in determining the size of the dose required to initiate infection.

7. Physiologic State of the Parasite

Physiologic states of parasites which attenuate them or increase their virulence contribute importantly to their ability to initiate infection. For example, tubercle bacilli are more infective if delivered in a fresh aerosol than in dried dust. In contrast, the rickettsiae that cause Q fever *(Coxiella burnetii)* are notoriously resistant to drying and can retain infectivity in dust particles for extended periods with little loss of infectivity.

8. Virulence Potential of the Parasite

The term *virulence* is difficult to delineate because it must be defined in the context of a given host-parasite relationship. Nevertheless, every pathogen has evolved highly specialized properties that endow it with the ability to overcome the innate resistance of the host and allow it to implant, multiply, and precipitate disease. Virulence of strains of organisms within a species often varies greatly, but in only a few instances are the principal factors that determine virulence known. As the pathogen proliferates following the establishment of an infection, all of the forces of immunity may come into play; only when the parasite overcomes these factors will inapparent infection progress to overt disease.

K. THE EVOLUTIONARY BASIS OF MICROBIAL PARASITISM

1. General Considerations

Since the evolution of parasitism is a slow, continuing process spread over eons of time, its study during such short periods of decades or centuries is difficult. Nevertheless, it is reasonably certain that when a microbial pathogen meets a new host population, evolution is invariably in the direction of "perfect parasitism", (mutualism). Perfect parasitism is exemplified by the constant residents of the normal microbial flora that benefit the host in diverse ways, such as by: opposing colonization by pathogens, providing vitamins, attacking certain ingested foodstuffs to provide host nutrients and stimulating immunity to pathogens as the result of carrying Ags that cross-react with Ags of pathogens.

It is probable that the original source of the present-day microbial parasites of man and animals was principally the pool of saprophytes in nature and moreover, that when saprophytes first established themselves in animal hosts, most of them probably behaved as imperfect parasites that harm their hosts (pathogens). The evolution of such a pathogen and its host toward perfect parasitism involves a loss of virulence on the part of the pathogen and an increase in resistance on the part of the host.*

2. Evolution of Pathogenism

Pathogens are imperfect parasites that harm their hosts but rarely convey any benefit, exceptions being to incite cross-reactive immunity (yaws conveys cross immunity to syphilis) and to competitively discourage colonization by another pathogen (Chap. 30).

The evolutionary paths that various pathogens and their hosts may have

*Close DNA relationships exist between many microbial pathogens of man and soil saprophytes, e.g., members of the genus *Mycobacterium.*

taken to arrive at present states of adaptation are difficult to visualize and are largely speculative; a few examples are given below.

Corynebacterium diphtheriae is an example of a noninvasive pathogen that has only evolved to the point where it can survive superficially on respiratory mucosae and even then only intermittently. The production of toxin and disease may be regarded as a fortuitous event due to chance lysogeny by a corynephage carrying the tox gene and not as a property that contributes to permanent persistence of the organism in man, its sole natural host.

Streptococcus pneumoniae is an extracellular pathogen that has adapted to the point *where it can resist all* of the forces of extracellular defenses except phagocytosis induced by opsonic Abs; consequently the period of disease production is short (albeit sometimes lethal) and is arrested as soon as free Abs appear in the serum. Permanent persistence of the organism in man depends on intermittent carriage fueled by an occasional case of active disease.

Mycobacterium tuberculosis is a facultative intracellular pathogen that has adapted to the point where it can resist all forces of extracellular and intracellular defense *except intracellular killing by activated macrophages.* Essentially all exposed individuals develop lifetime infections, the majority of which are silent. The precise intracellular or extracellular site(s) at which organisms persist in silent infections is not known. Persistence of *M. tuberculosis* in man, its sole natural host, rests on the shedding of organisms by individuals with *active disease, either acute or chronic. Unfortunately chronic disease often remains undiagnosed.*

Obligate intracellular parasites have developed extensive adaptations. Since conditions within phagolysosomes of either professional or nonprofessional phagocytes are hostile, these parasites have evolved so that they either obstruct phagosome-lysosome fusion and grow in the phagosome or destroy the phagosomal membrane to escape and grow in the cytoplasm.

One example, *Chlamydia trachomatis,* the cause of trachoma, obstructs phagosome-lysosome fusion and grows in the phagosome. The organism is maintained in man, its natural host, by producing chronic disease, the carrier state and latent infection with recrudescent disease. The organism has undergone another adaptation, namely, it has acquired the capacity to assume two unique intracellular forms, one a reproductive form and the other a spore-like form that survives well extracellularly and serves to infect new host cells.

Another example, *R. prowazekii,* the cause of epidemic typhus, destroys the phagosomal membrane and grows in the cytoplasm. The organism is maintained in man, its sole natural host, by individuals with latent and recrudescent disease who serve as reservoirs for transmission by human lice as vectors.

There is a final category of pathogens which evidently grow extracellularly but appear to be controlled by the environment generated by cellular immune mechanisms. In this group, *Actinomyces israelii* multiplies extracellularly, yet immunity appears to be of a cellular nature. This naturally raises the question as to whether a local CMI reaction can create a hostile microenvironment for an extracellular parasite. This principle could apply to the immune response against many types of pathogens, including certain fungi that grow extracellularly.

Whereas given pathogens have probably evolved to achieve increasingly lower virulence by selection involving mutation, episomal-induced genetic change or transduction, host populations have evidently adapted toward increased resistance as the result of selective killing of the most susceptible hosts by the pathogen. However, host-pathogen balances are so delicate and overall immunity so complex that it is extremely difficult to pinpoint the small changes that are responsible for increased resistance to a given pathogen. Some success along this line has been achieved in animal models such as, the comparative studies of Lurie who selected and bred strains of rabbits with respective high and low resistance to tuberculosis (Chap. 27).

As the result of the evolution of pathogenism, most pathogens that have man as their natural host, commonly produce inapparent or self-limiting infections; only a few produce serious infections that are lethal for a substantial proportion of their hosts.

REFERENCES

Beaman, L., and Beaman, B.L.: The role of oxygen and its derivatives in microbial pathogenesis and host defense: Annu. Rev. Microbiol. *38*:27, 1984.

Beaver, P.C., and Jung, R.C.: Animal agents and vectors of human disease. 5th Ed. Philadelphia, Lea & Febiger, 1985.

Brown, R.W.M., and Williams, P.: The influence of environment on envelope properties affecting survival of bacteria in infections. Annu. Rev. Microbiol. *39*:527,1985.

Brubaker, R.R.: Mechanisms of bacterial virulence. Annu. Rev. Microbiol. *39*:21, 1985.

Burnet, F.M., and White, D.O.: Natural History of Infectious Disease. 4th Ed. Cambridge, University Press, 1972.

Curnutte, J.T., and Boxer, L.A.: Clinically Significant Phagocytic Cell Defects: Current Clinical Topics in Infectious Diseases, Vol. 6, (eds.) Remington, J.S., and Swartz, M.D., New York, McGraw-Hill Book Co., 1985.

Eidels, L., et al.: Membrane receptors for bacterial toxins. Microbiol. Rev. *47*:596, 1983.

Golden, A., et al.: Pathology: Understanding Human Disease. Baltimore, Williams & Wilkins, 1985.

Middlebrook, J.L. and Dorland, R.B.: Bacterial toxins: cellular mechanisms of action. Microbiol. Rev. *48*:199, 1984.

Mims, C.A.: The Pathogenesis of Infectious Disease. New York, Academic Press, Inc., 1982.

Moulder, J.W.: Comparative biology of intracellular parasitism. Microbiol. Rev. *49*:298, 1985.

O'Hanley, P., et al.: Molecular basis of *Escherichia coli* colonization of the upper urinary tract in BALB/c mice. Gal-Gal pili immunization prevents *Escherichia coli* pyelonephritis in the BALB/c mouse model of human pyelonephritis. J. Clin. Invest. *75*:347, 1985.

Sarov, I., et al.: Specific Serum IgA Antibodies in the Diagnosis of Active Viral and Chlamydial Infections. *In* New Horizons in Microbiology. (eds.) A. Sanna and G. Morace. Amsterdam, Elsevier Science Publishers, 1984, p. 157.

Schoolnik, G.K., et al.: Bacterial Adherence and Anticolonization Vaccines. Current Clinical Topics in Infectious Diseases, Vol. 6. (eds) Remington, J.S. and Swartz, M.D. New York, McGraw-Hill Book Co., 1985.

Smith, T.: Parasitism and Disease. New York, Hafner Publishing, 1963.

9

STAPHYLOCOCCUS

Most normal human beings carry large numbers of staphylococci and related organisms, both in the nose and on the skin. The relatively nonpathogenic, opportunistic *Staphylococcus epidermidis* is almost always found among the normal flora of the skin and mucous membranes of the respiratory and alimentary tracts, whereas the pathogen *Staphylococcus aureus* is a transient member of the microbial flora. A third species, *Staphylococcus saprophyticus,* which is free-living in nature but can also colonize the skin, recently has been found to cause urinary tract infections in humans. Distinguishing characteristics of these 3 species are summarized in Table 9–1. These Gram-positive cocci, which are members of the family *Micrococcaceae,* were given their genus name *Staphylococcus* because they tend to occur in grape-like clusters (Greek *staphyle* = cluster of grapes). Members of the genus *Staphylococcus* are the only organisms in the family *Micrococcaceae* that are medically important.

Even though *Staph. aureus* and *Staph. epidermidis* are often carried by healthy individuals, under certain circumstances they can cause life-threatening diseases. Because of their presence on body surfaces, they are in a position to invade whenever host defenses are even slightly impaired. Consequently, they are common causes of both traumatic and surgical wound infections, as well as more superficial skin infections. In this regard, a frequent and characteristic staphylococcal lesion is the boil or furuncle in the skin. However, the organisms may invade and infect virtually any tissue, causing such diverse diseases as osteomyelitis, septicemia, pneumonia, and enterocolitis.

A. MEDICAL PERSPECTIVES

Staphylococci have been recognized as a cause of pyogenic diseases since the earliest days of microbiology. During the 1870s, Koch, Pasteur, and other pioneer microbiologists described these Gram-positive cocci in pus. Although physicians of that era did not consider that they carried potentially pathogenic

Table 9–1. Characteristics that Distinguish the Three Major Species of the Genus *Staphylococcus*

	aureus	*epidermidis*	*saprophyticus*
Production of coagulase(s)	+	–	–
Acid from mannitol (anaerobic)	+	–	–
Production of alpha toxin	+	–	–
Protein A in cell wall	+	–	–
Ribitol phosphate type teichoic acid in cell wall	+	–	+
Glycerol phosphate type teichoic acid in cell wall	–	+	v
Anaerobic growth and fermentation of glucose	+	+	–
Novobiocin sensitivity	+	+	–

Adapted from Baird-Parker: Ann. N.Y. Acad. Sci., *236*:7, 1974.
v = variable, + = >90% strains positive, – = >90% strains negative.

organisms and were the source of organisms that infected patients causing severe and often fatal postsurgical and postpartum infections, they were well aware of the extremely high postsurgical mortality rate due to infection. In fact, at the University of Aberdeen a large sign in the operating room warned patients: "PREPARE TO MEET THY GOD." Alexander Ogston, a young surgeon at Aberdeen, accepted this state of affairs without question until he visited Edinburgh and was persuaded by Lister that wound infections could be prevented, an amazing and almost heretical concept in those days when pus formation was considered to be an essential stage in wound healing. Ogston was so impressed by Lister's teachings that, upon returning to Aberdeen, he tore down and burned the sign, instituted successful aseptic techniques in surgery, and went on to make notable contributions to the study of the staphylococcal etiology of wound infections.

Although the use of aseptic techniques greatly reduced the incidence of iatrogenic (physician-induced) infections, staphylococcal diseases continued to occur with high frequency, and many of them were fatal in the preantibiotic era. In the 1940s, when penicillin first became available, hopes were high that the end of staphylococcal diseases was finally in sight. At that time, most staphylococci were susceptible to penicillin, and it was assumed that these infections would soon be totally under control.

Unfortunately, it was not long before penicillin-resistant strains of staphylococci emerged, at first in the hospital environment, and later throughout the community. It soon became apparent that staphylococci had a greater ability to develop resistance to antibiotics than virtually any other pathogen. New antimicrobial agents that are effective against staphylococci are constantly being discovered; however, there is the ever-present threat that resistant strains of the organism will be selected more rapidly than new agents can be developed. Consequently, efforts are being made to achieve a better understanding of the biology of these organisms, the pathogenesis of the diseases they produce, and the mechanisms of acquired immunity to them. Nevertheless, it seems likely that staphylococcal diseases will continue to be a major medical problem for the foreseeable future.

B. PHYSICAL AND CHEMICAL STRUCTURE

The staphylococci are facultatively anaerobic, Gram-positive spheres about 0.8 to 1.0 μm in diameter. In pus, they occur singly, in pairs and irregular clusters, and occasionally in short chains seldom more than 4 cocci in length. The characteristic grouping of grape-like clusters is most marked among organisms grown on solid media, and results from random division in multiple planes, with failure to separate after each division. In common with most Gram-positive bacteria, they tend to become Gram-negative in old cultures.

The staphylococcal cell wall contains a peptidoglycan backbone and species-specific teichoic acids. In addition, *Staph. aureus* possesses an antiphagocytic surface component called protein A, which is linked to the peptidoglycan layer but may also be partially released extracellularly. Protein A has a strong affinity for the Fc portion of immunoglobulin G molecules and can, therefore, bind these molecules to the cell surface, while leaving their specific Ab-binding sites (FAb portions) externally exposed and intact. For this reason, protein A evidently has an anti-opsonic effect and interferes with phagocytosis. The phenomenon of interaction of protein A with the Fc portion of IgG has been used to prepare suspensions of formalin-fixed, heat-killed *Staph. aureus* cells coated with IgG specific for diagnostic surface Ags of other pathogenic bacteria. These commercially available, antibody-coated *Staph. aureus* cells can be used to rapidly "coagglutinate" and, therefore, identify serotypes or serogroups of other pathogenic bacteria such as pneumococci, streptococci, meningococci, *Hemophilus influenzae,* etc.

Some strains of *Staph. aureus* have an antiphagocytic polysaccharide capsule; however, the ability to produce a capsule does not appear to be a requirement for virulence in humans.

Most strains of unencapsulated *Staph. aureus* produce a surface component called "clumping factor" or "bound coagulase." Clumping factor converts fibrinogen to fibrin on the bacterial surface; this reaction is responsible for the clumping of *Staph. aureus* when the bacteria are suspended in plasma.

All strains of *Staph. aureus* contain a major species-specific polysaccharide A in the cell wall. The antigenic determinant of this polysaccharide is the N-acetyl glucosaminyl ribitol unit of teichoic acid. On the other hand, the species-specific Ag in strains of *Staph. epidermidis* is composed of glycerol teichoic acid (Polysaccharide B).

A specific phage receptor site present in the peptidoglycan-teichoic acid complex is expressed on the surface of *Staph. aureus* cells. Differences in susceptibility patterns to an international set of more than 20 phages are reflected by lysogenic immunity patterns and are of value in typing strains for epidemiologic studies. Of the 4 main groups identifiable by phage typing, strains of groups I and III are most often responsible for hospital infections with antibiotic-resistant staphylococci (Table 9–2).

C. GENETICS

Detectable mutations include gain or loss of the ability to form pigment or various extracellular products, changes in colony morphology, and changes in phage susceptibility. However, in terms of human disease, the most important genetic changes are those leading to antibiotic resistance. Antibiotic resistance in *Staph. aureus* is plasmid-mediated. Resistance to penicillin is caused by production of penicillinase, a plasmid-coded enzyme that hydrolyzes the beta-lactam ring of penicillin. *Staph. saprophyticus* differs from *Staph. aureus* and *Staph. epidermidis* in that it is novobiocin resistant.

Table 9–2. Lytic Groups of Staphylococcus Typing Phages Included in the Internationally Agreed Set of Typing Phages

Lytic Group									
I	29	52	52A	79	80				
II	3A	3B	3C	55	71				
III	6	7	43E	47	53	54	75	77	83A
IV	42D								
Not allotted			81	187					

D. EXTRACELLULAR PRODUCTS

Pathogenic staphylococci produce a variety of biologically active extracellular substances, some of which are listed in Table 9–3.

E. CULTURE

Simple media support the growth of staphylococci over a wide range of temperatures (15 to 40°C) and pH (4.8 to 9.4). Although *Staph. aureus* is a facultative anaerobe, growth is better under aerobic conditions, and aerobic culture on blood agar is usually employed for identification. The large, smooth, frequently beta-hemolytic colonies may be golden yellow due to the production of carotenoid pigments; however, pigment formation is extremely variable among various *Staph. aureus* isolates, and is influenced by the growth conditions. *Staphylococcus epidermidis,* which lacks pigment, forms smooth, white colonies which are usually non-hemolytic.

F. RESISTANCE TO PHYSICAL AND CHEMICAL AGENTS

Staphylococci are among the most resistant of the nonsporulating bacteria and can survive many adverse environmental conditions enroute from host to host. They are highly resistant to extremes of temperature and drying, and survive in dried pus and sputum for days to weeks; moreover, they can withstand moist heat as high as 60°C for 30 minutes. Staphylococci are markedly resistant to phenols, to many other disinfectants, and to high salt concentrations (7.5–10% NaCl).

G. EXPERIMENTAL MODELS

Rabbits and mice are most often used for experimental studies of staphylococci; guinea pigs are used less frequently. The cocci are relatively nonpathogenic for laboratory animals under natural conditions, but may be lethal when injected intravenously or intraperitoneally.

Kittens and monkeys are susceptible to the effects of staphylococcal enterotoxins and have been used for their study.

H. INFECTIONS IN MAN

The intermittent carrier rate of *Staph. aureus* is estimated to be about 30 to 50% and the anterior nares are the usual habitat of the bacteria in carriers. The fact that individuals can carry and shed pathogenic *Staph. aureus* over prolonged periods indicates that they and most of their contacts have a substantial degree of immunity to the organism. It is only when normal host defenses are lacking or are compromised that *Staph. aureus* is likely to invade. Thus, disease most often occurs at a site of local injury, such as a burn, abrasion, or other wound. Abnormalities which result in systemic immunodepression and predispose to staphylococcal diseases include diabetes, leukemia, renal failure, immunosuppressive therapy, and immunodeficiency diseases involving defects in humoral Ab responses.

Enterocolitis. The colonization of staphylococci on integuments is normally restricted by the antibiosis exerted by members of the normal flora, However, following intensive oral administration of broad-spectrum antibiotics, many members of the normal flora of the intestine are decreased in number and their

Table 9–3. Some Biologically Active Extracellular Products of *Staphylococcus aureus**

Product	Activity	Other Properties
Free coagulases	Enzymes that clot citrated plasma by converting prothrombin to thrombin, which converts fibrinogen to fibrin	
Staphylokinase	Enzyme that degrades fibrin clots by converting plasminogen to the fibrinolytic enzyme, plasmin	
Cytolysins (toxins that disrupt plasma membranes of many mammalian cells)	All are hemolytic	
α	Narrow hemolytic spectrum (most active against rabbit erythrocytes)	Causes spasm of vascular smooth muscle; dermonecrotizing; lethal
β	A sphingomyelinase C with a narrow hemolytic spectrum (most active against sheep erythrocytes)	Activated by Mg^{2+}
γ	Narrow hemolytic spectrum	Two basic proteins acting in concert; inhibited by agar and cholesterol
δ	Broad hemolytic spectrum	Peptide surfactant; inhibited by phospholipids
Leukocidin	Kills leukocytes	Two interacting proteins; heat-labile
Enterotoxin	Emetic	Five heat-stable (100°C, 30 min) serotypes
Epidermolytic toxin ("exfoliatin")	Cleavage of desmosomes in stratum granulosum of epidermidis	Produced most often by organisms of phage group II
Toxic shock syndrome toxin 1 (TSST1)	Pyrogenic, lethal hypotension, rash	Associated with strains that cause toxic shock syndrome
Hyaluronidase	Enzyme that degrades hyaluronic acid	Also called spreading factor
Lipases	Enzymes that degrade lipids	
Proteinases	Enzymes that degrade proteins	
Penicillinase	Enzyme that splits β-lactam ring of penicillin	

*All products listed are proteins.

antagonism may be lost, thus allowing extensive growth of endogenous *Staph. aureus* and the production of pseudomembranous enterocolitis.

Furuncles and Carbuncles. These infections are among the most common staphylococcal diseases that occur in human beings. Hair follicles, sebaceous glands and sweat glands commonly become infected resulting in superficial skin infections. If such infections invade the subcutaneous tissue a suppurative lesion commonly referred to as a *boil* or *furuncle* develops. If the infection spreads to the deep tissues and forms multiple foci of infection it is called a *carbuncle.* Carbuncles are usually limited to the upper back and neck. Superficial skin infections are usually painless, whereas furuncles and carbuncles can be painful due to the tissue pressure exerted by the suppurating lesion. Furuncles usually peak in 4 to 6 days and then drain spontaneously, which relieves the pain and allows host defenses to eradicate the remaining organisms and promote healing. A few individuals are prone to develop repeated attacks of furunculosis usually caused by the same phage type of *Staph. aureus.*

Acute osteomyelitis. Bone disease caused by *Staph. aureus* is most common in young boys under the age of 12 years. In this disease, blood-borne organisms, which are usually derived from a skin lesion, localize, preferentially in the metaphysis of growing bones; local bone trauma commonly predisposes to such infections.

Clinical symptoms of acute osteomyelitis include pain in the area of bone involvement, fever, chills and muscle spasm. If the infection occurs near a joint, staphylococcal *pyoarthrosis* (arthritis) is a possible complication.

Approximately 50% of cases of bacterial arthritis *(pyoarthrosis)* are caused by *Staph. aureus.* Staphylococcal arthritis may follow orthopedic surgery in cases of osteomyelitis, local staphylococcal skin infections or the injection of staphylococcal-contaminated material into the joint.

Septicemia and Endocarditis. Bacteremia, which can lead to septicemia, may occur in staphylococcal infections, particularly from infections of the skin, genitourinary tract and the lungs. Staphylococcal septicemia produces fever, chills and a generalized toxic condition. A frequent complication is bacterial endocarditis, which can result in valve destruction. Bacteremia can also lead to metastatic abscesses which may involve subcutaneous tissues, kidneys, and brain.

Impetigo. Certain strains of *Staph. aureus* can cause bullous impetigo which is a highly communicable superficial skin disease most often seen in infants and children; phage type 71 strains are most often involved. Since these strains produce exfoliatin, the disease can be viewed as a localized form of scalded skin syndrome.

Other infections commonly caused by *Staph. aureus* are *paronychia* which involves soft tissue around nails and *styes* which involve the eyelids.

Food poisoning. Staphylococcal food poisoning is not an infectious disease, but instead an intoxication resulting from ingestion of contaminated food containing preformed toxin. The source of contamination is usually a food handler who is a carrier of *Staph. aureus.* The organisms grow rapidly, even at room temperature, and dangerous concentrations of enterotoxins can accumulate in foods within a few hours. Since the toxins are highly heat-resistant, heating does not render enterotoxin-containing food safe for consumption. After ingestion, symptoms of nausea, vomiting, and diarrhea become apparent within 1 to 6 hours. The illness is self-limiting and complete recovery usually occurs within a day or two.

Scalded skin syndrome (toxic epidermal necrolysis, Ritter's disease). This disease is produced by strains of *Staph. aureus* that synthesize and excrete a protein toxin called "epidermolytic toxin" or "exfoliatin." The disease can occur in immunocompromised adults but is most common in children younger than 5 years of age. The signs and symptoms of the disease (erythema, bullae formation, and desquamation of the skin epithelium) are believed to be caused by toxin-induced cleavage of the desmosomes linking the cells in the stratum granulosum of the epidermis.

Toxic shock syndrome. This disease is produced by strains of *Staph. aureus* that synthesize and excrete a protein toxin called "toxic shock syndrome toxin 1" (TSST1). The disease was first described in 1978 and is characterized by the rapid onset of fever, hypotension, and skin rash, and delayed epidermal desquamation on the palms and soles. Although children, men, and nonmenstruating women may develop the disease, the vast majority of recent cases have occurred during menstruation in women who have been using tampons containing polyester foam or polyacrylate rayon fibers. Recently, these materials have been found to enhance in vitro production of the causative toxin because they bind and reduce the content of magnesium ion in the culture medium. The ability of the fibers to function in the same way in vivo could explain the association of the disease with the use of tampons containing the synthetic fibers.

Staphylococcus epidermidis, a normal inhabitant of the skin, occasionally can cause endocarditis. It is the most common cause of infection on or around prosthetic heart valves. *Staph. epidermidis* is the second most common cause of infections of total hip replacements after *Staph. aureus.* Certain strains of *Staph. epidermidis* produces slime which favors their adherence to prosthetic biomaterials. The slime appears to protect the organism from normal host defenses as well as antibiotics. It is also involved in infections of the prostate and urinary tract of elderly males.

Staphylococcus saprophyticus like *Staph. epidermidis,* is also a classical opportunistic pathogen in compromised hosts. It is reported to be the etiologic agent of 10 to 20% of urinary tract infections in young women.

Pneumonia. Staphylococcal pneumonia is extraordinarily rare; however, it represents a serious infection because of a mortality rate that can reach 50%. Infants less than 12 months old account for up to 75% of cases. Children with cystic fibrosis or measles, patients with viral influenza and patients receiving immunosuppressants have an increased risk of developing staphylococcal pneumonia.

I. MECHANISMS OF PATHOGENESIS

Staphylococcal lesions tend to remain localized. It has been suggested that localization may be due to coagulase which causes fibrin to form in the lesion. However, coagulase negative staphylococci also produce localized lesions so this explanation is doubtful. The inflammatory response to staphylococcal infection mobilizes large numbers of neutrophils which leads to the suppurative response. As the abscess develops it erodes to the surface of the skin, ruptures and the pus drains to the surface. Following drainage the lesion undergoes the normal healing process.

The various extracellular products of *Staph. aureus,* some of which may contribute to its virulence, are listed in Table 9–3.

As stated earlier, both the capsule possessed by some strains and protein A are distinctly antiphagocytic. In addition, most strains produce a leukocidin, which is toxic for human neutrophils and macrophages.

Staphylococcal *alpha-toxin* (alpha hemolysin) is cytolytic for human macrophages (but not neutrophils), epithelial cells, and many other types of cells. It produces necrosis when injected into rabbit skin and is lethal for mice and rabbits when injected intravenously.

Although some of the staphylococcal toxins are thought to contribute to the establishment of infection and the disease process, the use of antitoxin or killed vaccines has not been generally effective for protecting against disease, presumably because the organism possesses other important virulence factors not represented in the vaccines.

It is thought that *hypersensitivity reactions,* especially of the delayed type, contribute to local tissue damage in staphylococcal infections. For example, repeated infections induced in the skin of rabbits produce local lesions of increasing severity. This state of increased susceptibility can be passively transferred with lymphoid cells but not with serum, the hallmark of delayed-type hypersensitivity states.

A variety of staphylococcal enzymes may function to allow survival of the organisms on integuments and invasion of tissue. For example, lipases produced by staphylococci may permit the organisms to survive the antibacterial action of lipids on the skin. Also, free fatty acids generated by lipase activity can produce localized skin tissue damage.

Enterotoxins, which are formed by about a third of *Staph. aureus* strains, are the major factors in the pathogenesis of staphylococcal food poisoning. They are heat-stable, trypsin-resistant protein exotoxins. Although the manner by which ingested toxin acts has not been fully elucidated, it appears to stimulate the emetic center in the brain. Five antigenic types of enterotoxin have been described and the production of one type of enterotoxin (type A) has been shown to be induced by a temperature phage. Most enterotoxin-producing strains belong to bacteriophage groups III and IV.

In addition to the well-documented role of *Staph. aureus* enterotoxins in producing staphylococcal food poisoning, the epidermolytic toxin and TSST1 are believed to play important roles in the pathogenesis of scalded-skin syndrome and toxic shock syndrome, respectively (see Sect. H).

J. MECHANISMS OF IMMUNITY

It is apparent that the forces of innate immunity which contribute to protection against *Staph. aureus* are markedly amplified by specific acquired immunity. Humoral antibodies are important in protection, as evidenced by the fact that patients with immunoglobulin deficiencies are especially prone to develop staphylococcal infections. Opsonization probably makes a major contribution to humoral immunity by promoting phagocytosis by the neutrophil, a phagocyte essential to antistaphylococcal immunity.

The abscess, with its barrier of fibrin, phagocytes, and granulation tissue around the area of suppuration, is important in localizing infection. The dangers involved in breaking the barrier around a staphylococcal lesion are readily apparent, as evidenced by the observation that systemic infections may follow the "squeezing" of a staphylococcal abscess. Although delayed hypersensitivity reactions may contribute to local tissue injury, they are probably beneficial from an overall standpoint in that they tend to localize the disease and thus discourage life-threatening systemic spread of organisms. This does not exclude the possibility that an exaggerated delayed sensitivity reaction could cause severe tissue damage or interfere with immunity.

K. LABORATORY DIAGNOSIS

The presence of typical single or clustered Gram-positive cocci in pus from lesions, or in blood or other body fluids, is suggestive of staphylococcal infection. Identification of *Staph. aureus* is made by cultural characteristics and demonstration of *free or bound coagulase* production. Recently, various commercial biochemical test systems and a rapid latex agglutination test have been introduced for the rapid identificaiton of *Staph. aureus.* The latex slide tests consist of latex particles coated with plasma that contains fibrinogen to detect bound coagulase (clumping factor) and immunoglobulin G for detection of protein A. Nutrient agar or trypticase soy agar supplemented with blood is often used for culture because it is a rich medium which demonstrates the possible hemolytic properties of the organism. *Catalase production,* a trait of staphylococci, is a useful characteristic to distinguish staphylococci from the catalase-negative streptococci which may be similar morphologically on stained smears. Selective solid media containing a high concentration of NaCl (5 to 7.5%) or containing phenylethyl alcohol are often used to isolate staphylococci from mixtures of organisms in fecal specimens and urine specimens. A common high-salt-containing selective medium for staphylococci is mannitol-salt agar. It contains 7.5% NaCl which is extremely inhibitory to most other bacteria. In addition, the medium can be used to differentiate *Staph. aureus* from *Staph. epidermidis. Staph. aureus* usually ferments mannitol and causes the pH indicator (phenol red) in the medium to turn yellow, whereas, *Staph. epidermidis* usually does not ferment mannitol (Table 9–1). *Staph. saprophyticus* may be differentiated from other staphylococci by virtue of its resistance to novobiocin. Also, it grows poorly if at all under anaerobic conditions and, unlike other staphylococci, it does not ferment glucose well.

Staphylococcal food poisoning is often diagnosed on the basis of clinical evidence alone. Usually the food involved is heavily contaminated with typical Gram-positive cocci which can be seen in direct smears. Cultures yielding large numbers of *Staph. aureus* are confirmatory. Also, the enterotoxins in the food can be detected by gel diffusion techniques or by radioimmunoassay with specific antiserum.

L. THERAPY

Penicillin G is the drug of choice for the treatment of penicillin-susceptible staphylococcal infections; unfortunately most strains produce penicillinase and are therefore resistant to penicillins G and V and to ampicillin. Therefore, pending the results of antibiotic susceptibility tests, a penicillinase-resistant, semi-synthetic penicillin or cephalosporin often is recommended (Chap. 6).

Prior to the development of antibacterial chemotherapeutic agents, the only effective treatments for staphylococcal infections were surgical drainage of suppurative lesions and correction of underlying predisposing causes. Today, *drainage of lesions* is even more important than formerly because chemotherapeutic agents do not diffuse readily into areas of suppuration nor do they act effectively in such areas. In the case of drugs such as penicillin, which are only effective against growing organisms, therapeutic failure probably rests in part on the fact that many organisms in purulent exudates are not growing. Also, the activity of certain chemotherapeutic agents is destroyed by the products of tissue necrosis.

M. RESERVOIRS OF INFECTION

Although various domestic animals, such as cattle and horses, may carry and may become infected with *Staph. aureus,* the human carrier (especially the nasal carrier) is essentially the only source of human infections. The carrier state is usually established in the newborn infant within the first few days of life and becomes intermittent with age.

The factors that determine the carrier state are not known. However, infants who possess maternal IgG Abs but lack or possess only small quantities of Abs of other Ig classes have a higher carrier rate than adults. This suggests that Abs of classes other than IgG, or the forces of cell-mediated immunity, or both, contribute to immunity in adults. The most dangerous carriers are food handlers and those in frequent contact with susceptible individuals (such as medical attendants in contact with patients). In the hospital environment, the most dangerous carriers (because of their frequent contact with patients) are found among the hospital personnel, especially individuals employed in operating rooms, surgical wards and nurseries.

N. CONTROL OF DISEASE TRANSMISSION

Control of transmission of staphylococcal infection is currently centered about hospitals and clinics where susceptible individuals are concentrated. Successful control demands great vigilance by the entire personnel of the institution, under the guidance of experts. The measures used include rigorous hygiene, detection of carriers, control of contacts, control of airborne and fomite spread, and tracing sources of infection by means of phage-typing and antibiotic-resistance patterns.

Control of staphylococcal food poisoning depends principally on proper food handling. Prompt refrigeration of food is essential to prevent growth of staphylococci and production of enterotoxin. Efforts must be made to detect carriers among food handlers and to remove them from situations where they could contaminate food. However, adequate refrigeration of food is the most effective control measure and is usually sufficient to prevent outbreaks of the disease. It should be emphasized that large containers of food cannot be cooled adequately because of the time required to decrease the temperature throughout the food mass. Staphylococci can grow and secrete toxin as the food cools; therefore, large quantities of foods that support staphylococcal growth should be prepared immediately before use.

REFERENCES

Easmon, C.S.F., and Adlam, C.: Staphylococci and Staphylococcal Infections. Vol. I and II. New York, Academic Press, 1984.
Gristina, A.G., et al.: Adherent bacterial colonization in the pathogenesis of osteomyelitis. Science *228*:990, 1985.
Hirsch, M.L., and Kass, E.H.: An annotated bibliography of toxic shock syndrome. Rev. Infect. Dis. *8*:S1, 1986.
Hovelius, B., and Mardh, P.: *Staphylococcus saprophyticus* as a common cause of urinary tract infections. Rev. Infect. Dis. *6*:328, 1984.
Kloos, W.E., and Jorgensen, J.H.: Staphylococci. *In* Manual of Clinical Microbiology, 4th Ed. Lennette, E.H., et al: (eds). Washington, D.C., American Society for Microbiology, 1985.
Parisi, J.T.: Coagulase-negative staphylococci and the epidemiological typing of *Staphylococcus epidermidis.* Microbiol. Rev. *49*:126, 1985.
Rogolsky, M.: Nonenteric toxins of *Staphylococcus aureus.* Microbiol. Rev. *43*:320, 1979.
Seligman, S.J.: Current concepts of *Staphylococcus aureus* infection with emphasis on treatment. Compr. Ther. *9*(5):27, 1983.
Shantz, E.M.: Pansorbin *Staphylococcus aureus* Cells: Review and Bibliography of the Immunological Applications of Fixed Protein A-Bearing *Staphylococcus aureus* Cells, Calbiochem Brand Biochemicals, La Jolla, Calif., 1984.
Sheagren, J.N.: *Staphylococcus aureus:* The persistent pathogen. N. Engl. J. Med. *310*: 1368, 1984.
Verhoef, J., and Verbrugh, H.A.: Host determinants in staphylococcal disease. Annu. Rev. Med. *32*:107, 1981.

10

STREPTOCOCCUS

Streptococcus pyogenes

The genus *Streptococcus* is one of 5 genera of the family *Streptococcaceae;* it is comprised of a large number of species of saprophytes together with pathogenic and nonpathogenic parasites several of which are among the normal flora of the mouth and intestine of man. They are Gram-positive, nonsporing, nonmotile bacteria that vary widely in their O_2 requirements. Streptococci are widespread in nature; some species are important in lactic acid fermentations, whereas others selectively parasitize and infect different animal hosts. Their taxonomy is complex, difficult and confusing; a complete and practical scheme of classification has not been devised. On a broad basis strains of streptococci are arbitrarily divided into those which are *alpha-hemolytic,* those which are *beta-hemolytic,* and those which are *nonhemolytic.** Other species vary in their hemolytic activity, some strains being hemolytic and others nonhemolytic. Division on the basis of hemolytic activities is unfortunate because the criterion of hemolysis varies widely among strains within a species and shows little correlation with other properties. Some species comprise strains all of which are nonhemolytic. All β-hemolytic streptococci are pyogenic but not all pyogenic streptococci are β-hemolytic.

Current serologic classification of streptococci, excluding those of the *viridans* and pneumococcus groups, was established by Lancefield; it is based on distinctive group and type Ags. Serologic groups, which are designated by capital letters, are distinguished on the basis of antigenic differences in cell wall carbohydrate. Groups D and N Ags are located in the membrane. Serologic types within Group A, which number about 70, are separated on the basis of antigenic differences of a protein in the cell wall designated "M protein" (Sect. B); they are designated by Arabic numbers.

Properties of some of the *Streptococcus* species that cause human infections are shown in Table 10–1. The two most important human pathogens are *Streptococcus pyogenes* and *Streptococcus pneumoniae. This chapter will be largely devoted to a full discussion of Strep. pyogenes. Streptococcus pneumoniae* was formerly classified as *Diplococcus pneumoniae.* Since *Strep. pneumoniae* differs markedly from other streptococci in many respects, including bacterial characteristics as well as host-parasite interactions, it will be discussed in a separate chapter. *Streptococcus pyogenes* is the only species in serologic group A and is often referred to as the "group A" streptococcus.

STREPTOCOCCUS PYOGENES

A. MEDICAL PERSPECTIVES

Streptococcus pyogenes has always been a constant cause of serious diseas[e] in man. Few organisms produce a wider variety of toxins or elicit a broade[r] spectrum of lesions and sequelae. Clear descriptions of the distinct clinica[l] entities caused by streptococci, scarlet fever and erysipelas, appeared in th[e] medical literature centuries before Billroth (1874) described "chains of globula

---

*Many *Streptococcus* species containing strains that produce alpha hemolysis ar[e] sometimes grouped under the heading "viridans streptococci" because of a greenis[h] color which develops in the zone of partial hemolysis. The term "beta hemolysis" i[s] used to designate the clear-zone of hemolysis which results from the complete lysis [of] red blood cells. The old term, gamma hemolysis, which was sometimes used to designat[e] *no hemolysis* should be dropped.

[illegible]perties of Streptococcal Species that Cause Human Infections

Group & Species	*Other Designations*	*Usual Hemolytic Activity*	*Lancefield Group*	*Principal Habitat(s)*	*Major Human Diseases*
pyogenes	Group A	β	A	Man	Pharyngitis Glomerulonephritis Rheumatic fever
agalactiae	Group B	β,α,NH[1]	B	Man, cattle	Meningitis Sepsis of neonates
suis	RST group	β	R,S, or T	Swine, dog	Meningitis
equisimilis	None	β	C	Man, animals	Rare
equi	None	β	C	Horses (strangles)	Rare
zooepidemicus			C	Animals	Rare
equinus	None	α or NH	D	Feces of horses	Rare
bovis	None	α or NH	D	Feces of mammals	Endocarditis
faecalis	Enterococci	α or NH	D	Feces of mammals	Endocarditis
faecium	Enterococci	α or NH	D	Feces of mammals	Endocarditis
durans	Enterococci	α or NH	D	Feces of mammals	Rare
Species of G	Group G	β	G	Man	Neonatal sepsis Puerperal fever
sanguis	Viridans streptococci	α or NH	Most strains have group H Ag		Endocarditis
milleri (angiosus)	Viridans streptococci	α or NH	Variable or none	Oral cavity of man	Endocarditis Abscesses in internal organs
mutans	Viridans streptococci	α or NH	Variable or none	Oral cavity of man	Dental caries Endocarditis
mitis	Viridans streptococci	α or NH	Variable or none	Oral cavity of man	Endocarditis
salivarius	Viridans streptococci	α or NH	Variable or none	Oral cavity of man	Endocarditis
pneumoniae	Pneumococcus	α	None	Man	Lobar pneumonia Meningitis Otitis media

[1]NH = nonhemolytic

microorganisms" in the lesions of erysipelas and Fehleisen (1882) reproduced erysipelas in volunteers with pure cultures of streptococci. *Rheumatic fever* (RF) and *glomerulonephritis,* serious sequelae of streptococcal infection, and the once deadly *puerperal fever* (commonly caused by *Strep. pyogenes*) were also recognized as clinical entities more than a century ago. The fascinating essays of Oliver Holmes and Semmelweis on the infectious nature of puerperal fever stand as classics in medical literature and are well worth the attention of any student of medicine. Lying-in hospitals during those prehygienic times were often referred to as "houses of death." The staggering death toll exacted throughout history by secondary streptococcal pneumonia during epidemics of influenza ending in the major 1918 pandemic can never be forgotten.

Streptococcus pyogenes is no longer the dreaded pathogen it was as late as the early decades of this century, largely because of the advent of effective antibiotics. It is especially notable that in developed countries today streptococcal infections and their sequelae are not only less common, but less severe than during the last century. The precise reasons for these changes, which began before the advent of chemotherapy, are not known but probably include evolutionary changes in host and parasite as well as better nutrition and living conditions. *The most serious sequela of streptococcal infection in the U.S.A. is heart disease resulting from RF.* A decade ago new cases were estimated to range from 50,000 to 100,000 annually; fewer occur today. However, prior to the introduction of effective chemotherapy, streptococcal disease with its sequelae stood with tuberculosis and syphilis as the three most serious and important infectious diseases in the U.S.A.

Although the future importance of infectious diseases cannot be predicted with any degree of certainty, there is no reason to believe that streptococcal diseases will return to their former high incidence or severity or become unmanageable, particularly so long as antibacterial drugs remain effective as prophylactic and therapeutic agents. Moreover, *there is the prospect that effective "vaccines" of M protein may be developed,* albeit the large number of M types poses a problem.

Good examples of the effectiveness of penicillin against streptococci are the marked reduction in the incidence of repeated attacks of RF afforded by long-term penicillin prophylaxis and the marked lowering of the fatality rate in cases of streptococcal meningitis treated with the drug.

B. PHYSICAL AND CHEMICAL STRUCTURE

Streptococcus pyogenes is a commonly capsulated, spherical to oval organism about 1 μm in diameter arranged in chains of varying length. The cells elongate in the axis of the chain prior to division. Since the cells usually fail to separate after division, the chain grows in length with persisting cell wall material linking the cells together. Adherence is strongest immediately after division which leads to what appear to be chains of "diplococci." Since post-fission separation of cells results from enzymatic cleavage of cell wall materials, Abs against the surface M protein of the cell, which serve to mask the cell-wall substrates responsible for chaining, can prevent cleavage. This provides the basis for the "long-chain" serologic test used for detecting specific serum Abs against M proteins.

The outermost layer of *Strep. pyogenes* is a *capsule composed of hyaluronic acid,* a nonantigenic substance which is antiphagocytic. Capsules are better maintained in vivo than in vitro. Strains which produce little hyaluronidase and possess large capsules form relatively large "mucoid" colonies on blood agar which, as they age and dry, transform to flat wrinkled colonies (matt colonies). In contrast, strains which produce large amounts of hyaluronidase and accumulate little or no capsular material form small glossy colonies.

Streptococcus pyogenes forms short, fine fibrillae (fimbriae) composed of M protein and lipoteichoic acid; the latter serving to effect adherence to host cells by contributing to the hydrophobicity of the bacterial surface and binding to fibronectin on host cells (Fig. 10–1).

The cell wall of *Strep. pyogenes* contains M, T and R proteins. They can be detected on the bacterial surface by agglutination tests with appropriate antisera, and when extracted in the soluble form can be identified by precipitin tests. Since the T and R Ags occur independently of M Ag and have not been shown to be medically important, they will not be discussed. Some of the M protein combines with cell wall components by primary chemical bonding.

The M protein (mol wt 50 to 60,000), which is highly resistant to acid and heat, is of paramount importance because it is the sole "protective immunogen" and is the major determinant of virulence. Its antiphagocytic property is even stronger than that of hyaluronic acid; moreover it precipitates fibrinogen and is toxic for platelets and secondarily for neutrophils which aggregate with the injured platelets. The M protein on the cell surface is readily accessible for linking with specific Ab which tends to cancel out any protection against phagocytosis afforded by M protein. Moreover, most human sera contain an unidentified relatively heat-stable factor that counteracts the antiphagocytic

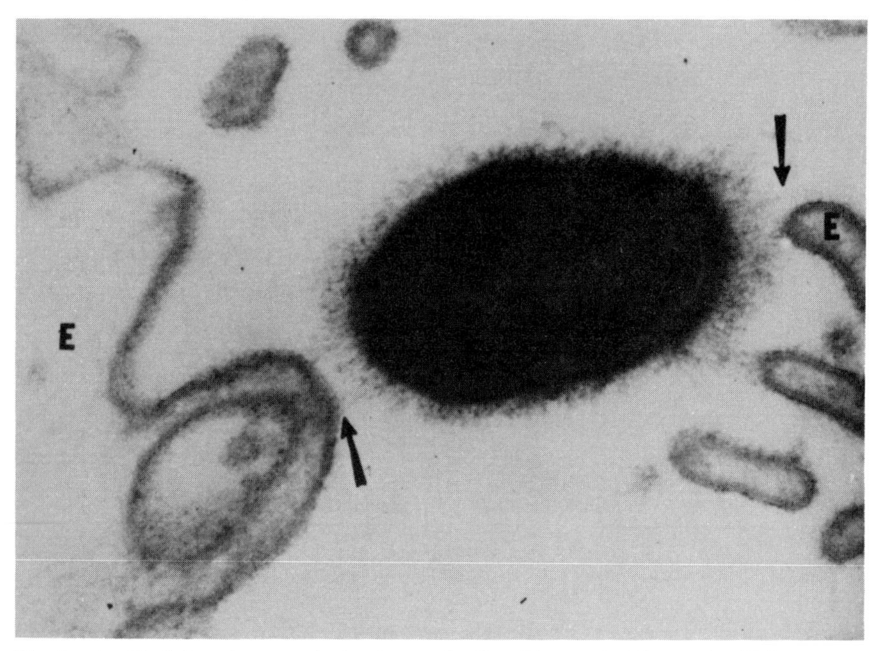

Fig. 10–1. Fimbriae of a group A streptococcal cell making contact (arrows) with human oral epithelial cells (E). (From: Beachey, E.H., and Ofek, I.: J. Exp. Med. *143*:759, 1976.)

activity of the hyaluronic acid capsule. Destruction of M protein by trypsin treatment, which does not kill the organism, greatly reduces its resistance to phagocytosis. Stripping the cell of both hyaluronic acid and M protein abolishes all resistance to phagocytosis. Whereas most strains of *Strep. pyogenes* produce but traces of proteinase during the last stages of growth, some strains produce such large amounts of this extracellular enzyme that they destroy their own M protein and certain other proteins as well. Since the proteinase is highly active at 37°C but not at 22°C, the capacity of strong proteinase-producers to synthesize M protein can be tested by incubating the cultures at 22°C to allow accumulation of the protein to detectable levels. The extent to which M protein may augment adherence of organisms to host cells effected by lipoteichoic acid coexisting in fimbriae is controversial; the two agents may act as a complex. In any event, adherence may be another attribute of M protein contributing to invasiveness and virulence. Since both anti-lipoteichoic acid Abs and anti-M Abs of the class, secretory IgA, can block the adherence of streptococci to the mucosa of the URT, they may play an important role in immunity and resistance to carriage and infection by opposing colonization and invasion of tissues. The M proteins of the different serologic types of streptococci are antigenically distinct and immunity is type specific.

The cell wall also contains the "group specific" polysaccharide ("C-carbohydrate" Ag) which because of its cross reactivity with heart valve glycoproteins could play a part in the pathogenesis of RF. It is tightly bound to the underlying mucopeptide layer of the inner cell wall lying adjacent to the cytoplasmic membrane and protects it against lysozyme action (Fig. 10–2). The cell wall mucopeptide shows in vitro toxicity for neutrophils and cultured kidney cells and produces dermonecrosis. Recent animal work has shown that cell wall fragments of *Strep. pyogenes* containing the C-carbohydrate-mucopeptide complex are singularly resistant to digestion by phagocytes and can persist in tissues to produce chronic rheumatic fever-like lesions as well as arthritis (Sect. H-8).

The cytoplasmic membrane contains Ags which cross-react with heart muscle, caudate nucleus neurons, glomerular basement membrane, and transplantation-like Ags that speed allograft rejection in guinea pigs (Sect. H). Lipoteichoic acid, an amphiphile, found in most Gram-positive bacteria originates in the cell membrane and extends through the cell wall. Secreted lipoteichoic acid can bind to M protein and act as an adhesin to promote adherence of *Strep. pyogenes* to the mucosa of the URT. Except for lethal toxicity, lipoteichoic acid has all of the properties of lipopolysaccharides of Gram-negative bacteria. Anti-lipoteichoic acid Abs can block adherence of *Strep. pyogenes* to mucosal epithelium. Hence lipoteichoic acid, as well as M protein, might serve as a protective immunogen.

Highly pleomorphic forms of streptococci appear in aging cultures and wall-less organisms (protoplasts) develop and grow in hypertonic media containing penicillin and lysozyme. Although such protoplasts multiply in vitro and have been observed in tissues of patients, their possible role in disease production is unknown.

C. GENETICS

Genetic variation is common among group A streptococci and involves many traits, including the synthesis of extracellular products, the capsule, and cell-

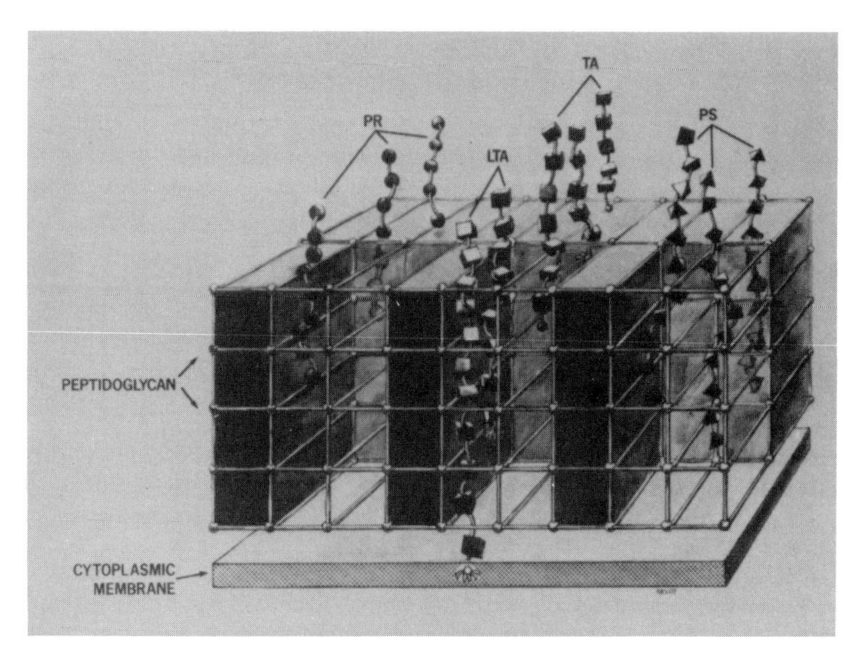

Fig. 10–2. Diagrammatic representation of the streptococcal cell wall. Note the antigenic polymers within the peptidoglycan structure. PR = protein, PS = Polysaccharide, TA = Teichoic acid, LTA = lipoteichoic acid. (Adapted from Secretory Immunity and Infection. New York, Plenum Publishing Co., 1978, pp. 761–762.)

wall components, and colonial morphology and drug resistance. Variations in the production of hyaluronic acid capsules and M protein appear to be independent genetic traits. However, since both hyaluronic acid capsules and M protein are antiphagocytic and account for virulence, differences in virulence involve variations in one or both of these substances. For example, those organisms which possess an abundance of both substances are highly virulent, those deficient in one or the other substance are of intermediate virulence, and those deficient in both substances are of lowest virulence. Repeated culture in vitro favors overgrowth by noncapsulated variants and contrariwise, low virulence organisms that lack hyaluronic acid capsules can be selectively suppressed by mouse passage to obtain more virulent capsulated mutants. In like manner, low virulence strains lacking M protein can be converted to M protein-containing strains of higher virulence. The natural selective pressure of immune forces is evident in man; for example, in the face of developing immunity carried organisms tend to lose their M protein and virulence, whereas epidemic strains which pass rapidly from one susceptible person to another are rich in M protein and are of high virulence. Whether the loss of M protein during carriage may result from high proteinase-producing mutants appears not to have been thoroughly investigated.

Whereas resistance to certain chemotherapeutic agents, such as the sulfa drugs, develops readily, strains resistant to penicillin seldom emerge as the result of penicillin therapy.

The production of the erythrogenic toxins responsible for scarlet fever is the outcome of lysogenization of streptococci by temperate bacteriophages, as in the case of Corynebacterium diphtheriae. However, unlike *C. diphtheriae,* which produces but a single toxin, four antigenically distinct erythrogenic toxins called A, B, C, and D are produced by *Strep. pyogenes,* presumably under the influence of different phages. Some strains produce both A and B toxins simultaneously. The quantity of erythrogenic toxin(s) produced by different strains of streptococci varies greatly, but whether variations in toxic strength occur is not known. Lysogeny is common among group A streptococci, and it is probable that future studies will reveal that additional attributes of pathogenicity and virulence are controlled by phage.

D. EXTRACELLULAR PRODUCTS

Group A streptococci produce at least 20 antigenic extracellular substances, including many enzymes, a number of which have not been identified. Only those that show evidence of being medically important will be considered.

1. Erythrogenic Toxins

The erythrogenic toxins of group A streptococci mentioned above are low molecular weight proteins. Some group C and G strains may produce toxins similar to if not identical with group A erythrogenic toxins. Erythrogenic toxins are destroyed by boiling and when injected into the skin of the "nonimmune" subject cause a delayed local erythema, which peaks at 24 hours (positive Dick test).* A scarletinal rash is produced when a large dose of the toxins is administered systemically to a nonimmune subject. The toxins produce no reaction in an immune subject who possesses neutralizing antitoxin (negative Dick test). Although these toxins have many biologic activities, their only known contribution to the pathogenesis of streptococcal disease is the production of a scarletinal rash. Whereas man is exquisitely sensitive to the toxin, most animals are highly resistant.

2. Streptolysin S and Streptolysin O

Two beta-hemolytic products, streptolysin S, a peptide, and streptolysin O, a protein, have been identified; the first is produced by essentially all strains and the second by the large majority of strains of *Strep. pyogenes.* Streptolysin O is oxygen sensitive, but streptolysin S is not. Consequently, streptolysin O only produces hemolysis deep in a blood agar plate, whereas streptolysin S produces hemolysis around surface colonies as well. Both hemolysins are highly toxic for neutrophils and macrophages and streptolysin O is a general cytotoxin and cardiotoxin. Streptolysin O elicits a neutralizing Ab, designated antistreptolysin O (ASO), but streptolysin S is nonantigenic. The levels of ASO in serum have proved to be a valuable index of a recent streptococcal infection, particularly with respect to the diagnosis of RF. These Abs appear in the serum

*There is uncertainty about the basis of the Dick reaction; some evidence indicates that it is not a straightforward toxin-antitoxin regulated event but also involves delayed hypersensitivity.

of about 70% of patients with pharyngitis and peak at about 3 to 5 weeks after infection.

3. Nicotinamide Adenine Dinucleotidase (NADase)

Nicotinamide adenine dinucleotide (NAD), a respiratory enzyme cofactor, is hydrolyzed by NADase. The enzyme is produced by some strains of *Strep. pyogenes,* including strains of a number of types such as 3, 4, 6 and 12. Streptococcal NADase, which is active only when bound to the cell wall, can kill phagocytes. Whether a phagocyte succumbs undoubtedly depends on the number of streptococci present within the cell.*

4. Streptokinase

Streptokinase is a nontoxic enzyme produced by essentially all strains of *Strep. pyogenes* and by group C and G organisms as well. Several antigenically-distinct streptokinases exist.

By acting on a proactivator of the fibrinolytic system, streptokinase brings about the conversion of blood plasminogen to plasmin, a proteolytic enzyme which digests fibrin. Neutralizing Abs to streptokinase become demonstrable in the serum in about 75% of cases of streptococcosis but do not appear to block the spread of infection in tissues.

5. Deoxyribonuclease

Streptococcal deoxyribonuclease (DNase or streptodornase) is a nontoxic enzyme produced by all strains of group A streptococci and by organisms of several other groups as well. Four antigenically-distinct DNases are recognized, A, B, C and D; except for type B, DNase neutralization by specific Ab is irregular. Some strains of *Strep. pyogenes* can produce all 4 serologic types of the enzyme simultaneously. The enzyme contributes to the liquefaction of exudates rich in DNA. In fact, commercial mixtures of streptokinase and streptodornase have been used with limited success for "enzymatic debridement" of wounds and abscesses in order to clear them of fibrin and pus.

6. Hyaluronidase

Streptococcal hyaluronidase is produced in varying amounts by some strains of *Strep. pyogenes,* most notably by strains 4 and 22. Although production of the enzyme in vitro is usually difficult to demonstrate it is evidently produced readily in vivo since increasing titers of neutralizing Abs almost always appear in convalescent sera. Since hyaluronidase attacks the hyaluronic acid of the ground substance in connective tissue, it could conceivably promote the spread of infection within tissues.

7. Proteinase

Streptococcal proteinase, a cathepsin-like enzyme with broad substrate specificity, is produced in vitro in varying amounts by certain strains of group A streptococci but rarely by strains of the other serologic groups. It is formed

*One report suggests that streptolysins rather than NADase may be the leukotoxin studied.

autocatalytically from a precursor under reducing conditions. Low pH (below 6.8) and high incubation temperature (37°C) favor its production.

8. Other Products

Several additional extracellular factors alleged to represent distinct molecular entities have been reported. Their importance in pathogenesis is controversial. They are lymphocyte mitogen, nephrotoxin, cardiohepatic toxin, a cell sensitizing factor (SF) that sensitizes red cells to agglutination and lysis by serum, and a nephritis-strain-associated protein.

E. CULTURE

Streptococcus pyogenes produces an abundance of lactic acid which lowers the pH of culture media and tends to limit growth. Rich, well-buffered media, particularly those containing blood or serum are favored. Some strains require CO_2 for primary isolation. A chemically-defined medium has recently been formulated for *Strep. pyogenes.*

F. RESISTANCE TO PHYSICAL AND CHEMICAL AGENTS

Streptococcus pyogenes is "average" in its resistance to most physical and chemical agents. It is moderately resistant to drying but is destroyed by pasteurization and is readily killed by the common disinfectants. The organism is susceptible to a wide variety of chemotherapeutic agents. Resistance develops readily to sulfonamides and certain antibiotics, but, fortunately, not to penicillin, the therapeutic drug of choice.

G. EXPERIMENTAL MODELS

It is only on rare occasions that *Strep. pyogenes* causes natural disease in wild animals, such as voles and field mice, but when inoculated many strains are pathogenic for the laboratory mouse, guinea pig and rabbit. Virulence may be enhanced by animal passage and in the case of the mouse is frequently attended by an increase in M protein and capsule size. The mouse is commonly used for virulence tests, but unfortunately mouse virulence does not necessarily parallel M protein production or human virulence. The pathogenesis of the sequelae of streptococcal infections in man is poorly understood despite an enormous literature on the subject. No animal models have been developed that closely simulate RF or acute poststreptococcal glomerulonephritis in man.

H. INFECTIONS IN MAN

Streptococcus pyogenes maintains itself in nature as a carried organism in the URT of man, its only constant natural host. The carrier rate in healthy subjects is usually less than 10%, but varies with season and other circumstances, being higher in winter than in summer and highest when URT infections are most prevalent in a population. During epidemics of streptococcal pharyngitis the organisms may be *encapsulated* and are usually rich in M protein; as epidemics decline the M protein content of infecting organisms declines.

The organism has high invasive powers and exhibits a strong tendency to spread to the blood via lymphatics or directly as a cellulitis. It has great versatility as a secondary invader, particularly in the case of viral infections, and can be a serious secondary lung invader in chicken pox, influenza, pertussis, and measles. The two most common sites of infection are the nasopharynx and the skin.

Nosocomial infections due to *Strep. pyogenes* remain a constant threat and can occur whenever hygienic practices are relaxed; the sources of infection include both patients and their attendants.

1. Pharyngitis

Streptococcus pyogenes is the most common cause of bacterial sore throat (pharyngitis). It is usually contracted from a carrier or an infected subject via droplet infection, direct contact or fomites. "Strep throat" is most frequent between the ages of 5 to 15 years but is rare in infants; in most instances the infection is mild and benign. Despite the protection afforded by maternal Ab, infections can occur in the newborn, especially in the poorly-protected umbilical stump. Since innate immunity is not fully protective, the infant becomes highly susceptible to infection after maternally derived anti-M Ab has waned. With each succeeding infection (which averages about one each year during childhood) or carriage experience the individual develops specific immunity to organisms of the type involved; thus immunity is type specific and cumulative with advancing age. Contraction of a new infection only occurs providing the individual has not had previous experience with organisms of the type in question or if the organism is of unusually high virulence or immunity has waned. Reinfections with organisms of the same type are rare, and when they occur are usually mild or occult, presumably because of the anamnestic response.

The character of streptococcal pharyngitis or tonsillitis changes with age. *The first infections in infancy and early childhood* lack a well-defined onset and symptoms, show little tendency to localize, are accompanied by relatively mild inflammatory responses and run a protracted course. Even though pharyngitis in infants is mild, often presenting only a rhinorrhea, suppurative lymphadenitis or otitis media sometimes occur.

Infections in older children and adults are usually mild and in some 20% of cases pass unnoticed. Severe infections show a greater tendency to localize and are characterized by a sudden and intense inflammatory response; they usually heal without suppurative complications. This change in the character of infection with age is thought to occur because with succeeding infections the individual becomes increasingly sensitive (both immediate and delayed types) to somatic Ags common to all group A streptococci, irrespective of type.

When a child develops a streptococcal sore throat, it is usually observed that at least one or more of the other members of the household (sometimes including adults) contract the disease. The more virulent the organism, the greater the chance of spread.

Severe streptococcal pharyngitis and viral pharyngitis are often mistaken one for the other. Severe streptococcal pharyngitis is commonly heralded by the rapid development of a sore throat of varying intensity (usually without significant cough), fever, leukocytosis, headache, malaise, tonsillitis, swollen tender cervical lymph nodes, exudate rich in neutrophils, and often abdominal pain, nausea and vomiting; definitive diagnosis rests on laboratory findings (Sect. E).

2. Scarlet Fever

Most cases of scarlet fever are due to *Strep. pyogenes* and most follow pharyngitis. In a limited number of instances of streptococcal pharyngitis an added complication develops, a skin rash, the hallmark of scarlet fever. The rash is commonly generalized and after several days is usually followed by desquamation which is especially evident in the palms of the hands and soles of the feet. Whether the patient with a streptococcal infection develops the scarletinal skin rash depends on whether the particular infecting strain is endowed with the capacity to produce one or more of the four different erythrogenic toxins and whether the individual is immune to scarlet fever by virtue of possessing specific antitoxic Abs. About 25% of individuals possess such Abs, presumably as the result of carried organisms or previous infection. Specific antitoxin, if present, can prevent the skin rash (scarlet fever) but fails to convey immunity against streptococcal infection, a property limited to anti-M Ab. Erythrogenic toxins are no longer regarded to produce serious injury and, except for the rash, the scarlet fever patient is not clinically in greater jeopardy than the patient with streptococcal pharyngitis without the rash. Immunization with erythrogenic toxin and the use of antitoxin for therapy are no longer advocated and the Schultz-Charlton blanching reaction produced in scarletinal skin by injecting convalescent serum is seldom used for diagnosis.

The decreasing severity of streptococcal infections with associated scarlet fever in the last century has been remarkable; whereas the death rate due to streptococcal infections attended by scarlet fever in Great Britain was over 900 per million in 1861–65, it fell to some 29 per million by 1921–25 and now stands near zero. This change may represent, in part, an example of evolution of host and/or parasite. Today scarlet fever is principally of interest because it is an example of the unique circumstance in which bacteriophages regulate toxin production by a pathogen.

3. Peritonsillar Abscess

The tissues involved in streptococcal pharyngitis are commonly confined largely to the throat mucosa and local lymphoid structures. In severe cases confluent exudates may accumulate and cover the affected area and the organisms may pass to the cervical lymph nodes and produce an abscess. Involvement of peritonsillar tissues with abscess formation (quinsy) or spread to retropharyngeal tissues, with the attending danger of airway obstruction and thrombophlebitis, probably results most often from a mixed infection with *Strep. pyogenes, Bacteroides* species and anaerobic streptococci. Infection sometimes spreads through non-lymphoid tissues of the floor of the mouth with associated abscess formation in the neck, a condition called *Ludwig's angina.* Rarely, group A organisms cause a "doughnut lesion" on the palate.

4. Spread to Other Organ Systems

Organisms may also spread readily from the pharynx into the paranasal sinuses, to skin to produce *impetigo,* or directly through the eustachian tube to produce middle ear infection *(otitis media);* spread may also extend to the air cells of the mastoid bone to produce *mastoiditis. Meningitis* may result from invasion through the mastoid bone to the meninges.

Rapid spread from any local site to the blood via lymphatics may occur and may lead to metastatic lesions in joints, bones, endocardium, etc.

Streptococcal bronchopneumonia, often with accompanying bacteremia, usually develops as a complication of some other pulmonary disease, particularly influenza; untreated cases carry a high mortality rate. Because of the tendency to spread through lymphatic vessels, even in retrograde fashion, the organisms readily reach the pleura; hence *pleuritis* with copious pleural exudates commonly develops. *Empyema* (pus in the pleural cavity) occurs in about 40% of cases and bacteremia is frequent. Neonatal sepsis or meningitis may sometimes occur due to *Strep. pyogenes* in the nursery environment.

5. Puerperal Fever

Puerperal fever (childbed fever) is a postpartum infection of the uterus which frequently leads to septicemia and death. It is caused by *Strep. pyogenes* in about 80% of cases; most of the remaining cases are caused by anaerobic streptococci, and *Clostridium perfringens,* which frequently colonize the vagina. A substantial number of cases of puerperal fever result from mixed infections which makes therapy difficult. Infection with *Strep. pyogenes* originates from organisms shed from the respiratory tract of the patient in some 25% of cases and from attendants in the remainder of cases. Without treatment the case fatality rate from disease due to anaerobic streptococci is about 35 to 40% and from disease due to *Strep. pyogenes* about 60%. Deaths from puerperal fever are now only a small fraction of the numbers that occurred before 1935.

6. Infections of the Skin

The skin is the second most common site of streptococcal infection where two types of lesions occur, *pyoderma* and *erysipelas.* Spread may sometimes extend to produce paronychia or vaginitis. Skin and wound infections are of longer duration than pharyngeal infections and are more frequently accompanied by local lymph node involvement and bacteremia. For unknown reasons "skin strains" of streptococci, which are largely limited to the recently discovered *higher-numbered* types, are more prone to produce skin infection than other strains. Most skin strains isolated from skin lesions during epidemics lack capsules and fail to produce mucoid colonies. Nevertheless, type specific immunity slowly develops and reinfection with the same type is probably rare.

Pyoderma includes the streptococcal form of *impetigo contagiosa* which is seen frequently in epidemic form among children and infants, especially in underprivileged families and particularly during summer in warm, humid climates. Trauma of skin, insect bites and poor hygiene are predisposing factors. Most skin infections arise from endogenous organisms inhabiting healthy skin of the patient, possibly as the result of colonization. Also, flies and other insects often contribute to transmission by serving as vectors of organisms in open lesions. Following skin infection the organisms may spread to the URT and scarletinal rash develops on occasion. The typical vesicular, crusted lesions of *impetigo contagiosa* due solely to the streptococcus are distinct from the *bullous lesions of impetigo caused by the staphylococcus.* However, the staphylococcus commonly superinfects the lesions produced by the streptococcus. *Nephritis but not RF* sometimes follows streptococcal impetigo. Recent outbreaks of nephritis following impetigo have been reported in the tropics.

Erysipelas, the second form of skin infection, usually involves the face and on rare occasions results from group C streptococci. The disease shows some tendency to spread from person to person. It is most common in infancy and middle age and for unknown reasons tends to recur at the same site in the same individual over a span of months to years. Facial erysipelas is usually strictly limited to skin of the face and is thought to result from endogenous throat-derived organisms which probably often enter through abrasions in skin about the nostrils and mouth; there is usually a preceding pharyngitis. The onset is sudden with fever and violent chills; the lesions take the form of spreading cellulitis of facial skin characterized by marked erythema and edema and by sharp advancing margins where the lymph channels are packed with organisms. Erysipelas may also develop in other parts of the body due to organisms invading through burns, natural abrasions of the skin or surgical wounds. Spontaneous recovery usually occurs after about a week or more, but without treatment there is danger of spread to the blood stream and the meninges.

7. Poststreptococcal Glomerulonephritis

"Bright's disease," a form of acute hemorrhagic glomerulonephritis, was first described by Bright in 1836 as a frequent sequela of scarlet fever. The disease is produced by relatively few types of *Strep. pyogenes* and only by certain strains within a type.* Outbreaks within families and schools due to a single strain are common and can follow infection in the pharynx, the skin, or elsewhere. During epidemics the attack rate is 10 to 15% and the latent period may extend from a few days to 10 to 20 days or longer. The death rate is highest in adults and has been variously estimated at 5 to 10% or less; most patients who die succumb to acute disease within a year. *The disease commonly follows type 12 infections, which are primarily pharyngeal, and type 49 infections which occur primarily in skin.* The attack rate in different epidemics due to pharyngeal infection with organisms of a single type can vary greatly (0 to 17% with type 12); this occurs because some but not all type 12 strains are nephritogenic. Moreover, the disease has been shown to follow either skin or pharyngeal infection with type 49 organisms. A few strains that belong to types 1, 2, 3, 4, 6, 18, 25 and types 55, 57 and 60 have been shown to be nephritogenic. Recently, streptococcal strains isolated from patients with poststreptococcal glomerulonephritis have been shown to secrete a nephritis-strain-associated protein (NSAP). This protein is *not* produced by strains isolated from individuals with other types of streptococcal disease. The disease is most common in children over 2 years of age but infants are also affected. Hematuria, edema, hypertension and sometimes gastrointestinal and CNS symptoms do not appear until about a week after the beginning of infection. Recurrent attacks are rare, presumably because of the limited number of existing nephritogenic strains and because exposed individuals often possess type specific immunity to nephritogenic strains due to prior experience with non-nephritogenic strains of the same type. In the face of epidemics M protein immunization under the shelter of penicillin prophylaxis should be effective.

The pathogenesis of poststreptococcal nephritis is puzzling and there is no

*Glomerulonephritis results from group C streptococci on rare occasions.

convincing evidence that any known streptococcal component plays a key role in its pathogenesis (Chap. 8). Since Ag-Ab complexes have been found in the serum of patients with glomerulonephritis, many investigators believe that the condition is an "immune complex disease." The concept that the major offending Ag is M protein is difficult to accept since both nephritogenic and non-nephritogenic strains produce M protein. Others believe that it results from Abs to streptococcal glycoprotein Ags that cross-react with related Ags of the glomerular basement membrane; such Ags may be limited to the "nephritis prone" individual. More recently Abs produced against NSAP have been demonstrated to bind to the glomeruli of renal biopsies from patients with the disease. These Abs did not bind to biopsies from non-streptococcal glomerulonephritis patients. This indicates that NSAP is localized in the glomerulus during the disease process. In addition, glomerulonephritis patients produce Ab to NSAP. It still remains to be clarified whether NSAP and/or the anti-NSAP Ab produced by the patient are involved in pathogenesis.

Studies on serial renal biopsies taken in different stages of the disease have thrown considerable light on its histopathology. Morphologically *the disease resembles experimental immune complex disease* produced with foreign proteins; serum C levels are sometimes low and typical "grainy deposits" are alleged to contain globulins. Streptococcal membrane Ag and C3 are frequently seen on the endothelial aspect of the glomerular basement membrane and in the mesangial matrix.* The initial glomerular lesions are generalized and involve endothelial and mesangial cells, but resolving lesions are focal in distribution. Although the acute disease usually undergoes spontaneous resolution, some 10 to 20% of patients, especially those with the greatest glomerular damage and depressed renal function, either die within a year of renal failure or progress to a chronic form of fibroepithelial proliferative disease characterized by extensive epithelial cell proliferation and crescent formation. Whereas the chronic lesions may gradually heal in some cases, they persist for years in others and carry an uncertain prognosis.

8. Rheumatic Fever

Rheumatic fever is a serious delayed sequela unique to human infections with *Strep. pyogenes.* Contrary to an old belief, RF is by no means rare in the tropics. It continues to be an important cause of heart disease in children, especially girls; the attack rate in the U.S.A. has not changed during the past 3 decades. Its pathogenesis remains one of the most challenging unsolved problems in medicine and has given rise to a multitude of hypotheses and a maze of conflicting opinions. Rheumatic fever is acute in nature and usually terminates within weeks. Patients are uniquely prone to develop recurrent attacks and progressive heart valve damage whenever they subsequently contract *Strep. pyogenes* pharyngitis, however mild it may be. *The disease occurs following infections limited largely if not exclusively to the URT* and affects only some 3% (range 0.3 to 20%) of infected individuals, a figure which appears to vary somewhat depending on the virulence of the infecting strain and the magnitude of the immune response. The onset of RF is delayed for about 1 to 4 weeks or more after infection (which passes unnoticed in 40 to 60% of

*Alternatively, deposits may occur on the epithelial aspect of the membrane.

patients) and commonly includes signs and symptoms reflecting widespread involvement of the heart and joints. The cardiac lesions are the most sinister and it has been aptly said that the disease "licks the joints but bites the heart." Symptoms may include fever, malaise, leukocytosis, chorea, epistaxis, subcutaneous nodular lesions, erythema marginatum, migratory polyarthritis and sometimes abdominal pain and vomiting. In decades past, aching in the extremities of children due to RF was so common that it was often mistakenly dismissed as "growing pains"! In many cases of RF medical attention is not sought and in others the disease is misdiagnosed; subsequent rheumatic heart disease may remain as the only indication of a preceding attack of RF. Permanent cardiac damage, particularly of the mitral valve, often occurs, which sets the stage for later serious infection of the damaged valves by other microbes, especially viridans group streptococci that produce subacute bacterial endocarditis (SBE). The subcutaneous nodules of RF tend to develop at sites subjected to frequent nonspecific injury, such as areas over elbows and shins. The hallmark of microscopic cardiac lesions of RF is the "Aschoff body," a DH-type lesion consisting of macrophages, epithelioid cells and giant cells surrounded by lymphocytes, plasma cells and occasionally some neutrophils; acute myocarditis can be fatal.

Since RF can follow infection with essentially all types of *Strep. pyogenes* (except possibly certain "skin strains"), repeated attacks and repeated heart valve damage commonly occur over a period of years, each due to a streptococcus of a different type.* There is no convincing evidence that an anamnestic response with shortening of the latent period takes place upon succeeding attacks of RF. Mere carriage of organisms without an immune response to streptolysin O does not induce recurrences. The initial infection usually subsides completely and the organisms are often cleared from the throat before the onset of RF. Since streptococci cannot be cultured from the blood of rheumatic patients or rheumatic lesions, the development of RF evidently depends on events that take place subsequent to infection. Epidemics of RF follow in the wake of epidemics of "Strep pharyngitis," especially when young people are grouped together as in military barracks.

Any theory on the pathogenesis of RF must take into account certain truths, namely that the disease (1) is only observed in human beings and only follows infection with *Strep. pyogenes,* (2) follows URT infections but seldom if ever follows skin infections, (3) follows infection with strains representing most, if not all, types of *Strep. pyogenes,* (4) is usually accompanied by the development of a high ASO titer and a long-persisting anti-group A polysaccharide titer, (5) develops in a small percentage of individuals within a population; although there is probably a genetic component in susceptibility to RF this issue remains controversial, (6) does not occur in infants, (7) does not recur in the rheumatic subject after each and every episode of streptococcal pharyngitis and (8) certain cell wall components persist in tissues.

One favored theory is that RF represents an allergic response to some streptococcal Ag(s), possibly of membrane origin. Presumably its rarity in children under the age of 3 years is because they have not reached full immunocompetence and have not had time to develop hypersensitivity to streptococcal

*Whether pharyngitis due to "skin strains" can lead to RF is not known.

products. The long latent period of a week or more, the abundance of lymphocytes and plasma cells in the lesions, and the close resemblance between RF and known allergic diseases of both the immediate and delayed types are compatible with the allergic theory. However, the precise events that occur, and the Ags, Abs, immune complexes and/or immune cells involved have not been identified.

Another favored theory is that the disease is due to cross-reactive Abs which are stimulated by a streptococcal Ag but react with tissue Ag. This is based on the observation that Abs which are reactive with both streptococci and human heart Ags, especially sarcolemmal Ags, are present in higher titer in the sera of patients who develop RF than in control subjects. However, the possible role which such cross-reactive Abs may play in the development of RF lesions is unknown. In addition, a cross-reaction with caudate nucleus neurons has been demonstrated in patients with rheumatic chorea. The sera of these patients contain an Ab which binds to the neurons as well as the streptococcal membrane. The Ab appears to be present at highest concentration during periods of choreiform movements.

Heart valve damage (particularly of the mitral valve) resulting directly from RF or from complicating bacterial infection of such damaged valves (usually SBE) is often life threatening (see section on Streptococci of Other Species).

Schonlein-Henoch purpura, an anaphylactoid phenomenon, and the nodular skin lesion, *erythema nodosum,* sometimes follow *Strep. pyogenes* infections.

I. MECHANISMS OF PATHOGENESIS

In common with most pathogenic bacteria, the mechanisms which enable *Strep. pyogenes* to invade and injure the host are largely obscure. Since the majority of streptococci are killed within minutes after ingestion by human phagocytes, especially neutrophils, their notable ability to resist phagocytosis by virtue of surface M protein and, to a lesser degree, surface hyaluronic acid, is without doubt the major determinant of virulence. Moreover, virulence for man, particularly as it relates to pharyngeal infection, correlates well with M protein production. Other probable factors of virulence directly concerned with the persistence and sometimes the growth of engulfed organisms are leukotoxin which often kills the phagocyte, and the cell-wall mucopeptide complex which resists digestion. Another potential virulence factor is the adhesion, lipoteichoic acid of fimbriae, which causes adherence to host cells.

Streptococci that infect skin wounds are often nonencapsulated and of low virulence, whereas those that produce pharyngitis are usually capsulated and of high virulence. The reason for this paradox is not known. Presumably factors other than capsules and M protein are the dominant factors determining virulence of skin strains. The reasons for the marked tendency of streptococci to spread within tissues and to produce injury are not known but could rest, in part at least, on the independent or cooperative activities of one or more of the following substances: DNase, hyaluronidase and leukotoxic and tissue-toxic components, streptolysin S, streptolysin O, NADase, M protein and the cell-wall mucopeptide complex.

The pathogenesis of sequelae considered in detail in Section H, is obviously different in RF and glomerulonephritis since the two diseases rarely occur simultaneously in the same patient. Although their pathogenesis is not understood, it may well rest on a combination of toxic and immunologic effects involving more than one streptococcal component.

J. MECHANISMS OF IMMUNITY

Immunity to *Strep. pyogenes* depends primarily on the acquisition of type specific opsonins. Whereas innate immunity involving surface phagocytosis may often serve to prevent invasion by streptococci of lesser virulence, it is ineffective against large numbers of highly virulent organisms. In contrast, acquired immunity is usually fully protective and repeated attacks with organisms of the same type seldom occur unless the initial infection is arrested by chemotherapy. Whether there is any significant degree of cross immunity between types is uncertain. Once induced, anti-M Abs persist for essentially a lifetime, probably as the result of frequent carrier contact; in the face of infection they are undoubtedly rapidly reinforced by the anamnestic response. Evidently total acquired immunity is gained in a cumulative manner as the result of succeeding encounters with streptococci of new types, be they carried organisms or infecting organisms.

Macrophages from infected rabbits have been shown to carry cytophilic Abs specific for C polysaccharide. However, nothing is known about the possible role of macrophages and DH in either immunity or lesion development in human streptococcal disease.

K. LABORATORY DIAGNOSIS

Streptococcal pharyngitis is diagnosed on the basis of both clinical and laboratory findings, including culture and serologic identification of *Strep. pyogenes* in smears. Other infections, including diphtheria, infectious mononucleosis and particularly infections due to adenovirus, can simulate streptococcal pharyngitis and make its diagnosis difficult. Virus pharyngitis of short duration is frequent, consequently penicillin treatment of misdiagnosed viral pharyngitis is sometimes mistakenly assumed to represent arrest of a streptococcal infection. A negative culture test is even more meaningful than a positive culture and serves to exclude streptococcal pharyngitis in more than half of the suspected cases. Since 5 to 10% of the population are nasopharyngeal carriers of *Strep. pyogenes, a positive culture only establishes the possibility that the infection is due to Strep. pyogenes.*

The specimen should be collected with a swab passed over the tonsils and pharynx, including exudate if present, and taking care not to touch the tongue, lips or uvula; anterior nasal cultures are sometimes needed. Because many strains of *Strep. pyogenes* require CO_2 for initial isolation, prompt culture in an atmosphere of 90% air and 10% CO_2 is advised using blood agar (preferably sheep blood) and a combined pour-streak plate method to permit colony counting and to avoid overlooking beta-hemolytic colonies. Incubation is carried out for 18 to 24 hours at 35°C and the plates are examined for relative numbers of beta-hemolytic streptococci. The catalase test and culture on low concentrations of bacitracin are used to rule out staphylococci (which are catalase positive) and, presumptively, streptococci of other groups, which unlike most strains of

Strep. pyogenes are rarely inhibited by low concentrations of bacitracin.* Selective-enrichment culture techniques are often useful.

Since beta-hemolytic streptococci of various Lancefield groups may be carried in the throat, final group identification with specific antisera is important. This is most effectively accomplished with a direct slide test using specific fluorescein-tagged Ab applied to cultured organisms from broth inoculated 2 to 5 hours previously with swabbed material. The identification of *Strep. pyogenes* in material from wound and skin infections is complicated by the fact that these are often "mixed infections;" anaerobic culture may be helpful in such cases. Rapid tests (20 to 30 minutes) based on chemical or enzymatic extraction and serologic identification of group A Ag from throat swabs are promising new diagnostic procedures.

Identification of organisms in blood, spinal fluid and pus from internal sources is usually easily made with the fluorescent Ab test.

Diagnosis of the sequelae, acute glomerulonephritis and RF, rests largely on clinical grounds and, since living organisms are not directly involved, throat cultures can only detect carriage. Very few patients with sequelae have seen a physician for treatment of a prior sore throat. Titrations of various Abs are of value as indicators of antecedent streptococcal infections that may have escaped clinical notice, and for assessing the immune response. Antistreptolysin O (ASO) titration serves to rule in or deny the possibility of an URT streptococcal infection and has been particularly useful in the diagnosis of RF. Titers above 250 to 300, especially if they are rising, are highly suggestive of a recent pharyngeal infection with *Strep. pyogenes.* Titers of ASO do not rise as the result of carriage. Skin infections seldom engender marked rises in ASO and anti-NADase titers but instead, stimulate high titers of anti-DNAase B and anti-hyaluronidase Abs. Because nonspecific host responses, such as increased sedimentation rate and C-reactive protein, occur in RF, their absence is useful for eliminating the disease as a possibility.

L. THERAPY

The objectives of therapy are to cure the infection, avoid complications and sequelae and prevent transmission of infection to others by preventing carriage. Penicillin is the drug of choice, but lincomycin or erythromycin can be used if the patient is allergic to penicillin. Since drug resistance develops to the tetracyclines and sulfonamides, they should not be used. In some instances additional measures, such as treatment with cephalosporin acid derivatives and surgical drainage, may be necessary to halt infection. Whereas penicillins given at the onset of pharyngitis or within a few days thereafter reduces the chances of developing RF, the prevention of acute glomerulonephritis demands treatment at the onset of infection; treatment should be continued for about 2 weeks.

M. RESERVOIRS OF INFECTION

Human carriers and infected subjects are the principal sources of infection. Except for the cow, animals are not important reservoirs of *Strep. pyogenes.*

*An occasional group A strain may be nonhemolytic or alpha-hemolytic and bacitracin-resistant. Anaerobic incubation promotes β-hemolysis and is a useful adjunct to conventional culture.

Numerous milk-borne epidemics occurred before wide-scale pasteurization of milk was adopted. Usually the milk was contaminated by infected milk handlers but in some instances the teats and udders of the cows became infected from milkers and then served as massive reservoirs of milk contamination. Consumption of food in which group A streptococci have grown can occasionally lead to massive outbreaks of streptococcal pharyngitis.

N. CONTROL OF DISEASE TRANSMISSION AND PREVENTION OF SEQUELAE

Transmission of infection is favored by close contact and crowded living conditions, by nasal rather than pharyngeal carriers, and by moist droplets rather than dust particles. Chronic asymptomatic anal carriers are also a reservoir of infection. Prevention of disease transmission involves carrier control and the usual hygienic precautions taken with patients with infectious diseases, particularly within hospitals with their highly susceptible patient populations.

The overall carrier rate for *Strep. pyogenes* is usually under 5% in summer; in winter it is about 10% but on occasion may reach 50% or higher, especially during epidemics of URT infection. Carriage is frequent after 1 year of age and decreases during late childhood and puberty to reach an average rate of less than 5% in adults. Convalescent carriage, which varies directly with virulence, usually lasts for only a few weeks, during which time the original infecting organisms are eliminated leaving relatively avirulent M protein-deficient nontypeable organisms as the last survivors. Evidently this change results from the selective effect of the developing immune state on the carried organisms. Penicillin is effective for arresting carriage and its mass use can quickly reduce the carrier rate and halt small-group epidemics.

Among the most important and irrevocable outcomes of streptococcal infection are the sequelae. Consequently, principal control efforts center on prevention and early arrest of infection.

In the case of known rheumatic subjects, continuing long-term prophylactic treatment with penicillin has been used successfully for preventing repeated attacks of RF; the incidence of attacks being reduced from 40 to 70% to 4%. Benzathine penicillin given monthly by the intramuscular route is preferred. In the absence of cardiac involvement treatment should be continued through college age but if cardiac symptons develop, treatment should be continued for indefinite periods beyond 5 years.

In the case of non-rheumatic subjects with pharyngitis the question as to how far the therapeutic and prophylactic use of penicillin should be carried is a matter of great concern and debate. Even though penicillin given early in streptococcal pharyngitis dampens the specific immune response and leaves the patient susceptible to reinfection with organisms of the same type, most physicians elect to treat all such patients rather than run the risk of complications and sequelae.

In the case of carriage, which induces immunity without the threat of infection, it is probably best to withhold treatment unless the carrier poses a special danger to others.

Assuming that M protein is not directly involved in the pathogenesis of sequelae, a logical alternative approach to penicillin prophylaxis in special risk groups would be to immunize them while under the shelter of penicillin prophylaxis using a pool of M proteins representing types of streptococci which are the most common causes of streptococcal disease in the population.

STREPTOCOCCI OF OTHER SPECIES

Streptococci belonging to other serologic groups pathogenic for man, e.g., those of groups B, C, D, G, and O, are frequently carried by man; however, infections are relatively rare and are usually mild. For example, only about 1.5% of illnesses due to streptococcal infection in children are due to streptococci of groups other than group A. On rare occasion isolates of pathogenic streptococci have been reported that do not belong to any of the known Lancefield groups.

Streptococci of groups D and O are especially worthy of attention because they are the most common causative agents of SBE. The organisms attack heart valves that are defective or have been damaged in various ways; since RF is the most common cause of valve damage, SBE occurs most frequently on valves injured by this disease. The organisms, which are common members of the normal flora, reach the valves from endogenous sources, such as the mouth and gut, via the blood stream to which they continuously gain entrance in small numbers; three common representative species are *Strep. sanguis, Strep. mitis* and *Strep. faecalis.* If untreated, SBE is almost always fatal and irreversible damage often occurs before treatment is instituted. Bacteriologic diagnosis and antibiotic resistance testing are important. Repeated blood cultures are often necessary for diagnosis and therapeutic management. Since the organisms are highly resistant to antibiotics, therapeutic success, which can be achieved in about 90% of cases, demands prolonged and rigorous treatment with suitable antibiotics or combinations of antibiotics. Penicillin and streptomycin is the combination of choice for treating SBE caused by the enteric streptococci.

Enterococci of Group D *(Strep. faecalis)* are a frequent cause of urinary tract infections. Their ability to grow in 6.5% NaCl broth is an aid for distinguishing them from other streptococci.

Anaerobic streptococci which are now classified in the genus *Peptostreptococcus* are also worthy of special mention because of their frequent participation in mixed infections, especially wound infections, and their high resistance to penicillin (Chap. 18).

Streptococcus agalactiae (group B streptococcus) deserves special consideration because of the increasing frequency of serious infections in neonates, especially meningitis and sepsis, caused by this organism. Some 1 to 3 cases per 100,000 births occur with a mortality of 30 to 60%. Perinatal infection can be divided into two groups; *early-onset disease* within the first 7 days of life usually results from vaginal organisms carried by the mother, whereas *late-onset* disease arising after the 7th day usually results from nursery sources, including other babies and the nursery staff. The vaginal carriage rate in women is about 30%.

Of the various therapeutic regimens, combinations of one of the penicillins and an aminoglycoside are often used; blood transfusion is sometimes employed.

It is alleged that neonatal infection is prone to occur in the offspring of mothers whose serum is deficient in protective Abs. Since group B polysaccharides vary structurally and can be divided into serotypes Ia, Ib, Ic, II and III, it has been proposed that Ab-deficient pregnant women be vaccinated with a pool of these Ags; the potential value of such a vaccine remains to be determined.

Dental Caries. Four species of the genus *Streptococcus* are associated with dental caries in man *(Strep. sanguis, Strep. mitis, Strep. salivarius* and *Strep. mutans).* It appears that *Strep. mutans* is the dominant species that initiates decay affecting smooth enamel. This pathogenetic characteristic is probably dependent on its ability to adhere to teeth and generate large bacterial plaque accumulations; *Strep. mutans* can synthesize high MW dextrans from sucrose which facilitate adherence and plaque formation. Caries occurs when lactic acid accumulates under the plaque in sufficient concentrations to dissolve the enamel.

REFERENCES

Baker, C.J., and Kasper, D.L.: Group B streptococcal vaccines. Rev. Infect. Dis. *7*:458, 1985.
Bland, E.F.: Declining severity of rheumatic fever. N. Engl. J. Med. *262*:597, 1960.
Clewell, D.B., et al.: Conjugative transposons and the dissemination of antibiotic resistance in streptococci. Annu. Rev. Microbiol. *40*:635, 1987.
Ellen, R.P., and Gibbons, R.J.: M protein-associated adherence of *Streptococcus pyogenes* to epithelial surfaces: prerequisite for virulence. Infect. Immun. *5*:826, 1972.
Fox, E.N., et al.: Primary immunization of infants and children with group A streptococcal M protein. J. Infect. Dis. *120*:598, 1969.
Ferrieri, P., et al.: Natural history of impetigo. I. Site sequence of acquisition and familial patterns of spread of cutaneous streptococci. J. Clin. Invest. *51*:2851, 1972.
Ferrieri, P., et al.: Natural history of impetigo. Etiologic agents and bacterial infections. J. Clin. Invest. *51*:2863, 1972.
Ginsburg, I.: Mechanisms of cell and tissue injury induced by group A streptococci: relation to poststreptococcal sequelae. J. Infect. Dis. *126*:294, 419, 1972.
Holmes, O.W.: Medical Essays: 1842–1882, Repr. of 1983 ed. lob. bldg. Arden Lib., 1977.
Husby, G., et al.: Antibodies reacting with cytoplasm of subthalamic and caudate nuclei neurons in chorea and acute rheumatic fever. J. Exp. Med. *144*:1094, 1976.
Ofek, I., et al.: Cell membrane-binding properties of group A streptococcal lipoteichoic acid. J. Exp. Med. *141*:990, 1975.
Ofek, I., et al.: Oxygen-stable hemolysins of group hemolysis of group A streptococci. VII. The relation of the leukotoxic factor to streptolysin S. J. Infect. Dis. *122*:517, 1970.
Quinn, R.W.: Dick test results: 1969–1971. J. Infect. Dis. *126*:136, 1972.
Read, S.E., and Zabriskie, J.B.: Streptococcal Diseases and the Immune Response. New York, Academic Press, Inc., 1980.
Sanders, W.J., and Rammelkamp, C.H.: Pneumonia caused by the group A streptococcus. Seminars in Infectious Diseases *5*:36, 1983.
Schoolnik, G.K., et al.: Bacterial adherence and anticolonization vaccines. Current Clinical Topics in Infectious Diseases, Vol. 6, (eds), Remington, J.S. and Swartz, M.D., New York, McGraw Hill, 1985.
Shibi, A.M.: Effect of antibiotics on adherence of microorganisms to epithelial cell surfaces. Rev. Infect. Dis. *7*:51, 1985.
Slaughter, F.G.: Semmelweis the Conqueror of Childbed Fever. New York, Collier, 1950.
Unny, S.K., and Middlebrook, B.L.: Streptococcal rheumatic carditis. Microbiol. Rev. *47*:97, 1983.
van de Rijn, I., et al.: Serial studies on circulating immune complexes in post-streptococcal sequelae. Clin. Exp. Immunol. *34*:318, 1978.
van de Rijn, I., and Kessler, R.E.: Growth characteristics of group A streptococci in a new chemically defined medium. Infect. Immun. *27*:444, 1980.
van de Rijn, I., and Fischetti, V.A.: Immunochemical analysis of intact M protein secreted from cell wall-less streptococci. Infect. Immun. *32*:86, 1981.
van de Rijn, I., et al.: Group A streptococcal antigens cross-reactive with myocardium. J. Exp. Med. *146*:579, 1977.
Villarreal, H., Jr., et al.: The occurrence of a protein in the extracellular products of streptococci isolated from patients with acute glomerulonephritis. J. Exp. Med. *149*:459, 1979.

11

STREPTOCOCCUS PNEUMONIAE

Among *Streptococcus* species *Strep. pneumoniae* is unique in many ways both with respect to its intrinsic properties and the infections it produces. It is commonly referred to as the "pneumococcus" and was formerly classified in the genus, *Diplococcus.*

Streptococcus pneumoniae is a pyogenic extracellular pathogen and is frequently carried in the URT of man, its principal natural host. Pneumococci have been isolated irregularly from horses, cattle, dogs, monkeys, guinea pigs, and rabbits. Although epizootics of pneumococcal disease occasionally occur in laboratory animals such as rats, guinea pigs, and monkeys, it is notable that such outbreaks tend to be narrowly limited to certain serologic types (type 19 for the guinea pig and type 2 for the rat). No animal is known to be an important reservoir of pneumococcal infection for man.

Streptococcus pneumoniae was discovered independently by Pasteur and Sternberg in 1881. It is classified on the basis of numerous serologic types originally designated by Roman numerals.

A. MEDICAL PERSPECTIVES

Prior to the development of therapy with specific antiserum in the early 1930s and later with drugs, the pneumococcus was a major cause of death; in hospitalized patients some 95% of all cases of pneumonia were due to the pneumococcus. The organism produces infections in various organs and tissues, the most common and serious being bronchopneumonia, (patchy pneumonia) and lobar pneumonia; the latter involves entire lobes of the lung or major portions thereof. Whereas pneumococcal bronchopneumonia is largely confined to the very young, the aged, and the debilitated, epidemic lobar pneumonia occurs most frequently in healthy young adults, especially males. In decades past it took a terrible toll of lives in this group; for those in the prime of life Osler called it "Captain of the Men of Death." The case fatality rate ranged from 20 to 40% or more and exceeded 50% in the case of infection due to organisms of type 3. Current therapeutic practices have reduced the case fatality rate to about 5 to 10%, and except for type 3 infections, uncomplicated cases, if treated early, can usually be saved. However, the incidence of pneumococcal infection has not decreased, because diagnosis is frequently delayed and because complicating disease is often present; as a consequence, pneumococcal infections still remain a major cause of serious illness and death. Deaths from pneumococcal pneumonia alone in the U.S.A. are estimated to number about 25,000 annually and pneumococcal pneumonia continues to be the most frequent of the bacterial pneumonias; its numbers approach some ½ million annually.

It is of singular importance that although treatment of patients with pneumococcal meningitis or bacteremic pneumonia may be "bacteriologically successful," i.e., most or all of the organisms are killed within a matter of hours, such patients often die later, apparently of delayed toxemia. Consequently, future possibilities for lowering mortality rates must hinge not only on early diagnosis and early treatment, vaccines and chemoprophylaxis, but on devising therapeutic means for counteracting the toxemia involved.

The discovery of capsule serotypes by Neufeld and Handel in 1909 was followed by one of the most remarkable team achievements in medical history, namely, the elucidation of the major virulence mechanism involved in pneumococcal infection. This was accomplished by a succession of investigators during the 1920s and 1930s at the Rockefeller Institute for Medical Research, including Dochez, Avery, Heidelberger, Goebel, MacLeod, and McCarty. They made major contributions toward establishing that the antiphagocytic capsule of the pneumococcus is the major determinant of virulence and that acquired immuity results from specific opsonizing Ab directed at the particular polysaccharide making up the capsule of the infecting organism. *Thus the type-specific capsular polysaccharide was found to be both the principal virulence factor and immunogen.* In addition, their studies on the phenomenon of Griffith's transformation of serologic types of pneumococci led to the discovery that DNA can serve as a transforming agent and opened the field of molecular genetics. The findings of this group of investigators provided the basis for specific serum treatment of lobar pneumonia instituted in the 1930s. Ironically, but fortunately, this long-awaited advance in therapy was quickly supplanted by sulfonamide treatment and later by treatment with penicillin and other antibiotics.

B. PHYSICAL AND CHEMICAL STRUCTURE

Streptococcus pneumoniae is typically capsulated, Gram-positive, lance-shaped cocci usually positioned in pairs with tips pointing outward. In vivo the organism may occur singly and in chains of varying length as well as in pairs. In common with other pathogenic cocci it is nonsporing and nonmotile.

Aside from the unexplained toxicity it exerts during infection, the most notable attribute of the pneumococcus is its enormous capsule which swells upon the addition of specific antiserum, a phenomenon called the Quellung reaction (Fig. 11–1). Based on antigenic differences in their capsular polysaccharides, 94 distinct serologic types of pneumococci have been described. The capsular material is soluble in water and hence is sometimes referred to as "soluble specific substance" (SSS). Capsule size is directly related to the rate of polysaccharide synthesis because capsular material readily passes into solution in the surrounding menstruum; hence capsules shrink as cultures age and synthesis slows. There is limited antigenic relatedness between the capsular polysaccharides of some of the different pneumococcal types and polysaccharides from various other sources including the genera *Escherichia, Klebsiella* and *Haemophilus.* For example, type 14 polysaccharide cross-reacts with Abs specific for the blood group A substance.

The cell wall of the pneumococcus contains a Forssman-like carbohydrate substance (C-substance) which is species specific and M proteins beneath its capsule which are analogous to but distinct from the M proteins of *Strep. pyogenes.** Since neither C nor M antigens of the pneumococcus exert appreciable antiphagocytic activity, anti-M antibodies and anti-C antibodies play no significant role in immunity.

An abnormal beta-globulin called "C-reactive protein" is present in the serum of a variety of patients with acute inflammatory diseases; it is so-named because it reacts with pneumococcal "C substance" to form a precipitate. The possible significance of the occurrence of this protein in disease is not known. The

*The C-substance (C) of the pneumococcus should not be confused with complement which is likewise designated by the capital letter, C. The M protein plays no part in the virulence nor is it a protective immunogen.

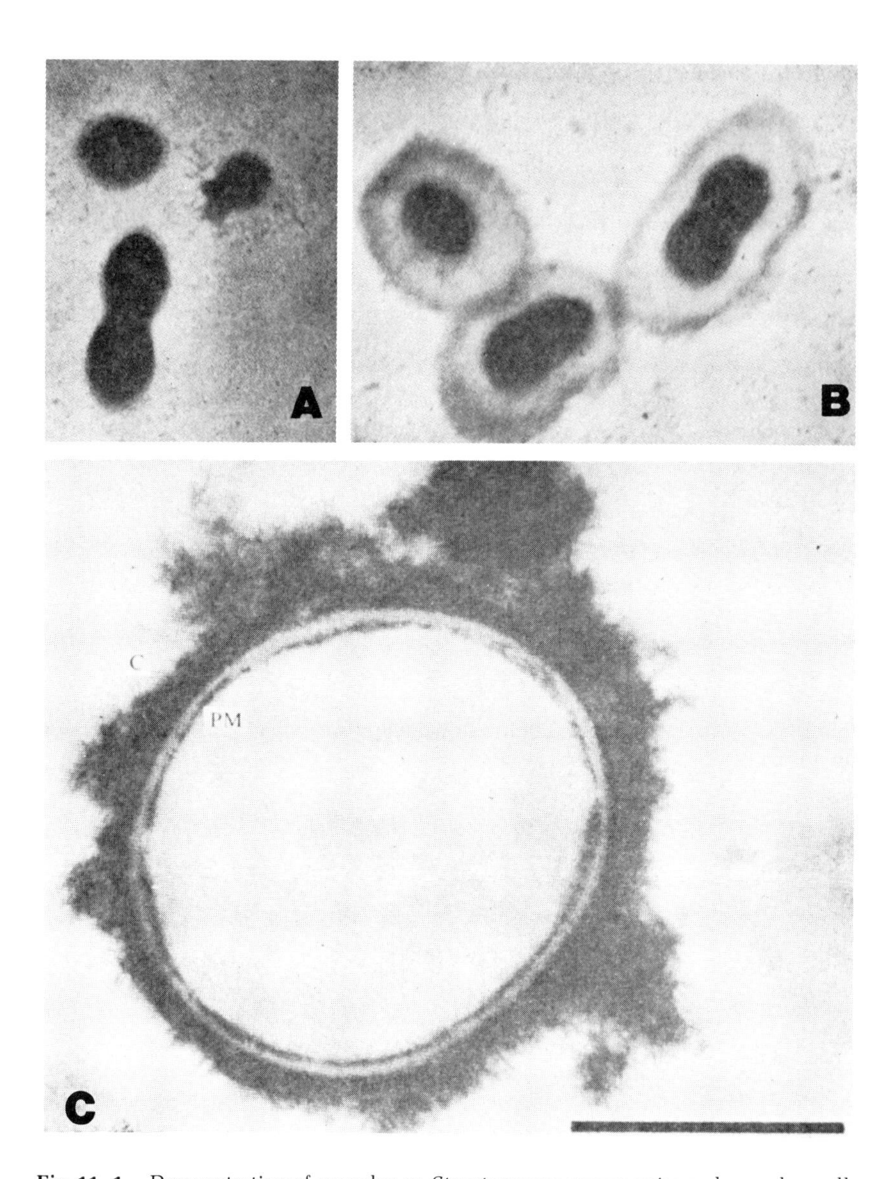

Fig. 11–1. Demonstration of capsules on *Streptococcus pneumoniae* and capsule swelling in the presence of specific anticapsular antibody. *A, S. pneumoniae* Type 1 (note normal capsule) × 10,500; *B, S. pneumoniae* Type 1 incubated 3 minutes with specific anti-Type 1 capsular antibody (note swelling of capsule) × 10,500. (Mudd, S., Heinmets, F., and Anderson, T.F.: J. Exp. Med. *78*:327, 1943.) *C, S. pneumoniae* stained with ruthenium red depicting the capsule (C) and plasma membrane (PM). Marker represents 0.5 μm. (Springer, E.L., and Roth, I.L.: J. Gen. Microbiol. *74*:21, 1973.)

pneumococcus contains lipoteichoic acid which inhibits an autolytic enzyme produced by the organism; it belongs to the Forssman group of Ags.

C. GENETICS

One type of genetic variation among pneumococci is concerned with the amount of capsular polysaccharide produced, a property which generally correlates directly with virulence, types 3 and 8 being the most virulent. Most encapsulated strains tend to produce colonies with smooth surfaces and outlines (S colonies), whereas strains lacking capsules tend to produce rough colonies (R colonies). A special form of extreme roughness occasionally occurs due to extensive chaining of organisms and may sometimes be seen among capsulated as well as noncapsulated strains. Mutation from S to R and from R to S occurs and can be readily demonstrated. Since one or a few S organisms can infect and kill a mouse, mutation from R to S is easily demonstrated by injecting a large dose of R organisms into the animal's peritoneum. Whereas the R organisms are rapidly phagocytized and killed, the few S mutants present soon grow and can be recovered in pure culture. The selective environment of the mouse peritoneum accounts for the increases in capsule production and virulence attending mouse passage.

A culture of pneumococci liberates DNA which can penetrate other pneumococci, including those of a different strain, and impose constitutional genetic change in the recipient. The first DNA-induced change studied was concerned with the transformation of a rough noncapsulated strain of the pneumococcus to a smooth capsulated strain. For example, when DNA extracted from a capsulated type 3 organism was added to a noncapsulated strain previously derived from a type 2 strain, the noncapsulated strain was transformed to a capsulated type 3 strain. Strains possessing more than one polysaccharide type in their capsules occur naturally on rare occasions, presumably as the result of partial transformation. Transformation has since been shown to involve many other genetic traits, including the quantity of capsular substance produced and resistance to therapeutic drugs. Transformation has been demonstrated experimentally in the mouse and presumably occurs naturally in human beings. Additional tools for genetic studies comprise pneumococcal plasmid and bacteriophages. An event of grave concern is the recent emergence of penicillin-resistant strains of type 19A pneumococci in South Africa, the basis of resistance being the production of a penicillin-binding protein.

D. EXTRACELLULAR PRODUCTS

Growing pneumococci produce H_2O_2 and liberate an oxygen-sensitive hemolysin which causes α hemolysis on blood agar, as well as capsular polysaccharide and the enzymes neuraminidase and hyaluronidase. The latter plays no part in virulence or pathogenesis.

The capsular polysaccharide is produced in great abundance and, being soluble, reaches the blood in such large amounts that it can often be detected in the serum and urine of patients with lobar pneumonia (Sect. K). However, it is not toxic for animals, even when given in large doses.

Autolysis of organisms releases a *"purpura-producing factor"* which causes local hemorrhage when injected into the skin of rabbits. It has also been observed that high concentrations of *pneumolysin,* a unique intracellular toxin possessing β hemolytic and general cytotoxic activity, are dermotoxic and lethal. Some strains are reported to produce a *leukocidin* in small amounts. Neuraminidase is lethal for weanling mice and at present is a strong candidate for explaining the irreversible toxicity and delayed death that sometimes results from human pneumococcal infections following a favorable response to therapy; it is antigenic and induces the production of neutralizing Abs.

E. CULTURE

The pneumococcus requires relatively complex media for growth. It is a facultative anaerobe lacking catalase and peroxidase; consequently it is commonly grown on media containing red blood cells, a rich source of catalase, in order to prevent the accumulation of toxic levels of the metabolite, hydrogen peroxide which they produce. The organism also stores best in media rich in blood and can be maintained at 4°C in such media for months. Its growth range is 25° to 42°C. The pneumococcus produces large amounts of lactic acid which is growth inhibitory unless highly buffered media are employed. The organism is notorious for the readiness with which it undergoes spontaneous lysis due to activation of autolytic enzymes; when the phase of exponential growth is passed, the organism rapidly becomes Gram-negative and lyses. This results in loss of turbidity in liquid cultures and shrinking of the central portions of old colonies on solid media which lends to them a centrally dimpled appearance. Although the mechanisms that promote autolysis are not known, surface-active agents, such as bile salts, trigger lysis and are the basis of the so-called "bile solubility tests" used in laboratory diagnosis to aid in differentiating pneumococci from α-hemolytic streptococci. Any suspected organism which is "bile-insoluble" is usually not a pneumococcus. The media commonly employed for growing pneumococci include blood agar and beef infusion broth containing 10% serum or blood and a reducing agent such as thioglycolate. When grown aerobically on the surface of blood agar, virulent capsulated strains produce medium-sized (0.5 to 3 mm) round, glistening colonies which are initially dome shaped and become surrounded by a greenish-brown zone of partial hemolysis containing red cell ghosts (α-hemolysis). Small amounts of CO_2 stimulate growth, and some strains have been reported to require 20% CO_2 for primary isolation. Organisms that produce unusually thick capsules, particularly those of type 3, form relatively large mucoid colonies.

F. RESISTANCE TO PHYSICAL AND CHEMICAL AGENTS

Pneumococci are "less than average" with respect to resistance to most physical and chemical agents. They are resistant to bacitracin but are susceptible to a number of other antibiotics and soaps (Sect. L). Rare strains have been found in South Africa that are multiresistant to several antibiotics. The pneumococcus is highly susceptible to ethyl-hydrocupreine hydrochloride, whereas the α-hemolytic streptococci are not. This is the basis of the "optochin test" for differentiating between the two organisms. When growth ceases pneumococci readily undergo spontaneous autolysis; they are also uniquely sensitive to "solubilization" with bile salts which accelerate natural spontaneous autolysis.

G. EXPERIMENTAL MODELS

Injected organisms are pathogenic for mice, rats, rabbits, dogs, and sub-human primates, the mouse and rabbit being the most susceptible. The pneumococcus is singularly virulent for the mouse which is commonly used for virulence tests; one to 5 organisms given intraperitoneally usually cause death within 16 to 48 hours. Organisms of a few types, including capsulated type 14, have limited virulence for the mouse which serves to illustrate that with certain types at least, important virulence factors other than the quantity of capsular polysaccharide exist and that virulence factors vary with the host species. Indeed the chemical nature of the polysaccharide, as determined by type, is important as indicated by the observation that strains which cause the most serious infections are restricted to unique types that fail to activate the alternative C pathway and thus escape phagocytosis until specific Ab is formed. Evidently the type as well as the quantity of capsular polysaccharide produced is a major determinant of virulence.

Disease closely simulating lobar pneumonia in man has been produced in dogs and subhuman primates. Birds, such as pigeons, are highly resistant because they possess natural opsonins, presumably acquired by natural exposure to cross-reacting Ags.

H. INFECTIONS IN MAN

Currently 80% of all bacterial pneumonias are caused by the pneumococcus and 75% of cases of pneumococcal pneumonia in adults are due to 12 types; namely, types 1, 2, 3, 4, 5, 7, 8, 12, 14, 18, 19, and 23 (50% are due to types 1 and 2 alone); in children the most common types are 1, 6, 14, 19, and 23 (also 5 and 7 in earlier reports). The prevalence of types causing disease varies with time and geographic area. The high-numbered types are, in general, less virulent, and bacteremia seldom occurs in infections caused by types above type 33. *In type 3 pulmonary infections with associated bacteremia the mortality exceeds 50% even with treatment!* In a small percentage of cases, mixed infections with more than one type of pneumococcus or with the pneumococcus and *Klebsiella pneumoniae* occur. Such cases obviously present problems in both diagnosis and management.

Most pneumococcal pneumonias result from organisms carried by the individual in his respiratory tract; its failure to produce pharyngitis remains an unsolved mystery. The lung is the most common site of infection; predisposing causes include: measles, respiratory virus infection, leukopenia, asthma with excessive mucus, alcoholism, nutritional cirrhosis, fatigue, dusty atmosphere, toxic gases, congestive heart disease, anatomic pulmonary disease, sickle cell anemia, and primary defects in immunocompetence. Pneumococcal pneumonia often leads to serious complicating infections at other sites, either by direct extension (empyema, pericarditis) or via the bloodstream, with resulting endocarditis, meningitis, or arthritis; endocarditis and meningitis are almost always fatal unless treated. Meningitis can sometimes result from direct extension of infection from the mastoid or from organisms invading through anatomic defects, such as those caused by basal skull fractures experienced even years earlier. Primary otitis media, an unusually common infection in young children, and mastoiditis are often due to the pneumococcus. Another type of infection, peritonitis in female children, evidently occurs as the result of direct invasion of the organisms from the vagina, where they are sometimes carried. Pneumococcal pneumonia presents as either lobar pneumonia or bronchopneumonia, the latter being most frequent in predisposed individuals, such as very young, very old, or debilitated persons.

Lobar pneumonia is the only pneumococcal infection that will be discussed in detail. Over 95% of cases of lobar pneumonia are of pneumococcal causation and presumably result most frequently from highly virulent indigenous or carried organisms to which the individual lacks adequate specific acquired immunity.* Most cases of pneumococcal pneumonia arise in the wake of viral influenza or the common cold, the majority being among otherwise healthy adults over 40 years of age; it was formerly most common in young men. Since late therapy often fails, early diagnosis and treatment are of the utmost importance. The disease is sudden in onset and is commonly heralded by chills, leukocytosis, high plateau-type fever, prostration, nausea and vomiting, cough, tenacious rust-colored sputum (often streaked with blood), arthralgia, and frequently, stabbing unilateral pleural pain. Excessively sticky sputum is usually due to infection with type 3 pneumococci or *K. pneumoniae* with all of its potential metastatic consequences; it occurs in about 25% of cases and is a serious prognostic sign. Hypoxia from impaired respiratory exchange is often so extreme as to require oxygen; congestive heart failure is a common complication. Icterus due to the combined effects of excessive red cell destruction and depressed liver function from anoxia sometimes occurs and is also a grave prognostic sign. In untreated patients who survive, a sudden improvement of symptoms may occur (recovery by "crisis"), usually about the 7th or 8th day of illness (range 5 to 10 days), or the disease may gradually subside (recovery by "lysis"). Chemotherapy can, of course, produce artificial crisis (Sect. M). The organisms that initiate infection are thought to invade directly through the bronchial tree. At onset, the alveoli become filled with fluid rich in rapidly dividing pneumococci but largely lacking in host cells. Although the cause of sudden excessive edema is not known, the suggestion has been made that it may be an allergic response to proteins of the pneumococcus. The organisms spread peripherally by way of intercommunicating bronchioles and from alveolus to alveolus through communicating channels in alveolar walls (pores of Cohn) opened by the stretching effects of excessive fluid in the alveoli. *The stage of edema* is quickly followed by fibrin accumulation and infiltration of blood cells, especially red cells, lending the appearance and texture of liver (stage of red hepatization). During these early stages the organisms multiply to reach enormous numbers. Subsequently, neutrophils infiltrate in large numbers, red cells lyse, and many pneumococci are phagocytized (stage of gray hepatization). The pleura becomes coated with leukocytes and fibrin and frank empyema sometimes results. In patients who recover promptly the organisms are largely eliminated by phagocytosis before the next stage (stage of resolution) occurs, during which fibrin is digested and neutrophils are replaced by a heavy influx of infiltrating macrophages. The latter help phagocytize the remaining

*Lobar pneumonia is defined as "an inflammation of the lung in which the distribution of lesions involves one or more lobes or large portions of lobes, but not others."

pneumococci, red cells, dying neutrophils, and debris; some semifluid debris may be coughed up. Fluid is absorbed, and the lung returns to normal.

During resolution, the rapidity and completeness of restoration of normal histology and function of the lung without scarring has long been a source of wonderment; it probably rests on (1) lack of serious injury of alveolar walls by the organism,* (2) rapid lysis of fibrin by enzymes liberated by dying neutrophils, thus thwarting the ingrowth of fibroblasts, and (3) good blood supply and lymphatic drainage, which maintain adequate nutrition. In some instances resolution does not proceed to completion, and persisting fibrin is replaced by scar tissue which converts the affected lung into a dense semielastic tissue, a process termed "cornification." Chemotherapy tends to favor fibrosis, possibly by killing organisms, thus minimizing the infiltration of neutrophils which contribute the enzymes responsible for fibrin digestion. In some cases of lobar pneumonia, especially involving children, alveolar walls may be ruptured by the extensive exudate with resulting emphysema.

Shock in pneumococcal pneumonia is extremely difficult to manage and if present for any extended time is usually fatal. Early shock with its fall in temperature (pseudocrisis) may be mistaken for crisis. Although the death rate in pneumococcal pneumonia is highest in cases in which bacteremia occurs (about 18% even in treated cases), there is also, in general, a direct correlation between mortality and the mass of lung involved. The metastatic consequences of bacteremia are meningitis, pericarditis and endocarditis. The death rate in penicillin-treated patients with bacteremia rises from 7% in the 12 to 49 year age group to 28% in patients over 50 years of age! The overall fatality rate in treated patients is 5 to 10%.

It is of singular interest that resolution of pneumococcal infection at other sites is less effective than in the lung and such lesions often require surgical drainage. Penicillin treatment of abscesses often fails, because in such lesions many of the organisms are not growing and phagocytosis is minimal, most of the leukocytes being dead.

I. MECHANISMS OF PATHOGENICITY

Pneumococcal pneumonia, indeed pneumonia due to many infectious agents, commonly results from a breakdown of one or more of the normal defense barriers of the lower respiratory tract, including the cough and epiglottal reflexes, the mucociliary escalator, the lymphatic drainage of the bronchi and bronchioles, and the macrophage defense of the alveoli. Lobar pneumonia evidently results from the aspiration of heavily contaminated secretions from the URT into the alveoli rather than from inhalation of contaminated sputum droplets or dust particles. During viral respiratory disease the URT secretions are usually excessive and are heavily contaminated with carried organisms. Since the secretions are thin in consistency, aspiration and drainage are apt to carry them to the terminal bronchioles and alveoli. In this regard, it is of singular interest that the initial infection commonly begins in the lower lobes where aspirated fluids tend to gravitate. Prior inflammation accompanied by accumulation of edema fluid in alveoli is particularly predisposing to infection,

*An exception to lack of injury sometimes occurs in type 3 infections in which there may be necrosis of alveolar walls.

presumably because the fluid serves as a good medium for bacterial growth and hampers the phagocytic activities of resident alveolar macrophages. Experimental evidence supports this concept; normal lungs of animals are more resistant to intratracheal challenge than edematous lungs.

It has been suggested that atelectasis may predispose to infection in both animals and man. Once infection begins, the influx of edema fluid into the alveoli is intense and carries the organisms centrifugally into adjacent alveoli through the opened channels between alveoli and communicating terminal bronchioles. As the result of bronchial spread, large areas of complete consolidation involving whole lobes or portions of lobes develop, which is in contrast to the patchy distribution of lesions seen in bronchopneumonia. Resident macrophages are apparently unable to arrest the unopsonized organisms by phagocytosis, and growth of pneumococci may not be contained by neutrophils until sufficient specific Ab accumulates to neutralize the large amounts of free solubilized capsular Ag present and accomplish specific opsonization. Agglutination of the organisms by Ab may help prevent their spread, but is a weak restraining force at best. Maturation of the lesion through its various stages occurs earlier at the initial site of infection than in peripheral areas of spread characterized by edema; consequently all stages of lesion development may be seen in the same lung until after final arrest takes place. Small numbers of organisms probably reach the blood in all cases of the disease, evidently by way of lymphatics draining the alveoli, but are only detected by blood culture when appreciable numbers are shed into the circulation. Persisting bacteremia indicates that organisms have not only reached the circulation but are growing in intravascular foci and have overwhelmed the forces which normally clear the blood of particulates. Early pleural involvement is common, probably as the result of lymphatic spread. Infections of the adjacent pleural and pericardial cavities may occur and represent serious complications; empyema occurs in about 5% of cases.

Despite the fact that patients with pneumonia show evidence of marked "toxemia," no causative toxin has been implicated with certainty. Pneumococcal neuraminidase is toxic for experimental animals, and the possibility that it might be an important and possibly the sole agent responsible for toxemia in human infections has not been ruled out (Sect. D).

Occasionally, degeneration of renal tubules occurs, but the reason for this is not known. It is of interest that immune complex disease involving renal glomeruli does not occur despite the probability that enormous amounts of soluble Ag-Ab complexes are formed. This raises questions as to the localization pattern of these complexes and their capacity to fix C.

J. MECHANISMS OF IMMUNITY

Since immunity is specific and lasts for decades, second attacks with organisms of the same serologic type are rare. Both macrophages and neutrophils readily kill ingested pneumococci. Hence, the critical immune forces operating in the alveolus against invading pneumococci involve primarily opsonin-dependent phagocytosis. Whereas small amounts of natural complement-dependent polyspecific opsonins for pneumococci occur in most normal individuals, neither they nor the forces of surface phagocytosis serve to meet any substantial challenge on the part of invading organisms.

Healthy individuals lacking detectable specific anticapsular Ab can carry highly-virulent organisms without developing lung infection. Since a few of the carried pneumococci must occasionally reach alveoli, macrophages, which are essentially the only phagocytes in normal alveoli, must effect the destruction of such organisms largely by the alternative C pathway and surface phagocytosis. They thus represent the effector cells of nonspecific antipneumococcal defense of the alveolus of the normal lung. In contrast, under the abnormal conditions existing at the time of infection, resident alveolar macrophages, unless assisted by substantial amounts of specific Ab, are evidently incapable of coping with the large numbers of pneumococci present at the initiation of a progressive infection.

Once infection is established, macrophages evidently contribute less to total phagocytic defense than neutrophils, which mobilize, ingest, and kill pneumococci with remarkable facility in the presence of specific opsonins. However, maximum killing requires that both the alternate and classical C pathways are intact. Consequently *the availability of adequate amounts of specific Ab is usually the critical factor determining the outcome of infection.* Enormous amounts of Ab are probably required to combine with available free Ag to the point where opsonization can occur. *Thus death or recovery may be likened to a race between the ability of the host to produce specific Ab and the ability of the organisms to produce capsular polysaccharide.* This is attested by the observations that free opsonins usually appear in the blood at or near the time of crisis, that the formerly used treatment with specific antiserum was effective, and that individuals lacking in the capacity to form Ab, such as patients with agammaglobulinemia, are highly susceptible to pneumococcal infection. The fate of Ag, the classes of Abs formed, their functions, and the dynamics of Ag-Ab interactions in lobar pneumonia have not been fully elucidated. With respect to the Abs produced and their functions, it has been claimed that in the mouse protection test IgM Ab is 100,000 times more effective than IgG Ab. Although this could explain why precipitin and agglutinin titers do not correlate well with mouse protection tests, there is no assurance that Abs of different immunoglobulin classes behave similarly with respect to pneumococcal immunity in the mouse and man. The alleged role of IgA Abs in pneumococcal immunity in the mouse needs further investigation. It is conceivable that high avidity IgA Abs could assist by neutralizing circulating capsular polysaccharide thus leaving IgG or IgM Abs available for opsonization. It appears that IgA is not capable of functioning as an opsonin.

The continuing production of anticapsular Abs for decades after infection may occur because pneumococcal polysaccharide is not readily destroyed by tissue enzymes; hence immunogenic amounts remain in the body for long periods. A disturbing finding is that the polysaccharide readily produces immunologic tolerance, at least in the mouse. The possible implications of this property of the polysaccharide with respect to human infections and vaccination are not known. Tolerance obviously does not occur in patients who pass crisis and live or perhaps even in patients who die.

The basis of the Francis reaction is of theoretical interest. This is an immediate type of skin reaction induced by the intradermal injection of purified pneumococcal polysaccharide. It depends on the presence of free antipneumococcal Ab in serum and becomes positive about the time of crisis in patients who recover, but is seldom positive in those destined to die. The test is positive in convalescent patients and in some normal individuals. For some unknown reason it may be negative in some patients who develop serum Ab but who die later without undergoing crisis.

K. LABORATORY DIAGNOSIS

Since early chemotherapy in pneumococcal pneumonia is critical, rapid methods for making a presumptive diagnosis should be used. Although most cases of pneumococcal lobar pneumonia can be diagnosed easily by clinical findings and simple laboratory tests, others are difficult to diagnose, the most notorious being patients with infected pulmonary infarcts. Pneumonias due to staphyloccci, *Streptococcus pyogenes,* and *Klebsiella pneumoniae* must always be considered. It is especially important to differentiate between pneumonia caused by the pneumococcus and *K. pneumoniae* since the causative organisms differ markedly in their susceptibility to antibiotics. Differential diagnosis should include consideration of coexisting infectious disease and noninfectious conditions, such as congestive heart failure, pulmonary infarction, and atelectasis.

Whenever lobar pneumonia is suspected, sputum specimens should be promptly stained and cultured, and blood should be inoculated into thioglycolate medium and plated on blood agar (Sect. E). Usually a Gram stain of deep sputum will permit a tentative diagnosis so that chemotherapy can be instituted promptly; however, it is impossible to distinguish between the pneumococcus and certain strains of α hemolytic streptococci by the Gram stain alone. If antisera are available, the capsular swelling test on sputum provides a rapid and reasonably reliable test for early diagnosis.* When positive, the detection of free pneumococcal Ags in blood or urine by modern techniques is clear evidence of pneumococcal infection. One method is counterimmunoelectrophoresis using a multivalent antiserum.

The sputum should come *from deep in the respiratory tract* and should be collected in a sterile jar at the bedside by the physician. Throat swabs, transtracheal aspiration, or lung puncture may be necessary in children. Swabs should be transported to the laboratory in a sterile medium to prevent drying, and cultures should be incubated at 35 to 37°C in a CO_2 incubator, since 10% of newly isolated strains will not grow without added CO_2.

Blood agar is commonly used for streak plating. Colonies are small and should be observed with a dissecting microscope. Bile solubility tests and the optochin test are useful for identifying the pneumococcus.

The presence of pneumococci in blood or deep lung secretions secured by transtracheal or lung puncture is virtually certain evidence that the pulmonary infection is of pneumococcal causation. Their presence in blood in appreciable numbers carries an unfavorable prognosis. Because of the high frequency of pneumococcal carriers, the presence of pneumococci in sputum is only of significance when the organisms are in great abundance and are accompanied by large numbers of neutrophils. Pneumococci in pus tend to become Gram-negative.

*Antisera reactive against 84 types are available at the State Serum Institute, Copenhagen, Denmark.

In difficult cases repeated examinations of sputum should be made using Gram, acid-fast, and negative stains. In some instances the mouse virulence test on sputum may be useful as a check on pathogenicity and to provide organisms for typing. The mouse inhibits the growth of most bacteria in sputum except the pneumococcus, which can usually be isolated in pure culture from the peritoneum or heart blood of the injected animals. *Illness may become apparent as early as 5 to 8 hours.* A few strains of pneumococci have limited virulence for mice and produce a more protracted infection.

L. THERAPY

Although specific antisera and sulfa drugs were formerly used to treat pneumococcal pneumonia, penicillin preferably in high dosage has become the therapeutic agent of choice, and strains resistant to penicillin are rare. Other chemotherapeutic agents are less useful because resistant mutants commonly emerge. Alternative drugs include erythromycin, cephalosporin and chloramphenicol. Tetracyclines are not recommended because strains resistant to these drugs often develop.

Most cases of pneumococcal pneumonia respond dramatically to penicillin treatment, often with cessation of bacteremia within a few hours and the passing of crisis within 24 hours. Penicillin has reduced the overall mortality from about 30% to 5 to 10%. If fever persists in the face of penicillin treatment, it is probably "drug fever" due to hypersensitivity to penicillin rather than to the pneumonia itself. Despite successful antipneumococcal treatment, a few patients develop severe and often fatal secondary infections with organisms such as the staphylococcus.

Early treatment is of the utmost importance in order to stop the spread and growth of the organisms short of the point where toxemia is irreversible. Late treatment, especially of patients with bacteremia or meningitis, often fails despite rapid killing of the organisms, and it has been postulated that some unknown lethal substance or effect is produced by growing organisms which causes irreversible lethal injury. Obviously there is urgent need for better understanding of the cause of toxemia, for without such information there is no rational basis for improving therapy and little hope that the death rate can be reduced much below the present level of 5%.

M. RESERVOIRS OF INFECTION

The principal, if not the sole, natural reservoir of human infections is man.

N. CONTROL OF DISEASE TRANSMISSION

The possession of type-specific Ab by normal persons unquestionably conveys strong protection against infection and lessens the chance that the specific organism can establish the carrier state; individuals known to be immune to given types seldom become carriers of those types. Moreover, most carriers eventually become immune to the type(s) they carry and tend to free themselves of the carried organisms. Most normal individuals have low levels of specific Abs to a number of types, presumably as a result of carrier experience. The types of carried organisms change constantly, carried types being lost and new types acquired by chance through droplet transmission from other carriers. Convalescent patients seldom carry the causative type organism more than 2 to 3 weeks.

Acquisition of the carrier state by a nonimmune individual is apparently largely a matter of chance and commonly results from droplet transmission from other carriers rather than from patients. The carrier rate at any given time may range from 10 to 30%. It is estimated that 40 to 70% of individuals are carriers at some time during each year and that during the peak period in late winter and early spring carrier rates often exceed 40%. Except for type 3, the carrier rate of highly virulent types is normally low. When the carrier rate in a population increases and when the carried organisms are highly virulent, the risk of an epidemic of lobar pneumonia is great, especially if an epidemic of a predisposing respiratory infection, such as viral influenza, occurs at the same time. *Whether or not an individual develops lobar pneumonia hinges on whether he (or she) is carrying a virulent strain of organism to which he is not immune at a time when his resistance becomes depressed.*

Since infection is commonly endogenous, effective control of pneumococcal pneumonia cannot be achieved by measures designed to prevent transmission of the disease from patients to healthy individuals, and stringent isolation of patients is only warranted when highly susceptible contacts are at risk, such as individuals with congestive heart failure, other pulmonary disease, debilitating illnesses, or immunosuppressed states. Potential methods for control of pneumococcal pneumonia are (1) prevention of predisposing illnesses, such as by vaccination against viral influenza, (2) active immunization with polyvalent pneumococcal vaccine containing specific polysaccharides representing the types commonly responsible for the disease in the geographic area, and (3) chemoprophylaxis with penicillin.

Both influenza vaccines and polyvalent pneumococcal vaccines are useful measures for protecting certain highly susceptible individuals and for preventing or halting epidemics of lobar pneumonia. Although a polyvalent vaccine containing polysaccharides of 14 of the most common infecting serotypes was licensed in 1977 the overall protection rates reported have ranged from 30 to 80%. Hence the practical value of the vaccine is controversial and further tests are necessary; the vaccine is probably worthwhile for high-risk groups, such as the elderly, and persons whose antimicrobial defenses are seriously impaired, but whose humoral response is intact.

REFERENCES

Bolan, G, et al.: Pneumococcal vaccine efficacy in selected populations in the United States. Ann. Intern. Med. *104*:1, 1986.

Clemens, J.D., and Shapiro, E.D.: Resolving the pneumococcal vaccine controversy: Are there alternatives to randomized clinical trials? Rev. Infect. Dis. *6*:589, 1984.

Hosea, S.W., et al.: The critical role of complement in experimental pneumococcal sepsis. J. Infect. Dis. *142*:903, 1980.

Kearney, R., and Halliday, W.J.: Immunity and paralysis in mice: Serological and biological properties of two distinct antibodies to type III pneumococcal polysaccharide. Immunology *19*:551, 1970.

Kelly, R., and Greiff, D.: Toxicity of pneumococcal neuraminidase. Infect. Immun. *2*:115, 1970.

Klainer, A.S.: Pneumococcal pneumonia. Semin. Infect. Dis., *5*:3, 1983.

Kumar, A. et al.: Discrepancies in fluorescent antibody, counterimmunoelectrophoresis, and neufeld test for typing of *Streptococcus pneumoniae.* Diagn. Microbiol. Infect. Dis. *3*:509, 1985.

MacLeod, C.M.: Prevention of pneumococcal pneumonia by immunization with specific capsular polysaccharides. In *Infectious Agents And Host Reactions.* (ed. Mudd, S.), Philadelphia, W.B. Saunders Co., 1970.

Winkelstein, J.A., et al.: Activation of C3 via the alternative complement pathway results in fixation of C3b to the pneumococcal cell wall. J. Immunol. *124*:2502, 1980.

12

HAEMOPHILUS

The genus *Haemophilus* (family *Brucellaceae)* comprise small rod-shaped pleomorphic Gram-negative parasites that are nonsporing, nonmotile aerobes which can behave as facultative anaerobes. They parasitize man and various animals as their natural hosts and commonly colonize the URT and/or urogenital tract in such a high proportion of their hosts that most species are considered to represent part of the normal flora. However, in opportunistic situations some of these species can readily adapt as facultative pathogens to cause serious infections. Indeed there is an element of opportunism in essentially all infections produced by *Haemophilus sp.*

Some of the species are highly fastidious and require hemin and nicotinamide adenine dinucleotide (NAD) for growth. The important potential pathogens for man include *H. influenzae, H. parainfluenzae, H. aphrophilus, H. aegyptius,* and *H. ducreyi.** Several species are among the normal nasopharyngeal flora of man; one of these species, *H. hemolyticus,* produces beta hemolysis and may be confused with colonies of β-hemolytic streptococci. Several species are animal pathogens. For example, a venereal disease in rabbits is caused by *H. caniculus; H. suis* acts synergistically with swine influenza virus to produce influenza in swine. Properties of some species of *Haemophilus* are shown in Table 12–1.

Table 12–1. Properties of Some Species of the Genus *Haemophilus*

Species	*V Factor Requirement*	*X Factor Requirement*	*CO_2 Requirement*	*Hemolysis on Blood Agar*	*Reservoir*
H. influenzae (capsulated)	+	+	–	–	Man
H. influenzae (noncapsulated[x])	+	+	–	–	Man
H. aegyptius	+	+	–	–	Man
H. haemolyticus	+	+	–	+	Man
H. parainfluenzae	+	–	v[xx]	–	Man
H. aphrophilus	–	+	+	–	Man
H. ducreyi	–	+	+	+	Man (genital)
H. haemoglobinophilus (canis)	–	+	–	–	Dogs (genital)
H. caniculus	+	+		+	Rabbit (genital)
H. suis	+	+	–	–	Swine

[x]whereas capsulated strains comprise types a through f, noncapsulated strains are non-typable. X Factor = hemin; V Factor = NAD.
[xx]variable

**Haemophilus aegyptius* is clearly related to *H. influenzae* by DNA homology but is distinct in several other respects; some microbiologists classify it as a variant of *H. influenzae.*

HAEMOPHILUS INFLUENZAE

A. MEDICAL PERSPECTIVES

Haemophilus influenzae was first isolated by Pfeiffer in 1892 from several patients suffering from viral influenza; subsequently it was referred to as "Pfeiffer's bacillus." It is of special interest that *H. influenzae* was erroneously considered to be the cause of viral influenza until the 1918 influenza pandemic, which accounts for the species designation. In retrospect, *H. influenzae* probably plays a minor role in viral influenza as a secondary invader. The naming of *H. influenzae* was based on faulty data and has caused much confusion; accordingly, it is necessary to emphasize to students that *H. influenzae is not the cause of influenza.*

Shope's discovery (1931) that swine influenza is caused by a virus acting in concert with *H. suis* raised the possibility that synergism might also exist in the human influenza system. However, no data have been obtained to support this possibility.

The role of *H. influenzae* as an important cause of meningitis in children was described by Slawyk in 1899 and its importance in obstructive bronchitis in infants and children was recognized by Lemierre in 1936. *Haemophilus influenzae* continues to be a frequent cause of a variety of serious infections in young children, because of their poor immune response and the increasing emergence of drug-resistant organisms.

B. PHYSICAL AND CHEMICAL STRUCTURE

Haemophilus influenzae is typically a small, plump (0.5 × 0.7 × 1.3 μm) rod, but, in vivo, often displays pleomorphism and grows in short chains resembling filaments.

Principal surface Ags comprise a serologically variable endotoxic lipopolysaccharide (LPS) Ag, a large number of serologically distinct surface membrane (M) protein Ags that vary from strain to strain (strains fall into 8 protein Ag patterns) and a capsular polysaccharide which varies structurally and serologically from strain to strain. For example, the capsule of one group, type b, is composed of unique repeating polyribosylribitol phosphate (PRP) subunits. Capsules are present on some strains but not others. Capsulated strains of *H. influenzae* fall into 6 serologic types, a through f, which can be identified by the classical capsular swelling reaction with specific antisera or by other serologic reactions, such as the precipitin-ring reaction around colonies grown on agar media containing specific antiserum. It is of singular interest that some 95% of all cases of invasive disease are caused by strains of type b, the most virulent type. The capsular Ags of *H. influenzae* cross-react with Ags from many sources, including those of pneumococci, streptococci, coliforms and staphylococci. With respect to biochemical activities, strains of *H. influenzae* fall into 5 biotypes.

C. GENETICS

Virulent clinical isolates are unstable and readily lose their capsules on subculture, thus indicating that noncapsulated mutants quickly dominate. Accordingly, the rate of spontaneous S→R mutation is high. Experimentally, capsular types of organisms have been altered by DNA transfer utilizing the principle of transformation. For example, streptomycin-resistant strains occur naturally and can be induced with DNA. Plasmids are responsible for the transfer of antibiotic resistance.

D. EXTRACELLULAR PRODUCTS

Haemophilus influenzae does not produce any known soluble toxins. It does, however, liberate capsular polysaccharide during infection which usually accumulates in detectable quantities in the blood and spinal fluid during the acute phase of the disease; soluble capsular polysaccharide probably neutralizes the early Abs formed, thus blocking the opsonic action of the first Ab made and delaying the effector limb of specific acquired immunity.

E. CULTURE

Haemophilus influenzae is highly fastidious and requires a culture medium with a rich base, such as brain-heart infusion plus blood. For example, rabbit defibrinated blood (5 to 10%) is added to a hot (80° to 90°C) enriched melted agar medium which disrupts the RBCs and "chars" the hemoglobin giving the plates a chocolate appearance. The colonies on chocolate agar are small and colorless, they resemble small droplets of moisture. Whereas capsulated organisms form smooth glossy colonies, noncapsulated organisms form colonies which are granular or rough in appearance. Capsules can be demonstrated most readily by culture in broth for 6 to 8 hours or culture on Levinthal's medium for 6 to 12 hours, at which time colonies of capsulated organisms exhibit iridescence when viewed obliquely with transmitted light. *Haemophilus influenzae* requires the heat-stable hemin (X factor) and heat-labile NAD (V factor) that are supplied by the blood in "chocolate" agar. The heating of the blood in the preparation of chocolate agar liberates hemin from the RBCs and also inactivates enzymes in fresh blood that can degrade NAD.

When streak plates made on sheep blood agar are cross streaked with *Staph. aureus,* colonies of *H. influenzae* growing in the near vicinity of the staphylococcal growth are larger than those growing at a distance. This is referred to as the satellite phenomenon and is due to extra V factor supplied by the staphylococci; this property can be an aid in the identification of *H. influenzae.*

The pH optimum for growth is 7.6, and some strains grow better on primary isolation in the presence of 5 to 10% CO_2 than without CO_2; they are facultatively anaerobic and can grow at 35 to 40°C.

F. RESISTANCE TO PHYSICAL AND CHEMICAL AGENTS

Haemophilus influenzae is highly susceptible to desiccation and to the common disinfectants. It is readily killed by exposure at 55°C for 30 minutes.

G. EXPERIMENTAL MODELS

Strains of type b *H. influenzae* produce a fatal infection in mice if the inoculum is suspended in mucin and injected intraperitoneally (i.p.); large i.p. doses with mucin produce toxic death. Experimental bronchopneumonia and meningitis have been produced in monkeys. Capsulated type b strains are much more virulent for the young rat than their noncapsulated counterparts; one capsulated organism inoculated i.p. can cause death.

H. INFECTIONS IN MAN

Type b *H. influenzae* is the most virulent type; type f, is the second most virulent type. Type b causes 95% of the acute infections involving deep tissue invasion, including meningitis, primary segmental or lobar pneumonia, laryngotracheitis, epiglottitis, cellulitis and pyoarthrosis (purulent arthritis). Brief discussions of type b infections of the above categories are presented below. Most of the invasive infections occur in immunologically immature children under 3 years of age and are only observed in adults with severe systemic immunodeficiencies, especially humoral defects.

Haemophilus influenzae is carried in the URT of 20 to 50% of normal persons but only about 5% of carried organisms are capsulated, although this figure rises during epidemics. The source of infection is usually endogenous. Although noncapsulated and hence nontypable strains of *H. influenzae* are less virulent than capsulated strains, they often produce local infections in individuals whose systemic immune defenses are low or whose local defenses are severely compromised. Such local infections may be acute, subacute or chronic; they include otitis media and chronic sinusitis.

Noncapsulated strains may sometimes participate in chronic mixed infections that are based on underlying conditions which compromise local antimicrobial defenses. Such mixed infections include chronic bronchitis, bronchiectasis and chronic secondary bronchopneumonia.

Certain other species of *Haemophilus* that are not discussed in this chapter appear to play opportunistic roles similar to those of noncapsulated *H. influenzae.*

Haemophilus influenzae, even in the capsulated state, is of such low virulence that it fails to infect normal individuals with full immunologic competence. Clinical infections occur only when the individual's antimicrobial defenses are compromised at either the systemic level or the local level or both. It is of singular interest that as the decay of maternal Ab progresses, systemic susceptibility (usually evident at 3 months of age), as indicated by invasive disease, such as meningitis, varies inversely with age (Fig. 12–1, Sect. J). Sometimes a genetic component may be involved. For example, the incidence of meningitis in Alaskan Eskimo infants under 2 years of age may range as high as 7%.

When infants carrying protective maternal Ab are excluded, the age populations fall into 3 age-related susceptibility groups based on systemic susceptibility alone, namely a high susceptibility group (ages 3 months to 3 years), a moderately susceptible group (ages 3 years to 8 years) and a low susceptibility group above 8 years of age. *On the basis of local susceptibility resulting solely from impairment of local defenses* only 1 significant group exists, namely, persons over 8 years of age.

The above considerations explain in large measure the incidence and nature of infections in various age groups. Since systemic immune forces strongly oppose tissue invasion, type b infections in the *low susceptibility group* comprising older children and adults are few and tend to remain localized. In contrast, most type b infections in the high-susceptibility group (ages 3 months to 3 years) comprising individuals severely lacking in systemic immunocompetence are invasive.

Lastly, in the *moderately susceptible* group (age 3 years to 8 years) comprising children partially lacking in systemic immunocompetence, localized type b infections and invasive type b infections tend to be more nearly equal in number.

Needless-to-say noncapsulated strains rarely cause invasive infections even in the high susceptibility group, but can cause localized infections in this group or any group, providing local defenses are severely compromised.

Acute pharyngitis is common in young children and may persist for several days unless the patient is treated; the throat is inflamed and hyperemic.

Acute epiglottitis progresses rapidly with abscess formation and extreme edema; in a matter of hours it can lead to respiratory obstruction that demands emergency tracheotomy and is often accompanied by severe septicemia.

Acute laryngotracheitis is a rapidly progressing infection characterized by a croupy-cough and intense edema that causes occlusion of the trachea. Tracheotomy is often necessary to prevent death, which can occur within 24 hours after the onset of symptoms.

Acute secondary pneumonia may sometimes occur in children as a serious superimposed complication of pertussis or rubeola or as a sequela of measles or viral influenza.

Acute bronchiolitis in children is characterized by a persistent non-productive cough, wheezing and dyspnea. It is a serious, rapidly progressive infection that can lead to early death unless treated promptly.

Meningitis due to H. influenzae is the most common cause of bacterial meningitis in children between the ages of 3 months and 2 years but is progressively less frequent during the remaining years of childhood. About 8,000 cases occur in the U.S.A. annually. It rarely occurs before the age of 3 months because of the protection afforded by persisting maternal IgG Abs. The disease is usually preceded by *H. influenzae* pneumonia; meningitis may occasionally occur as the result of head injury, an event which may also occur in older age groups. The symptoms of *H. influenzae* meningitis are usually typical of classical bacterial meningitis, the exception being meningitis in young infants who may only exhibit swelling of the fontanels. The disease is rapidly progressive and is almost invariably fatal without early treatment. Consequently an early diagnosis is critical and therapy should be initiated as soon as a presumptive diagnosis can be made. Chemotherapy has reduced the mortality to about 5 to 10%. However, a number of patients saved by treatment exhibit residual brain damage.

Cellulitis usually occurs in periorbital tissues as an extension of acute pharyngitis. Bacteremia frequently develops and metastatic spread to distant sites is common.

Pyoarthrosis (purulent arthritis) of infants due to *H. influenzae* is usually associated with infection at another site and accompanying bacteremia.

Influenzal segmental or lobar pneumonia is principally a disease in young children (peak incidence 6 months to 1 year of age) that can lead to meningitis. It is usually preceded by coryza and can pursue an acute, subacute or, occasionally, chronic course with fever of about 100° to 103°C, cough, pleuritic pain and other pulmonary symptoms. The disease is occasionally preceded by one of the childhood viral diseases. Bacteremia is almost always present and granulocytosis of varying degrees occurs. If neutropenia eventuates, the disease is almost always fatal. Isolation of type b organisms from sputum or the detection of capsular Ag in body fluids provides a good presumptive diagnosis; isolation of organisms from blood, pleural fluid or lung aspirates provides a definitive diagnosis. Supportive treatment and chemotherapy (Sect. L) can reduce the mortality rate to 5% or less.

I. MECHANISMS OF PATHOGENESIS

The mechanisms responsible for injury in *H. influenzae* infections are obscure. No extracellular products of the organism are known to be injurious. Fever probably results from endogenous pyrogens of leukocytes. Although invasiveness may rest primarily on the antiphagocytic activity of the capsule, this activity may not be a simple act of physically repelling the phagocyte but, instead, of blocking or subverting the activities of C that promote phagocytosis and the killing of Gram-negative bacteria by Ab (Sect. J and *Fundamentals of Immunology* Chaps. 2 and 22).

The virulence of most strains of *H. influenzae* is low and there is an element of opportunism in the infections they produce. Although the capsule is probably a virulence factor for all capsular types of *H. influenzae*, it is unlikely that relative virulence among types depends solely on capsule size but additionally, involves structural differences in the polysaccharide that block or subvert the antibacterial activities of C in a fashion not unlike that of the pneumococcus. For example, invasive strains of type b are reported to be uniquely unable to *activate* the alternative C pathway and, in consequence, thus evade the nonspecific phagocytosis generated by such activation. In the case of type b organisms other virulence factors appear to reside in outer membrane protein subtype Ags; Abs specific for these Ags are also protective.

Adhesins that can effect adherence to host cells and colonization are probably present on some capsulated as well as noncapsulated strains of *H. influenzae*, but the nature of the adhesin(s) remains obscure.

Haemophilus influenzae type b LPS could be a virulence factor because it has been noted to paralyze the cilia on respiratory epithelium; in so doing, it probably lowers local antimicrobial defense by slowing bacterial clearance.

Since *H. influenzae* is the only *Haemophilus sp* that produces IgA1 protease, this enzyme could conceivably serve as a virulence factor by neutralizing the adherence blocking action of secretory IgA Ab. However, with respect to adherence to host cells and colonization on mucosae it has been reported that, whereas noncapsulated strains display strong adherence most capsulated strains do not; this could account for the observation that most localized infections are caused by noncapsulated strains as contrasted to capsulated strains, which are invasive.

J. MECHANISMS OF IMMUNITY

One of the most singular aspects of immunity to invasive *H. influenzae* infections, such as meningitis, is that the relation between age and susceptibility to infection correlates inversely with the bactericidal powers of blood (Fig. 12–1) and with the titer of anticapsular Abs in serum as well. Whereas the high resistance of the newborn is due to passively acquired maternal Abs, which perforce are of the IgG class, later resistance, which develops progressively after the age of 3 years, may involve protective Abs of other Ig classes as well.

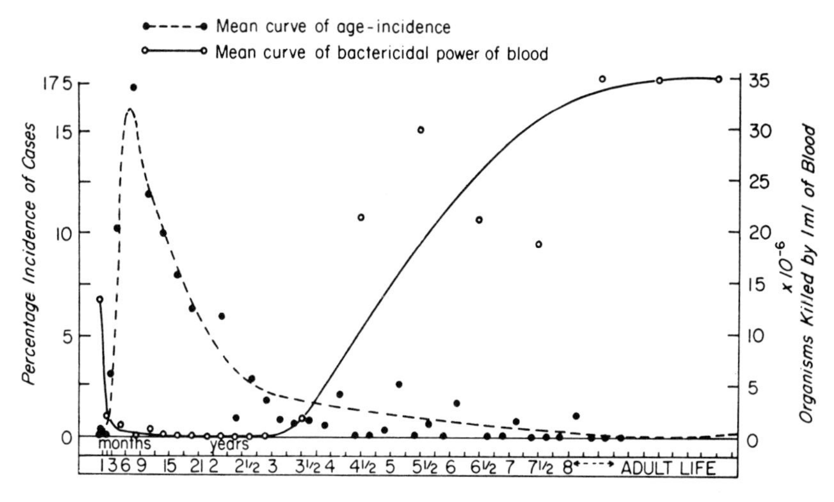

Fig. 12–1. The relation of age incidence of influenzal meningitis to the bactericidal power of human blood at different ages against a smooth meningeal strain of *H. influenzae.* (Adapted from Fothergill, L.D. and Wright, J., J. Immunol. *24*:281, 1933).

The bactericidal activity of blood after 3 years of age evidently rests on the combined effects of opsonization and C-dependent bacteriolysis, which potentially could involve either IgG or IgM Abs or both.

It has also been observed that in the resistant-age group above 8 years of age there is an inverse correlation between age and the anticapsular Ab response in patients convalescing from type b infections and in subjects vaccinated with PRP (Sect. B).

Another observation is that in a certain few individuals of the resistant-age group (persons over 8 years of age) who lack anti-capsular type b Abs, resistance correlates directly with the level of Abs to somatic Ags.

Acquired immunity to H. influenzae is strictly humoral in nature and may be effected principally, but not entirely by anticapsular Abs. An important defense role for Abs against somatic Ags is indicated by the observation that such Abs from patients are highly protective in animal models. Anticapsular Abs can be generated by either infecting organisms or organisms of the normal flora, be they of the same strain or alternatively, organisms of other strains, species or genera that produce cross-reacting Ags. *Indeed it is probable that the acquisition of resistance with age is due largely to cross-reacting Ags of other carried organisms rather than to type b carried organisms;* the type b carriage rate is only about 5%.

The effector mechanisms of acquired immunity to invasive *H. influenzae* infections are probably multiple, complex and ill-understood. Potentially they include Abs of various Ig classes and differing specificities that may act in various ways, either independent of or dependent on C. For example they could block adherence of organisms to host cells or serve as opsonins or bactericidins. Since type b organisms are the most common cause of serious infections and capsular Abs are the most protective, the present discussion will be centered on the mechanisms by which type b anticapsular Abs of different Ig classes may act. Although it is conceded that opsonic activity of anticapsular IgG Abs is a major force in defense, the opsonic role of IgM Abs, which perforce require C to act, appears to be weak. The defense role played by these 2 classes of Ab acting directly as lysins and cidins is even more uncertain. Although C-dependent bacteriolysis can be demonstrated in cell-free in vitro systems, doubts that bacteriolysis occurs in vivo have been expressed. Influenzal LPS type b comprises 3 antigenic specificities but the extent to which anti-LPS Abs may be protective is not known. Whether adhesins exist that generate protective anti-adhesin Abs is also not known. Antibodies specific for outer membrane proteins can protect animals against experimental infection. Other potentially protective Abs comprise Abs specific for the antigenic types of IgAl proteases.

K. LABORATORY DIAGNOSIS

A *presumptive diagnosis* of infection due to *H. influenzae* is made on the basis of clinical findings and examination of specimens. The most useful specimens include blood, pleural fluid, urine, cerebrospinal fluid (CSF) and aspirates from lesions. Because carriage rates of *H. influenzae* are high, tests on throat swabs and sputum specimens are not significant. Smears should be stained by Gram's method using carbol-fuchsin for counterstaining since the organism has some tendency to retain the Gram-stain.

Capsular swelling tests provide serologic confirmation of cultural identification. The capsular swelling test, in which antisera specific for capsular polysaccharide are used, is helpful for identifying *H. influenzae* in specimens such as spinal fluid, sputum and exudate. Refined tests for specific type b *H. influenzae* Ag, which is often shed into body fluids in detectable amounts during infection, also promises to be a valuable adjunct for early definitive diagnosis of most *H. influenzae* infections. Present tests for specific type b capsular Ag include counterimmunoelectrophoresis, latex particle agglutination, and the ELISA procedure. The preparation of purified Ags, such as PRP, and Abs such as anti-PRP monoclonal Ab, should make these tests highly sensitive and reliable.

Culture should be carried out on enriched "chocolate" agar or a comparable medium. *Haemophilus influenzae* can be differentiated from related species on the basis of lack of hemolysis on blood agar plates and its CO_2, X and V factor requirements (Table 12–1). Demonstration of the satellite phenomenon on blood agar may be helpful (Sect. E).

L. THERAPY

Chemotherapy of *H. influenzae* infections is becoming progressively more difficult because of the rapid emergence of drug-resistant organisms, including strains carrying either chromosomally mediated or plasmid-mediated resistance. Depending on geographic location, the incidence of ampicillin-resistant strains has been reported to vary from 5 to 55%. Strains resistant to chloramphenicol and tetracycline have also been reported.

Chemotherapy of invasive *H. influenzae* infections varies depending on the site(s) and nature of infection. Some of the various antimicrobial agents used, either singly or in various combinations include penicillin G, ampicillin, amoxicillin, methicillin, sulfonamides, erythromycin, cefamandole, chloramphenicol, gentamicin, tetracycline, streptomycin, trimethoprim-sulfamethoxazole and the newer cephalosporins. *Treatment should be initiated early* and altered appropriately as soon as in vitro sensitivity tests are completed. Chloramphenicol or a combination of chloramphenicol and ampicillin is the treatment of choice for systemic infections.

It is hoped that therapeutically useful antisera may soon be produced (Sect. N).

M. RESERVOIRS OF INFECTION

Man is the only natural host and reservoir of *H. influenzae.*

N. CONTROL OF DISEASE TRANSMISSION

Spread of *H. influenzae* infection is by droplet transmission from infected individuals or carriers to susceptible persons. There appears to be an increasing number of adults who lack protective Abs specific for *H. influenzae,* indicating a lack of exposure or failure of response to the carried organism or cross-reacting relatives; this could be the result of the widespread use of antibiotics.

Because of the increasing incidence of type b infections, much effort is being devoted to the production of protective vaccines and antisera. However, trials with a purified type b PRP vaccine have been disappointing because most infants in the high-risk group under 18 months of age do not form Abs against PRP and are not protected against systemic infection. Whether the unresponsiveness of this age group is due solely to immunologic immaturity or whether other factors are involved, such as immunologic tolerance, is not known. The addition of an adjuvant to the PRP vaccine could conceivably overcome unresponsiveness. One promising vaccine is a type b polysaccharide-protein complex designed to enhance the anti-PRP response. Other possible vaccines include, membrane protein Ags, alone or combined with adjuvant and PRP conjugated with diphtheria toxoid.

Since anti-PRP Abs can confer passive protection against *H. influenzae* type b, they may be of both therapeutic and prophylactic value for protecting infants or adults whose antimicrobial defenses are compromised. The need for passive protection is increased because of the rapid emergence of drug-resistant strains of type b *H. influenzae* and the increasing incidence of type b infections. Human monoclonal anti-PRP Abs have been produced that protect rats against experimental sepsis with *H. influenzae* type b. Indeed the production of monoclonal Abs may be extended to involve other Ags; such Abs would be highly useful for identifying and purifying Ags as well as for diagnosis, prophylaxis and therapy.

HAEMOPHILUS AEGYPTIUS

Haemophilus aegyptius was first observed by Koch in smears from infected eyes of patients in Egypt (1883). In 1887, Weeks of New York isolated the microbe in pure culture; since then it has been known as the Koch-Weeks bacillus. *Haeomphilus aegyptius* resembles *H. influenzae* in many ways, including DNA relatedness and the mutual presence of some cross-reactive Ags as well as the growth requirement for X and V factors. However, the definitive work of Pittman and Davis clearly established that it should be classified as a separate species.

Haemophilus aegyptius is responsible for a highly communicable form of a purulent conjunctivitis commonly referred to as "pink eye." In contrast to *H. influenzae,* it is not virulent for mice even if suspended in mucin. It has been shown to produce typical conjunctivitis in human volunteers but fails to infect the common laboratory animals.

Laboratory diagnosis depends on culturing and identifying the organism. As a rule, topical sulfonamides are effective chemotherapeutic agents although ampicillin and chloramphenicol are the logical drugs of choice in severe infections.

HAEMOPHILUS PARAINFLUENZAE

Haemophilus parainfluenzae requires the V but not the X factor for its growth. It should probably be regarded as a member of the normal flora, which under special circumstances can cause subacute bacterial endocarditis. The organism may be involved in some primary respiratory infections, but is more often a secondary invader that plays a minor role in such infections. Some strains of *H. parainfluenzae* are capsulated.

HAEMOPHILUS APHROPHILUS

This organism is commonly present in the normal respiratory tract. It is an uncommon cause of primary pneumonia but may be involved as a secondary invader. In addition, it may produce bacterial endocarditis on rare occasions.

HAEMOPHILUS DUCREYI

Haemophilus ducreyi, the cause of chancroid or soft chancre was discovered by Aguste Ducrey in 1889. Chancroid is an acute, localized, chancre-like lesion found on the external genitalia; it is basically a painful ulcer with a ragged and irregular margin that is commonly tender and swollen. In addition, the inguinal nodes may be enlarged and painful and may undergo suppuration (buboes).

Haemophilus ducreyi is a short, plump, Gram-negative rod that may exhibit bipolar staining. In stained smears of exudates from lesions the organisms usually appear singly, in small clusters and in long "school of fish-like" lateral columns situated between inflammatory cells or shreds of mucus. Only occasional organisms are found intracellularly. *Haemophilus ducreyi* can be cultured with some difficulty on enriched blood agar incubated at a slightly reduced oxygen tension. Its DNA relatedness to other *Haemophilus* sp. is distant or absent.

Chancroid is more frequent in the tropics than elsewhere and is more common in Blacks than in Caucasians. It occurs predominantly in female prostitutes and males. The organism is usually transmitted by sexual intercourse; it can be acquired from sexual partners who have no clinical evidence of an active infection. About 2 to 5 days after contact with an infected partner, a localized ulcer appears at the site of initial infection. Mild systemic symptoms may appear as a consequence of regional lymph node involvement, such as headache, fever, malaise or anorexia.

Laboratory diagnosis involves examining a stained smear of the exudate taken from under the edge of the lesion and demonstrating a preponderance of typical Gram-negative rods. Exudate or scrapings from lesions should be streaked on chocolate agar containing vancomycin to inhibit contaminants and incubated at 35°C under 5% CO_2; other media may also be useful. Colonies may not appear short of a week or more of incubation. The majority of patients with chancroid can mount a positive delayed-type skin reaction to killed *H. ducreyi* injected intradermally. However, the skin test is of limited value in diagnosis because specific sensitivity persists for years even after the disease has been cured. The nature of the lesions and mode of transmission demand that the possibility of syphilis be ruled out by appropriate serologic tests and darkfield examination. An agglutination test using specific rabbit antiserum is useful for identifying isolates.

The drug of choice for treating chancroid is sulfisoxazole (Gantrisin). Chloramphenicol and the tetracyclines are also effective chemotherapeutic agents. If buboes are large, they should be aspirated during treatment to prevent spontaneous rupture.

Experimental infections can be produced in monkeys, rabbits, and chick embryos.

REFERENCES

Ambrosino, D.M., et al.: Passive immunization against disease due to *Haemophilus influenzae* type b; concentrations of antibody to capsular polysaccharide in high-risk children. J. Infect. Dis. *153*:1, 1986.

Denny, F.W.: Effect of a toxin produced by *Haemophilus influenzae* on ciliated respiratory epithelium. J. Infect. Dis. *129*:93, 1974.

Eskola, J. et al.: Antibody levels achieved in infants by a course of *Haemophilus influenzae* type b polysaccharide/diphtheria toxoid conjugate vaccine. Lancet *1*:1184, 1985.

Hunter, K.W., Jr.: Human monoclonal antibacterial antibodies: Protection against *Haemophilus influenzae* type b by antibodies to the capsular polysaccharide. *New Horizons in Microbiology,* eds. Sanna, A. and Morace, G. Amsterdam, Elsevier, 1984.

Schreiber, J.R., et al.: Functional characterization of human IgG, IgM and IgA antibody directed to the capsule of *Haemophilus Influenzae* type b. J. Infect. Dis. *153*:8, 1986.

Smith, A.L.: *Haemophilus influenzae* Pneumonia. Semin. Infect. Dis. *5*:56, ed. Weinstein, L. and Fields, B.N., New York, Thieme-Stratton Inc., 1983.

13

BORDETELLA

The genus *Bordetella* (Family: *Brucellaceae*) contains 3 species of medical importance, namely, *Bord. pertussis, Bord. parapertussis* and the animal pathogen, *Bord. bronchiseptica. Bordetella parapertussis* is now considered to be a variant of *Bord. pertussis* (Table 13–1).

The organisms belonging to the genus *Bordetella* are small, ovoid, non-motile (except *Bord. bronchiseptica*), nonsporing, Gram-negative rod-shaped aerobes that often exhibit metachromatic bipolar staining; they commonly parasitize the respiratory tract of animals and man. Bordetella pertussis, the causative agent of *pertussis* (whooping cough), is the most important species; *Bord. parapertussis* produces a comparatively mild disease resembling whooping cough. *Bordetella bronchiseptica* produces occasional respiratory and wound infections in humans.

BORDETELLA PERTUSSIS

A. MEDICAL PERSPECTIVES

The first, classic description of pertussis was written by de Baillou in 1578 as follows:

> The lung is so irritated so that every attempt to expel that which is causing trouble, neither admits the air nor again easily expels it. The patient is seen to swell up, and as if strangled, holds his breath tightly in the middle of his throat . . . For they are without this troublesome coughing for the space of four or five hours at a time, then this paroxysm of coughing returns, now so severe that blood is expelled with force through the nose and through the mouth. Most frequently an upset of the stomach follows.

Bordetella pertussis was first isolated by Bordet and Gengou in 1906. Naturally acquired pertussis is a highly contagious disease; the attack rate among nonimmunized children exposed to the disease is approximately 90%. However, an effective vaccine is available and vaccination, antimicrobial drugs and

Table 13–1. Characteristics of *Bordetella pertussis, Bordetella parapertussis* and *Bordetella bronchiseptica*

Characteristics	*Bord. pertussis*	*Bord. parapertussis*	*Bord. bronchiseptica*
Motility	–	–	+
Strict aerobes	+	+	+
Catalase	d	+	+
Oxidase	+	–	+
Ability to catabolize carbohydrates	–	–	–
Glutamic decarboxylase activity	–	–	+
Urease	–	+	+
Sensitive to:			
Bacitracin, thionine, pyronin	–	–	–
Erythromycin, chloramphenicol, streptomycin	+	+	+
Safranine	+	+	–

d = 11 to 89% positive

From *Bergey's Manual of Systematic Bacteriology,* Vol. 1, Edited by Holt and Krieg, Williams & Wilkins, 1984.

proper management has reduced the mortality rate in the USA from 12.5 (1920) to 0.3 (1950) per 100,000. Mass immunization was initiated in 1940. Currently, about 2,000 cases of pertussis are reported in the USA each year. Deaths dropped from 1100 in 1950 to 12 in 1970 most of which were in infants under 1 year of age. Pertussis remains as a major cause of death in the developing world. For example, in Africa alone the estimated death toll for 1985 was 1.5 million.

B. PHYSICAL AND CHEMICAL STRUCTURE

Fresh virulent isolates of *Bord. pertussis* are capsulated and fimbriated (piliated). However, on culture they undergo phase changes with associated loss of capsules and fimbriae etc. (Sect. C). Several components of *Bord. pertussis* are of potential importance in pathogenesis and immunity.

Historically *Bord. pertussis* was observed to exert several unique effects on host tissues. Although the bacterial component(s) that cause these effects were not known, they were assigned hypothetical labels according to the effects produced, namely; a *histamine sensitizing factor (HSF)* which increases host sensitivity to histamine, *a pancreatic islet-activating protein (IAP)* that alters carbohydrate metabolism with resulting hypoglycemia, *a lymphocytosis promoting factor (LPF)* that elicits polyclonal T cell activation and traps lymphocytes in the circulation with resulting lymphocytosis (it is a potent T cell mitogen in vitro) and *a toxic hemagglutinin (HA)* that agglutinates RBCs in vitro and acts as an adhesin to bind the HSF-LPF-IAP-HA molecule to host cells. *It is now known that all of these 4 activities are due to a single toxic heat-stable envelope protein.* This multifactor protein possesses tissue toxicity and behaves as an exotoxin. In the mouse model specific Ab against the HSF-LPF-IAP-HA molecule blocks adherence to host cells and protects against lethal challenge with virulent organisms. It has potential as a protective immunogen for man.

Another surface protein with in vitro hemagglutinating activity is *fimbrial hemagglutinin* (FHA) that binds selectively to receptors on the cilia of respiratory epithelium. Anti-FHA Abs in immune serum are protective, evidently because they block adherence of infecting bacteria to host epithelium; local secretory IgA Abs should be even more protective.

Another toxic component is a cytoplasmic *heat-labile toxin (HLT)* that is only liberated on cell disruption and is a weakly antigenic protein that can produce dermonecrosis and death in mice. This toxin is alleged to be 10 times more abundant in phase I than in phase IV organisms. Anti-HLT Abs are not protective.

The cell walls of *Bord. pertussis* contain *heat-stable endotoxic* LPS that is similar to but distinct from the classical endotoxin of the *Enterobacteriaceae.* Anti-LPS Abs effect C-dependent bactericidal activity *in vitro* but do not protect mice against intracerebral challenge with virulent organisms.

The numerous serologically distinct heat-labile surface proteins of Kaufmann (K Ags) serve to distinguish between *Bord. pertussis, Bord. parapertussis* and *Bord. bronchiseptica* (Table 13–2). The K Ags bear no relation to the severity of disease and are not protective Ags. However, anti-K agglutinin titers above 1:320 correlate with immunity to pertussis and are valuable measures of such immunity. Although the protective Ag(s) for man remains to be identified, fimbrial hemagglutinin is the leading candidate.

Table 13–2. Differential and Common Antigens of Phase I Strains of the Species of the Genus Bordetella

Antigen	*Bord. pertussis*	*Bord. parapertussis*	*Bord. bronchiseptica*
K-antigens			
Common to genus:			
Factor 7	+	+	+
Species specific:			
Factor 1	+	−	−
Factor 14	−	+	−
Factor 12	−	−	+
O-antigen, common	+	+	+
Filamentous hemagglutinin	+		+
Toxins:			
HLT[a]	+	+	+
Pertussis[b]	+	−	−
LPS[c]	+	+	+

[a]HLT = heat-labile (56°C) dermonecrotic toxin
[b]Pertussis = Exotoxin (true toxin) (HSF-LPF-IAP-pertussigen), heat labile (80°C)
[c]LPS = lipopolysaccharide
From *Bergey's Manual of Systematic Bacteriology,* Vol. I, Edited by Holt and Krieg, Williams & Wilkins, 1984.

Smooth colonies, capsules, and virulence characterize phase I organisms but the nature of the capsule is uncertain and no particular Ag is known to characterize S-form organisms. Any role that capsules might play in pathogenesis and immunity remains to be revealed.

It has been reported that a heat-stable somatic protein, designated O Ag, is present in cells and culture fluids of *Bord. pertussis.* This Ag is not a protective Ag.

C. GENETICS

Fresh isolates from patients in the catarrhal stage of pertussis are in the smooth-colony, capsulated, fully virulent phase (phase I); phase IV is a term used for the rough avirulent form. Intermediate phases, designated II and III, are also recognized. If virulent strains are subcultured repeatedly, these phase variations regularly occur, demonstrating that a virulent capsulated strain readily mutates to noncapsulated variants that ultimately outgrow the capsulated parent strain in vitro. These in vitro phase changes, which represent a typical S→R mutational shift from virulence to avirulence, also occur in the course of human infection (Sect. H).

D. EXTRACELLULAR PRODUCTS

Bordetella pertussis produces an exotoxin with multiple activities (Sect. B). The organism also produces an adenylate cyclase that is reported to depress phagocyte function.

E. CULTURE

Bordetella pertussis is a relatively fastidious, strict aerobe that grows well on the glycerine-potato-blood agar originally devised by Bordet and Gengou. In contrast to *H. influenzae,* it does not require the X and V factors, hemin and nicotinamide nucleoside. The colonies are smooth, glistening and dome-shaped; they tend to be opaque or pearl-like and produce a fuzzy type of hemolysis. Phase I colonies are mucoid and tenacious.

F. RESISTANCE TO PHYSICAL AND CHEMICAL AGENTS

Bordetella species have greater than average susceptibility to the common disinfectants but resist drying for 5 days or more. They are readily killed by heating at 55°C for 30 minutes.

G. EXPERIMENTAL MODELS

Experimental infections have been produced in chimpanzees, monkeys, dogs, rabbits, rats, mice and ferrets. Intranasal instillation of *Bord. pertussis* in mice results in an interstititial pneumonia; intracerebral challenge is also used. Infections of chimpanzees produce a lymphocytosis and a characteristic paroxysmal cough simulating human pertussis.

H. INFECTIONS IN MAN

Pertussis can occur at any age, but is principally a disease of childhood; it can be divided into 3 stages, namely, the catarrhal stage, the paroxysmal stage and the convalescent stage. In unimmunized populations most individuals become infected during infancy and early childhood.

The *catarrhal stage,* which lasts about 7 to 14 days, begins about 10 to 15 days after contact with an infected person; it is characterized by coryza, sneezing, a mild cough and a low-grade fever. During this stage, phase I organisms predominate; as immunity is acquired the less virulent phase II and III organisms begin to emerge. Toward the end of this stage the leukocyte count commonly rises to about 20,000 to 30,000 (range 12,000 to 200,000) with >60% lymphocytes; lymphocyte levels exceeding 40,000 have been recorded.

The *paroxysmal stage* (patients with uncomplicated disease are afebrile) usually lasts between 1 to 6 weeks and is characterized by a violent, repetitive cough. The coughing commonly forces the air out of the lungs; this is followed by an inspiratory "whoop," due to the narrowed glottis and the presence of thick, mucoid sputum; vomiting commonly follows the whoop. Small localized areas of necrosis appear in the bronchial epithelium and patchy interstitial pneumonia usually develops. Paroxysmal cough is probably due in part to local irritation of the mucosa and in part to CNS effects of circulating toxin. The viscous, ropy, mucinous bronchial secretions are extremely difficult to expel and cause local obstruction leading to areas of emphysema and atelectasis.

Although the *convalescent stage* normally lasts for about 2 to 3 weeks, it may last for months with a gradual decline in the severity of coughing; true carriage does not follow recovery.

Bronchopneumonia, due to secondary infection, occurs in 1 to 10% of patients. The organisms most frequently responsible for secondary infection include *Strep. pneumoniae, Staph. aureus, H. influenzae* and *Strep. pyogenes;* infection due to *Strep. pneumoniae* is the major cause of death in pertussis.

Neurologic symptoms, which may appear at the peak of the paroxysmal stage, occur most frequently in infants and young children with bronchopneumonia. In a study of a series of 35 hospitalized patients, Byers and Rizzo reported that 17% of infants developed some permanent CNS damage. Schachter found that 27.5% of a series of 200 children with uncomplicated whooping cough exhibited subsequent retardation; even a higher percentage displayed personality changes. Other serious neurologic complications include convulsions, coma, paralysis, blindness and psychic disorders. About 50% of children who develop coma do not survive. Late sequelae are epilepsy, spastic paralysis and mental retardation.

I. MECHANISMS OF PATHOGENESIS

When considering mechanisms of pathogenesis, it is important to remember that factors that enable microbes to invade and produce disease do not necessarily serve as protective immunogens. One major factor contributing to the virulence of *Bord. pertussis* is evidently the adhesin, fimbrial hemagglutinin, which selectively attaches to receptors on the cilia of epithelial cells of the respiratory tract. The adherent organisms then grow and kill their ciliated targets, primarily by the local liberation of toxin(s). Some of the liberated toxin enters the circulation to produce systemic effects, such as lymphocytosis and CNS symptoms. *Adherence* is apparently a necessary stage in pathogenesis since specific Abs induced with purified FHA (Sect. B) can block adherence of organisms to cultured cells and can protect animals against either aerosol or intracerebral challenge with live organisms.

The multifactorial exotoxin molecule (Sect. B) is evidently largely responsible for the systemic alterations seen in pertussis. Purified exotoxin can elicit all of the activities inherent in the molecule, such as histamine sensitization and lymphocytosis. Moreover, Abs specific for the multifactorial molecule can prevent the systemic effects of purified Ag and can protect against challenge with live organisms as well.

The heat-stable LPS endotoxin of the cell wall may contribute to pathogenesis even though anti-LPS Abs are not protective in the mouse model. The roles which other bacterial components may play in pathogenesis are more controversial.

It has been proposed that some type of neurologic alteration caused by toxin may be responsible for the whoop because the paroxysmal cough persists long after the organisms have been eradicated. The observation that pertussis vaccine produces β-adrenergic blockade in mice, thus making the lung hyperreactive to injury, supports the neurologic hypothesis.

Bordetella pertussis possesses many potent biologically active components that could contribute to pathogenesis. However, the relative contributions of these substances remain to be elucidated.

J. MECHANISMS OF IMMUNITY

Active acquired immunity following pertussis is not permanent. However, the anamnestic response of waning immunity can attenuate subsequent infections to the point where they do not resemble whooping cough; some 20% of present-day cases are mild infections of this sort. Indeed, it is the consensus that repeated subclinical or mild infections probably act as boosters to maintain immunity, thus creating the false impression that immunity is life-long after only a single bout of whooping cough. It is noteworthy that, although passive immunity in newborn infants due to maternal Ab is comparatively weak, it can apparently attenuate the disease in young infants to the point where the symptoms consist of apnea and choking without whoop. The low level of immunity in neonates evidently reflects the low level of protective Abs in serum of adults in general. Immunity may wane to a low level with age, and, as a consequence, the disease may recur in a severe form in the aged.

The FHA Ag present in the cells of virulent phase I organisms is probably the most important immunogen. The heat-stable exotoxin is also a potential immunogen.

The anti-adhesin effect of sIgA Abs probably contributes to immunity. This is in accord with the observation that anti-FHA Abs in immune serum protect the mouse against aerosol as well as intracerebral challenge with live organisms; human convalescent serum is also protective.

K. LABORATORY DIAGNOSIS

Material for nasopharyngeal culture is obtained by employing a special nasal swab made from a flexible wire with a small cotton swab at one end. The swab is passed through the nose to the pharyngeal wall and held in place until the patient coughs. The swab is withdrawn, passed through a drop of penicillin or cephalosporin solution to inhibit contaminants and placed on a Bordet-Gengou agar plate. The drop is then streaked in the conventional manner. Characteristic colonies appear after 2 days' incubation at 37°C. An agar plate in which the patient coughs into an exposed agar plate (cough plate) is less reliable. Final identification can be made by employing a specific agglutination test with appropriate antiserum. Organisms are most abundant in the early phases of pertussis.

A direct identification can be made by staining the organisms in nasopharyngeal smears with specific fluorescein-labeled Ab.

L. THERAPY

Erythromycin is the drug of choice for the treatment of patients with severe pertussis; chloramphenicol and tetracyclines are also used. In most instances, the clinical response to antimicrobial treatment is slow and questionable. However, the bacterial phase of the disease can be arrested by antimicrobial therapy and convalescent shedding can be reduced from weeks to days. Persons with mild forms of whooping cough do not require specific treatment except to arrest shedding of organisms.

Hyperimmune human globulin injected early in the course of the illness can attenuate the disease and is recommended for infants under 2 years of age.

It is highly important to replace water and salt loss in patients who suffer severe and frequent vomiting. Prompt detection and treatment of complications, such as secondary bacterial infections of the lungs or middle ear, are of utmost urgency.

M. RESERVOIRS OF INFECTION

Man is the sole natural host and the only known reservoir of infection.

N. CONTROL OF DISEASE TRANSMISSION

Bordetella pertussis is endemic throughout the world; epidemics tend to occur in cycles ranging from 2 to 5 years when critical numbers of susceptible children accumulate in a population. *Long-term chronic carriers of Bord. per-*

tussis have not been observed. It is difficult to explain the maintenance of this pathogen in the population based only on clinically recognized infections. Presumably, many mild cases of the disease, especially in adults, go unrecognized and serve to cause nursery and family outbreaks; they also maintain the organism in the population.

Active immunization of all infants and children with the triple vaccine DPT (diphtheria-pertussis-tetanus) is 90% effective against pertussis. Killed phase I cells of *Bord. pertussis* are included in this vaccine. Immunization should be started in children at 2 months of age, followed by a booster dose of vaccine at 1 year of age; subsequent booster doses should be given at 3 and 5 years of age. The degree of protection tends to parallel the serum agglutinin titers at the time of exposure. Vaccination presents a minor risk because of the possibility of severe or fatal *vaccine encephalopathy.* Pittman has estimated that the incidence in the USA is about 1 fatal reaction in 5 to 10 million doses of vaccine administered. The production of a safer and more effective vaccine is urgently needed. *It is of singular interest that Bord. pertussis has marked adjuvant activity and enhances the Ab response to the other Ags in DPT.* Chemoprophylaxis with erythromycin may be useful for preventing infection in susceptible contacts exposed to patients.

BORDETELLA PARAPERTUSSIS

This Gram-negative rod resembles *Bord. pertussis* except that it develops larger colonies. *Bordetella parapertussis* is the cause of a mild whooping-cough-like disease that, as a rule, does not require specific therapy. The organism possesses somatic Ags that cross-react with Abs of *Bord. pertussis* and *Bord. bronchiseptica.* Apparently DPT vaccine does not immunize against either *Bord. parapertussis* or *Bord. bronchiseptica.*

BORDETELLA BRONCHISEPTICA

This organism is a common cause of sporadic and epidemic pulmonary infections in rabbits and guinea pigs. Colonies of *Bord. bronchiseptica* are small, round and glistening. In contrast to the other species of *Bordetella,* this organism is motile.

Based on cross-immunity experiments, *Bord. bronchiseptica* is more closely related to *Bord. pertussis* and to *Bord. parapertussis* than the latter are related to each other. *Bordetella bronchiseptica* produces a whooping-cough-like disease in children on rare occasions.

REFERENCES

Additional standards: Pertussis vaccine. Federal Register *33*:1818, 1968.

Bordet, J. and Gengou, O.: Le microbe de la coqueluche. Ann. Inst. Pasteur *20*:731, 1906.

Byers, R.K. and Rizzo, N.D.: A follow-up study of pertussis in infancy. N. Engl. J. Med. *242*:887, 1950.

de Baillou, G.: Whooping cough. In *Classic Descriptions of Disease,* 2nd Ed. ed. R.H. Major. Springfield, Charles C Thomas, 1939.

Hewlett, E.L.: Selective primary health care: Strategies for control of disease in the developing world. XVIII. Pertussis and Diphtheria. Rev. Infect. Dis. *7*:426, 1985.

Holt, L.B: The pathology and immunology of *Bordetella pertussis* infection. J. Med. Microbiol. *5*:407, 1972.
Munoz, J., et al.: Mouse-protecting and histamine-sensitizing activities of pertussigen and fimbrial hemagglutinin from *Bordetella pertussis.* Infect. Immun. *32*:243, 1981.
Muse, K.E., et al.: Scanning electron microscopic study of hamster tracheal organ cultures infected with *Bordetella pertussis.* J. Infect. Dis. *136*:768, 1977.
Nagel, J.: Isolation from *Bordetella pertussis* of protective antigen free from toxic activity and histamine sensitizing factor. Nature *214*:96, 1967.
Roop, II, R.M. et al.: Virulence factors of *Bordetella bronchiseptica* associated with the production of infectious atrophic rhinitis and pneumonia in experimentally infected neonatal swine. Infect. Immun. *55*:217, 1987.
Schachter, M.: Le pronostoc neuropsychologique des enfants d'une coqueluche precoce non compliquee. Proxis *42*:464, 1953.
Tuomanen, E.: Piracy of adhesins: Attachment of superinfecting pathogens to respiratory cilia by secreted adhesins of *Bordetella pertussis.* Infect. Immun. *54*:905, 1986.

14

NEISSERIA

Neisseria meningitidis

Neisseria gonorrhoeae

The genus *Neisseria* comprises a number of aerobic and facultatively anaerobic nonpathogenic and pathogenic species that parasitize animals and man. Several nonpathogenic species are among the normal flora of mucosae, principally in the URT of man, including *N. sicca, N. subflava, N. flavescens, N. mucosa* and *N. lactamica.*

Only two species of the genus *Neisseria* are important pathogens of man, their only natural host, *N. meningitidis* and *N. gonorrhoeae.*

On rare occasion certain neisseriae, especially *Branhamella catarrhalis* and *N. subflava,* produce opportunistic infections, including meningitis and septicemia; *B. catarrhalis* and *N. sicca* have been reported to cause endocarditis. Other genera of the family Neisseriaceae include: *Acinetobacter, Branhamella, Moraxella,* and *Kingella.*

NEISSERIA MENINGITIDIS

A. MEDICAL PERSPECTIVES

Few microbes produce a higher mortality than *N. meningitidis* and none kills more quickly; death often occurs, within a few hours after the first symptoms appear.

Cerebrospinal meningitis is the most common of the serious infections produced by *N. meningitidis,* and explosive epidemics of the disease have probably occurred for untold centuries. Despite its neurologic nature and epidemicity, meningococcal meningitis was not recognized as a clinical entity until Vieusseux presented his classic description of the disease in 1805. The causative organism was first isolated by Weichselbaum in 1887.

Epidemics of meningococcal meningitis occur most frequently under circumstances in which large numbers of individuals, especially from different geographic regions, are brought together under crowded and stressful conditions. For example, major epidemics have occurred repeatedly among recruits in military camps of the USA during this century. Epidemics are especially common and mortality rates are unusually high in certain geographic areas, such as the "meningitis belt" of Northern Africa, where over half a million cases were reported between 1939 and 1962. In the Brazillian epidemic of 1971–1974 the attack rate reached 65 per 100,000 and involved a serogroup shift from group C to A. During epidemics the mortality rate in untreated cases averages 70% and ranges from 20 to 90%. Carriers account for spread of the disease, and mounting carrier rates to high values often presage the onset of epidemics. Both case and carrier rates are highest in males; carriage lasts for days to months.

Treatment of meningococcal disease with specific antiserum, which was introduced by Flexner in 1913, reduced the average mortality from 70 to about 50%. The introduction of sulfonamide treatment some 50 years ago was a marked advance in therapy and further reduced the mortality to about 10%. Initially, sulfonamides were also highly effective for eliminating carriers and for mass prophylaxis to arrest epidemics. However, after a time increasing numbers of "carried" and "case" strains of sulfonamide-resistant meningococci arose. The widespread occurrence and persistence of sulfa-resistant strains has seriously limited the usefulness of sulfadiazine for therapy and carrier control.

Although penicillin has provided a highly effective substitute drug for therapy, it is relatively ineffective for eliminating the carrier state, probably because the concentrations of the antibiotic achieved in the tears and mucous secretions that bathe the nasopharynx are too low.

Whereas meningococcal infections continue to be of public health importance, it is unlikely that they will ever be as serious a threat as they have been in the past, and there is high hope that better control may soon be achieved through the use of meningococcal immunizing agents and new drugs for controlling carriers.

B. PHYSICAL AND CHEMICAL STRUCTURE

Neisseria meningitidis is a nonsporing, nonmotile, Gram-negative, piliated, oblong diplococcus arranged with its long axis parallel to the line of division. The opposing surfaces of the two organisms are flattened and slightly concave which gives the cells a kidney-bean shape; its dimensions are approximately 0.8 × 0.6 μm.

The organisms contain metachromatic granules and stain readily with the usual stains; in aging, autolyzing cultures they tend to become swollen and distorted and may fail to stain.

Although in vivo grown organisms of groups A and C present capsules demonstrable by the capsular swelling test, group D organisms usually lack an easily demonstrable capsule. Capsulation is irregular in cultured group B organisms, apparently because the capsular substance is readily depolymerized by enzymes of the organism.

The medically important Ags of *N. meningitidis* consist of a capsular polysaccharide in close association with a macromolecular complex in the outer membrane which is comprised of endotoxic lipopolysaccharide (LPS) and membrane proteins. Whereas, the capsular polysaccharide is the basis for 9 *serogroups,* namely A, B, C, D, 29E, X, Y, Z, and W135, the membrane protein is the basis for some 18 *serotypes* that are heterogeneously distributed among the various *serogroups.* Some strains of *N. meningitidis* are not groupable. The endotoxin of the membrane complex is antigenic. A number of the Ags of *N. meningitidis,* including somatic nucleoproteins and carbohydrates, cross-react with various species of *Neisseria* and other bacteria. Serogrouping and serotyping are valuable for studies on pathogenesis, epidemiology and vaccine production. It is notable that the ratios of organisms of various serogroups and serotypes constantly shift within human populations, both with respect to carriage and disease production.

Serogroup and serotype Ags engender IgM, IgG, and IgA Abs; whereas all can act as opsonins, only IgM and IgG Abs possess C-dependent bactericidal activity. The protective interrelationships of these Abs are not clear; allegedly IgA Abs may sometimes block the bactericidal activities of Abs of other classes. Severe complement deficiencies can predispose to infection.

Apparently group B polysaccharide is unique in being a sialic acid polymer consisting principally of n-acetyl-neuraminic acid. Whereas polysaccharides of groups A and C purified with cationic detergent are highly immunogenic for man, group B capsular material is a poor immunogen that is identical to the K1 capsular polysaccharide of *E. coli.*

The capsules of meningococci are antiphagocytic and the capsule is an important factor in virulence; its antiphagocytic effects are neutralized by specific Ab which serves to opsonize the organism.

Marked chemical and biologic heterogeneity exists among various endotoxins and there is some evidence that meningococcal endotoxin may be more complex than the endotoxins of the rod-shaped Gram-negative bacteria. The lipid A moiety, which appears to represent the active component of LPS, is antigenic.

C. GENETICS

Repeated subculture of smooth (S) capsulated strains of *N. meningitidis* on artificial media selects for rough (R) noncapsulated avirulent mutants which soon dominate in the culture. Inoculation of mice with rough organisms artificially coated with mucin reverses the selective process and virulent organisms can be recovered from the animals.

Meningococci readily develop resistance to drugs, including sulfonamides and streptomycin. Transformation of meningococci with DNA can be accomplished.

D. EXTRACELLULAR PRODUCTS

A number of extracellular enzymes are produced but no secreted extracellular product is known to be important in pathogenesis. The organism produces an enzyme, indophenyl oxidase, which is the basis of the oxidase test used for detecting colonies of neisseriae. Meningococci produce a heat-labile (65°C for 30 minutes) enzyme(s) which causes autolysis in aging cultures. Although they are not extracellular products, surface substances shed to the exterior during growth include capsular polysaccharides and outer membrane components. The latter are exteriorized when the outer membrane forms and sheds vesicles from the bacterial surface.

E. CULTURE

On primary isolation *N. meningitidis* is fastidious in its nutritional requirements; it grows well on chocolate agar, but is highly sensitive to trace metals and fatty acids present in most culture media. Consequently agents, such as serum or starch, are often added to culture media to bind and neutralize these toxic components. Stock strains tend to be nutritionally less fastidious than freshly isolated strains and many of them can grow on synthetic media. The organism prefers aerobic conditions and grows poorly, if at all, under strict anaerobic conditions. Most strains grow best under added CO_2 for primary isolation and proliferate best at pH values of 7.4 to 7.6. The organism grows over a temperature range of 25 to 42°C (optimum 37°C); however, on primary isolation it is especially sensitive to sudden reductions in temperature.

For primary isolation chocolate agar or a modified Thayer-Martin selective medium called "Transgrow medium" can be used. The latter serves as both a transport and growth medium for both *N. meningitidis* and *N. gonorrhoeae.* It permits transport at ambient temperatures for 2 to 4 days or longer, especially if incubation is carried out overnight before transport. The medium selects against common contaminants, including nonpathogenic neisseriae. Following growth on laboratory media, the organisms tend to die rapidly; hence continued maintenance in culture requires frequent transfer or use of a "maintenance medium" such as the medium of Levine and Thomas, which allows survival for 1 to 2 months.*

On blood agar the colonies are nonhemolytic, elevated, moist, smooth and bluish-gray in color; they range from 1 to 5 mm in diameter. Rough, smooth and mucoid colonial types occur depending on the presence and amount of capsular polysaccharide produced, the rough colonies being formed by organisms lacking capsules.

Like all neisseriae, the colonies of *N. meningitidis* show a positive oxidase reaction with the oxidase reagents dimethyl- or tetramethyl-paraphenylene diamine. The treated colonies first turn pink and then black due to oxidation of the dye; organisms in treated colonies die after a few minutes.

F. RESISTANCE TO PHYSICAL AND CHEMICAL AGENTS

Neisseria meningitidis is notoriously susceptible to destructive physical and chemical agents. It is highly sensitive to disinfectants and many chemotherapeutic drugs and is easily killed by desiccation and heat. It dies within a few days at 0°C but can be preserved at extremely low temperatures.

Although the neisseriae have the major characteristics of most Gram-negative bacteria, i.e. carry endotoxin in their cell walls and are susceptible to destruction by Ab and C, they possess a characteristic in common with many Gram-positive bacteria, namely susceptibility to penicillin.

G. EXPERIMENTAL MODELS

No experimental models closely resembling disease in man have been developed. The 15-day-old chick embryo is perhaps the best model; irrespective of the site of inoculation the organisms produce a bacteremia followed by elective localization in the lungs and CNS. The mouse model is commonly used but is highly artificial; fatal infection can be produced, providing the inoculum is suspended in mucin. Apparently the mucin serves to protect the organisms against phagocytosis until they have had time to synthesize capsular polysaccharide sufficient to afford maximum protection against phagocytosis.

H. INFECTIONS IN MAN

Although *N. meningitidis* often colonizes the nasopharynx of man, *it seldom, if ever, causes clinically evident nasopharyngeal infection* and rarely causes primary pneumonia or primary pericarditis.

Between sporadic epidemics the carrier rate ranges from 5% or less to about 20%. The disease is most common in late winter and early spring; in civilian populations large-scale epidemic disease peaks in cycles of about 10 years. Because of the decay of maternally derived protective Ab, morbidity and mortality rates are highest among infants between the ages of 4 months to 1 year. Mortality rates are also high in extreme old age.

Infection is usually acquired as the result of recent close association with a healthy carrier and results primarily from droplet infection. Infection occurs almost invariably within closed populations, such as the family and military groups, and can affect most any organ. Clinical disease appears about 2 to 3

*It is important that laboratory technicians keep abreast of modifications in culture media and techniques.

days to a week after exposure. Apparently, and for unknown reasons, colonizing organisms occasionally transgress the mucosa of the nasopharynx to reach the bloodstream and/or the CNS to produce disease. Although chilling and fatigue have been presumed to predispose to infection, this is uncertain, and the possible routes and mechanisms responsible for invasion are poorly understood. Patients often give a history of a preceding sore throat but whether this ever represents a meningococcal pharyngitis is debatable. Instead, such sore throats may represent virus infections that are predisposing. Overcrowding, especially in sleeping quarters, contributes importantly to the chances of developing meningococcal infection; whether this rests on the "quantity of exposure" is not known. Whereas most investigators believe that in case of anatomic defects, infecting organisms can sometimes pass directly from the nasopharynx to the CNS without entering the bloodstream, it is the consensus that in the absence of anatomic defects invasion of organisms from the nasopharynx is commonly initiated via lymphatics and moreover, that the resulting bacteremia, however transient it may be, is the usual source of organisms producing meningitis.

Bacteremia may be acute or may last for weeks to months; it may be progressive or intermittent and complications may or may not develop. The reasons for such wide variations in the course of meningococcal disease are obscure but probably depend to a substantial degree on differences in the development of specific acquired immunity. Since serum levels of group specific Ag appear to reflect both the severity of meningococcemia and the stage and intensity of the immune response, future studies on serum Ag may throw light on this problem. In a small percentage of cases the initial mild bacteremia is rapidly followed by the development of fulminating meningococcemia with attending widespread thromboembolic metastatic lesions that sometimes require skin transplants and digit amputation.

Fulminating meningococcemia begins abruptly with chills, headache and dizziness and often terminates in early circulatory collapse and death in some 70% of untreated cases. The metastatic lesions show perivascular infiltration with leukocytes and are often hemorrhagic; in the skin they are present as large irregular geometrical purpuric spots which are especially striking (Fig. 14–1) and commonly yield positive smears and cultivable organisms. The tendency to hemorrhage is presumed to be due to meningococcal endotoxin which causes excessive activation of the clotting mechanism with resulting depletion of blood clotting factors (consumption coagulopathy). In severe cases, especially in infants, thrombocytopenia and extensive hemorrhage into the adrenal glands may occur with associated adrenal insufficiency (Waterhouse-Friderichsen syndrome) and death. Hemorrhage into the pituitary may also occur and contribute to early death.

Mild acute or subacute meningococcemia, a frequent type of infection, is highly responsive to chemotherapy. It may have an insidious onset but is usually heralded by the sudden development of irregular or intermittent fever, leukocytosis, chills and malaise. Other initial symptoms may include skin rash, acute mono- or poly-arthritis, nausea and conjunctivitis. The rash may be mild or severe and may resemble the "rose spots" of typhoid fever. Smears of capillary blood may sometimes reveal intracellular Gram-negative diplococci, as do smears of exudates from skin lesions. Skin petechiae may form which de-

Fig. 14–1. Fatal meningococcemia in an infant. (Courtesy Dr. George Ray.)

velop into ulcers. Some patients recover spontaneously after weeks or months, whereas others develop sequelae or complications, such as meningitis.

On rare occasion a chronic form of meningococcemia develops which often leads to purulent synovitis or arthritis. Culture of organisms from the blood is difficult in such patients; if left untreated, they may eventually develop other complications, such as meningitis, endocarditis or pericarditis.

Although *meningitis, the most frequent serious meningococcal disease,* is commonly thought to represent a complication of bacteremia, it is certain that the organisms occasionally invade the meninges directly through anatomic defects, such as those produced by head injury. The disease begins with symptoms reflecting meningeal involvement and generalized infection or a combination of both. The histopathology consists of a fibrinopurulent type of inflammation which may block the circulation of CSF and compress the brain. Various signs of inflammation of the meninges develop and coma may follow within a few hours. Many complications of meningitis arise, including cerebral thrombosis, brain abscess, transient or permanent paralysis, hydrocephalus and altered cerebration. Relapses are frequent and may occur several times during a period of 1 to 2 years. The mortality ranges from 2 to 6% despite chemotherapy.

Differential diagnosis of meningococcal infection should include consideration of a number of diseases such as typhus, typhoid fever, subacute bacterial endocarditis, rheumatic fever, diabetic coma, and uremia.

Although attendants seldom contract infection from patients, it is common practice to regulate the patient's contacts until 48 hours of chemotherapy have passed and to monitor known contacts for evidence of carriage and developing illness. Attempts should be made to eliminate convalescent carriage of organisms before discharging patients (Sect. N).

I. MECHANISMS OF PATHOGENESIS

Epidemiologic studies have established that "epidemic strains" of meningococci exist. Presumably they are more invasive and "toxic" than other strains, but the factors responsible for these attributes have not been fully elucidated. *Neisseria meningitidis* is usually destroyed promptly after engulfment by professional phagocytes but can adhere to, enter, persist within, and kill endothelial cells by shedding endotoxin.

The basis of the profound toxemia of meningococcal infections may be more complex than is generally believed. A protein (P substance), which is toxic for animals, has been reported but its role in the toxemia of infections in either animals or man has not been established. Endotoxin probably plays the major role in pathogenesis by producing vascular damage, consumption coagulopathy and circulatory collapse through mechanisms common to those of the Shwartzman reaction. It has been reported that endotoxin produces lesions in animals similar to those attending natural disease in man and that anti-endotoxin protects animals against lethal doses of either endotoxin or living meningococci. Cultured isolates from patients with invasive disease are reported to liberate more endotoxin than strains isolated from carriers.

The capsule is an important determinant of virulence and anticapsular Ab affords group-specific protection against infection by acting as an opsonin as well as a bactericidin.

The ability to form iron-binding "siderophores" for scavenging iron may be another attribute of virulence; for example, pathogenic but not nonpathogenic *Neisseria* species can scavenge iron from transferrin to support in vivo growth.

J. MECHANISMS OF IMMUNITY

Immunity to meningococci is complex and ill-understood. However, herd resistance is relatively high due to natural protective Abs generated by carriage. Acquired immunity is a major force of resistance and the titer of serum bactericidal activity is strongly correlated with resistance to infection. For example, maternal IgG Abs provide substantial humoral immunity to young infants until 4 to 6 months of age when they become more susceptible than at any time during life; indeed most childhood cases of meningitis occur between 4 months and 1 year of age and about half of all cases takes place before the age of 5 years. It is probable that the course of infection is largely determined by developing forces of specific acquired immunity. Antibodies with agglutinating, precipitating, complement fixing, opsonic and bactericidal activities arise early (5 to 15 days) in the course of infection. Most of these Abs are directed against surface Ags, including the *capsular immunogens,* which engender specific *serogroup* immunity. The relative role that *serotype* Abs may play in immunity has not been fully elucidated; however, in the case of group B organisms they totally account for the bactericidal Abs in serum. Antibody-containing sera afford specific protection to both mice and men. All of the various Abs and Ab activities that may convey protection to either mouse or man are by no means clear.

Although IgA Abs may block bacteriolysis by IgM and IgG Abs they may protect by opsonizing organisms for ingestion by monocytes. These paradoxical activities of IgA Abs are the basis for one hypothesis to account for epidemics, namely, that strains of *E. coli* carrying Ags that cross-react with *Neisseria* cyclicly parasitize the gut and stimulate cross-reactive IgA Abs that block IgG and IgM anti-meningococcal serogroup Abs. The observation that the production of the IgA-splitting meningococcal protease correlates with strain virulence adds to the enigma of interpreting protective immunity.

Whether anti-endotoxins afford substantial protection against invasion by meningococci or injury resulting from infection is not known. In any event anti-endotoxin has not been found useful for therapy despite favorable results in animals.

It is tempting to assume that secretory IgA may contribute to immunity to nasopharyngeal carriage but this is pure speculation. Carriage of non-groupable strains of meningococci stimulates cross-reactive bactericidal Abs against numerous but not all meningococci of other groups. The identity of the cross-reactive Ag(s) involved is not known. In this regard it is of interest that children are frequent carriers of the relatively avirulent non-groupable strains and thus acquire broad cross-reactive immunity to meningococci. Whereas some individuals may remain chronic carriers of meningococci for years, the carrier state is usually limited to weeks, presumably because of the early development of specific Abs to the carried organisms. After 2 weeks of carriage most individuals become resistant to infection by the organisms they carry. Protective Abs can be demonstrated in the sera of some 80% of healthy individuals. They are probably initiated, broadened and reinforced throughout life by intermittent exposure to various carried organisms. It is assumed but not proved that immunity acquired by carriage does not selectively promote a shift in carriage from typable to nontypable organisms.

K. LABORATORY DIAGNOSIS

Diagnostic procedures include examination and culture of spinal fluid, blood, nasopharyngeal secretions and other specimens as indicated.

In routine bacteriologic diagnosis it is common practice to culture 10 ml of fresh blood in 10 volumes of an appropriate liquid medium, such as tryptose phosphate broth, and to spread 0.1 ml of blood on a plate of chocolate agar or Thayer-Martin selective medium. In cases of fulminating septicemia, and sometimes in lesser states of septicemia, the organisms can be demonstrated in a drop of capillary blood or the buffy coat using the Gram stain. Repeated blood culture may be needed on occasion. Incubation is carried out under 10% CO_2 and the broth is observed and examined for growth of Gram-negative diplococci during a week of incubation. The oxidase test is applied to colonies on plates. Oxidase-positive colonies of characteristic morphology are picked promptly to avoid killing, and the organisms are identified by fermentation tests and agglutination tests with known antisera. *Neisseria meningitidis* can be distinguished from *N. gonorrhoeae* and other common species of neisseriae by the fermentation of different sugars and the ability of the latter to grow on nutrient agar at the temperature of 22°C.

Fresh specimens of spinal fluid are examined using capsular swelling tests and precipitin tests with known antisera and by culture and direct staining for typical Gram-negative diplococci. Since organisms are abundant in spinal fluid, direct smears of spinal fluid sediment will often reveal them, both free and within phagocytes. The organisms tend to occur in large numbers within certain neutrophils but not others. Cultures are often positive before leukocytes appear in spinal fluid and reduction in the glucose content of the fluid occurs. Organisms which frequent the respiratory tract and which may be mistaken for *N. meningitidis* are *B. catarrhalis, N. sicca* and members of the genera *Moraxella* and *Veillonella.*

Positive bacteriologic findings on blood and spinal fluid are highly significant indicators of meningococcal disease but positive nasopharyngeal cultures only indicate carriage of the organisms.

L. THERAPY

If organisms have been killed by drug therapy or rapid diagnosis is indicated, resort can be made to special immunologic methods for detecting specific Ag; the newer Ab tests for specific polysaccharides include tests on serum, spinal fluid, synovial fluid and urine using counter-current immunoelectrophoresis, latex agglutination and coagglutination by Ab and staphylococcal protein A.

Although it is often desirable to determine the susceptibility of an infecting organism to an antimicrobial drug before initiating therapy, early treatment of acute meningococcal disease is so crucial that the presence of typical organisms on smear or identification of specific Ag in body fluids justifies the initiation of chemotherapy before drug sensitivity tests are completed. Certain drugs that do not pass the normal blood-brain barrier readily can be used in meningococcal meningitis because the permeability of the blood-brain barrier increases during inflammation. Consequently, during about the first 3 days of effective treatment the barrier remains moderately permeable to drugs which would not pass the normal barrier. Although sulfadiazine readily passes the blood-brain barrier, it is no longer the drug of choice because of the ubiquity of sulfa-resistant strains of meningococci; instead massive doses of penicillin are used.* A combination of sulfadiazine and penicillin G may be advisable if the infection is severe and the organism is not sulfa-resistant. Still other alternative drugs are the newer cephalosporins. Drug treatment should be initiated promptly and if endotoxin shock, adrenal insufficiency or intravascular coagulation is suspected, appropriate supportive treatment should be added. Recurring symptoms of infection should be monitored for superinfection with another organism, as well as for changes in drug resistance of the infecting meningococcus.

M. RESERVOIRS OF INFECTION

Man is the only known reservoir of meningococcal infection.

---

*Chloramphenicol is the alternative drug of choice in cases of penicillin allergy.

N. CONTROL OF DISEASE TRANSMISSION

As stated earlier, meningococcal infection is usually acquired by recent close association with healthy carriers.

Because carrier rates and morbidity rates of *endemic* meningococcal disease are low, preventive measures directed at carriers or patients are relatively ineffective during endemic periods. This is in contrast to *epidemic* disease, which appears to rest on both a high incidence of carriers of virulent organisms and a high percentage of corresponding "susceptibles" in the population; under these circumstances regulation of carriers is a highly effective preventive measure.

During epidemics, chemoprophylaxis with sulfonamides among carriers and close contacts of carriers continues to be recommended provided sulfa-susceptibility is first established. However, because of the current prevalence of sulfa-resistant strains, chemoprophylaxis must often be carried out with rifampin or minocycline singly or in combination. Rifampin is of temporary and limited value because of the rapid emergence of rifampin-resistant strains and minocycline often causes serious vestibular dysfunction.

Great strides in prophylactic immunization are being made despite the low immunogenicity of group B polysaccharide and the poor response of infants to serogroup Ags. Licensed group A vaccines are protective for all persons over 3 months of age and group C vaccines for all persons over 2 years of age. They are not recommended for routine use in young children but are advised in the face of epidemics. Other serogroup vaccines are nearing perfection. Membrane proteins freed of endotoxin also show promise as vaccines, especially when combined with capsular polysaccharides including the weakly antigenic group B polysaccharide.

Protective antitoxin can be produced against the core glycolipid of endotoxin. This glycolipid vaccine protects animals against endotoxin shock, but whether the vaccine or the antitoxic Abs it generates will prove useful in man is uncertain. Children with agammaglobulinemia can be protected with pooled human gammaglobulin.

NEISSERIA GONORRHOEAE

A. MEDICAL PERSPECTIVES

Although gonorrhea undoubtedly occurred among early civilizations, it was not clearly described until the 13th century when Guielmus de Saliceto recognized the disease as a distinct clinical entity of venereal origin.*

For centuries gonorrhea was considered to be an early stage of syphilis, an opinion which was erroneously reinforced by the famous experiment of John Hunter, who in 1767 acquired syphilis by inoculating himself with exudate from a patient who allegedly had only gonorrhea but who evidently had *both syphilis and gonorrhea.* It was not until the work of Hill (1790), Bell (1792) and Ricord (1831–1860) that syphilis and gonorrhea were clearly recognized as distinct diseases. Neisser discovered the causative organism of gonorrhea in

*The term "gonorrhea" was first used by Galen in the first century to designate another disease.

1879 and Loeffler cultured it in 1882. Until recent decades bacteriologic diagnosis was usually made by smear alone and therapy was of little or no value. The marked effectiveness of sulfonamide therapy introduced in 1935 was short-lived because of the rapid emergence of resistant strains, and it was not until 1943 that penicillin was shown to be a highly valuable therapeutic agent.

However, in spite of effective chemotherapy and improved methods for diagnosis, the disease is probably as prevalent today as at any time in history. Again, as is so often true, the opportunities for controlling a disease with a highly effective chemotherapeutic and/or prophylactic agent have come to naught in the present instance because of changing social mores, complacency, ignorance and public indifference. Indeed it has become almost axiomatic that development of a highly effective agent that can control a disease commonly leads to a loss of interest on the part of research workers and lack of concern on the part of the medical profession and the public. Today gonorrhea is a worldwide pandemic disease of staggering proportions that has plateaued since 1975; over 25% of cases are among teenagers. The peak incidence is in the 20- to 24-year age group but is shifting downward. The threat the disease poses destroys what little solace might be taken from the fact that gonorrhea is seldom life-threatening. The mounting number of cases with serious complications presents a major public health problem.

The incidence of gonorrhea is difficult to assess because the actual number of cases is at least 5 to 10 times the number reported by physicians; the estimated annual incidence in the USA is 3 million. Also the disease is usually self-limiting and many cases go undiagnosed. There are many reasons for the continued high incidence of gonorrhea, some related to the inherent nature of the disease, such as the marked frequency of asymptomatic infections and reinfections, some related to the inadequacy of methods for detecting carriers, and others related to social attitudes and sexual promiscuity. It appears that short of a sharp reversal in social attitudes and sexual behavior, the disease will be controlled only by new approaches, such as better methods for detecting carriers and the development of an effective vaccine.

B. PHYSICAL AND CHEMICAL STRUCTURE

Neisseria gonorrhoeae is piliated but is otherwise similar to *N. meningitidis* with respect to morphology and staining properties. Irregular "giant forms" are frequently present in aging cultures and a decrease in stainability occurs in vivo within a few hours after initiation of penicillin therapy. The gonococcus possesses a capsule. Capsules rapidly diminish on subculture, a change which is associated with conversion from smooth (S) to rough (R) colonial growth. Subculture also results in the loss of pili and virulence. Although the gonococcus possesses *weakly antigenic capsular polysaccharide* Ags that are related to the K-like polysaccharides of group B meningococci, they have not been well studied. Capsules are meager but can usually be demonstrated by special methods, including the use of India ink, capsular swelling and electron microscopy.

Piliation is a major determinant of virulence and there are at least 4 serotype pilus proteins (Fig. 14–2).

Other Ags include the outer membrane Ags comprising LPS (endotoxin) and proteins. Outer membrane protein Ags are divided into 3 major groups based

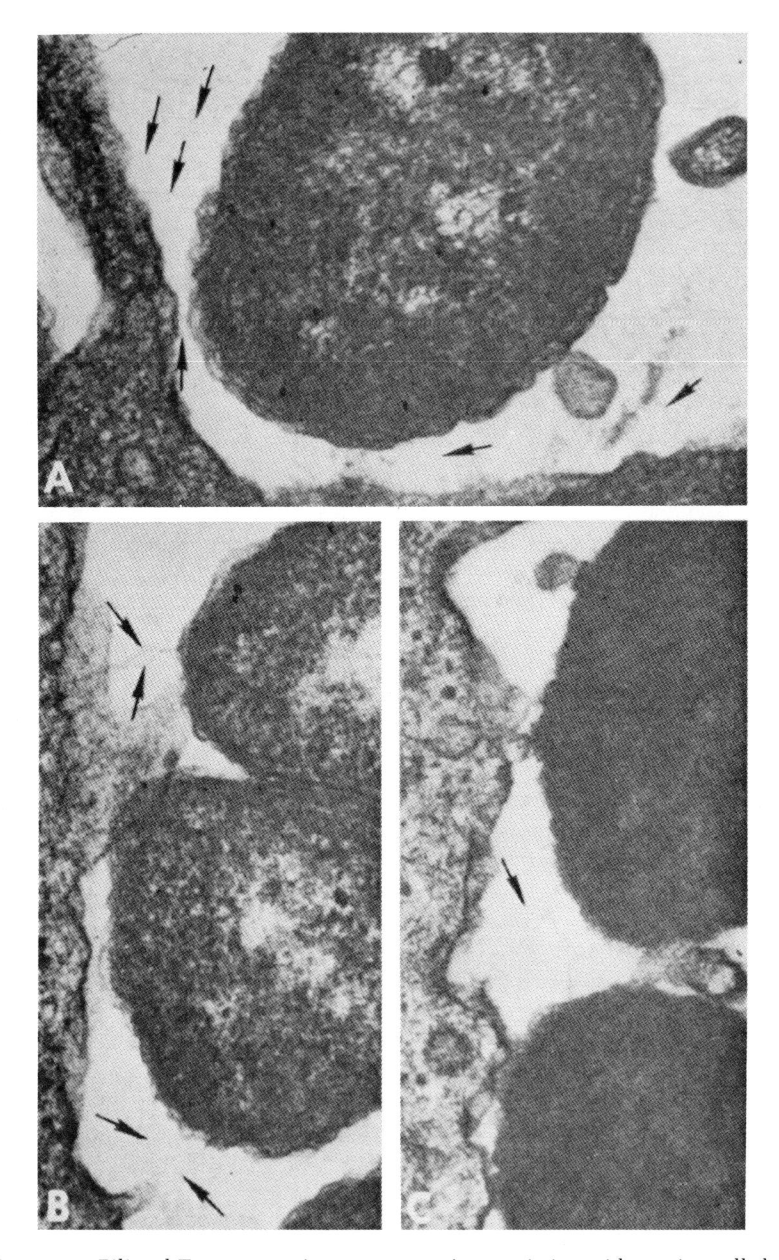

Fig. 14–2. Piliated T-2 gonococci are numerous in association with amnion cells both after exposure to the cells in suspensions (A) or as monolayers (B and C). Pili can be seen extending from the gonococci (arrows) toward the plasma membranes of amnion cells in these thick sections. In some instances (A and B) several pili course from each organism to the amnion cell surface. × 80,000 (A and B) and × 60,000 (C). (From Swanson, J.: J. Exp. Med. *137*:578, 1973.)

on physical properties, Protein I, Protein II and Protein III (PI, PII and PIII) which have been serologically subdivided into 9 or more serotypes. Protein III is common to all strains of gonococci but PI and PII proteins vary among different strains.

There are structural differences in LPS from different strains of gonococci. However, serologic differences between various polysaccharide moieties have not been determined. The lipid moiety (lipid A) is antigenically distinct.

Precise studies on other Ags have not been done. Nevertheless, tests that reflect total patterns of antigenicity have been conducted; cross absorption tests using formalin-killed organisms of various strains and cross-absorbed antisera have identified about a dozen distinct strains involving *3 major serologic, classification systems.* The confusing serology of the gonococcus sorely needs mending by use of the precise tool, monoclonal Abs.

As a reflection of differing cell-surface characteristics, *5 colonial types* of *N. gonorrhoeae* have been described. The small virulent types, T1 and T2, are piliated; they predominate in the exudates of acute infections and appear in primary culture on suitable media. On repeated subculture on liquid media they soon convert to avirulent nonpiliated types T3, T4, and T5, which are poor in surface K-like Ag. Type 1 can only be maintained in culture by cultivation on special media. Autoagglutinability varies among colonial types and appears to depend in large part on the presence of peculiar regions on the surface of the organism that promote adherence to adjacent organisms.

Another procedure for colony typing based on nutritional needs (auxotyping) has permitted division of strains into *over 30 auxotypes;* auxotyping is especially useful for epidemiologic studies.

Virulent organisms that produce disseminated gonococcal infection (DGI) resist the bactericidal effects of normal serum (Sect. I); they are usually sensitive to penicillin and commonly belong to a single auxotype.

C. GENETICS

Gonococci can be easily transformed with naked transforming DNA, especially if they are piliated. It is probable that gonococci often transform in vivo by exchanging DNA; indeed the close homology of the DNA of meningococci and gonococci, which permits in vitro interspecies transformation, makes it theoretically possible that this event could happen in vivo. Transformation with plasmid DNA is relatively inefficient but has been well-studied, especially with relation to penicillinase (β-lactamase) production by strains with unusually high resistance to penicillin. Penicillinase-producing strains (about 1% of isolates in the USA) tend to be resistant to certain antibiotics other than penicillin. The chance that interspecies passage of gonococcal plasmid DNA may eventually occur to establish penicillin-resistant strains of *N. meningitidis* is an unfortunate possibility. Antibiotic resistance among non-β lactamase-producing strains of gonococci is due to the additive effects of mutations at several loci. Three genetic loci that regulate serum resistance to Ab and C (sac loci) have been identified.

Reliable genetic studies have shown that reversible piliation results from reversible rearrangement of chromosomal genes. A gene has been cloned that specifies the H8 outer membrane protein common to *N. gonorrhoeae* and *N. meningitis;* its use has provided information on the immune response to the

H8 Ag in patients with meningococcemia and in patients with disseminated gonococcal infection. By using cloned DNA and plasmid ligation an IgA1 protease-deficient strain of *N. gonorrhoeae* has also been derived, although its virulence has not been tested in humans.

It is reasonable to predict that information now being gathered with modern techniques, especially studies involving recombinant DNA, monoclonal Abs and lymphocyte subsets, will soon clarify the mechanisms of pathogenesis and immunity to the gonococcus and thus provide more rational bases for the control and treatment of gonococcal infections.

D. EXTRACELLULAR PRODUCTS

Extracellular products produced by the gonococcus include a weak hemolysin, an oxidase which is important in the oxidase test and a protease that cleaves IgA1 but not IgA2.

E. CULTURE

The growth requirements of the gonococcus are similar to but more exacting than those of the meningococcus. The gonococcus requires CO_2 and grows best at 35° to 36°C over a pH range of 7.2 to 7.6. Some strains do not grow at 37.5°C and growth ceases below 30°C or above 38.5°C; this temperature sensitivity was the basis of artificial fever therapy attempted several decades ago. Blood cultures are seldom positive even in disseminated infections. Mucosal swabs or scrapings should be streaked promptly; after 48 hours of incubation at 36° to 37°C on a selective medium, such as Thayer-Martin medium or "Transgrow medium" containing added antimicrobics, small oxidase-positive colonies appear as raised, translucent, finely granular, mucoid structures about 5 mm in diameter. If delay of more than a few minutes up to 4 hours before culture is involved, a nutritive or nonnutritive transport medium, such as Stuart's medium, should be used. Speciation is accomplished by fermentation studies; glucose is the only sugar fermented. The organism dies out on culture media within 3 to 4 days but can be preserved for weeks by covering the culture with paraffin oil. Preservation for years can be accomplished by storage at −70°C.

F. RESISTANCE TO PHYSICAL AND CHEMICAL AGENTS

The gonococcus is highly susceptible to most chemicals, including the common disinfectants. Its high susceptibility to silver nitrate is the basis of the common practice of preventing ophthalmia neonatorum by placing drops of a 1 to 2% silver nitrate solution in the eyes of newborn infants. The organism is usually killed by sunlight and drying within 1 to 2 hours. For susceptibility to chemotherapeutic agents see Sections C and L.

G. EXPERIMENTAL MODELS

Good experimental models of gonorrhea in man are greatly needed; it is only recently that a disease closely simulating human gonorrhea has been produced in the expensive model, the chimpanzee. Consequently, most of the research on pathogenesis has been accomplished by the use of human volunteers. Infection can be produced in the chick embryo, in the anterior chamber of the rabbit's eye, and in mice injected intraperitoneally with mucin-coated organisms.

H. INFECTIONS IN MAN

Gonorrhea is the *most frequently reported infectious disease* in the USA. About 30% of cases develop serious complications and chronic disease lasting many years. Infections are almost invariably of venereal origin and the lesions are usually confined to the urogenital tract. The incubation period is usually 2 to 8 days (range 1 to 30 days).

In the male, the onset of disease is almost always symptomatic and abrupt and is characterized by frequent painful urination and a yellowish mucopurulent discharge from the urethra, which usually prompts the individual to seek medical attention. *Neisseria gonorrhoeae* has a predilection for parasitizing columnar and transitional (but not squamous) epithelium. The organisms transgress the epithelial layer (Sect. I) to reach subepithelial tissues where they grow extracellularly for a short time and engender an acute inflammatory lesion characterized by heavy infiltration with neutrophils, lymphocytes, plasma cells and mast cells. Fever does not develop so long as the infection is limited to the urethra. Spread of the disease, which is by direct extension via lymphatic vessels, often results in prostatitis, *epididymitis* and sometimes sterility. Most *cases undergo spontaneous cure without serious complications.* Occasionally secondary invaders, such as *Staph. aureus,* supplant the gonococcus and prolong the urethritis. In some cases columnar epithelium may be destroyed and replaced by resistant stratified squamous epithelium; sloughing epithelium and the *formation of scar tissue may cause obstruction* or narrowing of ducts. This tendency to obstruct ducts often results in *retention cysts or abscesses.* Strictures of the urethra may lead to urinary tract infection with other organisms. Other less frequent complications are discussed below. In about half of the cases of prostatitis, spread to the rectum and resulting proctitis occurs. In a small percentage of cases the individual may remain infectious for months to years due to asymptomatic chronic disease.

In the female, primary infection usually involves the urethra, cervical glands, Skene's glands or Bartholin's glands. The infection spreads externally to the rectum in about half of the cases. It may involve the fallopian tubes and lead to acute *purulent salpingitis or chronic salpingitis;* the latter predisposes to ectopic pregnancy, obliterative fibrosis and sterility. Pelvic inflammatory gonococcal disease is sometimes complicated by intercurrent infection with other microbes, such as anaerobes or *Chlamydia trachomatis. Gonorrhea in the female (either chronic or acute) is mild or asymptomatic in about 75% of cases* and usually involves the cervix. Consequently discovery of the female "carrier" is most often the result of a search for sources of infection to account for disease detected in the male.

Chronic disease (often undiagnosed) *with vague pelvic and abdominal symptoms and serious complications occurs in about 15% of infected females.* Occasionally spread occurs to produce perihepatitis (Fitz-Hugh-Curtis syndrome).

In some cases of gonorrhea, principally those with a chronic pelvic focus, the organisms *invade the bloodstream* and often produce disseminated infection involving lesions in the tendon sheaths, joints, heart valves and skin, etc. but rarely the meninges. Septic death may occur in some cases of disseminated gonorrhea.

Various *sexual practices* present different patterns of disease, proctitis being common in males. Oral-genital practices may lead to pharyngitis and parotiditis as well as genital lesions; pharyngeal carriage has been reported.

Gonococcal bacteremia is usually accompanied by chills, fever and malaise and in about 50% of the cases small *dermal infarcts* due to septic emboli are seen. These infarcts usually occur in the extremities; they often reveal gonococci on stained smears but seldom contain cultivable organisms. The majority of patients with severe bacteremia, accompanied by chills, fever and skin lesions, complain of *polyarthritis* and/or *tendosynovitis* of the wrist or ankle. Less than 50% of patients with arthritis yield positive blood cultures. It is extremely difficult to culture the organism from the synovial cavity, either because synovial fluid may be sparse or absent, or limited to the joint capsule and membrane. The clinical picture of gonococcal arthritis without positive bacteriologic findings can be mistaken for rheumatic fever; however, arthritis patients will respond to trial antibiotic therapy within 2 to 3 days if the disease is caused by the gonococcus. Immune complexes have been reported to contribute to the pathogenesis of late gonococcal arthritis.

Endocarditis, a life-threatening complication, may result not only from gonococcal bacteremia, but may cause bacteremia to persist. Consequently if heart murmurs are heard, adequate and prolonged chemotherapy is in order. The rare cases of gonococcal *meningitis* that occur are often mistaken for meningococcal disease and treated as such.

The newborn may contract infection of the eyes (ophthalmia neonatorum) during passage through the infected birth canal. Before laws were enacted requiring prophylaxis with silver nitrate for all newborn infants (Credé treatment) the disease caused 12% of all blindness but now affects only 0.1 to 0.2% of all newborns and is easily controlled by antibiotic therapy. Another infection contracted at birth is the highly-destructive disease, neonatal gonococcal arthritis. Gonococcal *conjunctivitis* or *iridocyclitis* sometimes occurs in adults, particularly in patients with gonococcal arthritis. Conjunctivitis is a tragic complication if not treated promptly, since it can rapidly lead to corneal ulceration, perforation and blindness.

Gonococcal vulvo-vaginitis sometimes occurs in prepubescent females, usually as the result of sexual abuse or, on rare occasion, from moist fomites or by other means. Indeed, like tuberculosis, gonococcal infection should be viewed as a household disease; infection of any mucous membrane can occur as the result of direct contact with an infection site or with freshly contaminated fingers, etc. or fomites such as wash cloths, towels, drinking cups or eating utensils, etc.

I. MECHANISMS OF PATHOGENESIS

The mechanisms concerned in the virulence and pathogenesis of gonococcal disease have not been fully elucidated.

The gonococcus produces the virulence factor, IgA1 protease. Destruction of secretory IgA1 Ab at the locus of infection is presumed to permit the organism to persist on the mucosal surface.

Pili also represent a virulence factor and most piliated organisms are virulent. Pili, together with an associated surface component, called leukocyte association factor, effect adherence of organisms to phagocytes and mucosal epithelial cells. Piliated gonococci also promote endocytosis by epithelial cells and then grow within and destroy these cells by shedding endotoxin (Fig. 14–3). Adherence to neutrophils may sometimes cause the phagocytes to degranulate after engulfment; degranulation may be followed by bacterial growth and cell destruction. The phagocytosis-impairing activity of pili, especially with regard to macrophages, is neutralized by anti-pilus Ab.

Phagocytosis by neutrophils plays a variable role in immunity to different strains of gonococci because of the multiple factors concerned. As compared with serum-resistant strains that characterize disseminated infections, serum-sensitive strains that characterize local infections are more chemotactic for neutrophils and are more readily opsonized by Ab and C.

The extent to which capsules oppose phagocytosis and to which respective Abs may promote phagocytosis is not known; however, capsular swelling by Ab can be demonstrated. In any event an association exists between capsulation and the severity of disseminated disease.

Since *serum resistance* contributes to the virulence of members of a number of genera of microbes, including the gonococcus, a clear understanding of the mechanisms underlying serum resistance is of great importance. The anti-gonococcal Abs in normal human serum that effect C killing of serum-sensitive gonococci are presumed to be generated by Ags of microbes of the normal flora that cross-react with gonococcal Ags. Serum resistance is not an all or none phenomenon and serum-resistant gonococci can be killed with specific hyperimmune rabbit serum. Whereas, isolates from patients with DGI tend to be serum-resistant, isolates from patients with local venereal lesions tend to be sensitive. The extent to which serum Abs in various patients with disseminated infections effect C killing of the infecting organisms has not been determined. However, their contribution to immunity is probably important because patients with C deficiencies involving C9 are highly prone to develop disseminated disease.

Recent investigations indicate that the difference between serum-resistant and sensitive gonococci rests with structural differences in cell surface components concerned with the binding and activation of C; whereas exposure to anti-gonococcal Abs and C results in the binding of C5b-9 components by both resistant and sensitive gonococci, the complex is bound in different molecular configurations such that C-activation occurs on sensitive but not resistant gonococci. The precise surface components responsible for serum resistance have not been determined.

Endotoxin shed by engulfed gonococci through vesiculation contributes to pathogenicity by killing host cells in which the organisms grow. Anti-LPS Abs contribute to protection in man by acting as bactericidins. Although little endotoxin reaches the circulation to produce systemic endotoxin shock, anti-endotoxin Abs can protect experimental animals against endotoxin shock induced by injecting large numbers of killed or living gonococci. Since endotoxin is present equally in both virulent and avirulent gonococci, this agent is obviously not a key to virulence. The use of pyocins for recognizing sugar residues and monoclonal Abs for identifying epitopes should soon reveal the relation of LPS structure to pathogenesis and immunity.

The gonococcus plays the role of an intracellular as well as an extracellular parasite with all of its implications, e.g., inapparent infection, chronic carriage and escape from some of the usual forces of defense.

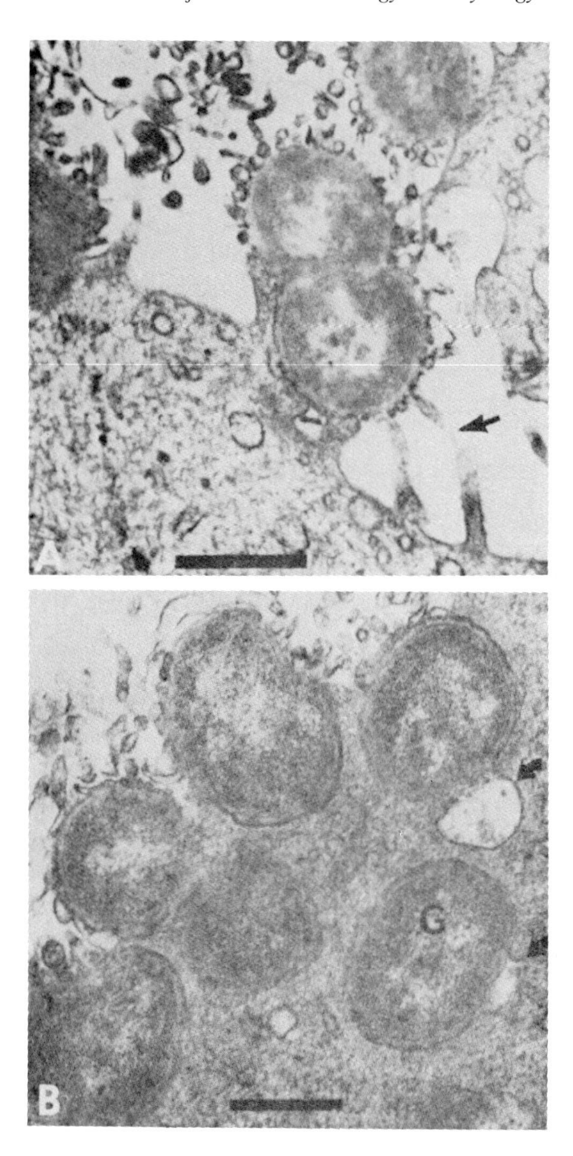

Fig. 14–3. Adherence and embedding of gonococci into the epithelial cells of the urethra. A. Processes from a urethral epithelial cell appear to be twisted over and attached to an adjacent gonococcus (arrow). The bar represents 500 nm. B. Electron micrograph showing gonococci deeply embedded in a urethral epithelial cell. Note the membrane of the host cells (arrows) around the gonococcus (G). The bar represents 500 mm. × 36,000 (A) and × 30,750 (B). (Ward, M. E. and Watt, P. J.: J. Infect. Dis. *126*:603–604, 1972.)

J. MECHANISMS OF IMMUNITY

Immunity to the gonococcus is multifactorial, unusually complex and ill-understood.

Innate immunity is evidently weak or nonexistent because all individuals appear to be equally and fully susceptible to primary infection.

Acquired immunity is likewise weak and is probably largely if not solely humoral; it arrests most acute infections but frequently does not suffice to cope with chronic infection. Multiple Ag-Ab systems and anti-virulence factors undoubtedly participate in acquired immunity in many diverse ways. The Ab activities concerned probably include bactericidal, opsonic, anti-pilus and anti-IgA1 protease effects.

A substantial part of acquired immunity is strain specific as attested by the observation that volunteers who possess anti-gonococcal Abs specifically resist reinfection with their original infecting organisms. Moreover, sequential reinfections with antigenically different strains often follow spontaneous cure of a prior infection. The durability of strain-specific immunity has not been accurately determined.

The marked susceptibility to reinfection with the gonococcus was amply documented many years ago in the diary of the famous James Boswell, "a compulsive patron of prostitutes," who recorded some 19 separate attacks of gonorrhea between 1760 and 1790. Evidently he paid dearly for his sexual exploits for the records indicate that he finally died from complications of the disease.

New knowledge on anti-gonococcal immunity could help solve two great needs, an effective vaccine and a more suitable method for detecting carriers.

K. LABORATORY DIAGNOSIS

Fresh exudate from acute lesions in a patient with a clinical history of gonorrhea commonly yields positive smears for typical Gram-negative diplococci. The organisms are largely intracellular and tend to be present in large numbers in some phagocytes in the smear but not others (Fig. 14–4). *A presumptive diagnosis by smear is fairly reliable in the case of the male but not the female; in the latter case the smear should be confirmed* by culture. In the hands of experts, false-positive microscopic tests, due principally to organisms of the Mima-Herellea group, can usually be avoided by care and the use of proper techniques.

Bacteriologic diagnosis of self-treated cases and of chronic gonorrhea, especially in the female, often fails because the organisms are few in number and are extremely difficult to find, even by culture. When only a single specimen is examined, the diagnosis may be missed in 20 to 40% of cases. Repeated examination of specimens taken over a period of days or weeks from the glands of Skene and Bartholin and from the cervix and rectum greatly reduces the number of missed cases. The use of suppository tampons for collecting specimens promises to improve the success of culture.

Specimens should be promptly inoculated after collection, preferably on Thayer-Martin medium containing selective antimicrobics or Transgrow medium and incubated under 10% CO_2 at 35° to 36°C. Stuart's medium is another good transport medium. Blood, synovial fluid and spinal fluid should prefer-

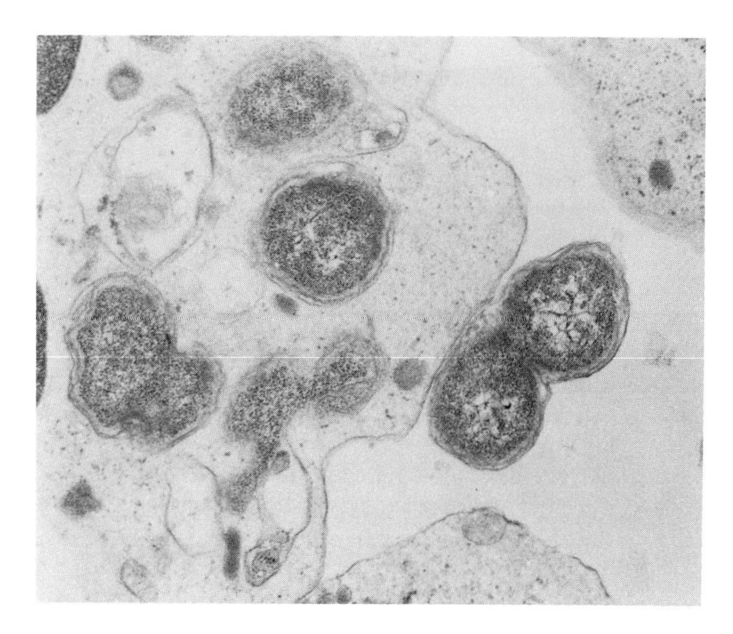

Fig. 14–4. *Neisseria gonorrhoeae.* A section of a polymorphonuclear leukocyte from a human exudate. A typical diplococcus is seen near the cell surface as well as various intracellular neisseria. ×25,000. (Courtesy of Eva S. Leake.)

ably be cultured for a week in a thin layer of ascitic fluid broth under 10% CO_2. Although detection of oxidase-positive colonies of typical morphology on Thayer Martin medium affords a presumptive diagnosis of gonorrhea, final identification of *N. gonorrhoeae* rests on sugar fermentation tests and/or tests with specific antisera. Effective mass control of gonorrhea demands rapid but accurate bacteriologic diagnosis. Whereas the direct fluorescent Ab (FA) method applied to fresh exudate usually succeeds with acute exudates, the "delayed" FA reagent method in which the specimen is cultured for 16 to 20 hours before applying FA is even more successful than routine culture and is useful in cases of chronic disease.

The immune response in gonorrhea is weak and no useful diagnostic skin test has been developed. However, a new test for anti-pilus Ab in serum promises to be useful for detecting asymptomatic patients and chronic carriers.

Simple but accurate methods for detecting asymptomatic infection are sorely needed.

L. THERAPY

The drug of choice for most cases of gonorrhea is penicillin. Whereas most strains were formerly sensitive to about 0.1 units/ml of penicillin, it is now common to find mutant strains for which the minimum inhibitory concentration is 2 to 4 units/ml. Consequently, the recommended "one-shot" dose of penicillin for females is now 4.8 million units of procaine penicillin G with probenecid to prolong effectiveness. Parenteral administration of penicillin is usually best because it assures that the patient receives the full dosage of the drug. A few plasmid-transformed strains of Far-East origin that produce penicillinase (TEM-type β-lactamases) are highly resistant to both penicillin and tetracycline. The penicillin-resistant strains possess a 4.4 md plasmid that mediates β-lactamase production. In some areas penicillinase-producing *N. gonorrhoeae* (PPNG) account for up to 5% of the cases of gonorrhea. The drug of choice for penicillinase-producing strains is spectinomycin. Some of the third generation cephalosporins such as cefizox are also effective for treating patients infected with penicillin-resistant gonococci and patients who are allergic to penicillin. A course of oral tetracycline is effective for tetracycline sensitive strains. Unfortunately, many strains of gonococci have developed resistance to tetracycline. Other drugs that can be used in various regimens designed for selected cases with respect to allergy and drug resistance include ampicillin and amoxicillin.

The high frequency of mixed infections with two or more sexually transmitted microbes such as chlamydial infections has led to the expanding use of new broad-spectrum antimicrobics and treatment regimens.

Since organisms within cells and abscesses are not readily reached by many chemotherapeutic agents, unusually high doses must often be given to be effective. Also cases of proctitis and pharyngitis may be refractory because penicillinase produced by other bacteria is present in the area of infection. Whether multiple-drug resistance, which is on the increase, will ever progress to the point where drugs will no longer be therapeutically valuable cannot be forecast. Cure is evaluated by weekly cultures continued for 2 to 3 weeks after terminating treatment.

Since some 3% of patients with venereal disease have both gonorrhea and syphilis, penicillin treatment often aborts coexisting cases of early syphilis. Needless to say, all patients with gonorrhea should be subjected to serologic tests for syphilis before treatment and periodically for 6 months thereafter.

M. RESERVOIRS OF INFECTION

Man is the only reservoir of the gonococcus. The organism is most commonly harbored in the genitalia and rectum. The incidence of disease and carriers is high among prostitutes, averaging some 20%. *Long-term carriage is more frequent in women than men.* Whereas patients sometimes contract the disease from the acutely infected mate, infection most commonly results from contact with the *asymptomatic chronic carrier* who is often unmindful or unaware of ever having been infected; currently, approximately 2 million such carriers exist in the USA. The unawareness of the chronic carrier was well documented by Boswell, who contracted the disease from a "safe" girl friend who believed herself innocent because she had no lover for 6 months before becoming intimate with Boswell. Lacking the benefit of today's knowledge, Boswell doubted her fidelity; he wrote: "There is scarcely a possibility that she could be innocent of this crime of horrid imposition, and yet her positive asseverations have really stunned me." Since approximately 1% of all females have asymptomatic gonorrhea, it is obvious that the female carrier constitutes the major reservoir of disease transmission; the relative importance of the asymptomatic male remains to be determined but is appreciable.

N. CONTROL OF DISEASE TRANSMISSION

The fallacious assumption that highly effective antimicrobial agents can be used to effectively control an infectious disease irrespective of the nature of its transmission and the human elements involved is well illustrated in the case of gonorrhea. No suitably effective practical measures for preventing gonorrhea in the exposed individual have been developed despite the numerous gadgets, ointments, chemical irrigants, etc. that have been tried over the centuries without success. Condoms probably afford the best protection. Penicillin, if given in adequate dosage at or near the time of exposure, is an effective prophylactic and should be employed for all individuals known to have had sexual contact with an active case of the disease. Treatment of the patient and his or her contacts should be continued until cure is effected since sexual relations are apt to continue despite advice to the contrary.

Efficient case finding is a critical but delicate matter which should be reserved for experts. In theory, if a good case-finding team is at hand, the rapid identification and prompt treatment of all of the sexual contacts that the patient has had within a week before infection, regardless of whether they present symptoms, should go a long way toward controlling the disease. Of course, contacts of contacts should also be considered since some of the original contacts may have reached the infective stage within a week.

Filling the great need for a simple test to detect carriers and a protective vaccine is now a distinct possibility (Sect. C).

REFERENCES

Andersen, B.M. and Solberg, O.: Endotoxin liberation and invasivity of *Neisseria meningitidis.* Scand. J. Infect. Dis. *16*:247, 1984.

Cannon, J.G. and Sparling, P.F.: The genetics of the gonococcus. Annu. Rev. Microbiol. *38*:111, 1984.

Greenwood, B.M.: Selective primary health care: Strategies for control of disease in the developing World. XIII. Acute bacterial meningitis. Rev. Infect. Dis. *6*:374, 1984.

Joiner, K.A., et al.: Complement and bacteria: Chemistry and biology in host defense. Annu. Rev. Immunol. *2*:461, 1984.

Joiner, K.A.., et al: Monoclonal antibodies directed against gonococcal protein I vary in bactericidal activity. J. Immunol. *134*:3411, 1985.

Joiner, K.A., et al.: Bactericidal but not nonbactericidal C5b-9 associated with distinctive outer membrane proteins in *Neisseria gonorrhoeae.* J. Immunol. *134*:1920, 1985.

Macrina, F.L.: Molecular cloning of bacterial antigens and virulence determinants. Annu. Rev. Microbiol. *38*:193, 1984.

Martin, J.E. and Lester, A.: Transgrow, a medium for transport and growth of *Neisseria gonorrhoeae* and *Neisseria meningitidis.* Health Service and Mental Administration Health Reports. *86*:30, 1971.

Ober, W.B.: Boswell's gonorrhea. Bull. N.Y. Acad. Sci. *45*:587, 1969.

Perine, P.L., et al.: Epidemiology of the sexually transmitted diseases. Annu. Rev. Public Health *6*:85, 1985.

Plaut, A.G.: The IgA1 proteases of pathogenic bacteria. Annu. Rev. Microbiol. *37*:603, 1983.

15

CORYNEBACTERIUM

The genus *Corynebacterium* of the family *Corynebacteriaceae* contains many species parasitic to animals and to man, but the only pathogen of major importance for man is *Corynebacterium diphtheriae,* the cause of diphtheria. The virulent form of the organism is notable for the powerful heat-labile exotoxin it secretes, which serves as the principal agent responsible for its pathogenicity, virulence and immunogenicity. *Corynebacterium diphtheriae is maintained in man, its only natural host, by a small but relatively constant population of chronic carriers who commonly harbor the organism in the nose or nasopharynx* (Sect. D). Certain nonpathogenic corynebacteria (often referred to as "diphthcroids") are commonly found on the mucous membranes of the upper respiratory tract (URT), urogenital tract and eyes. On occasion certain of them may assume the role of opportunists. Two nonpathogenic species which are notable because they often confuse the diagnosis of diphtheria, are *C. hofmannii,* a frequent inhabitant of the URT, and *C. xerosis,* which is often present on the conjunctivae. As might be expected, several species of *Corynebacterium* are pathogenic for animals. The focus in this chapter will be on *C. diphtheriae* and the infections it produces.

A. MEDICAL PERSPECTIVES

No serious bacterial disease of man is more thoroughly understood and controlled than diphtheria, largely because the disease is a simple toxemia uncomplicated by invasion of organisms beyond the local lesions they produce on external integuments and because artificial immunization produces strong immunity which commonly persists for months to years.

Diphtheria, a world-wide disease, was first accurately described as a clinical entity by Bretonneau of Tours in 1821. However, clear references to the disease are found among the earliest writings of civilized man. The Greeks recognized diphtheria as a disease of the throat and tonsil, which often suffocates its victims; the Spanish called it "morbus suffocans." Bretonneau recognized the *false membrane* as being highly characteristic of the infection and consequently named the disease *diphtheritis* after the Greek word "diphthera" meaning leather, skin or membrane. He observed that diphtheria is communicable and proposed that it is caused by a specific germ. However, the causative organism was not observed until 1883 (Klebs) and not fully established as the causative agent until 1884 (Loeffler). Based on his observation that, although the growth of organisms in infected guinea pigs was limited to the site of inoculation, damage occurred in distant organs; Loeffler (1884) reasoned that the organism must produce a toxin. The truth of this postulate and the role of antitoxin in immunity was established when Roux and Yersin (1888) discovered the toxin in culture filtrates and Frankel, von Behring and Kitasato (1890) discovered that the sera of immunized animals contained a neutralizing substance (antibody) which protected against the toxin. This Nobel prize-winning work was the first evidence of Ab formation as a phenomenon.

Prior to the advent of antitoxin therapy and mass immunization, diphtheria was a dreaded disease which regularly took a huge toll of lives, especially infants and young children among whom morbidity and mortality are highest. Even in advanced countries it was the single leading cause of death among children. Sometimes whole families were wiped out. For example, during an epidemic in the Atlantic Coast States in 1735–1740 some 20% of all children under 15 years of age died of the disease.

In the decades before immunization procedures were introduced, periodic pandemics occurred, which sometimes swept the world. For example, in Spain the year 1613 is known as the "year of the garrotillo," and a world-wide pandemic originated in France in 1850 which began its decline in 1885 but did not terminate until about 1941. The mortality rate tends to be high during pandemics, but its relation to the cause of pandemics is not known. The cause(s) of pandemics is an enigma but could be related to evolutionary changes in the organism and the host population (Sect. N).

Advances, including the introduction of antitoxin treatment in 1891, the Schick test for measuring immunity in 1913, and immunization with the toxin-antitoxin complex (von Behring, 1913) markedly improved the control of diphtheria; however, progress was slow and it was not until after the highly effective immunogen, diphtheria toxoid, developed by Ramon in 1923, was perfected for use as a vaccine that the important contribution of mass immunization to control the disease was fully realized (Sect. N). Toxoid is prepared by detoxifying toxin with formaldehyde. The antigenicity of toxoid can be enhanced by treatment with alum to form alum precipitated toxoid, or combining it with pertussis vaccine and tetanus toxoid to form the vaccine complex (DPT). Since 1940 mass *immunization of children in advanced countries has become routine, and diphtheria seldom occurs* (Sect. N). Moreover, antitoxin treatment of the disease reduces the average mortality from about 18 to 5%. A few elderly Americans can recall school epidemics in the early 1900s that took death tolls as high as 10 to 20%. The conquest of diphtheria has been painfully slow but nevertheless stands as one of the greatest achievements in medical science.

Sporadic outbreaks of diphtheria still occur in the USA, involving a few hundred cases and fewer than 100 deaths annually, including an increasing number in adults. Nevertheless, there is little reason to expect that the disease will become a major health problem again so long as mass immunization of children is continued and good epidemiology is practiced. Until better world control of diphtheria is effected, it is probable that the disease will never be completely eradicated from any large population for an extended period of time. Consequently, *relaxation or abandonment of any of the current public health practices would lead to serious epidemics.* The future outlook for control of diphtheria in the world at large is far less promising than it should be, and in certain underdeveloped countries the disease remains a major medical problem.

B. PHYSICAL AND CHEMICAL STRUCTURE

Corynebacterium diphtheriae is a slender, highly pleomorphic, Gram-positive, nonsporing, nonencapsulated, nonmotile, nonacid-fast rod-shaped organism which varies from 2 to 6 μm in length and from 0.5 to 1.0 μm in diameter. When grown on a suboptimal medium, the organisms are often club shaped and sometimes pointed; they tend to appear in pallisades or clumps arranged in angular aggregates resembling Chinese letters. When stained with Loeffler's alkaline blue or toluidine blue, cytoplasmic granules of polymerized polyphosphates stain metachromatically and present a reddish hue. This staining behavior lends a characteristic beaded appearance to the organisms.

Since the corynebacteria, mycobacteria and nocardiae are closely related many of their structural components are similar. For example, the O polysaccharides are cross-reactive. They also produce similar cytotoxic cord factors. Serologically heterogeneous proteins called K Ags are present in the outer cell wall of *C. diphtheriae.* They are alleged to contribute to adherence to host cells and to induce antibacterial immunity.

C. GENETICS

When cultured on selective-differential media containing potassium tellurite, 3 principal colonial biotypes are seen. They were initially termed "gravis," "intermedius," and "mitis" to designate differences in the severity of disease produced (Sect. E); however, the relationship between virulence and biotype is not constant and the identification of biotypes is of epidemiologic value but of little diagnostic value.

A singularly significant aspect of the genetics of *C. diphtheriae* was introduced by the remarkable and biologically important discovery of Freeman in 1951, namely, that *exposure of non-toxigenic strains of C. diphtheriae to corynephage derived from toxigenic strains converted them to toxigenicity, a process called lysogenic conversion.* Diphtheria toxin production by toxigenic strains of *C. diphtheriae* rests on parasitism by those closely-related phages of the β phage family that carry the *structural gene for toxin production (tox gene);* most β phages carry the tox gene and most of them can induce lysogenic conversion. Beta phage carrying the tox gene is a temperate phage that can either initiate the *lysogenic cycle* in which entering linear DNA circularizes and integrates with the bacterial chromosome to yield latent virus (prophage β) or alternatively, initiate the *lytic cycle* characterized by phage DNA replication, phage maturation and bacterial lysis by endolysin (Chap. 3).

Since toxin production by corynebacteria can vary depending on mutations involving either the host bacterial cell and/or phage, it is no surprise that nonconverting β family tox mutants occur, certain of which fail to effect lysogenic conversion because the tox gene remains unexpressed or because the tox gene is so altered that the gene product is toxin-like but nontoxic. Other tox mutations of β phage may lead to complete loss of the tox gene. Such tox deficient mutants can regain conversion power by recombining with related converting phages. In some instances a single cell may become infected with 2 different phages which is usually followed by the elimination of one or the other.

Another possibility of change is that nontoxigenic bacterial mutants may arise that carry the tox gene but do not express it because the bacterial cell somehow suppresses the synthesis or secretion of toxin.

For unknown reasons various strains of *C. diphtheriae* infected with β phage differ in their capacity to form toxin. It is also notable that some strains of other species of *Corynebacterium,* namely, *C. ulcerans* and *C. pseudotuberculosis (C. ovis)* are naturally infected with β phage and secrete diphtheria toxin in small amounts. Experimental lysogenic conversion can also be accomplished in these 2 species with β phage from *C. diphtheriae.* Moreover, most strains of the 2 species are infected with another phage and secrete a different toxin called "ovis" toxin.

The consequences of the various activities of corynephages with respect to toxin production and virulence are not fully apparent. Lysogeny is known to occur in human populations but the role that this converison may play in the carriage of virulent organisms and the epidemiology and natural history of the disease remains to be determined.

The manner by which phage DNA induces toxin production is not known; however, toxin production is favored by certain growth conditions, especially submerged culture in alkaline liquid media (pH 7.8 to 8.0) containing a certain low level of iron ranging near the value of 100 μg/ml. Toxin production is greater the longer the period between phage induction and cell lysis; apparently low levels of iron prolong the induction-lysis period and thus promote increased toxin production. The slow-growing Park and Williams strain 8 (PW8) of *C. diphtheriae,* which can grow at lower concentrations of iron than other strains, is the most potent toxin-producer isolated to date, producing toxin in vitro equal to about 5% of its dry weight. *Despite its high in vitro toxinogenicity, the PW8 strain is avirulent for animals.* This implies that toxin is not the only determinant of virulence or that it is not produced in vivo by this strain of *C. diphtheriae.* Whether the in vivo iron concentration is unsuitable for toxin production by PW8 or whether avirulence rests on inability to colonize or some other basis is not known.

Toxin production does not block host cell division or require lytic growth of phage because the tox gene can be expressed when the phage is present either as a prophage, a vegetative replicating phase or a superinfecting nonreplicating exogenote* in immune lysogenic cells.

D. Extracellular Products

The only recognized extracellular secretion product of *C. diphtheriae* of medical importance is diphtheria toxin, a slow-acting exotoxin that inhibits protein synthesis of target cells (Sect. I). It is lethal for susceptible animals in quantities of 100 μg/kg. Diphtheria toxin is a heat-labile protein consisting of a single polypeptide chain having a molecular weight of about 61,000. Mild trypsin digestion of "intact toxin" yields "nicked toxin" (activated toxin), which is a complex composed of two polypeptide components (fragments), A (mol wt 21,150) and B (mol wt 40,000), linked by two disulfide bridges.

Animal models for quantitative measurement of toxin and antitoxin have been developed. *The minimum lethal dose* (MLD) of toxin is the smallest amount which will kill a 250-g guinea pig within 4 to 5 days. Twelve MLDs injected by accident have been observed to kill a child. Antitoxin is measured by its capacity to protect animals against either the local or the systemic effects of toxin. Currently the measurement of toxin and antitoxin is done by use of the international standard antitoxin stored at Copenhagen, Denmark.

In vitro precipitin (flocculation) tests are not suitable for measuring either toxin or antitoxin since toxin can spontaneously denature to form toxoid without losing its capacity to bind Ab; moreover the precipitating and neutralizing capacities of Abs do not correlate in different preparations.

---

*Incomplete genome derived from the donor cells.

Other products are neuraminidase and N-acetylneuraminic acid lyase; their potential contributions to colonization and invasiveness are obscure.

E. CULTURE

Corynebacterium diphtheriae is an aerobe and grows poorly under strict anaerobic conditions. It has simple nutritional needs and grows over wide ranges of temperature (15 to 40°C) and pH (7.2 to 8.0).

For many years Loeffler's coagulated serum medium was the medium used for laboratory diagnosis; however, potassium tellurite is an additional medium of great value since it inhibits most contaminating bacteria while allowing recognizable colonies of *C. diphtheriae* to develop. Tinsdale's agar is also an excellent medium for primary isolation. The color of the colonies on tellurite medium, which ranges from gray to black, is due to reduction of the tellurite to tellurium, a property shared by few of the other bacteria which may be present in the respiratory tract. The three major colonial types, gravis, intermedius, and mitis, can be separated on the basis of colony size, color and form. The gravis strains produce large, flat, rough dull-gray to black colonies, whereas the mitis strains produce small, smooth, convex mushroom-gray colonies. Diphtheroid colonies are shiny gray-black with grayish-white edges.

Corynebacterium diphtheriae can be distinguished from *C. ulcerans* and from two common nonpathogenic diphtheroid species, *C. xerosis* and *C. hofmannii,* by colonial appearance on tellurite medium and by biochemical reactions. Distinction between avirulent (nontoxigenic) and virulent (toxigenic) strains of *C. diphtheriae* demands virulence tests of toxin production (Sect. G) because avirulent strains occur which are otherwise indistinguishable from virulent *C. diphtherae.*

F. RESISTANCE TO PHYSICAL AND CHEMICAL AGENTS

Corynebacterium diphtheriae is "average" in its resistance to most physical and chemical agents. It is readily destroyed by most of the common disinfectants and by heating sufficient to destroy vegetative bacterial cells. The organism is sensitive to erythromycin, penicillin and several other antibiotics. However, it is highly resistant to drying and may remain alive and infective for days to weeks in dust.

G. EXPERIMENTAL MODELS

Laboratory animals vary widely in their susceptibility to infection with *C. diphtheriae* and to diphtheria toxin; the monkey, rabbit, guinea pig and pigeon being highly susceptible and the rat and mouse highly resistant. It is of singular interest that varied susceptibilities of different species of animals to diphtheria toxin correlate with the susceptibility of their respective cultured cells to the toxin in vitro.

Disease resembling diphtheria in man can be produced in animals by introducing the organism into the respiratory tract or skin. For example, if a large dose of virulent *C. diphtheriae* is injected intracutaneously or subcutaneously into a guinea pig, the organism will remain localized but the animal will die within 12 hours to several days (depending on dosage) due to the systemic effects of liberated toxin. Within a few hours the injection site becomes tender, edematous and hemorrhagic; degenerative changes occur in heart, liver and kidneys and most notably *the adrenals show intense congestion, which is usually accompanied by hemorrhage.* Low doses of toxin produce late paralysis.

Death following the injection of either the organisms or the toxin can be prevented by prior adminstration of specific antitoxin. *The demonstrated protective action of antitoxin in animals challenged by injecting live organisms constitutes the "animal virulence test."* Insight into the mechanisms by which toxin produces cell injury has been gained by studies on the action of toxin on cell cultures (Sect. I).

H. INFECTIONS IN MAN

Man is the sole host and principal reservoir of *C. diphtheriae* and pharyngeal infection is commonly acquired from a patient, a person with subclinical infection, or a carrier, via respiratory droplet passage or direct contact. Less frequently, transmission may result from insect bites (skin infection), contaminated milk, fomites or dust. Diphtheritic lesions of skin or of other sites may occasionally serve as sources of respiratory tract infection. Diphtheritic infections are rare in infants because of maternal Abs. They occur in all other age groups but are most frequent in young children.

Following an incubation period of 2 to 7 days, pharyngeal infection is usually heralded by a headache and a sore throat of slight to moderate intensity *but notably devoid of the erythema and pain characteristic of streptococcal pharyngitis. Prostration* is a singular finding and tends to be out of all proportion to the early low grade fever which is usually present. *Differential diagnosis should, most notably, include streptococcal infection, Vincent's disease, infectious mononucleosis, mycotic infection, agranulocytic angina and adenovirus infection with exudate.*

The general course of the disease varies greatly depending on microbial virulence, host resistance and whether respiratory tract obstruction occurs; the mortality in untreated cases varies markedly among different epidemics and has ranged from about 12 to 30% or more. Presumably the highest mortalities are due to organisms that produce the greatest amounts of toxin in vivo. Late symptoms of injury of cranial and peripheral nerves often develop. Whereas in severe cases death may occur within 24 to 48 hours, *in other cases the patient may live for several weeks to finally die of cardiac damage.* In individuals who have previously developed antitoxin, the disease tends to be mild and is frequently not recognized because of the anamnestic response. *Presumably the anamestic response alone in the initial absence of antitoxin can abort the development of disease when the infecting organism is of low virulence.*

Evidently the organisms first reach and colonize the surface of the mucosa of the throat before lesions appear. It is presumed that the small amount of toxin initially produced soon kills underlying epithelial cells and induces a local inflammatory exudate in which further growth of organisms is favored. With time the initial small yellowish-white lesions, which are usually on the tonsil, coalesce and darken; they may spread laterally to include wide areas over the tonsils, posterior pharynx and nares, uvula, soft palate, larynx and trachea. Infection may occasionally spread through the eustachian tube to the middle ear. The tough grayish-white adherent *"pseudomembrane"* which forms, consists of necrotic epithelium embedded in a fibrinous exudate rich in red cells and leukocytes. When hemorrhage occurs, the membrane may darken and if it is forcefully dislodged, points of hemorrhage appear and a new layer of exudate forms. Membranes which eventually loosen and separate naturally may cause sudden *respiratory tract obstruction.*

The cervical lymph nodes become swollen and tender and *in severe cases massive edema of the neck may develop (bull-neck).* It is notable that the *growth of the organisms tends to be largely restricted to the mucosal layer* and any spread of live organisms to the draining lymph nodes and blood, which may occasionally occur, is commonly either an early or terminal event. The reason(s) for such strict localization and failure of the organism to invade deep tissues and produce distant lesions is not known.

There are two principal causes of death, one the systemic effects of toxin and the other, local obstruction of the respiratory airway. The most life-threatening effects of the systemic toxemia are cardiac damage and depression of respiration. Cardiac failure (the major cause of death) results from fatty degeneration of the myocardium or peripheral circulatory collapse. Myocarditis usually becomes manifest after the 2nd week and death may result during the next 2 months or later. Paralysis of the soft palate, the most common symptom due to cranial, nerve damage, often appears after the 3rd week and leads to the characteristic regurgitation of fluid through the nose during attempts to swallow. Late complications, including paralysis of oculomotor, facial, pharyngeal, laryngeal, diaphragmatic, intercostal and peripheral muscles, may occur. Respiratory paralysis may be the cause of death in some instances.

The second major cause of death, obstruction of the respiratory airway, is due to either excessive edema or a detached pseudomembrane. Membranes deep in the throat or trachea (laryngeal diphtheria) pose the greatest threat. *Obstruction by a detached membrane is a fortuitous event and can occur with great suddenness; it cannot be forecast on the basis of clinical examination,* a point which cannot be overemphasized since emergency intubation or tracheotomy can often be lifesaving.

Patients who recover from diphtheria, including those who show symptoms of neurotoxic injury, usually show no permanent effects of the disease. Clinical recovery is not attended by prompt elimination of the organisms, and the convalescent carrier state, which may persist for 1 to 2 months or longer, constitutes a reservoir of infection for others (Sect. N).

Extrarespiratory infections may occur in sites such as the ear, conjunctivae, vagina, anterior nares, endocardium, umbilicus and skin wounds. Some of these extrarespiratory infections, such as endocarditis and skin infections, are caused by nontoxigenic as well as toxigenic strains of *C. diphtheriae. Skin diphtheria is largely confined to the tropics and subtropics* where it occurs with considerable frequency, and constitutes a major reservoir of infection. The rare case of extrarespiratory infection that occurs in Northern USA (less than 1% of all *C. diphtheriae* infections) is often misdiagnosed. Such infections usually begin at a site of minor injury and develop as a chronic spreading "punched-out" ulcer covered by a gray membrane. Although skin diphtheria is not as life-threatening as pharyngeal diphtheria, presumably because toxin is produced at a slower rate in skin or reaches the circulation less readily, *ulcers may persist for months and the disease often leads to peripheral neuritis* (missed cases of mild respiratory infections may likewise lead to late peripheral neuritis).

The pathogenesis of skin infections is not fully understood, and there is no good correlation between serum antitoxin levels and resistance to skin infection. Instead, *resistance to skin infection apparently results from the natural acquisition of immunity unrelated to antitoxic immunity. Also, skin infections often involve nontoxinogenic rather than toxinogenic strains.* During past decades, the peak incidence of skin diphtheria in Southern USA occurred in late summer and early fall, evidently because skin diphtheria is commonly associated with infected insect bites, many of which involve mixed infections with staphylococci and/or streptococci.

Apparently antitoxic immunity is less effective in opposing infection by *C. diphtheriae* in skin wounds and skin lesions of various sorts than in the throat.

The role which skin infections, with either toxinogenic or nontoxinogenic strains, play in herd immunity and epidemiology of respiratory tract diphtheria is not known.

Taken together, the above observations exemplify the subtleties of host-parasite relationships and invite the speculation that early in the evolution of parasitism by *C. diphtheriae* the organism may have been a nontoxinogenic opportunist parasitizing insect-bitten primitive man; later the fortuitous development of toxinogenicity by lysogenic conversion endowed the organism with enhanced powers of invasiveness and pathogenicity, especially in the respiratory tract.

I. MECHANISMS OF PATHOGENESIS

In the intact animal certain cells and tissues appear to be more susceptible to toxin than others, particularly peripheral nerves, pericardium, diaphragm, heart muscle, liver and kidneys. Present evidence indicates that whole toxin liberated by *C. diphtheriae* into body fluids is first converted to the activated complex by enzymes, after which the C-terminal fragment B of the complex promptly binds to receptors on the host cell plasma membrane and then somehow incites cleavage of the complex and delivery of the enzymatically-active N-terminal A fragment into the cytoplasm where it proceeds to inhibit protein synthesis (Fig. 15–1). Passage of fragment A through the cell membrane evidently requires time since antitoxin can block the action of toxin on cultured cells if added as late as 30 minutes to 1 hour after adding toxin, but not later. Thus variations in susceptibility of various mammalian cells may depend on whether they have suitable binding sites for the A-B toxin complex, and membrane activities which allow fragmentation and penetration of fragment A into the cells. Apparently only 200 to 300 toxin molecules are required to kill a cultured cell. Whereas a shutdown in protein synthesis by cells takes 1½ to 3 hours, morphologic evidence of injury does not appear for several hours. For considerations of pathogenicity at the organ and systemic levels see Section H.

Apparently nontoxinogenic strains may sometimes produce mild respiratory tract disease often with membrane formation but without systemic toxemia. Whether actual membrane formation in such cases may result from mixed infections is controversial. Indeed the basis of disease production by any strain or species of *Corynebacterium* that is not known to produce an exotoxin is uncertain but could rest with cord factor and/or other cell products. Why *C.*

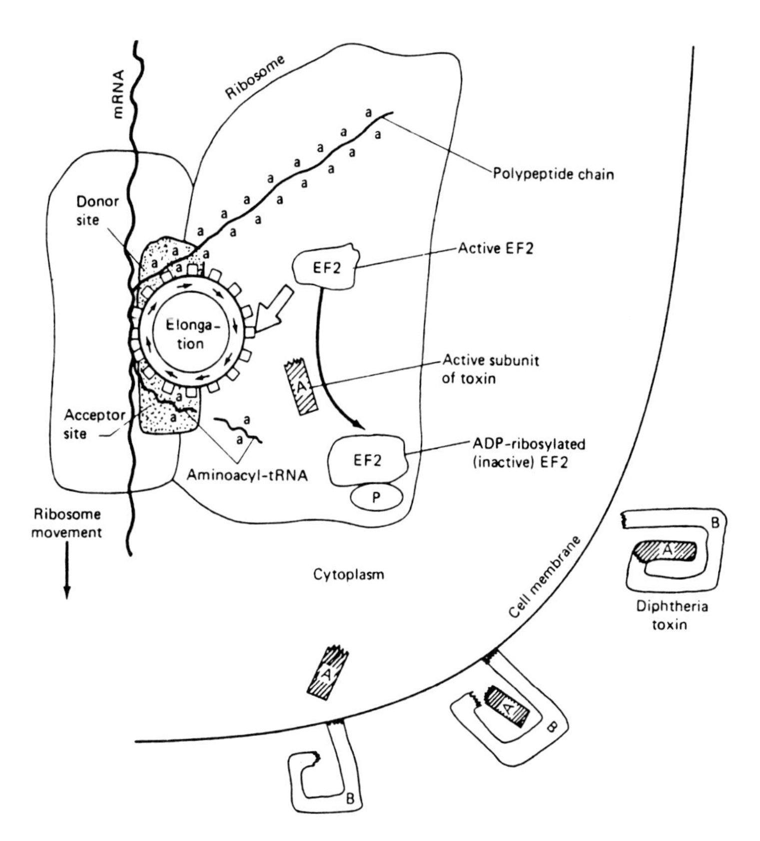

Fig. 15–1. Mechanism of action of diphtheria toxin. The diphtheria toxin molecule consists of a toxin-binding (B) peptide and a toxin active (A) peptide. The toxin A subunit enters the cell and catalyzes a reaction resulting in the inactivation of elongation factor 2 (EF2). Inactivation of EF2 stops the assembly of the polypeptide chain. (Adapted from: *Medical Microbiology,* Sherris, JC, (ed), New York, Elsevier, 1984.)

diphtheriae armed with its potent toxin fails to invade deep tissue beyond the mucosae remains a long unexplained mystery.

J. IMMUNITY

The dominant factor responsible for acquired immunity to respiratory diphtheria is antitoxin comprising Abs of the IgG and IgA classes; however, the relative roles played by antitoxin Abs of various classes of immunoglobulins are not known. Much of the antitoxin Ab may be non-precipitating. Evidently the power of the organisms to invade the respiratory mucosa and to injure depends largely on toxin, for antitoxin immunity is protective. However, total immunity to toxinogenic strains undoubtedly involves both antibacterial and antitoxic immunity and is never complete with respect to carriage. The nature of antibacterial immunity is unknown. In any event the immune response to K Ags probably contributes importantly to antibacterial immunity to *C. diphtheriae* although the mechanisms involved have not been elucidated.

The nature of immunity to skin diphtheria is less certain than the nature of immunity to respiratory tract diphtheria (Sect. H). Since skin infection occurs in an existing wound and usually involves other organisms, *factors which determine susceptibility and resistance to skin diphtheria are probably multiple and complex.*

Antitoxic immunity strongly opposes pharyngeal infection by *C. diphtheriae* and disease but the extent to which it may discourage carriage of virulent organisms is uncertain. Likewise, little is known about the carriage of nontoxigenic avirulent organisms in either immunized or nonimmunized populations or what, if any, influence they may have on carriage or infection by virulent organisms. In some studies the carriage rates of avirulent organisms have been found to be substantial, especially among strains of the mitis type. Healthy subjects exposed to carriers usually acquired carriage which terminates within a few days or weeks.

Although high herd immunity induced by mass immunization with diphtheria toxoid including booster doses every 10 years is accompanied by a low carrier rate of virulent organisms, it is uncertain whether such low carrier rates are due principally to antitoxic immunity per se or to a lesser chance of acquiring carriage because of a lowered incidence of disease and paucity of carriers in the population. Whereas virulent organisms can be carried in the face of antitoxic immunity, long-term carriage is rare. It is obvious that *the nature of immunity regulating carriage of either virulent or avirulent organisms is uncertain. Indeed the factors which oppose carriage may be more concerned with forces of "antibacterial" immunity involving Abs that block adherence to host cells and the antagonism afforded by the normal flora of the respiratory tract than with acquired immunity to toxin.*

Because immunity to respiratory tract diphtheria is due largely to antitoxin, susceptibility to the local effects of a small dose of toxin injected in the skin (the Schick test) provides a rough measure of immunity and is occasionally used for this purpose.

The *nonimmune individual who is not allergic to the toxin* will give a true Schick-positive reaction by responding to the Schick test dose or toxin (1/50 MLD for the guinea pig) with a late spreading inflammatory reaction of long duration, which is characterized by erythema, swelling and tenderness. The reaction usually appears by about 24 hours and peaks on the 5th to 7th day. In contrast *the immune individual who is not allergic to the toxin* will give a *"true Schick negative"* test by failing to react to the toxin. The average threshold level of antitoxin in serum that determines whether an individual will be Schick positive or negative is about 0.03 units/ml.

The Schick test is complicated by the fact that some individuals, especially older children and adults, may possess a delayed type of hypersensitivity to the toxin (immediate sensitivity is extremely rare). Consequently the Schick test is conducted by *injecting highly purified toxin* intradermally into the volar surface of one forearm and a *control toxoid preparation* into the opposite forearm.* In case the individual is *both immune and allergic,* allergic inflammation will develop at both sites within 12 to 18 hours but, because the true toxin reaction is lacking, will fade within 48 to 72 hours. This reaction in

*The use of highly purified toxoid preparations has largely eliminated allergic reactions due to contaminating antigens.

immune individuals is the so-called *"pseudoreaction."* In the occasional individual who is *nonimmune but allergic* to the toxin the early allergic reaction present at both sites will fade by 48 to 72 hours leaving a persisting toxin reaction *(Schick reaction)* at the toxin site. This reaction in the susceptible person is called the *"combined reaction."* Allergic reactions to toxin are rare in young children but increase in incidence with age and are lower in toxoid immunized populations than in populations not artificially immunized, presumably as the result of a higher carrier rate and more frequent inapparent infections in nonimmunized populations. *The Schick test often incites the production of antitoxin, especially if the anamnestic response is involved.*

Although the Schick test is only an approximate measure of existing immunity, it is an even less accurate measure of susceptibility to the actual disease since it does not fully allow for the anamnestic response, which in the face of infection can be profound, or exposure to highly virulent organisms, which can overwhelm some Schick-negative individuals. However, *in general, individuals who are Schick-negative are immune to diphtheria and those who are Schick-positive are susceptible* and need immunization. Since administration of the usual immunizing dose of toxoid can cause violent reactions in toxoid-allergic individuals, adults and older children to be immunized should first be tested for sensitivity to toxoid. The usual, although not infallible, procedure used is to administer the *Moloney test* with 0.1 ml of a 1:10 dilution of toxoid, and then to proceed cautiously with immunization, depending on the nature of the local response. *In toxoid-allergic individuals 3 weekly injections using the Moloney test dose of toxoid will usually serve to stimulate an anamnestic response and reawaken a good immunity.* Although Moloney-negative individuals can be given the full immunizing dose of toxoid, it is safer to give repeated small doses of toxoid.

K. LABORATORY DIAGNOSIS

In diphtheria the difference between recovery and death often hinges on the promptness of administering antiserum; delays involving hours or even minutes may be crucial. No rapid and reliable routine method for laboratory diagnosis has been developed to date. Consequently it is current practice to diagnose the disease on clinical grounds alone, which is most difficult in mild cases, and to use the laboratory only to judge the accuracy of clinical diagnosis in retrospect.

Because diphtheria is rare in the USA, present-day physicians misdiagnose an inordinate number of the cases they see; older physicians, however, can usually correct such misdiagnoses; however, it is often "too late."

Material used for laboratory diagnosis should be collected directly from the lesions and the nasopharynx by the physician before any therapy is begun and promptly transported to the laboratory for staining and culture (Sect. E). Since invasion of deep tissues is rare, blood culture is not worthwhile. A smear, stained with dilute fuchsin, should be made and examined to rule out the possibility of Vincent's disease, and a tellurite plate, a cystine-tellurite plate, a sheep blood agar plate and a Loeffler's slant should be inoculated and incubated at 35° to 37°C. *No attempt should be made to identify C. diphtheriae by examining smears* made from material taken directly from the lesions because the results are frequently erroneous and misleading even when negative.

The cultures should be observed after 15 to 24 hours at 37°C and, if negative, again at 48 hours. The blood plate should be examined for beta hemolytic streptococci, the tellurite plates for colonies typical of *C. diphtheriae* and Loeffler's medium for organisms typical of *C. diphtheriae;* colonies on cystine-tellurite medium vary from dark gray to black. Special procedures for rapid diagnosis, including the fluorescent Ab technique on direct smears, and toxin detection in saliva have been proposed but have not been found to be suitable for routine laboratory use.

Final identification of virulent *C. diphtheriae* demands a virulence test for toxinogenicity which is done with a pure culture derived from a tellurite colony. Cystine-tellurite is highly selective and is the most useful medium for primary isolation of *C. diphtheriae.* In some laboratories animal virulence tests have been replaced by the simpler in vitro double-diffusion precipitation-in-gel test of toxinogenicity devised by Elek, which serves to specifically identify toxin-producing organisms growing on potassium tellurite agar into which specific antitoxin is allowed to diffuse (Fig. 15–2). The test is only reliable when done with care using suitable antiserum, control strains and base medium. The classical animal virulence test consists of injecting the isolate subcutaneously into an antitoxin-protected guinea pig and an unprotected control. If the isolate is virulent, the protected animal will be spared but the unprotected control will die and yield postmortem findings characteristic of those produced by diphtheria toxin, especially hyperemia and hemorrhage of the adrenals.

L. THERAPY

If clinical evidence indicates that there is any reasonable chance that the patient has diphtheria, material for culture should be collected promptly and antitoxin treatment *initiated as soon as skin and ophthalmic tests for serum sensitivity can be completed (about 30 minutes).*

Administering antiserum of horse origin to an occasional nonsensitive patient who may not have diphtheria only entails the risk of developing serum sickness which is generally self-limiting, and not permanently harmful. The lethal activities of the toxin are so irrevocable that *no patient should be denied antiserum if there is any likelihood that the disease is diphtheria.*

Since toxin is quickly bound by tissues and soon rendered unavailable for neutralization by Ab, *the success of therapy demands early treatment* with an *adequate dose of antitoxin* (Table 15–1). Antitoxin treatment begun on the first day is highly effective but after the 4th day is of little or no value. The mortality among untreated patients varies greatly in different epidemics ranging from about 12 to 50%. Among all treated cases the average mortality is about 5%. Antitoxin, consisting of concentrated gamma globulin from immune horse serum, is usually administered intramuscularly (but never subcutaneously) as a single dose (100 U to 500 U/lb body weight) in order to ensure maximum immediate neutralization of toxin.* Another objective is to provide an excess of free Ab which will persist for many days beyond the time when a clinical cure is effected. A second injection of antitoxin may occasionally be needed. Patients who are sensitive to the antitoxic serum pose a special problem and must be either placed on "desensitization treatment," which consists of giving

*One unit of antitoxin neutralizes about 100 MLD of toxin.

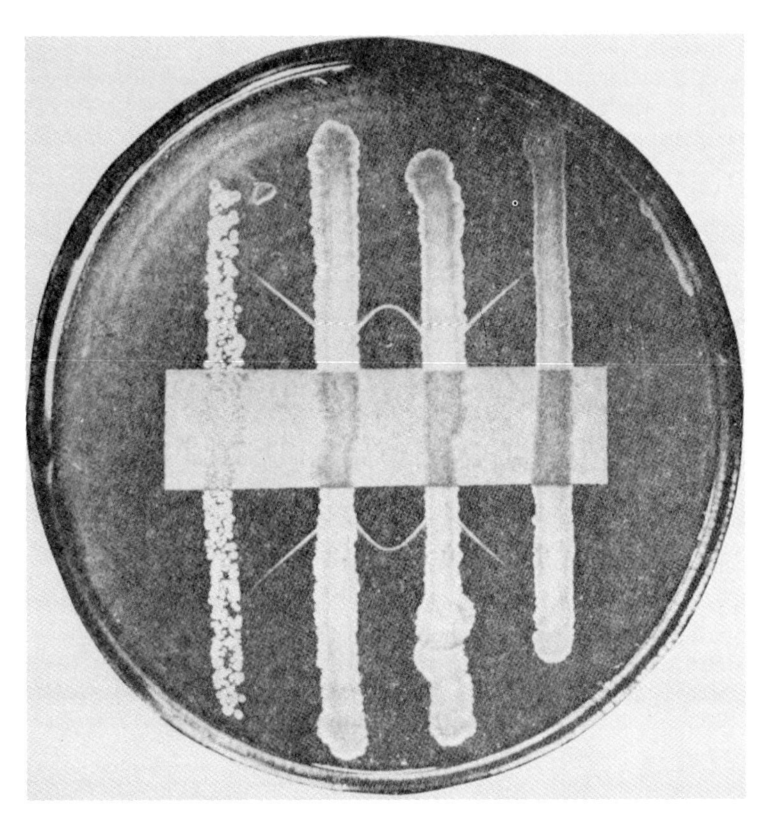

Fig. 15–2. Gel diffusion test for toxin-producing *Corynebacterium diphtheriae.* The organisms are streaked at right angles to a paper strip impregnated with diphtheria antitoxin. The outer two strains are avirulent, whereas the two inner strains are virulent. Note the precipitin arcs indicating lines of identity. A known virulent strain can be placed next to a suspected virulent strain to establish identity of a toxin-producing isolate. (E.O. King, M. Frobisher and E.I. Parsons. Am. J. Pub. Health *39*:1314, 1949.)

Table 15–1. The Influence of Delay in Antitoxin Administration on Mortality in Diphtheria

Antitoxin Given On	*Cases*	*Case Fatality (%)*
1st day of disease	225	0
2nd day of disease	1,441	4.2
3rd day of disease	1,600	11.1
4th day of disease	1,276	17.3
5th day of disease and upward	1,645	18.7

Russell, W.T.: The epidemiology of diphtheria during the last forty years. Med. Res. Council, Special Report. London No. 247, 1943.

the antitoxin cautiously in small increasing increments, a procedure which carries considerable risk of anaphylaxis, or giving antitoxins produced in some other animal, such as the goat, sheep, rabbit, or man.

Antibiotics such as erythromycin, penicillin, cephalosporins, and the tetracyclines have no therapeutic value against preformed toxins but tend to discourage continued toxin formation and complications due to other pathogens. In addition, antibiotic therapy tends to render the patient noninfectious for others by hastening the elimination of *C. diphtheriae* from the primary lesions and arresting convalescent carriage which otherwise often persists for 1 to 2 months or longer. Convalescent carriers pose a public health problem and should remain in isolation until carriage is terminated (Sect. N). *The exercise of vigilance to prevent or relieve respiratory tract obstruction is another matter of first order importance in therapy and cannot be emphasized too strongly.*

Future prospects for improving therapy are difficult to envision unless research on the mechanisms of toxin action should reveal some biochemical means for blocking injury by toxin after it has become fixed to cells.

M. RESERVOIRS OF INFECTION

The principal reservoirs of infection are human carriers and patients with the disease. On rare occasions animals have served as reservoirs; for example, unpasteurized milk from cows with teat lesions incited by hand milkers who were carriers has caused epidemics. Public health practices in advanced countries are highly effective for limiting and controlling reservoirs of infection and usually succeed in arresting epidemics.

N. CONTROL OF DISEASE TRANSMISSION

An understanding of the epidemiology of diphtheria in different populations demands consideration of numerous factors including the carrier rate, passive immunity in the newborn, opportunities for maintaining active immunity by artificial booster immunization or carrier contact and other possible factors.

Experience has shown that most infectious diseases are only effectively prevented when immunization procedures are introduced which strongly immunize a large majority of the population; in the case of diphtheria when 70% or more of the total population is immune, carrier rates drop, the chain of disease transmission is severely weakened, and its spread becomes limited and manageable. At this point the unimmunized in the herd are benefited because of lack of exposure.

Prevention of diphtheria in the USA currently rests principally on standard mass immunization of children, carrier regulation, and isolation of active cases. In addition, possible contacts are given booster immunization and sometime short-term prophylaxis with antibiotics. It is of singular interest that after mass immunization was introduced in one region in 1940, mortality rates dropped so sharply that deaths were reduced from some 100 per million to near zero by 1950.

Populations in which mass immunization is not practiced are typified by carrier rates of virulent organisms of about 5%, a high proportion of immune adolescents and adults, and frequent and sizable epidemics confined principally to children. The term "carrier" designates the long-term carrier. Convalescent carriers whose carriage extends beyond 10 weeks after infection are reclassified as "carriers". *Populations in which mass immunization is practiced are typified by low carrier rates* (a small fraction of 1%), a low proportion of immune adolescents and adults, and infrequent small epidemics, often among adolescents and adults.

In *"nonimmunized" populations* immunity is often incited naturally by carried organisms, especially by strains of low virulence, and is maintained in the majority of adults by frequent contact with carried organisms. Consequently many newborn infants in such populations (probably 90% or more) are well protected by maternally derived Ab. Moreover, because of the high carrier rate an appreciable number of such infants become actively immunized by carrier contact while they still possess a degree of protection due to maternal Ab. Those infants with maternally derived Ab who do not experience contact with carried organisms lose their passively acquired immunity after a few months and become highly susceptible to diphtheria. They constitute a large susceptible segment in such populations upon which epidemics feed.

In populations in which mass immunization of young children is practiced essentially all individuals remain immune for several years and sometimes through puberty into adulthood, and the case and carrier rates are low. Because of the low carrier rate, the booster effect of carrier contact on the maintenance of immunity among adults is largely lost. However, epidemics among adolescents and adults are infrequent in such populations because the reservoirs of infection, the carriers and cases of disease, are few. Since many mothers in such populations do not have appreciable serum levels of antitoxin, most infants lack passive immunity and need early immunization with toxoid combined with pertussis and tetanus immunogens (DPT vaccine).

Ideally, immunization should not only be instituted early in infancy (about 3 months of age) but should be continued into childhood and even adulthood with booster doses of toxoid given at appropriate intervals. Otherwise even the capacity to respond anamnestically may ultimately wane to the point where its effectiveness in the face of infection is largely lost. Substantial immunity is maintained for about 3 to 5 years after immunization. Immunity is not absolute and individual responses vary greatly.

Because continuing immunization into adulthood is not commonly performed, and herd immunity among adults and adolescents is low in advanced countries, *it is important to detect and control carriers and to take emergency prophylactic measures in the case of outbreaks of the disease.* This should include quarantine of patients, booster immunization and administration of antitoxin, or preferably antibiotics, to possible contacts. Cases of active disease should be isolated through convalescence until carriage of the organisms is eliminated; nasal carriage persists longer than throat carriage. Whenever chronic carriers of virulent organisms are discovered in a population, efforts should be made to eliminate carriage. Certain antibiotics, especially erythromycin, are useful for this purpose. Incubationary carriers are rarely detected and constitute especially dangerous reservoirs of infection. A better understanding of the role of phage on virulence conversion in nature and of the factors which determine the carrier state is needed in order to devise more effective methods for detecting and eliminating the carrier state.

Immunization of children is commonly accomplished with highly purified toxoid precipitated with alum or adsorbed to aluminum phosphate gel as adjuvant. The immunizing dose consists of 10 to 20 LF units in a 0.5 to 1.0 ml volume. The material is usually combined with pertussis vaccine and tetanus toxoid to form a pool of Ags (DPT). Immunization is begun at 6 to 8 weeks of age. Two or three doses of the Ag given 1 month apart followed by the first booster doses at 1 year of age and the second at the beginning of school usually provide adequate immunization until adolescence. Schick testing a few individuals before and after immunization serves as check on the adequacy of immunization. In mass-immunized populations allergy to toxin is infrequent in young children and Schick testing prior to immunization is not practiced. In adults allergy to toxin is more frequent and, as indicated in Section J, immunization of adults should be carried out on an individual basis after Moloney testing.

OTHER *CORYNEBACTERIUM* SPECIES AND CORYNEFORMS

Opportunistic corynebacteria are assuming increasing importance as the medical use of immunosuppressive drugs expands.

Some species of *Corynebacterium* other than *C. diphtheriae* have limited pathogenetic potential for man; they include members of the normal flora of the skin and mucosae of man and animals together with animal pathogens. Related organisms called "coryneforms", because of their morphologic resemblance to *C. diphtheriae* and potential pathogenicity for man, are included in this discussion. Among the medically most important species are: *C. ulcerans, C. pseudotuberculosis* and *C. minutissimum. Corynebacterium ulcerans* comprises strains, some of which carry β phage and produce small amounts of diphtheria toxin. They also produce other toxins that cause dermonecrosis and hemolysis. These toxins enable *C. ulcerans* to produce mild diphtheria-like infections.

CORYNEBACTERIUM PSEUDOTUBERCULOSIS (C. OVIS)

Strains carrying phage with a tox gene produce an exotoxin similar to but distinct from diphtheria toxin. The organism, an animal pathogen, produces ulcerative disease in horses and bovines and sometimes in man.

Corynebacterium minutissimum is the principal cause of *erythrasma,* a chronic localized infection of the stratum corneum characterized by reddish-brown scaly areas especially in the groin, axillae and toe webs. When examined with a Wood's lamp they display a coral-red fluorescence.

The list of other opportunists is long. A few of the most important species are: *C. xerosis, C. renale, C. equi, C. bovis,* and *C. haemolyticus.*

These organisms produce a wide range of infections in various mucosae and skin and at internal sites such as heart valves, ears and meninges. Most infections occur in immunodeficient individuals.

REFERENCES

Brubaker, R.R.: Mechanisms of bacterial virulence. Annu. Rev. Microbiol. *39*:21, 1985.
Buck, G.A., et al.: DNA relationships among some tox bearing corynebacteriophages. Infect. Immun. *49*:679, 1985.

Buck, G.A. and Groman, N.B.: Genetic elements novel for *Corynebacterium diphtheriae:* specialized transducing elements and transposons. J. Bacteriol. *148*:143, 1981.
Collier, R.J.: Diphtheria toxin: Mode of action and structure. Bacteriol. Rev. *39*:54, 1975.
Coyle, M.B., et al.: *Corynebacterium* species and other coryneform organisms, *Manual of Clinical Microbiology,* 4th Ed., eds. Lennette, E.H., et al., Washington, DC, Amer. Soc. Microbiol., 1985.
Groman, N.B.: Conversion of corynephage and its role in the natural history of diphtheria. J. Hyg. (Camb.) *93*:405, 1984.
Koppman, J.S. and Campbell, J.: The role of cutaneous diphtheria in a diphtheria epidemic. J. Infect. Dis. *131*:239, 1975.
Lipsky, B.A., et al.: Infections caused by nondiphtheria corynebacteria. Rev. Infect. Dis. *4*:1220, 1982.
Shibil, A.M.: Effect of antibodies on adherence of microorganisms to epithelial cell surfaces. Rev. Infect. Dis. *7*:51, 1985.

16

BACILLUS

Members of the genus *Bacillus* are sporeforming rods of the family *Bacillaceae* which grow best under aerobic conditions. Most species are saprophytic inhabitants of soil, water and decaying organic matter. Few are highly pathogenic for animals and only one, *Bacillus anthracis,* the cause of anthrax (Charbon, Milzbrand), is a pathogen for man. *Bacillus cereus* has been implicated in food poisoning. In addition, *B. subtilis* has been involved in bacteremias and eye infections in heroin addicts and in infections of immunocompromised patients.

A. MEDICAL PERSPECTIVES

Anthrax played an important role in the early development of medical bacteriology (Chap. 1). It was employed as a model by such investigators as Davine, Koch and Pasteur; Koch used *B. anthracis* to perfect his pure culture technique and to prove his famous postulates on the etiology of infectious disease, and Pasteur demonstrated the worth of his attenuated anthrax vaccine in a public field trial which brought him world acclaim.

Herbivores, ranging from goats to wild elephants, are common primary hosts of *B. anthracis;* infection is most frequent *in cattle* and occurs in "anthrax regions" where the soil commonly serves as a permanent reservoir of infection. Diseased herbivores and their products and remains serve as the principal sources of infection for carnivores and omnivores, including man. Being a sporeformer, the organism can remain alive and infectious on dried material for decades, including all sorts of animal products that have not been properly sterilized or disinfected. Anthrax has been the bane of domesticated bovines for centuries. It was probably among the "7 plagues of Egypt" at the time of the Pharaohs and *remains a threat to the livestock industry in areas where pastures with suitable soils have become contaminated.*

The current concept is that the organism is a *soil saprophyte* which, in its pathogenetic role, should be considered to be an opportunistic pathogen and *facultative parasite.* Infection of herbivores commonly results from spores. However, the relative importance of different routes of infection is uncertain and varies, depending on whether the animals are stable fed or pastured. In grazing animals, infection probably occurs most frequently through breaks in the skin (many of which are inflicted by biting insects) which are exposed to soil, dust, or water contaminated with spores. Infection probably occurs less often by ingestion or inhalation of contaminated forage foodstuffs and dust. Carriage of the organism does not occur and transmission by blood-sucking insects is infrequent. Direct animal-to-animal transmission among herbivores is uncommon; *the organism is not capable of perpetuating itself in nature, and is solely dependent on host-to-host transmission.* Although it has long been known that "anthrax regions" are established by contamination of the soils, diseased animals and their excreta and products, the manner by which the organism persists in such regions has been difficult to elucidate. Fields have been known to harbor the organism for 50 to 100 years under circumstances in which the chances of recontamination were highly unlikely. It was first proposed that spores may be able to survive in soil for decades without germinating and that long spore survival plus periodic recontamination by diseased animals may account for prolonged persistence of the organism in the soil. *Instead, it appears that prolonged persistence in soil depends on irregular periodic cycles of spore germination, vegetative growth and resporulation as determined by the microenvironment.* Prolonged persistence of the organism occurs only in soils having pH values near neutrality or above, particularly calcareous soils, and demands spore-vegetative cell-spore recycling. Recycling occurs under restricted conditions of temperature, moisture, pH and available organic matter. Such conditions are only met periodically in "incubator areas" which are often only small locales within a field. Water-killed grasses and decaying vegetation overlaid by silt deposits favor the growth of *B. anthracis.* Ponds, ephemeral streams, etc. provide ideal incubator areas when flood-drought cycles occur to cause growth and killing of vegetation and when minimum daily temperatures above 60°F occur to promote spore formation. Humidity above 80% and temperatures above 70°F favor spore formation. Consequently, within an "anthrax region" the production of crops of spores adequate to incite anthrax outbreaks tends to occur focally and sporadically, sometimes years apart. *It is highly probable that once the organism becomes established in the soil of a favorable region, it seldom if ever fails to persist indefinitely as a saprophyte unless incubator areas are eliminated by natural events or altered land management, such as soil drainage and tillage. Obviously it is important to maintain constant vigilance to avoid the establishment of new "anthrax regions" since, once established, they may remain indefinitely as a cardinal sin of pollution against both man and beast.*

Anthrax, which has been endemic in Africa, Asia and Europe for centuries, was spread to distant lands with the advance of Western civilization. Anthrax regions in the USA are distributed principally in areas of old Spanish and French settlements of the South and Southwest, along the routes of the early cattle drives from Texas to Montana and in the neighborhood of tanneries in the New England states.

Although good sanitary practices and vaccination have brought anthrax in developed countries under good control, large numbers of domesticated and wild animals and thousands of human beings (20,000 to 100,000) are afflicted each year in the world at large. Anthrax remains one of the major livestock diseases in the world and threatens to increase unless better control practices are adopted. It is currently a danger to the animals on the big game preserves of Africa where adding oral vaccines to the waterholes is being considered.

Anthrax is declining in the USA. During the period 1945 to 1952, 2,785 outbreaks and 14,708 deaths occurred among livestock. Small outbreaks continue to occur almost annually with livestock losses ranging from tens to hundreds. In 1968 a total of 165 cases of anthrax occurred among livestock on 34 farms in the state of California. However, human infections now are relatively rare in the USA. Four cases of inhalation anthrax occurred in a New Hampshire textile mill in 1957 and small sporadic outbreaks continue to arise.

In a recent annual report, 6 human cases of occupational anthrax were reported to the Centers for Disease Control. Four cases were acquired by unvaccinated employees exposed to imported goat hair, whereas 2 cases were associated with anthrax in cattle. *If all individuals at high risk were to be appropriately immunized with the vaccines available, the incidence of human anthrax in the USA should drop to near the vanishing point.* However, the establishment of new anthrax areas by contamination with imported animal products remains a continuing threat.

B. PHYSICAL AND CHEMICAL STRUCTURE

Bacillus anthracis is a large sporeforming Gram-positive nonmotile encapsulated rod averaging about 4 to 6 μm in length by 1 to 1.2 μm in width. Whereas cultured organisms grow in long chains and form central spores, organisms in infected hosts occur singly and in short chains and are devoid of spores. The capsule, a polypeptide composed of D-glutamic acid, is antiphagocytic and is one of the key materials in virulence, the other being a toxin. The formation of both toxin and capsules is favored by the presence of 5% CO_2 and media rich in serum. Other components of the cell are not of medical importance. Spores are formed in cultures, soil and tissue of dead animals. However, spores are never seen in tissues or blood of infected living animals.

C. GENETICS

Mutants of *B. anthracis* vary in their nutritional requirements, ability to form spores, colony morphology, virulence and resistance to drugs, bacteriophage and lysozyme. Repeated subculture tends to select for avirulence. Sporulation bears no relation to virulence because nonsporing mutants and sporeformers are equally virulent when injected into animals. The 48 different species of the genus *Bacillus* exhibit a wide range of G+C content of their respective DNA which is different from most other genera. Accordingly, there is remarkable diversity in phenotypic properties such as chemical composition, optimal temperature growth characteristics and secretion products; plasmids can probably act as mediators of toxin production.

D. EXTRACELLULAR PRODUCTS

Anthrax toxin has been separated into 3 distinct components which are thermolabile and appear to be proteins or lipoproteins. Factor I is designated edema factor (EF). The second component, Factor II, known as protective Ag (PA), induces protective Abs in guinea pigs. The third component (Factor III), which is essential for the lethal effect, is termed lethal factor (LF). Edema factor plus LF has no toxic effects. In contrast, LF plus PA is lethal but does not produce edema, whereas EF + PA only produces edema. A combination of EF, PA and LF exhibits the maximum toxicity. The triple component toxin complex causes edema and necrosis in rabbit skin and death of most animals when given intravenously. Factor II or PA is alleged to be the most immunogenic. Whereas PA alone is immunogenic, EF or LF given alone is not immunogenic. However, combinations of EF + PA, EF + LF, PA + LF as well as EF + PA + LF are immunogenic.

Recently it has been reported that PA + EF blocks phagocytosis of *B. anthracis* by human neutrophils. In addition, these two toxin components suppress phorbol myristate acetate and particle-induced chemiluminescence of neutrophils and concurrently increase the levels of cyclic AMP. These findings indicate that the toxins could markedly perturb phagocytic function.

E. CULTURE

Bacillus anthracis is aerobic and facultatively anaerobic. It grows best at a pH range of 7.0 to 7.4 over a temperature range of 12° to 45°C (optimum 35°C). The organism can be grown on simple media. Colonies are small, opaque, grayish-white and tough. They have fringed edges due to hair-like loops consisting of long, parallel chains of cells. Colonies characteristically exhibit the "medusa head" or "curled hair-lock" appearance due to the long chains of bacilli that grow in parallel. The colonies of encapsulated strains are smooth and those of nonencapsulated strains are rough. Spores are only formed in old aerobic cultures. The organism is proteolytic and fermentative.

F. RESISTANCE TO PHYSICAL AND CHEMICAL AGENTS

Whereas vegetative forms of *B. anthracis* possess the usual resistance of nonsporing bacteria, the spores are highly resistant to drying, heat (boiling 10 minutes) and most disinfectants (oxidizing agents are the most lethal). A temperature of 120°C for 15 minutes is normally used to inactivate the spores. The organism is susceptible to a number of antibiotics (Sect. L).

G. EXPERIMENTAL MODELS

Only an exceptional animal species is completely resistant to injected *B. anthracis,* the frog being one; epidemics involving many species have occurred in zoos. Whereas the white rat is markedly resistant, the mouse and guinea pig are highly susceptible species.

Herbivores, such as sheep and cattle, have been used as experimental models. Ingested spores survive stomach acid and pass through the intestinal epithelium to germinate and multiply in the submucosa and draining lymph nodes. In about 80% of cases the organisms reach the blood to produce septicemia and toxin-death, often within a few hours or even minutes after symptoms are first noticed (apoplectic anthrax). Indeed, except for possible cases of primary meningitis, death of both animals and man infected with *B. anthracis* is almost invariably a consequence of septicemia and the action of the toxin. Some strains or breeds of animals, belonging to species that are highly susceptible to natural infection, such as long-fleeced Algerian sheep, are unusually resistant to experimental infection. These resistant animals should be especially useful for studying the mechanisms of resistance to anthrax.

H. INFECTIONS IN MAN

Anthrax is usually established in man by spores present in animal products and seldom if ever by direct host-to-host passage of vegetative cells, such as by blood-sucking insects; carriage does not occur in either animals or man. Infection commonly results from spores entering through skin abrasions (95% of cases in the USA); on occasion spores enter through the respiratory tract by inhalation of dusts from hair, hides, ivory, bone meal, feces or wool (woolsorter's disease). In some regions of the world the organisms may enter the alimentary tract following ingestion of raw or inadequately cooked meat, milk, or blood products. The disease is most frequent among herdsmen, veterinarians, butchers, and industrial workers who handle animal products (industrial anthrax).

The *localized lesion of cutaneous anthrax, the malignant pustule,* is usually solitary; it is characteristically painless, necrotizing, seropurulent, ulcerous and hemorrhagic; the resulting *black eschar* (malignant pustule) is the basis of the term "anthrax" which is derived from the Greek word "anthrakos," meaning coal. It is surrounded by an area of intense nonpitting edema due to a gelatinous exudate in the tissue. The lesion heals spontaneously in about 80% of subjects without appreciable local lymphadenopathy. In about 20% the draining lymph nodes become so extensively involved that septicemia and death due to the toxin result within a few days.

Pulmonary anthrax is almost always fatal because septicemia leading to early toxin-death (24 hours) usually sets in before effective therapy can be instituted. Unless there is reason to suspect anthrax, the nonproductive cough and vague pulmonary symptoms commonly prompt the mistaken diagnosis of "viral pneumonitis." Apparently the inhaled spores do not remain in the alveoli but, instead, are transported by alveolar macrophages to the lymph nodes where they germinate and usually overwhelm lymph node defenses to incite a septicemia. Spores germinate within macrophages in vitro and presumably in vivo as well (Fig. 16–1). The resulting vegetative cells escape and multiply extracellularly. Hemorrhagic mediastinitis is a common complication. Pleurisy associated with precordial distress and with organisms in the pleural fluid occurs on occasion and can serve as a useful diagnostic clue.

Meningitis or pneumonia secondary to mediastinal lymph node involvement is rare. Subclinical infections may sometimes occur, as indicated by specific Ab responses, but chronic infection or carriage of the organisms in either animals or man has not been determined.

I. MECHANISMS OF PATHOGENESIS

Spores of B. anthracis are engulfed by macrophages but are not destroyed. The spores germinate to yield vegetative cells that escape and multiply extracellularly. Following skin infection a few vegetative organisms probably always reach the draining lymph nodes but seldom reach the blood in sufficient numbers to cause septicemia and death (20% of cases). In systemic disease the organisms are usually confined largely to lymph and blood channels and rarely enter the CNS. At death the bacilli may reach such enormous numbers in

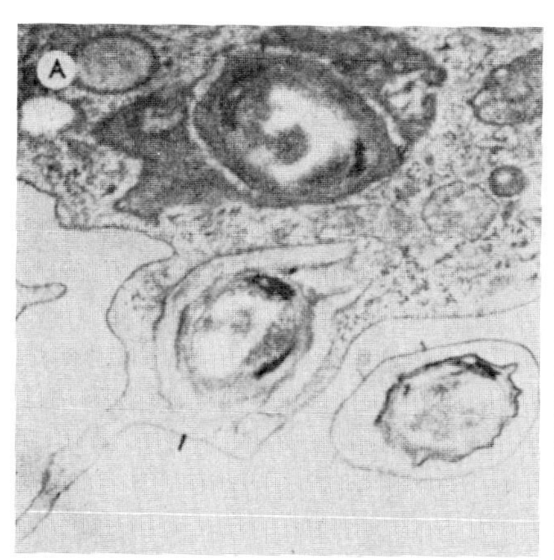

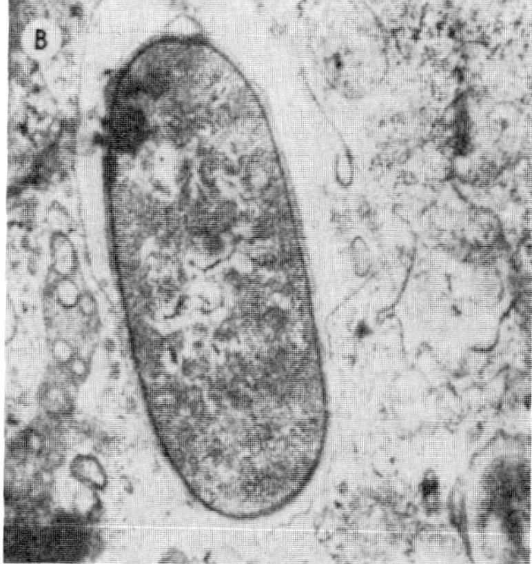

Fig. 16–1. Sections of alveolar macrophages containing *(A)* phagocytized anthrax spores, 17,000 ×. *(B)* Anthrax spore that germinated in the phagosome, 27,200 ×. (From Shafa, F., Moberly, B.J., and Gerhardt, P.: J. Infect. Dis. *116*:401, 1966. The University of Chicago Press.)

animals of certain species that blood capillaries, especially of the viscera, can become literally occluded with them.

The toxic basis of death in septicemia was not appreciated until the discovery of anthrax toxin in 1954. The mechanisms responsible for systemic toxicity are poorly understood. Evidently injury is irreversible because *late destruction of bacilli with antibiotics does not prevent toxic death.* Toxin seems clearly to have primary toxicity for the CNS, leukocytes and certain cultured cells. However, there is little evidence to indicate what cell types may be attacked to account for the edema, hemorrhage and necrosis of the local lesion, the malignant pustule, or the hemoconcentration, extreme anoxia, respiratory failure and renal failure of systemic shock.

Virulence is due to the combined effects of the toxin and the capsule; whereas both are antiphagocytic, the toxin has the additional pathogenetic activity of being toxic for other cells as well as phagocytes.

J. MECHANISMS OF IMMUNITY

The forces of immunity presented by different animal species appear to vary greatly. Compared to animals, man is moderately resistant. In susceptible animals and man the destruction of *B. anthracis* probably depends largely on the activities of neutrophils. It is questionable whether CMI and/or nonantibody anthracidal substances present in serum play a significant role in either innate or acquired immunity. Antigenic virulence factors of most bacteria are effective immunogens and, in theory, acquired immunity to anthrax should involve the combined effects of antitoxic and anticapsular Abs. However, protection by anticapsular Ab has only been demonstrated in the mouse. The role of antitoxin in protecting against infection is easily demonstrated; it evidently acts on toxin in two ways, namely, by blocking both its antiphagocytic effects and its "toxic effects."

The best animal vaccines are spore preparations made from attenuated nonencapsulated, toxinogenic strains plus alum-precipitated toxin. An alum-precipitated protective antigen (PA) vaccine protected 100% of the animals within a month after vaccination. However, immunity was temporary and only 52% protection was seen at the end of 3.5 months. A combination of the attenuated spore vaccine plus the precipitated PA vaccine provided a significant increase in both the level and duration of acquired immunity. The use of a living spore vaccine in South Africa has reduced the incidence of anthrax in cattle by over 99%. Although an effective alum-precipitated, pooled toxin is available for use in industrial plants to protect workers in high risk situations, better vaccines will doubtlessly be developed in the future as knowledge about immunogens and the nature of anthrax immunity advances. A better means of assessing anthrax immunity in man and more information on the possible role of anticapsular Abs in immunity are needed.

K. LABORATORY DIAGNOSIS

Blood should be examined by smear and culture in all subjects in whom anthrax is suspected. Wright-stained smears of exudates from early, but not from late, cutaneous lesions usually show an abundance of typical encapsulated organisms, whereas sputum from patients with pulmonary anthrax (which, locally, is largely restricted to pulmonary lymph nodes) seldom yields positive smears or cultures. In septicemia, Wright-stained smears of centrifuged sediment of blood cleared of red cells with 3% acetic acid often reveal the organism. Specimens should be cultured on a selective medium that will suppress contaminants; suspected outgrowth should always be checked for pathogenicity in guinea pigs or mice since other organisms, especially *B. cereus,* closely resemble *B. anthracis.* Characteristically, *B. anthracis* isolates produce rough colonies when grown in the absence of increased CO_2 but produce mucoid colonies when grown on a bicarbonate medium in an atmosphere of 5% CO_2. Other identification procedures include the string-of-pearl type of growth on nutrient agar containing 0.05 unit penicillin per ml, lysis by γ phage and fluorescent Ab techniques applied to direct smears of the lesions or cultured organisms.

L. THERAPY

Bacillus anthracis is highly susceptible to penicillin. Tetracycline can be used in patients sensitive to penicillin. Other potentially useful antibiotics are chloramphenicol, cephalosporins, and the aminoglycosides. However, in vitro testing for resistance should be done. The major problem in management is that diagnosis must be made early to effectively treat the disease. Cutaneous anthrax lesions should not be incised or drained because this enhances dissemination of the organism. In pulmonary anthrax, drug therapy and supportive measures are usually initiated too late to be of value.

M. RESERVOIRS OF INFECTION

Whereas soil commonly serves as a permanent reservoir of infection of herbivores, the only reservoir for man is the infected animal or its excreta or products.

N. CONTROL OF DISEASE TRANSMISSION

Anthrax is a reportable disease in both man and animals. Since carriage does not occur, carriers are not a problem. Control of the disease in animals by vaccination and sanitary practices is the first and major step for preventing infection in man. Vehicles that may contribute to infection in animals and man and to contamination of soil that may lead to the establishment of new anthrax areas are: fertilizers, animal feeds, animal products, offal and wastes, industrial wastes, fodders, hay, insects, carrion birds and wild animals. Insects that have fed on sick or dead animals are a special hazard to man. Also in recent years, cases of human anthrax have been traced to bone meal. Control measures include restrictions on the importation, movement, handling and processing of vehicles and animals from anthrax areas and disposal of infected animals by burning or deep burial in lime pits. It must be kept in mind that vegetative cells which escape from the infected animal via excreta, blood and exudates readily form spores. Unique examples of breaks in surveillance that have led to human epidemics are: a large outbreak during World War I due to imports of contaminated hair used to make shaving brushes and a recent small outbreak among factory workers engaged in cutting and polishing ivory.

BACILLUS CEREUS

This organism is similar to *B. anthracis* except it is commonly motile, β-hemolytic and is not susceptible to γ phage. *Bacillus cereus* causes two clinical types of food poisoning. One type has an incubation period of about 4 hours after eating and is characterized by severe nausea and vomiting. It can be mistaken for staphylococcal food poisoning. Foods such as fried rice have been implicated in epidemics. The second type of food poisoning has an incubation period of about 16 to 18 hours and is characterized by abdominal cramps and diarrhea; it can be confused with clostridial food poisoning.

The events leading to food poisoning are usually dependent on spores which survive the cooking process. The spores germinate and the vegetative cells multiply and produce enterotoxin; two types of enterotoxin have been described, one of which acts by stimulating adenyl cyclase production. Diagnosis is based on isolating *B. cereus* from the implicated food as well as the stools of the victims.

Bacillus cereus has also been implicated in opportunistic infections in patients carrying surgical implants (prosthetic devices) and in immunodeficient individuals, immunosuppressed transplant recipients, and patients with leukemia.

It is important to recognize that *B. cereus* unlike *B. anthracis* produces penicillinase and cephalosporinase and, as a consequence, is resistant to β-lactam antibiotics. The organism is commonly susceptible to aminoglycosides, vancomycin and clindamycin.

BACILLUS SUBTILIS

Bacillus subtilis is ubiquitous and abounds in air, dust, organic material and soil. It is a strict aerobe, is always motile and is a common contaminant in cultures in the laboratory. Like *B. cereus* it can cause opportunistic infections in immunosuppressed patients. It can produce septicemia in drug addicts exposed to contaminated street heroin. Infections due to *B. subtilis* respond favorably to β-lactam antibiotics.

OTHER *BACILLUS* SPECIES

Several *Bacillus* species can cause disease in insects and are commonly used in insect control. The most commonly used organism is *B. thuringiensis* which kills the larvae of *Lepidoptera.* It is of special interest that this organism forms toxigenic crystalline bodies during sporulation. When the spores germinate, the toxin is released by enzymatic action in the larval gut.

Another species, *B. stearothermophilus,* is used to evaluate efficiency of sterilization procedures by subjecting spores to routine sterilization procedures.

REFERENCES

Brachman, P.S.: Inhalation anthrax. Ann. N.Y. Acad. Sci. *353*:83, 1980.

Lincoln, R.E., and Fish, D.C.: Anthrax toxins. In *Microbial Toxins.* Vol. III. S.J. Ajl, et al., Eds. New York, Academic Press, 1970, p. 361.

McKendrick, D.R.: Anthrax and its transmission to humans. Cent. Ar. J. Med. *26*:126, 1980.

Metchnikoff, E.: Classics in infectious diseases. Concerning the relationship between phagocytes and anthrax bacilli. Rev. Infect. Dis. *6*:761, 1984.
Mikesell, P., et al.: Evidence for plasmid-mediated toxin production in *Bacillus anthracis.* Infect. Immun. *39*:371, 1983.
O'Brien, J., et al.: Effects of anthrax toxin components on human neutrophils. Infect. Immun. *47*:306, 1985.
Shafa, F., et al.: Cytological features of anthrax spores phagocytized *in vitro* by rabbit alveolar macrophages. J. Infect. Dis. *116*:401, 1966.
Sirisanthana, T., et al.: Outbreak of oral-oropharyngeal anthrax: an unusual manifestation of human infection with *Bacillus anthracis.* Am. J. Trop. Med. Hyg. *33*:144, 1984.
Tuazon, C.U., et al.: Serious infections from *Bacillus* species. J.A.M.A. *241*:1137, 1979.
Turnbull, P.C.B.: *Bacillus cereus* toxins. Pharmacol. Thera. *13*:453, 1981.
Van Ness, G.B.: Ecology of anthrax: anthrax undergoes a propagation phase in soil before it infects livestock. Science *172*:1303, 1971.

17

CLOSTRIDIUM

Tetanus

Clostridial Myonecrosis (Gas Gangrene), Anaerobic Cellulitis and *Cl. perfringens* Food Poisoning

Botulism

Clostridium difficile

The genus *Clostridium* includes the anaerobic members of the family *Bacillaceae;* it comprises a large number of species of Gram-positive spore-forming rods. Although most species lead only a saprophytic existence, some exist as both saprophytes and parasites. Many species can assume the role of *opportunistic pathogens* for man. Clostridia are present in soil, dust, sewage and mud and some species are often abundant in animal and human feces and sometimes in the female genital tract. A special attribute of the clostridia is that, in common with all members of the family *Bacillaceae,* the spores that they form are extremely resistant to physical and chemical agents and can survive in nature for long periods of time. Unlike the more susceptible vegetative cell they are highly resistant to many common disinfectants and some spores can survive 100°C. Consequently, the sterilization procedure of autoclaving at 120°C for 15 minutes or more is expressly designed to kill spores.

The pathogenic clostridia are characterized by their capacity to synthesize and secrete highly potent exotoxins which are important in the pathogenesis of the five diseases of man that they produce, namely, *tetanus, clostridial myonecrosis (gas gangrene), anaerobic cellulitis, Cl. perfringens food poisoning and botulism.* Although the clostridia are widely distributed in nature, infections due to these organism are comparatively rare. Lacerated external wounds, large or small, particularly puncture wounds contaminated with foreign materials such as soil, dust, feces, bits of clothing and fragments of glass, that often contain spores, favor the development of clostridial infections by providing anaerobic and other conditions that allow spore germination and bacillary multiplication. Injuries involving the intestines or uterus also frequently provide sites of clostridial infection.

It should be emphasized that irrespective of the cause, e.g., chemical or physical injury or lack of blood supply, any area of dead or dying tissue to which clostridial spores may gain access is a potential site of clostridial infection. Since the organisms cannot tolerate the aerobic environment of live tissue with an intact blood supply, they can only extend the initial lesions by secreting toxins that kill surrounding tissues, which they can then invade. Except for specific antitoxins, host defenses have little opportunity to act against organisms in the zone of infection. Phagocytes cannot migrate far from vital vascularized tissue to reach organisms in the zone of devitalized tissue, especially organisms contaminating foreign bodies. These facts emphasize the importance of prompt and thorough debridement and drainage of wounds, with removal of all devitalized tissue and foreign materials. The above considerations also make it obvious that the virulence factors of clostridia must rest with their extracellular products and that the defenses of the host must rest largely with antitoxic Abs. The fact that aerobes growing in a wound can promote the growth of anaerobes by creating anaerobic conditions also emphasizes the necessity of directing therapy at control of these organisms as well as the anaerobes themselves.

All of the medically important clostridia are obligate anaerobes that can be cultivated in meat infusion broth media containing a reducing agent, such as thioglycollic acid, or on agar plates incubated in an anaerobic jar (90% N_2 plus 10% CO_2). As a general rule they produce malodorous organic amines derived from their enzymatic action on proteins and peptides. Chopped meat and milk media are commonly used for obtaining primary cultures. A few clostridia are partially aerotolerant and will grow feebly in the presence of low tensions of O_2.

An abbreviated classification scheme of the medically important clostridia is presented in Table 17–1.

TETANUS

A. MEDICAL PERSPECTIVES

Tetanus, a horrifying clinical disease, was recognized by the early Greeks, and old drawings depict the lumbar spasm of tetanus.

The infectious nature of *tetanus* or *lockjaw* was not proven until 1884 when Carolo and Rattone produced tetanus in rabbits. Kitasato isolated the causative agent *Cl. tetani* in pure culture in 1889. In 1890, Von Behring and Kitasato succeeded in immunizing animals by injecting small amounts of toxin and Ramon developed the highly effective immunogen, tetanus toxoid, in 1923. Prior to the development of antitoxin, and especially toxoid, tetanus was a major medical problem, particularly during wars and in other high-risk situations. Tetanus, gas gangrene and other wound infections have taken a frightful toll of lives during wars; it is said that during the Civil War in the USA many more soldiers died of wound infections (largely tetanus and gas gangrene) than died in battle. Tetanus antiserum given the wounded saved many lives during World War I and mass immunization with toxoid during World War II reduced deaths from tetanus to essentially zero in the allied military forces! At the present time the estimated annual case rate of tetanus in the world at large exceeds 350,000 and the number of deaths 160,000. Currently in the USA the annual number of deaths from tetanus is less than 200, largely because of immunologic control. Tetanus is readily prevented by immunization, which represents one of the triumphs of immunology. *Since therapy is often of little or no benefit, every effort should be made to promote routine immunization with tetanus toxoid, including proper booster injections for those at high risk.* With adequate attention to active immunization and prophylactic passive immunization with antiserum, tetanus does not pose a future medical problem of any magnitude in the USA.

B. PHYSICAL AND CHEMICAL STRUCTURE

Clostridium tetani is a relatively slender Gram-positive rod about 3 to 5 μm in length and 0.3 to 0.7 μm in diameter. The young vegetative cells have peritrichous flagella. They develop swollen spores that are located terminally giving the cells the appearance of drumsticks or rackets (Fig. 17–1). The mature spores, which are spherical, do not stain with the Gram stain. Accordingly, developing spores appear as nonstainable refractile inclusions within the cells.

C. GENETICS AND VARIABILITY

Antigenic variability of somatic and flagellar Ags occurs. A common somatic Ag permits identification of *Cl. tetani* with fluorescein-labeled antisera. Strains of *Cl. tetani* have been differentiated into 10 types on the basis of flagellar antigens. However, all strains of *Cl. tetani* produce only one antigenic type of tetanus toxin.

Table 17–1. Laboratory Properties of Pathogenic Clostridia

	Nagler Reaction (Lecithinase)	*Spores*	*Motility*	*β-Hemolysis*	*Lactose Fermentation*	*Glucose Fermentation*	*H_2S*	*Indole*	*Urease*	*Nitrate Reduction*
C. perfringens	+	C	–	+	+	+	+	–	–	+
C. tetani	–	T	+	+	–	–	–	+	–	–
C. botulinum										
Types A,B,E,F	–	ST	+	+	–	+	+	–	–	–
Types C,D	–	ST	+	+	–	+	+	–	–	–
Type G	–	ST	+	+	–	–	+	–	–	–
C. novyi										
Type A	+	ST	+	+	–	+	+	–	–	+
Type B	+	ST	+	+	–	+	+	–	–	+
C. histolyticum	–	ST	+	+	–	–	+	–	–	–
C. septicum	–	ST	+	+	+	+	+	–	–	+
C. sporogenes	+	ST	+	–	–	+	+	–	–	+
C. sordelli	+	C	+	+	–	+	+	+	+	+

T, terminal; C, central; ST, subterminal.

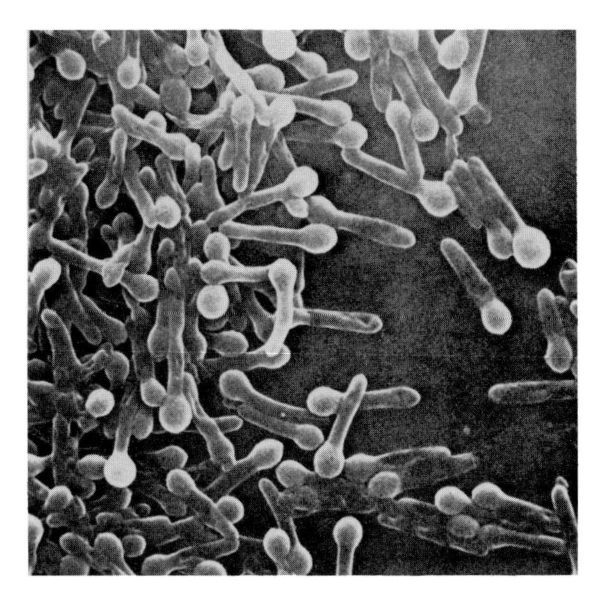

Fig. 17–1. Scanning Electron Micrograph of *Clostridium tetani*. Note terminal spores which give cells a drumstick appearance. From: *Atlas of Scanning Electron Microscopy in Microbiology*. Yoshi, Z. et al. Williams & Wilkins, Baltimore, 1976.

D. EXTRACELLULAR PRODUCTS

All virulent strains of *Cl. tetani* produce tetanus toxin which is a powerful heat-labile (5 minutes at 65°C) neurotoxin (tetanospasmin) liberated from the organisms mainly by cellular autolysis. Tetanus toxin consists of a single polypeptide chain with a MW of 150,000. Upon release the toxin molecule is nicked leaving 50,000 and 100,000 MW moieties. One mg will kill 50 to 75 million mice. It has been estimated that 0.13 μg is a lethal dose for man. It spreads from the local site of infection along motor nerves and via the blood.

The toxin binds with gangliosides in synaptic membranes and acts in the inhibitory synapses of nerves by blocking the normal function of the inhibitory transmitter. The toxin is transmitted intraaxonally against the flow. Upon reaching the nucleus it is transported to the inhibitory interneurons. At this point it inhibits the release of inhibitory transmitters. The inhibition of the release of inhibitory transmitter substances results in the more powerful muscles being dominant. In humans this is seen in the muscle spasms of the masseter muscles with trismus flexion of the upper extremities in conjunction with extension of the lower extremities causing arching of the back (opisthotonus). *Tetanolysin, another toxin, is distinct from "tetanus toxin";* it causes lysis of erythrocytes and kills neutrophils.

E. CULTURE

Clostridium tetani is an obligate anaerobe and grows under the strictest of anaerobic conditions. Nutritional requirements are complex and can be met by culturing on blood agar or cooked meat broth. Faint β hemolysis occurs on blood plates.

F. RESISTANCE TO PHYSICAL AND CHEMICAL AGENTS

Autoclaving at 120°C for at least 15 minutes is required for sterilization of spores. Spores are also highly resistant to most routine disinfectants.

G. EXPERIMENTAL MODELS

There are no routine animal models for tetanus, although laboratory animals are susceptible to tetanus toxin.

H. INFECTIONS IN MAN

Clostridium tetani is an opportunistic pathogen that initiates infection by chance. Tetanus spores can be readily isolated from soil samples, particularly in fertilized cultivated lands. It is important to emphasize that even trivial puncture wounds, especially if "dirty," are potential sites of tetanus infection; for example, tetanus can occur at injection sites in drug addicts and at the site of an insect bite. *Neonatal tetanus* (tetanus neonatorum) occurs when the severed umbilical cord stump (a debilitated tissue) becomes infected soon after birth. It is a major cause of infant mortality in underdeveloped countries; it is most effectively controlled in these regions by immunizing prospective mothers. *Since the organism does not invade healthy tissue having an adequate blood supply, tetanus infection remains localized in the wound area. The clinical symptoms are the result of the action of tetanus toxin which diffuses from the site of infection to reach the bloodstream.* High concentrations of toxin exist at the local site and continue to diffuse and reach the bloodstream for some time after growth of the organisms has been arrested by antimicrobial therapy. In general, the severity of tetanus is inversely proportional to the incubation period, which may vary from several days to many weeks, and to the onset period (interval between the first sign and generalized spasm). If the incubation period is less than 9 days, and the onset period is less than 48 hours, the attack will probably be severe. Tetanus is characterized by violent, painful convulsive contractions of voluntary muscles, usually beginning in the muscles near the infected wound (local tetanus) and the masseter muscles, followed by progressive involvement of other voluntary muscles throughout the body. Once observed the symptoms of severe tetanus are never forgotten; the rigid facial muscles create a sneering countenance and the painful contraction of voluntary muscles, especially of the legs and back, may be so violent that vertebrae are fractured. The patient assumes a rigid position with arched back, the weight being borne on head and heels (opisthotonos; Fig. 17–2). Fever, sweating and cardiovascular symptoms, which are usually present during the peak of severe illness, apparently result from overactivity of the sympathetic nervous system. Death usually results from spasms affecting the muscular apparatus involved in respiration. In a nonfatal case of tetanus the symptoms progress to a point and then gradually regress.

As a rule a poor prognosis is associated with a short incubation period between injury and the first seizure, injury close to the head, extremes of age and the frequency and severity of convulsions. For those patients who recover no permanent residual pathologic condition results unless fractures or pulmonary complications occur during the active disease.

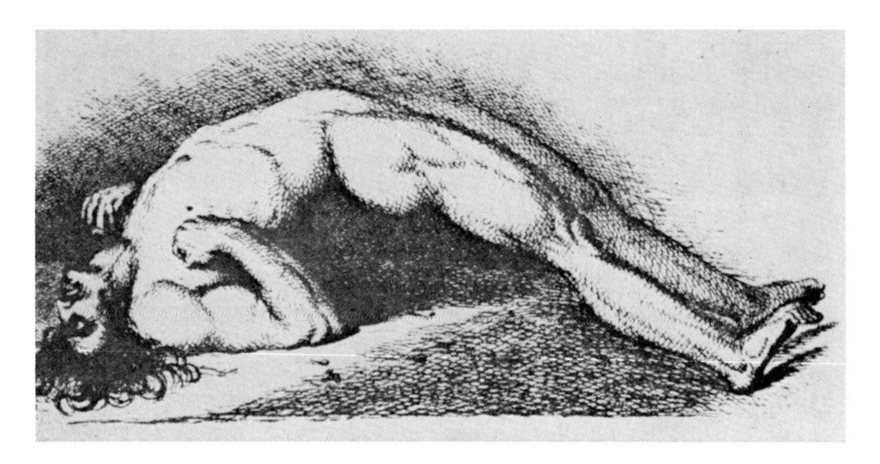

Fig. 17–2. Tetanus in a British soldier wounded in the battle of Corunna in 1809. Note the rigid position of opisthotonos and the "sardonic smile" of lockjaw involving spastic paralysis of muscles which persist in working against one another despite intense pain and exhaustion. Reproduced from a drawing by the Scottish surgeon Sir Charles Bell published in 1832 in his book *The Anatomy and Philosophy of Expression.*

I. MECHANISMS OF PATHOGENESIS

The toxin has a strong affinity for certain nerve tissues and once combined cannot be neutralized by antitoxin. It alters nerve function but no cytotoxicity or local lesions have been found. Ultimately the toxin is inactivated somehow because *patients who recover from tetanus have no apparent residual neurologic defects.* The action of tetanolysin, the cytotoxin, does not contribute importantly to pathogenesis.

J. MECHANISMS OF IMMUNITY

Tetanus toxin can be neutralized by specific Ab providing the Ab can react with the toxin before it reaches its target. Natural tetanus does not confer acquired immunity because the toxin concentration is apparently too small to be immunogenic. Accordingly, recurrent attacks are possible and patients should be immunized with tetanus toxoid at the onset of therapy.

K. LABORATORY DIAGNOSIS

The diagnosis of *tetanus* rests on clinical grounds; as a rule the fully developed disease is not confused with other diseases. The organism is sometimes difficult to isolate and its mere presence in a wound does not confirm or establish the existence of tetanus. Consequently wound culture is not done routinely.

L. THERAPY

There is no successful specific treatment for *tetanus.* Proper nursing care should include the placing of a padded tongue depressor between the teeth to prevent biting of the tongue, and removal of upper respiratory tract (URT) secretions by suction and postural drainage. Sedatives and muscle relaxants are important to prevent tetanic seizures. A centrally acting muscle relaxant such as diazepam and a sedative such as sodium thiopental form a useful combination. *Large doses of penicillin, tetracycline or some other effective antibiotic should be administered to all patients with tetanus.* Antibiotics not only act directly on *Cl. tetani,* but restrict the growth of contaminating aerobes which help to maintain anaerobic conditions favoring *Cl. tetani.* Active immunization with tetanus toxoid should also be started at the onset of therapy. Since antimicrobial chemotherapy does not interfere with the diffusion of preformed toxin from the site of infection, antitoxin should be administered although it may seem to be of limited value. Human tetanus-immune globulin is preferred over horse serum antitoxin. Careful debridement of the wound and the establishment of adequate drainage under the protective cover of antitoxin is highly important. Tracheostomy with nasogastric tube feeding, to prevent inhalation of pharyngeal contents and to permit artificial respiration, is another valuable therapeutic measure in severely ill patients. A combination of curare and intermittent positive pressure ventilation through a cuffed tracheostomy tube and anticoagulants may be indicated in the most severe cases. Patients with mild tetanus usually survive, but in those with more severe disease the best treatment only reduces the mortality from about 45 to 20%. A guide to tetanus prophylaxis in wound management is given in Table 17–2.

M. RESERVOIRS OF INFECTION

Clostridial spores are widespread in nature and are particularly prevalent in cultivated soil and feces.

N. PREVENTION OF DISEASE

Mass immunization with tetanus toxoid is the first and most effective step for preventing tetanus. Ideally, active immunization with alum precipitated toxoid should be performed during the first year of life, and a booster dose of toxoid should be given at 4 to 6 years of age and each decade thereafter. Preventive measures following injuries that predispose to tetanus include wound debridement, drainage, administration of antitoxin, a booster dose of toxoid and early chemoprophylaxis with drugs, such as penicillin or tetracyclines. *In all instances in which there is any reasonable chance that tetanus may develop, a booster dose(s) of toxoid should be given. Antibiotics alone are ineffective if given later than about 4 to 6 hours after injury.* If human hyperimmune antiserum is not available and a foreign antiserum (horse or bovine) to which the patient is sensitive is given, it may be rapidly rendered ineffective by immune elimination; there is also the problem of anaphylaxis and serum sickness.

CLOSTRIDIAL MYONECROSIS, ANAEROBIC CELLULITIS AND *CL. PERFRINGENS* FOOD POISONING

The most important clostridia responsible for myonecrosis (gas gangrene) and anaerobic cellulitis are *Cl. perfringens, Cl. novyi* and *Cl. septicum.* Three additional organisms, although of lesser importance, are *Cl. histolyticum, Cl. sordelli* and *Cl. fallax.* These clostridial species, which are referred to as histotoxic clostridia, produce a variety of toxins with different potencies. These

Table 17–2. Guide to Tetanus Prophylaxis in Wound Management

History of Tetanus Immunization (Doses)	*Clean, Minor Wounds*		*All Other Wounds*	
	*Tetanus and Diphtheria Toxoids (for Adult Use)**	*Human Tetanus Immune Globulin*	*Tetanus and Diphtheria Toxoids (for Adult Use)**	*Human Tetanus Immune Globulin*
Uncertain	Yes	No	Yes	Yes
0–1	Yes	No	Yes	Yes
2	Yes	No	Yes	No†
3 or more	No‡	No	No§	No

From Immunization Practices Advisory Committee, Centers for Disease Control: Ann. Intern. Med., *95*:723, 1981.
*For children less than 7 years of age, diphtheria and tetanus toxoids and pertussis vaccine adsorbed (or diphtheria and tetanus toxoids adsorbed, if pertussis vaccine is contraindicated) is preferred to tetanus toxoid alone. For persons 7 years of age or older, tetanus and diphtheria toxoids (for adult use) is preferred to tetanus toxoid alone.
†Yes, if wound more than 24 hours old.
‡Yes, if more than 10 years since last dose.
§Yes, if more than 5 years since last dose. (More frequent boosters not necessary and can accentuate side effects).

clostridia are widely distributed in nature and, as a consequence, contamination of wounds with more than one species is common.

Since *Cl. perfringens* is the most common and important species involved in the diseases covered in this section, it is given the most attention.

A. MEDICAL PERSPECTIVES

Clostridial myonecrosis (gas gangrene) is chiefly a disease of warfare although a significant number of civilian cases occur. The disease has always been an important medical problem and although improvements in prophylaxis and therapy have been made, it will undoubtedly continue to be of substantial concern for the foreseeable future.

B. PHYSICAL AND CHEMICAL STRUCTURE

Clostridium perfringens is an encapsulated, nonmotile, Gram-positive rod that occurs singly and in pairs. The organisms are 2 to 4 μm in length and 1 to 1.5 μm in width. Unlike *Cl. tetani* the spores occur in the subterminal position.

C. GENETICS AND VARIABILITY

Clostridium perfringens is classified into 5 serologic types (A–E) based on the toxins it produces (Table 17–3). Other species of clostridia involved in gas gangrene also produce many and varied toxins. Variants of *Cl. perfringens* that produce low levels of some of the toxins can be isolated.

D. EXTRACELLULAR PRODUCTS

Clostridium perfringens, type A, the most frequent and important agent of gas gangrene in man, produces a wide array of extracellular toxins that possess varying degrees of necrotizing and hemolytic activity. The most important toxin is the α toxin which is lecithinase C. Other products include collagenase, proteinase, hyaluronidase and deoxyribonuclease. Certain "food poisoning" type A strains produce a heat-labile enterotoxin; they usually produce only small amounts of α toxin. The other clostridial species also produce many different toxins which probably play important roles in the pathogenesis of anaerobic cellulitis.

E. CULTURE

Clostridium perfringens, type A, can be identified by anaerobic culture on agar plates containing egg yolk. Lecithinase C (α toxin) causes a precipitate to form around the colonies because it destroys the emulsifying action of lecithin. Specific Ab to α toxin can be incorporated in the agar to neutralize this effect and thus identify the toxin. Double zones of hemolysis, an inner zone of complete hemolysis and an outer zone of partial hemolysis, give a characteristic "target" appearance to *Cl. perfringens* colonies on blood agar.

F. RESISTANCE TO PHYSICAL AND CHEMICAL AGENTS

For information on this topic see chapter introduction.

Table 17–3. Toxins and Antigens of *Clostridium perfringens*[a]

	α	β	γ	δ	ε	θ	ι	κ	λ	μ	ν
A Gas gangrene, puerperal infection, septicemia, food poisoning:											
Classical	+ + + +	−	−	−	−	+ + −	−	+ +	−	−	+
Atypical	+ −	−	−	−	−	−	−	+ −	−	−	−
B Lamb dysentery, foal enterotoxemia, goats and sheep: hemorrhagic enteritis	+	+ + +	+	+ +	+ +	+	−	−	+ + +	+	+
C Sheep toxemia, calves and lambs: hemorrhagic enteritis	+	+ + +	+	+ +	−	+	−	+	−	−	+
Man: enteritis necroticans	+	+	+	−	−	−	−	−	−	−	+
D Sheep, lambs, goats, cattle: enterotoxemia	+ + +	−	−	+ +	+ + +	+ +	−	+ +	+ +	+ + −	+ + −
E Sheep, cattle (pathogenic?)	+ + +	−	−	−	−	+ +	+ + +	+ +	+ +	(+)	+ −

[a] + + +, produced by all strains; +, + +, produced in increasing quantities; + + −, + −, present in some strains only; −, not produced.
From: Willis, A.T. *Anaerobic Bacteriology: Clinical and Laboratory Practice.* 3rd Ed. London, Butterworth & Co. (Publishers) Ltd, 1977.

G. EXPERIMENTAL MODELS

Guinea pigs, pigeons and rabbits are the most susceptible animals to experimentally induced *gas gangrene.* They usually die within 12 to 48 hours. At death the injection site is purplish red and the adjacent muscles and subcutaneous tissues crepitate when pressed, due to bubbles of gas in the tissues. *Clostridium perfringens* can be typed by protecting infected mice with specific antitoxins.

H. INFECTIONS AND DISEASE IN MAN

Aside from simple wound contamination there are two types of clostridial infections, anaerobic cellulitis and true myonecrosis (gas gangrene). Since *Cl. perfringens* can be isolated from 80 to 90% of operative wounds and skin ulcers, it is important to recognize that the presence of clostridia does not indicate onset of disease.

Anaerobic cellulitis does not involve muscles and tends to remain localized. Germination of clostridial spores occurs in the wound where devitalized tissues exist and a low redox potential is established. Foreign material as well as growth of aerobic organisms can contribute to the lowering of the redox potential. These types of infection do not generally produce intense pain, toxemia or invasion of muscle. However, an infection initially manifesting a cellulitis can progress to a true myonecrosis.

Clostridial myonecrosis (gas gangrene) is a highly aggressive invasion of muscle. The infection perpetuates a cycle of progressive tissue injury. *Clostridium perfringens, Cl. novyi* and *Cl. septicum* are the most frequently recognized etiologic agents of myonecrosis; the role of other species as etiologic agents is less certain. Mixed infections with clostridia or with clostridia and certain pyogenic organisms are common. Only about 5% of patients with wounds and devitalized tissues potentially suited for the development of gas gangrene develop the disease.

The incubation period in gas gangrene may be as short as 4 to 6 hours and as long as 72 hours or more. Intense edema and pain develop in the wound. Moderate fever, early delirium and extreme prostration occur. Leukocytosis and thrombocytopenia are common findings and massive hemolysis leading to anemia and renal failure may be observed within 24 hours after onset.

If gas gangrene infection is of sufficient magnitude, enough toxin will be produced to incite a progressive cycle of tissue destruction and extension of the infection. The organisms ferment tissue carbohydrate and protein components leading to the accumulation of bubbles of hydrogen gas between muscle fascia and in other tissues, which can be detected by crepitation under pressure. Excessive gas promotes extension of the lesion by dissecting tissue and exerting pressure effects that interfere with the blood supply. Without treatment the death rate is essentially 100%.

Uterine infection represents a special type of clostridial myonecrosis which usually involves the gravid uterus. Clostridial infections of the uterus account for a significant number of cases of gas gangrene in a civilian medical practice. Such infections were largely the result of criminal abortions by nonmedical practitioners. Uterine myonecrosis is commonly associated with septicemia and intravascular hemolysis which lead to secondary renal failure. These infections progress rapidly and have a high mortality.

Clostridial septicemia results from invasion of the bloodstream, particularly in certain malignancies; *Cl. perfringens* and *Cl. septicum* are the usual etiologic agents and appear to initiate infection in the intestinal tract as the consequence of a progressive malignancy. Septicemia may also occur as the result of gastrointestinal or biliary tract surgery. Death may occur within 24 hours after the onset of clinical symptoms if the patient is left untreated.

Food poisoning caused by *Cl. perfringens* type A is more common than was formerly realized. It is due to ingestion of contaminated food containing preformed *Cl. perfringens* enterotoxin which causes hypersecretion by the small intestine. This type of food poisoning is similar to staphylococcal food poisoning; the incubation period is less than 18 hours. *Enteritis necroticans,* a rare but severe intestinal disease caused by atypical type C strains of *Cl. perfringens,* is usually diagnosed only during surgery or at necropsy. The β toxin of type C strains is responsible for the pathogenesis. Administration of Type C antitoxin to patients with enteritis necroticans has been reported to reduce the mortality rate.

I. MECHANISMS OF PATHOGENESIS

Gas gangrene is a progressive infection because the various toxins destroy tissue, thus creating an expanding anaerobic environment in which the organisms can spread. The primary lesion may extend to involve an entire extremity. Massive edema coupled with gas production within the tissues, toxemia and hemolytic anemia account for the pathologic picture. Alpha toxin probably plays the dominant role in the destruction of tissue.

J. MECHANISMS OF IMMUNITY

The only known acquired defense against gas gangrene results from Abs specific for extracellular toxins. Antibodies are probably incapable of neutralizing toxin that has fixed to cells. Passive immunization is probably of some benefit when antitoxin is given early, in conjunction with the other modes of treatment.

K. LABORATORY DIAGNOSIS

The diagnosis of *clostridial myonecrosis, clostridial cellulitis* or other clostridial infections such as *postabortal sepsis* rests largely on clinical findings. Bacteremia seldom occurs in myonecrosis but is common in postabortal sepsis. Since clostridia are present in small numbers in most external wounds, their isolation is of little significance in diagnosis of wound infections unless a quantitative approach is used. Gram-positive bacilli can sometimes be found in abundance in smears from biopsied muscle and identified by the fluorescent Ab method.

L. THERAPY

Early diagnosis of gas gangrene and prompt application of proper therapeutic regimens are important if success is to be achieved. The success of treatment is often tenuous and a few patients die even when treated by all the modes of therapy, namely, surgical debridement, antibiotics, antitoxins, hyperbaric oxygen and, in extreme cases, the old procedure, amputation. Amputation may sometimes succeed when other methods fail.

The most important approach for treating gas gangrene and less severe clostridial wound infections involves immediate extensive surgical debridement and open drainage. The administration of hyperbaric oxygen to create aerobic conditions in the wound provides another form of therapy which can produce prompt improvement in the patient. Penicillin is the antibiotic of choice. Polyvalent antitoxin appears to have limited value and has been discontinued as a method of treatment in many medical centers.

M. RESERVOIRS OF INFECTION

Clostridial spores are ubiquitous in nature and are particularly prevalent in soil and feces. Infection occurs on occasion due to lack of aseptic practices, such as the use of unsterilized surgical instruments.

N. PREVENTION OF DISEASE

Clostridial myonecrosis can only be prevented by guarding against traumatic injuries and instituting proper prophylactic procedures for patients at risk, such as wound debridement, drainage and chemoprophylaxis. Because of the large number of species of organisms and toxins involved, immunization with toxoids is not considered to be practical even among persons at high risk.

Anaerobic cellulitis is usually a mixed infection with more than one clostridial species, often including species such as *Cl. bifermentans, Cl. fallax* and *Cl. histolyticum;* it is a limited infection and does not involve muscles. The disease may represent an infection that has become stabilized as the result of host control of the infection which in turn limits the amounts of necrotizing toxins produced.

BOTULISM

Botulism is a rare, but severe and often fatal type of intoxication caused by eating food containing toxin produced by growth of the causative organism *Cl. botulinum.*

The disease is more common in animals than in man. For example, large outbreaks have been reported in cattle, poultry and ranch-raised mink. One of the principal concerns of wildlife management groups has been the massive outbreaks of botulism among waterfowl that feed in shallow waters rendered anaerobic at certain times by the high oxygen demand of aquatic flora.

A. MEDICAL PERSPECTIVES

The first case of human *botulism* described (van Ermengen, 1896) was caused by pickled ham contaminated with *Cl. botulinum.* Most of the cases of botulism in the USA result from eating home-canned vegetables and some fruits, whereas in Europe most cases are due to eating smoked, salted or spiced fish and meats. A total of 1,669 cases and 948 deaths due to botulism was recorded in the USA between 1899 and 1967; during the past decade an average of about 20 cases/year was reported.

The threat of botulism will continue in the future because heat-sterilization methods are always subject to errors. It is seldom that even a few months pass without the threat of botulism from large batches of improperly sterilized commercially canned foods.

B. PHYSICAL AND CHEMICAL STRUCTURE

Clostridium botulinum is a motile Gram-positive rod (3 to 8 μm long by 0.5 to 1.3 μm wide) that occurs singly or in short chains. Swollen spores generally appear in a subterminal position.

C. GENETICS

Clostridium botulinum comprises several types of organisms, each of which produces at least 1 to 8 serologically distinct forms of the exotoxin. The production of toxins C and D has been shown to depend on lysogeny by certain bacteriophages and some strains of *Cl. botulinum* have been shown to produce both A and F toxins.

D. EXTRACELLULAR PRODUCTS

Clostridium botulinum produces a powerful neurotoxin. Although classed as an exotoxin, botulinum toxin is not released into the medium until the organisms die and undergo autolysis. It is a classical exotoxin which is destroyed by boiling (100°C) for 10 minutes, but not by gastric juices; hence it passes the stomach to reach the small intestine where it is absorbed. The eight known serologically distinct toxins and the respective types of organisms that produce them are designated A, B, Cα, Cβ, D, E, F and G (Table 17–4). Less than 1 μg of a toxic polypeptide derivative (mol wt 150,000) is lethal for man and 1 mg will kill 20,000,000 mice.

Crystalline type A toxin is a protein with a MW of 900,000 made up of three constituents, namely, a neurotoxin, a hemagglutinin, and a nontoxic moiety.

Table 17–4. Animal Species Susceptible to Types of *Cl. botulinum*

Type	*Species*	*Sites of Outbreaks*
A	Human	United States, Soviet Union
B	Human, horse	United States, Northern Europe, Soviet Union
Cα	Birds, turtles	Worldwide
Cβ	Cattle, sheep, horses	Worldwide
D	Cattle, sheep	Australia, South Africa
E	Human, birds	Northern Europe, Canada, United States, Japan, Soviet Union
F	Human	Denmark, United States
G	No outbreaks recognized	

From Smith: *Botulism,* Springfield, Charles C Thomas, 1977.
A single toxin is produced by each type except for types C and D strains. Type C strains produce primarily C_1 (Cα) toxin but also small amounts of C_2 (Cβ), whereas type D strains produce mainly type D toxin but also small amounts of C_2. Prophages are responsible for the dominant toxins in types C and D.

E. CULTURE

It is important to stress the fact that growth of *Cl. botulinum* does not occur at pH values below 4.5. Consequently those home-canned foods which are highly acidic do not cause botulism.

Clostridium botulinum is a strict anaerobe that produces β hemolysis on blood agar except for strains of type G; four cultural groups are recognized.

F. RESISTANCE TO PHYSICAL AND CHEMICAL AGENTS

Canning techniques in which boiling water is used under ambient conditions (cold pack method) are not adequate for killing spores of *Cl. botulinum* which can withstand 100°C for 3 to 5 hours. This accounts for the high incidence of outbreaks following consumption of home-canned food.

G. EXPERIMENTAL MODELS

Laboratory animals are highly susceptible to botulinum toxin and are useful for detecting and identifying the toxin in food; the type of toxin in a food can be determined by demonstrating specific protection with a known antitoxin. Mice are commonly used for this purpose.

H. DISEASE IN MAN

It should be emphasized that botulism is not an infectious disease but, instead, is a severe intoxication resulting from the ingestion of improperly preserved food containing preformed toxin. Cases due to types A (62%), B (28%) and E (10%) are the most common; those due to F are rare. The E toxin is most often associated with improperly preserved fish.

The types of botulism seen in the USA can be classified into 4 categories. (1) Food-borne (most common) botulism, (2) infant botulism (due to ingestion of spores and subsequent multiplication), (3) wound botulism (a rare neuroparalytic disease associated with wounds) and (4) unclassified (those over 1 year of age with symptoms but no identifiable source of transmission).

After absorption in the intestine the toxin reaches susceptible neurons by way of lymphatics and the blood circulation. Symptoms usually begin within 12 to 36 hours after ingestion of food containing the toxin, but may appear as early as a few hours or as late as 8 days. The shorter the incubation time, the more severe the disease. Cranial nerve terminals are affected earliest and most severely. Symmetric descending weakness or paralysis occurs, but sensory and mental functions are not impaired. Hearing may be altered and vision is impaired, probably because of involvement of extraocular muscles. Diplopia develops and nystagmus and fixed dilated pupils may be present. No fever develops, but the mouth tends to become dry and painful. Double vision is followed by difficulty in swallowing and bulbar paralysis. Death usually results from respiratory failure, cardiac arrest, aspiration pneumonia or a combination of these events. The fatality rate is high, but varies with the type of toxin ingested; 32% of patients ingesting type A toxin succumb, whereas a 17% mortality occurs following ingestion of type B toxin and 40% for type E toxin. Recovery from botulism is slow but is usually complete without permanent damage.

Infant botulism results in an acute type of flaccid paralysis revealing weakness of head, face and throat muscles, including the trunk. The disease is caused by intestinal growth of *Cl. botulinum* and accounts for some 10% of "crib deaths," the leading cause of infant mortality in the developed world. Breast milk is relatively protective due to specific IgA Abs; however, these Abs are irregularly present. Death is caused by a paralyzed tongue or pharyngeal muscles that occlude the airway or by paralysis of the diaphragm and intercostal muscles. Peak incidence occurs at 1 to 2 months of age. The neurologic diseases likely to be confused with infant botulism are Guillain-Barré syndrome, myasthenia gravis and cerebrovascular accidents.

I. MECHANISMS OF PATHOGENESIS

Botulinal toxin is absorbed by mucous membranes throughout the gastrointestinal tract and ultimately blocks transmission in cholinergic nerve fibers. Crude toxin of type E strains is potentiated 10- to 1000-fold by trypsin digestion which presumably occurs in the GI tract. The toxin apparently interrupts neural impulses close to the point of final branching of terminal nerve fibrils short of the motor end-plate and prevents release of acetylcholine. Toxins have decreasing affinity for nerve tissue in the order A>E>B; as a result, type A toxin can only be detected in the blood for a few days after ingestion, whereas type B toxin persists for as long as 3 weeks.

J. MECHANISMS OF IMMUNITY

Immunity against botulism rests on neutralization of the toxin by antitoxic Ab and can be induced by either active or passive immunization. Antitoxin can neutralize free toxin but is ineffective against toxin that has been fixed to tissue. The lethal dose is less than the amount required for an Ab response. Consequently, no natural Ab is found in the sera of humans.

K. LABORATORY DIAGNOSIS

The diagnosis of botulism rests primarily on clinical findings. If the suspected food is available, helpful confirmatory evidence of botulism may be gained by injecting mice with extracts or saline emulsions. If the toxin is present, the mice will die within 24 hours but can be protected by type specific antiserum. Fresh serum from a patient with botulism is also toxic for mice; the toxin can be identified by mouse protection tests with A, B, E and F antitoxins.

L. THERAPY

Since death from *botulism* is a consequence of respiratory failure, *early tracheostomy and the use of a mechanical respirator are of utmost importance.* In addition, trivalent type A-B-E antitoxin should be administered promptly by the i.v. route.* Type E or bivalent type A-B antitoxin can be used if the causative type is known. Since the antisera are produced in horses, appropriate skin tests for horse serum sensitivity must be conducted prior to administration of antitoxin. Antiserum administration should never be delayed while waiting for laboratory results. Present evidence indicates that mortality is significantly decreased by proper administration of specific antitoxin in combination with

*Botulinal antitoxins are available at the USPHS Center for Disease Control, Atlanta, Georgia, and the Lederle Laboratories, Pearl River, New York.

supportive therapy. *Prophylactic administration of antitoxins is important and effective in those subjects who have eaten toxin-containing food but whose symptoms of botulism have not become apparent. The first case of botulism, in a group of individuals who have eaten contaminated food, becomes an important index case demanding that prompt measures be taken to protect the remaining exposed subjects.*

M. RESERVOIRS OF CONTAMINATION

The spores of *Cl. botulinum* are widespread in nature throughout the world and are found in many soils and muds. Since the organisms are so ubiquitous, all raw foods may be considered to be potentially contaminated. The different types of *Cl. botulinum* have distinctive geographic distribution patterns. Whereas type A is the most common type in North America, type B predominates in Europe. Type A organisms predominate in the soils of Western USA, but type B organisms predominate in Central and Eastern USA. The type E organism has been isolated from salmon on the West Coast and from fish of the Great Lakes. Type F *Cl. botulinum* has been found in salmon and marine sediments of the West Coast and in crabs from the York River of Virginia.

N. PREVENTION OF DISEASE

If food to be preserved is contaminated with botulinum spores, the environment in the food mass is anaerobic, and the pH is near neutrality or above, any spores that may survive the heat treatment may germinate, grow, and produce toxin.

The gas produced in canned food may or may not cause the can to bulge and a detectable foul odor may be evident. *Suspected food should never be tasted since only a short sojourn of a trace of food in the mouth, even if rejected, may allow absorption of enough toxin to be lethal.*

Botulism can be effectively prevented by properly canning and preserving foods. As an additional precaution, all home-canned foods that could support the production of botulinal toxin should be boiled for at least 10 minutes before consumption. Antiserum together with other prophylactic measures, including induced vomiting, gastric lavage and purgation, should be used whenever there is reasonable chance that toxin-containing food has been ingested. Although active immunization with pentavalent toxoid is effective, it is only employed for high-risk individuals, e.g., laboratory workers who handle the toxin. The toxoid has been used successfully for protecting animals against botulism.

CLOSTRIDIUM DIFFICILE

Clostridium difficile is an obligate anaerobe that is the etiologic agent of antibiotic-associated pseudomembranous colitis. It can be induced experimentally in hamsters and guinea pigs with penicillin or clindamycin which promotes selective growth of this organism. This organism produces two major toxins. Toxin B is cytotoxic for tissue culture cells. Toxin A, although more active in some systems, is less toxic for tissue cultures than toxin B. Although toxin can be detected in the stool of patients with this form of colitis, cytotoxin levels do not always correlate with the degree of disease. Pseudomembranous colitis can be treated with oral vancomycin.

REFERENCES

Ajl, S.J., et al.: *Microbial Toxins.* Vols. I, IIA, III. New York, Academic Press, 1970 and 1971.

Arnon, S.S., et al.: Infant botulism: Epidemiological, clinical and laboratory aspects. J.A.M.A. *237*:1946, 1977.

Arnon, S.S., Daus, K., and Chin, J.: Infant botulism: Epidemiology and relation to sudden infant death syndrome. Epidemiol. Rev. *3*:45, 1972.

Aronsson, B. et al.: Enzyme immunoassay for detection of *Clostridium difficile* toxins A and B in patients with antibiotic-associated diarrhoea and colitis. Eur. J. Clin. Microbiol. *4*:102, 1985.

Bartlett, J.G., et al.: Antibiotic-associated pseudomembranous colitis due to toxin-producing clostridia. N. Engl. J. Med. *298*:531, 1978.

Bizzini, B.: Tetanus toxin. Microbiol. Rev. *43*:224, 1979.

Cline, K.A., and Turnbull, T.L.: Clostridial myonecrosis. Ann. Emerg. Med. *14*:459, 1985.

Craig, J.M. and Pilcher, K.S.: *Clostridium botulinum* type F. Isolation from salmon from the Columbia River. Science *153*:311, 1966.

Eklund, M.W., et al.: Bacteriophage and the toxigenicity of *Clostridium botulinum* type D. Nature (New Biol.) *235*:16, 1972.

Feldman, R.A.: A seminar on infant botulism. Rev. Infect. Dis. *1*:607, 1979.

Finegold, S.M.: *Anaerobic Bacteria in Human Disease.* New York, Academic Press, 1977.

George, W.L., et al.: *Clostridium difficile* and its cytotoxin in feces of patients with antimicrobial agent-associated diarrhea and miscellaneous conditions. J. Clin. Microbiol. *15*:1049, 1982.

Hitchcock, C.R., et al.: Treatment of clostridial infections with hyperbaric oxygen. Surgery *62*:759, 1967.

Ho, K.L.: *Clostridium perfringens* meningitis. Int. Surg. *67*:271, 1982.

Immunization Practices Advisory Committee, Centers for Disease Control: Diphtheria, tetanus and pertussis—Guidelines for vaccine prophylaxis and other preventive measures. Ann. Intern. Med. *95*:723, 1981.

Kerr, J.: Current topics in tetanus. Inten. Care Med. *5*:105, 1979.

Kitamura, M., et al.: Interaction between *Clostridium botulinum* neurotoxin and gangliosides. Biochim. Biophys. Acta *628*:328, 1980.

Lewis, G.I.: *Biomedical Aspects of Botulism.* New York, Academic Press, 1981.

Libby, J.M., et al.: Effects of the two toxins of *Clostridium difficile* in antibiotic-associated cecitis in hamsters. Infect. Immun. *36*:822, 1982.

Mellanby, J., and Green, J.: How does tetanus toxin act? Neuroscience *6*:281, 1981.

Nelson, R.M., et al.: *Clostridium perfringens* bacteremia. Opportunist or killer? Am. Surg. *51*:301, 1985.

Raff, M.J., et al.: Spontaneous clostridial empyema and pyopneumothorax. Rev. Infect. Dis. *6*:715, 1984.

Sherertz, R.J., and Sarubbi, F.A.: The prevalence of *Clostridium difficile* and toxin in a nursery population. A comparison between patients with necrotizing enterocolitis and an asymptomatic group. J. Pediatr. *100*:435, 1982.

Simpson, L.L.: The origin, structure and pharmacological activity of botulinum toxin. Pharm. Rev. *33*:155, 1981.

Sugiyama, H.: *Clostridium botulinum* neurotoxin. Microbiol. Rev. *44*:419, 1980.

Sutphen, J.L., et al.: Chronic diarrhea associated with *Clostridium difficile* in children. Am. J. Dis. Child *137*:275, 1983.

Van Heyningens, S.: Tetanus toxin. Pharm. Ther. *11*:141, 1980.

Veronesi, R.: *Tetanus: Important New Concepts.* Amsterdam, Elsevier, Excerpta Medica, 1981.

Wilkins, T. et al.: Clostridial toxins active locally in the gastrointestinal tract. Ciba Found. Symp. *112*:230, 1985.

Willis, A.T.: *Clostridia of Wound Infection.* London, Butterworths, 1969.

Zedd, A.J., et al.: Nosocomial *Clostridium difficile* reservoir in a neonatal intensive care unit. Pediatr. Infect. Dis. *3*:429, 1984.

18

NONSPORULATING ANAEROBES

Bacteroides and *Fusobacterium*

Anaerobes, both sporulating and nonsporulating, are part of the complex normal microflora of most surface integuments of the body; the exceptions being the stomach, lower respiratory tract and upper urinary and genital tracts, which are essentially sterile (Chap. 7).*

The nonsporulating anaerobes that often serve as facultative pathogens in mixed infections, resulting from opportunistic situations, belong to several genera, some of the most frequent representative species being *Bacteroides fragilis, Fusobacterium necrophorum, Veillonella parvula, Peptostreptococcus putridus, Peptostreptococcus anaerobius,* and *Actinomyces israelii.* Additional nonsporulating anaerobic species that occasionally contribute to mixed infections include members of the genera *Campylobacter, Treponema, Borrelia, Dialister, Butyrivibrio,* and others (Chap. 8). Anaerobes of certain other genera are of limited or doubtful pathogenicity.

Only the nonsporulating anaerobes of greatest pathogenetic importance will be discussed in this chapter, namely, certain members of the genera *Bacteroides* and *Fusobacterium;* fusobacteria comprise a nondefinitive group, many species of which closely resemble *Bacteroides* species. For convenience, members of both genera are often lumped together as the *Bacteroides-Fusobacterium* (B-F) group.

The normal human adult excretes about 10^{11} bacteria per g of feces, the majority of which are pleomorphic, rod-shaped, Gram-negative, nonsporing anaerobes predominately of the genus *Bacteroides,* family *Bacteroidaceae.* These organisms are found as normal inhabitants of mucosae, principally in the lower intestinal tract and vagina. The fusiforms are most prominent in the mouth. Organisms of the B-F group frequently predominate in mixed infections in man involving 2 to as many as 10 species of microbes, including facultative as well as strict anaerobes and aerobes. The organisms invade as opportunists when local defenses are abrogated, as when gut injury liberates fecal contents into the abdominal cavity or when systemic immunity is depressed. The B-F group comprises many species, several of which participate in mixed infections. *Bacteroides fragilis* is the most frequent participant with *B. melaninogenicus* (a melanin producer) in second place. Other important participant species include *B. distasonis, B. ovatus, B. vulgatus, B. thetaiotamicron, F. necrophorum, F. nucleatum* and *F. mortiferum.*

BACTEROIDES AND *FUSOBACTERIUM*

Many of the characteristics of *Bacteroides* species are in common with *Fusobacterium* species. The most important differentiating characteristics of the *Fusobacterium* species are their ability to ferment glucose to butyric acid and their high susceptibility to penicillin G and certain other antibiotics.

A. MEDICAL PERSPECTIVES

Infections involving B-F species and other anaerobic bacteria of low pathogenicity often go undiagnosed. This results from a general lack of appreciation of the opportunistic potential of these organisms, as well as the failure to properly conduct anaerobic culture of specimens from such infections. The

*Some organisms classified among the anaerobes are actually microaerophiles.

increasing use of forms of immunosuppressive therapy that predispose to opportunistic infections, including infections with nonsporing anaerobes, makes it imperative that infecting organisms be recognized and identified because the diseases they cause are often fatal. One recent study of patients with bacteroidal bacteremia revealed that 15 of 39 (38%) died of their infections. With proper diagnosis and treatment the death rate should not exceed 5 to 10%.

The fact that most oportunistic infections concerned with the B-F group are *mixed infections* makes diagnosis and treatment extremely complex and difficult. Nevertheless, the future holds the promise that with time, simple and rapid methods for identifying etiologic agents in these infections will be developed and that the selection of effective chemotherapeutic regimens will become easier.

B. PHYSICAL AND CHEMICAL STRUCTURE

Lipopolysaccharide (LPS) is present in the cell walls of opportunistic pathogens of the B-F group. In addition most B-F species are nonmotile. The LPS of *B. fragilis* and most *Fusobacterium* sp. is of low toxicity and is chemically distinct from classic LPS of Gram-negative bacteria. Protein Ags of the outer and inner cell membranes have not been characterized. *Polysaccharide capsules* are present on *B. fragilis, B. vulgatus, B. ovatus, B. thetaiotamicron,* and *B. asaccharolyticus;* they are serologically species specific. The capsular polysaccharide of *B. fragilis* has been shown to be strain specific and antiphagocytic but its potential role in virulence is unknown. The specific antiserum recently developed for *B. fragilis* should prove useful for identifying the organism in clinical specimens.

C. GENETICS

Extensive DNA-DNA homology studies have shown that *Bacteroides* species are genetically distinct from each other and that these differences are reflected at the phenotypic level with respect to the serologic specificity of their capsular polysaccharides. Strains within species also show differences in DNA homology. Antibiotic-resistant strains of *B. fragilis* possess plasmids that carry resistance determinants to clindamycin, erythromycin and possibly tetracycline. Transfer of the determinants involves conjugation mediated by plasmids.

D. EXTRACELLULAR PRODUCTS

Although no extracellular products of the B-F group have been clearly shown to contribute to virulence or pathogenicity some of the numerous enzymes produced may play such roles (Sect. I). The mutagen, glyceryl-ester lipid, that is present in feces is a matter of concern because its incidence in various human populations correlates with the incidence of colon cancer. In one extensive study, production of mutagen was limited to 5 species of colonic anaerobes, all *Bacteroides* species. It is probable that a *heparinase* produced by certain *Bacteroides* species is an important cause of thrombophlebitis and septic embolism. *Fusobacterium necrophorum* may produce a leukotoxin.

E. CULTURE

Since anaerobes are present among the normal flora on the surface of external integuments, only specimens derived aseptically (usually by aspiration) from sources that are normally sterile can be expected to yield meaningful results. Members of the B-F group have simple nutritional requirements and ferment a number of carbohydrates. They are relatively tolerant of O_2 but culture should be prompt (within 20 minutes unless anaerobic transfer is effected); care must be taken to avoid excessive exposure of media to air before anaerobic culture in order to guard against the formation of toxic organic peroxide. Plates of blood agar, brain-heart infusion agar, hemolyzed blood-agar with menadione or other media, are streaked and incubated under anaerobic conditions. Plates should be observed daily since early colonies are more likely to be facultative than strict anaerobes. Liquid media should be freshly prepared or heated at 100°C for 20 minutes prior to use to drive off dissolved oxygen. Tubes of liquid media serve best if they are incubated in an anaerobic gas mixture containing H_2, CO_2, and N_2 or covered with a layer of melted petroleum jelly and paraffin to exclude contact with air. Thioglycollate broth or media containing chopped meat provide anaerobic conditions in the lower portion of the tube.

Most B-F species yield visible growth within 2 to 10 days. They often produce a foul odor as the result of their anaerobic metabolism, and gas formation may be demonstrable. Isolates should be identified as being either strict or facultative anaerobes.

F. RESISTANCE TO PHYSICAL AND CHEMICAL AGENTS

Relative susceptibility to O_2 is a notable characteristic of most species of the B-F group. Most *Bacteroides* sp. are highly resistant to penicillin and the cephalosporins, including tetracycline and cefotoxitin, and all strains are uniformly resistant to aminoglycoside antibiotics. Except for acquired resistance, most strains are susceptible to chloramphenicol, metronidazole and *clindamycin, the therapeutic drug of choice.* Most of the fusobacteria are sensitive to penicillin G and the older cephalosporins. Resistance of the B-F group to other physical and chemical agents generally parallels that of most Gram-negative, nonsporulating bacteria. Some *Bacteroides* sp. are attacked by bacteriophage but neither lysogeny nor transduction has been demonstrated.

G. EXPERIMENTAL MODELS

Organisms of the B-F group are not markedly pathogenic for laboratory animals, except for an occasional strain of *F. necrophorum.* Rabbits, rats and mice are used most frequently.

H. INFECTIONS IN MAN

Organisms of the B-F group can infect virtually any tissue in which anaerobic conditions exist. Thus, they are usually isolated from patients suffering from mixed infections, such as those complicating colectomy, colostomy, appendicitis, cholecystectomy, gunshot wounds, rectal abscess, and miscellaneous infections of the urogenital tract, puerperal or postabortal sepsis, necrotizing pneumonia, lung abscess and brain abscess. These organisms produce the typical symptoms of Gram-negative sepsis, including toxic shock in about 25% of cases. Infection with these anaerobes should be suspected whenever there is foul-smelling pus, a black discharge, necrotic tissue, or infection in a bite wound inflicted by either a human being or an animal. Thrombophlebitis occasionally develops near the site of infection, and emboli may be shed into the circulation.

Lung infections with B-F group organisms usually follow invasion of normal flora from the mouth and URT, and are prone to occur in the immunologically-compromised patient. The onset is often insidious, and symptoms such as malaise, cough and low grade fever may occur over a period of weeks to months before the physician is consulted. In such cases, an underlying malignant disease is often found to be the predisposing factor.

Fusospirochetal disease occurs when fusobacteria and spirochetes of the normal flora infect together. This may occur following injury to mucous membranes, as a result of other infections, or in systemically-immunosuppressed hosts. The most common example is ulcerative gingivostomatitis (trench mouth) and the associated condition, Vincent's angina, in which ulcerative lesions develop in the tonsillar areas. Trench mouth and Vincent's angina are usually associated with dietary deficiencies, particularly deficiencies of vitamins A and B. Fusospirochetal infections also produce lesions elsewhere, most notably pulmonary gangrene or abscess in debilitated elderly persons with chronic respiratory disorders. The bacteria that usually predominate in such pulmonary anaerobic infections are *Peptostreptococcus anaerobius, B. melaninogenicus* and *F. nucleatum.*

I. MECHANISMS OF PATHOGENESIS

Little is known about the mechanisms of pathogenesis of B-F infections although it is virtually certain that capsulation and endotoxins, however weak, play a major role, especially in those patients who develop toxic shock. The necrotizing nature of the lesions evidently reflects the activities of proteolytic enzymes and other extracellular products of the anaerobes. This concept is supported by the occurrence of foul-smelling putrid discharges that ostensibly result from proteolysis of tissues. Numerous enzymes of potential importance in pathogenesis include heparinase, collagenase, hyaluronidase, fibrinolysin, lecithinase, lipase, neuraminidase, elastase and others.

It is important to emphasize that mixed infections are singularly difficult to assess, diagnose and treat because the organisms involved vary markedly, both with respect to species and relative numbers. Assessing contributions to pathogenesis made by the different species involved is difficult because animal models do not closely simulate man, and the total virulence provided by the collective synergistic activities of the organisms in the mixtures vary; in addition, the patient's defenses are usually depressed. In accord with expectation, animal studies have shown that mixtures of anaerobes and facultative anaerobes, which alone are nonpathogenic, can act synergistically to produce severe infections; even mixtures of aerobes and anaerobes can complement each other in this respect.

J. MECHANISMS OF IMMUNITY

The normal immunologically uncompromised person has a high level of innate immunity against B-F organisms and never becomes infected in spite of being host to billions of normal flora organisms. It is only when defenses, *either local* or *systemic,* are abrogated that B-F organisms invade and produce disease. Little is known about the nature of immunity to organisms of the B-F group; phagocytosis is probably a first line of defense and C-dependent bacteriolysis and opsonophagocytic killing may occur.

K. LABORATORY DIAGNOSIS

Special care must be taken in collecting and transporting specimens to the laboratory, as well as in culture techniques. Precautions should be taken to avoid contamination of specimens with normal flora. Specimens should be collected aseptically, preferably by aspiration, and protected against undue exposure to O_2 until cultured (Sect. C).

Direct microscopic examination is often a valuable aid in diagnosis; both wet mounts and Gram-stained smears should be prepared. Specimens are inoculated heavily into chopped meat or some other anaerobic broth medium, streaked onto solid agar plates, and incubated anaerobically (Sect. E). Since mixed infections are the rule, pure isolates must be obtained from the primary cultures for identification based on colony characteristics, biochemical reactions, and patterns of susceptibility to antibacterial agents. A new microprocedure for determining preformed enzymes promises to greatly speed the identification of isolates.

Of course, aerobic culture of clinical specimens should also be carried out since aerobes are often present in mixed infections.

L. THERAPY

Drainage of lesions and surgical debridement to eliminate anaerobic foci are essential, and sometimes may be sufficient to limit the disease without other therapy. However, antibiotic treatment is important and necessary for serious infections, especially those associated with bacteremia. Although effective therapy depends on complete coverage of the various species causing the infection with due regard for their antibiotic sensitivity, early treatment is advised before species identification and sensitivity testing of cultured isolates is completed. Whereas the only guide to early therapy is smears made from clinical specimens, identification and antibiotic testing of cultured isolates permits modification of antibiotic regimens during later treatment.

Antibiotic sensitivity of B-F organisms varies with respect to both species and strains (Sect. F). Except for CNS infections and drug-resistant strains clindamycin is the drug of choice against *B. fragilis.* The drug of choice in CNS infections is metronidazole with chloramphenicol as an alternative. Many *Bacteroides* sp. produce β-lactamase and are resistant to the older penicillins and cephalosporins (Sect. F). However, some of the newer penicillins and cephalosporins are active against most *Bacteroides* sp.; one important organism, *F. nucleatum,* is highly sensitive to penicillin, ampicillin and cephalothin. Fully effective therapy in mixed infections must be broad enough to cover organisms that may be involved other than those of the B-F group.

M. RESERVOIRS OF INFECTION

The reservoirs of infection are the nonsterile surface integuments of the body (see introduction).

N. PROPHYLAXIS

Prevention of B-F infections is best achieved by avoiding or correcting predisposing conditions, for example, by using great care in treating patients with immunosuppressive agents, particularly with respect to the suppression of antibacterial immunity. Prophylactic treatment of patients with wounds that permit the entrance of normal flora into injured tissues is also important. Surgical subjects who run undue risk of developing mixed infection by normal flora, such as patients undergoing colonic resection, should receive 24 to 72 hours of prophylactic treatment with appropriate antibiotics beginning about 1 hour before surgery.

REFERENCE

Allen, S., et al.: Rapid identification of anaerobes. In *New Horizons in Microbiology.* Sanna, A., Morace, G., (eds.), Amsterdam, Elsevier Science Publishers, 1984, p. 233.

Bartlett, J.G.: Anaerobic Pneumonias. In *Semin. Infect. Dis.* Vol. 5, Weinstein, L., and Fields, B.N., (eds.), New York, Thieme-Stratton Inc., 1983, p. 131.

Finegold, S.M., et al.: International symposium on anaerobic bacteria and their role in disease. Rev. Infect. Dis. *6*:51, 1984.

Hofstad, T.: Pathogenicity of anaerobic Gram-negative rods: Possible mechanisms. Rev. Infect. Dis. *6*:189, 1984.

Joklik, W.K., et al.: *Zinsser's Microbiology,* 18th Ed., East Norwalk, CT, Appleton-Century-Crofts, 1984.

Rodloff, A.C., et al.: Inhibition of macrophage phagocytosis by *Bacteroides fragilis* in vivo and in vitro. Infect. Immun. *52*:488, 1986.

Salyers, A.A.: Bacteroides of the human lower intestinal tract. Annu. Rev. Microbiol. *38*:293, 1984.

Smith, L.D.: *The Pathogenic Anaerobic Bacteria,* 2nd Ed., Springfield, Charles C Thomas, 1975.

19

ESCHERICHIA COLI AND RELATED OPPORTUNISTIC ENTERIC BACTERIA

The normal human adult excretes, in the feces, approximately 100 billion *Escherichia coli* organisms daily, along with lesser numbers of other Gram-negative rods of the family *Enterobacteriaceae.* The members of this family are widespread in nature and range from obligate parasites to saprophytes with low parasitic potential. This family includes the genera *Salmonella, Shigella* and *Yersinia* whose members are commonly pathogenic (Chaps. 20, 21, 24). The purpose of this chapter is to discuss the opportunistic members of the family *Enterobacteriaceae.*

Escherichia coli and some of its close relatives are commonly referred to as "coliforms," because they are normal inhabitants of the intestinal tract. Included in the coliform group are members of the genera *Enterobacter, Edwardsiella, Klebsiella, Serratia, Citrobacter* and *Arizona.* In addition to the coliforms, most of which ferment lactose, the relatively nonpathogenic enteric rods include non-lactose-fermenting organisms of the genera *Proteus, Morganella,* and *Providencia.*

The prototype species of the coliforms, *E. coli,* has probably been studied more intensively than any other microbe. It is important as a biologic indicator of fecal contamination in water supplies because it is abundant in feces, durable in external environments and easily detected. Other highly pathogenic bacteria that are spread by fecal contamination, such as *Salmonella typhi,* are less durable in external environments and usually comprise only a small fraction of the total number of organisms excreted by an infected individual; because of the small numbers present *S. typhi* cannot be readily detected in water. As a consequence *E. coli* is a highly useful indicator of fecal pollution. If drinking water is polluted with *E. coli,* it is considered unsafe because of the potential risk of other enteric pathogens being present.

The first of the following sections centers on *E. coli* and some general properties of the coliforms; this is followed by separate discussions of the genera *Klebsiella, Enterobacter, Hafnia, Serratia, Proteus, Providencia, Morganella, Edwardsiella, Citrobacter* and *Arizona* (Table 19–1). Table 19–2 summarizes some important properties of these organisms.

ESCHERICHIA

A. MEDICAL PERSPECTIVES

Despite the fact that *E. coli* is a member of the normal flora and is not highly pathogenic, it is of great medical importance because of the frequency and potentially serious nature of the infections that it causes. The organism normally lives in the large intestine without causing apparent harm. However, because of its presence in feces, it often reaches and incites disease in other areas of the body, especially the urinary tract and peritoneum. In the decades before the advent of modern surgery and anti-microbial chemotherapy, many patients died of *E. coli* infections, such as peritonitis, attending rupture of the appendix and perforation of the intestine. Even today, such infections with this organism are often difficult to manage because of rapidly changing patterns of drug resistance, especially in nosocomial infections, and hence continue to be important causes of illness.

The increasing use of immunosuppressive therapy and antibiotics for a va-

Table 19–1. Tribes, Genera, and Species of the Opportunistic Members of the Family *Enterobacteriaceae*

Tribe	*Genus*	*Species*
Escherichieae	*Escherichia*	*coli*
Klebsielleae	*Klebsiella*	*pneumoniae* *ozaenae* *rhinoscleromatis* *oxytoca*
	Enterobacter	*cloacae* *aerogenes* *hafniae* *agglomerans*
	Serratia	*marcescens* *liquefaciens* *rubidaea*
Proteeae	*Proteus*	*vulgaris* *mirabilis*
	Morganella	*morganii*
	Providencia	*rettgeri*
Edwardsielleae	*Edwardsiella*	*tarda*
Salmonelleae	*Citrobacter*	*freundi* *diversus* *amalonaticus*
	Arizona	*hinshawii*

Table 19–2. Properties of the Coliforms and Related Enterobacteria

Genus	*Lactose Fermentation*	*Motility*	*Other Useful Characteristics*
Coliforms			
Escherichia	+	+	Metallic sheen on eosin-methylene-blue media; flat colonies
Enterobacter	+	±	Raised colonies; lack sheen on eosin-methylene-blue media
Klebsiella	+	−	Mucoid colonies due to capsules
Serratia	+	+	Some species produce red pigment
Citrobacter	Variable	+	
Arizona	Variable	+	Late lactose fermenters
Non-Coliforms			
Proteus	−	+	Some species exhibit swarming motility on solid media; rapidly degrade urea to yield ammonia
Providencia	−	+	Urea not degraded
Edwardsiella	−	+	

+ = most strains positive.
− = most strains negative.
± = some strains positive.

riety of conditions has led to an increased incidence of all kinds of opportunistic infections, including those caused by enteric bacteria. Such infections will undoubtedly remain a major medical problem in the future. In addition, it has been shown that some strains of coliforms are relatively virulent because they produce enterotoxins and that certain strains can cause epidemic enteritis, especially in infants. Thus, *E. coli* and its relatives have commanded increasing respect for their pathogenic capabilities as more is learned about them.

B. PHYSICAL AND CHEMICAL STRUCTURE

All members of the family *Enterobacteriaceae* have a similar structure (Fig. 19–1). They are Gram-negative, nonsporing rods, measuring about 1.0 to 2.0 μm × 0.5 μm. Motile organisms of this family possess peritrichous flagella. The flagellar protein (H Ag) is serologically distinct in various strains, even within a species. If a K Ag (capsular/envelope polysaccharide or protein) is

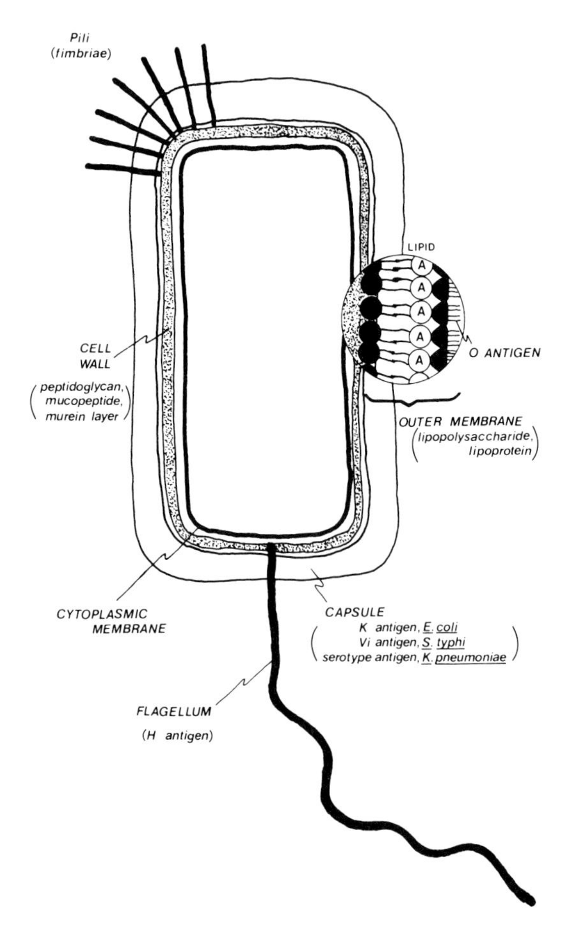

Fig. 19–1. Major cell wall antigens of the Gram-negative bacilli. See Figure 2–6 for details of the outer membrane.

present, it may mask the structural somatic (0) lipopolysaccharide Ag that is an integral part of the cell wall. Serotypes of *E. coli* are determined by the various combinations of O, K and H Ags. The K Ags are divided into 3 main types designated L, A and B. For example, the strain designated 0119:B4: H6, has somatic Ag 119, capsular K Ag B4 and type 6 flagellar Ag.

Escherichia coli produces 2 types of fimbriae or pili that are determinants of pathogenicity. The first type is the mannose-sensitive or common pili and the second type is referred to as mannose-resistant colonizing factor Ags (CFAs). The CFAs (I and II) bind enteropathogenic *E. coli* to human intestinal epithelium. It has been reported that mannose-sensitive and mannose-resistant pili occur on urinary tract isolates and appear to be necessary for binding of *E. coli* to uroepithelial cells. Purified CFAs are immunogenic and when used as experimental vaccines are effective in inducing Abs that block adherence and colonization.

C. GENETICS

Much of what is known about microbial genetics was learned from studies of *E. coli.* By using transduction and conjugation as tools, the "chromosome" of *E. coli* has been partially mapped.

Of particular medical importance is the ability of the enterobacteria to rapidly acquire multiple drug resistance through exchange of episomes or plasmids, such as resistance transfer factors (Chap. 4). Transmissible plasmids also control toxinogenicity, expression of CFAs, and the production of bacteriocins.* For example, colicins, which are bacteriocins formed by certain strains of *E. coli,* kill other strains selectively. Hence they can be used for strain typing, a procedure of value in epidemiologic studies.

The relationships between the enterobacteria are complex and there is often a thin dividing line between genera. It should be emphasized that these organisms are highly variable and cannot always be precisely classified. For example, modifications in surface Ags can result in new serologic types and acquisition of episomes commonly results in new biochemical traits such as H_2S production and sucrose fermentation as well as antibiotic resistance. Genes are also readily transferred between different species within a genus and between different genera. Therefore, it is not surprising to find that the taxonomy of the species in the family *Enterobacteriaceae* is constantly being modified.

D. EXTRACELLULAR PRODUCTS

Although enterobacteria produce a wide variety of enzymes, little is known about the roles of secreted enzymes in the pathogenesis of enterobacterial infections. It is thought that the bacteriocins produced are important in regulating the normal flora by the principle of antibiosis.

Certain strains isolated from human beings and animals produce enterotoxins. Both heat-stable (ST) and heat-labile (LT) enterotoxins have been isolated and purified. Some toxinogenic strains produce both stable and labile enterotoxins, whereas others produce only one of the enterotoxins. Toxigenicity is lost from many isolates upon culture although some strains remain stable en-

*Bacteriocins are protein substances which have bactericidal activity against other microorganisms which are frequently related to the producing organisms. It is common to refer to the bacteriocins produced by *E. coli* as *colicins.*

terotoxin producers for years. These observations suggest that stable maintenance of the gene may be restricted by plasmid incompatibilities or other unknown factors.

It is of interest that *E. coli* and *Vibrio cholerae* enterotoxins are immunologically cross-reactive and have identical or similar molecular mechanisms of action resulting in the activation of adenyl cyclase (Chap. 22). The known properties of heat-stable and heat-labile enterotoxins are summarized in Table 19–3.

E. CULTURE

Escherichia coli, as well as the other coliforms, have simple cultural requirements. They can be grown on a defined glucose-salts medium as well as on blood agar and eosin methylene blue (EMB) medium. *Escherichia coli* is facultatively anaerobic and readily ferments glucose, lactose and certain other sugars, producing both acid and gas. All of the coliform group have similar growth requirements. Biochemical and serologic tests are used to identify the various organisms.

F. RESISTANCE TO PHYSICAL AND CHEMICAL AGENTS

The usual physical and chemical agents employed for disinfection are generally effective against the enterobacteria; also, these organisms are readily killed by pasteurization and by chlorination. However, they can live for weeks to months at room temperature or below, and can survive for long periods in soil, water, or other suitable environments.

The enterobacteria are susceptible to a number of antibiotics, but resistant strains tend to emerge during antibiotic therapy due to the acquisition of plasmids that convey drug resistance.

G. EXPERIMENTAL MODELS

Rabbits and other laboratory animals are used to produce experimental urinary tract (UT) and bacteremic infections with the coliforms. Swine, rabbits and suckling mice also have been employed to study the toxigenic and virulence characteristics of *E. coli*. Certain strains that are prevalent in UT infections are virulent for young mice when injected intracerebrally, whereas others, such as those causing enteritis in infants, are avirulent for mice.

Table 19–3. Characteristics of *E. coli* Enterotoxins

Enterotoxin	*Molecular Weight*	*Immunogenicity*[b]	*Mechanism of Action*
Heat-labile (LT)	75–95,000	+	activates adenyl cyclase; increases intracellular c-AMP levels
Heat-stable[a] (ST)	2–3,000	–	increases intracellular c-GMP levels

[a]Remains active after 30 min at 60°C
[b]Elicits an antitoxic Ab response in experimental animals and humans

H. INFECTIONS IN MAN

The coliforms are the most common cause of UT infections, the majority of which result from ascending invasion by endogenous normal flora. Organisms enter the UT by natural means or may be introduced on catheters or other instruments. Their persistence in the UT is favored by anesthesia, paralysis, by any other agency that interrupts the normal voiding reflex, or by anatomic abnormalities that permit retention of urine.

Urine is an excellent culture medium and readily supports the growth of *E. coli* and many of the other bacteria that gain entrance into the bladder. The resulting cystitis may be self-limiting, or the infection may ascend further to reach the kidneys and cause pyelonephritis. The *E. coli* O serotypes most frequently found in UT infections are 02, 04, 06, 07, 025, 050 and 075.

Although UT infections occur in persons of all ages, they are much more common in females than in males for the following reasons. For example, in the female the orifices of the anus and the UT are proximal to each other, the urethra is short, and the infecting organisms may be mechanically introduced into the urethra during sexual intercourse. The UT infections may be acute, intermittent or chronic, symptomatic or asymptomatic; they can lead to permanent renal damage with all of its consequences; therefore, this problem continues to be a subject of intensive research.

It is convenient to classify UT infections into 4 categories: uncomplicated and complicated acute infections and asymptomatic and symptomatic chronic bacteriuria. Uncomplicated acute infections are characterized by production of the usual symptoms of cystitis (dysuria, pain on voiding) or pyelitis (flank pain, chills and fever) in patients without a history of prior UT infection. In these patients, endogenous *E. coli* is almost always the etiologic agent. Complicated acute infections occur in persons with anatomic abnormalities of the UT and tend to recur until the abnormality is corrected. In contrast to uncomplicated acute infections, which occur most often in women of childbearing age, complicated acute disease is seen most often in young men. In both conditions, however, *E. coli* is most often the etiologic agent. Asymptomatic bacteriuria occurs mostly in women without clear-cut UT disease; however, a careful history usually reveals that these patients have had some symptoms of UT infection in the past. The organisms isolated from them are often *E. coli*, but other species of enterobacteria are also frequently found. Asymptomatic bacteriuria is detected in about 5% of women in early pregnancy; unless treated, approximately 40% of these women develop acute symptomatic UT infections later in pregnancy or in the postpartum period. Chronic bacteriuria is most frequent in the elderly, the members of both sexes being equally affected. These individuals have repeated episodes of symptomatic infection, usually due to structural abnormalities, such as scarring or strictures, in the UT. Bacteria in the urine that cause infections of the UT must be distinguished from bacteria that normally contaminate urine during collection; this is accomplished by quantitative urine culture (Sect. K).

Endogenous coliforms can often be isolated from the tissues of an infected appendix removed by operation during attacks of apendicitis, from the infected peritoneum, or from any sort of abscess in areas exposed to fecal contamination, e.g., decubitus ulcers. Although these are often "mixed infection," *E. coli* usually predominates and is a major offender.

Escherichia coli is a major causative agent of neonatal meningitis and nosocomial Gram-negative systemic disease in immunosuppressed or debilitated hospitalized patients. Up to 80% of the invasive strains involved in neonatal meningitis possess the K1 capsular Ag which is immunologically cross-reactive with capsular Ags of *Neisseria meningitidis* groups B and C, *Streptococcus pneumoniae* types 1, 2, 4, 7, 10 and 23, and *Haemophilus influenzae* type b. It is thought that the *E. coli* strains associated with neonatal meningitis are usually contracted from the mother during delivery. Systemic infection with foci in internal organs, bacteremia, and septicemia due to *E. coli* are increasing problems in immunosuppressed and immunologically deficient patients. In general, the strains responsible for these infections reflect the normal flora of the patients. It is becoming clear, however, that strains possessing acidic polysaccharide capsular K Ags are usually the ones involved in these invasive infections. The reason for this correlation may be that encapsulated isolates tend to be resistant to the bactericidal activity of serum and, further, that they resist phagocytosis because of restricted complement activation.

A few strains of enteropathogenic *E. coli* are the cause of epidemics of infantile diarrhea in nurseries; serotypes 055, 0111 and 0127 are implicated most often. It is of singular interest that these serotypes rarely, if ever, cause UT infections.

There is overwhelming evidence that certain enterotoxin-producing strains of *E. coli* are responsible for many outbreaks of cholera-like disease ("turista" or traveler's diarrhea). Strains producing either heat-stable or heat-labile enterotoxin or both have been incriminated in a number of epidemics in both children and adults. Extensive efforts at vaccine development are, therefore, being aimed at creating mutants which are capable of colonizing the small bowel but which elaborate non-functional enterotoxins (natural toxoids) that are immunogenic but do not cause diarrhea. Such attenuated mutants administered orally would conceivably stimulate local IgA Abs that would prevent colonization and neutralize endotoxin as well.

Other *E. coli* enteric pathogens are represented by a group of "Shigella-like" *E. coli* strains which are referred to as *enteroinvasive strains.* These strains can invade the intestinal mucosa with subsequent destruction of the epithelial cells which results in a bloody dysentery. These strains apparently do not elaborate an enterotoxin.

Some data suggest that a change of environment permits the establishment of a new gut flora with a predominance of *E. coli* serotypes common in the new environment. Mild diarrheal illness may occur during the interval when the newly acquired serotypes are becoming established as part of the normal flora. This could explain some cases of the common "traveler's diarrhea."

I. MECHANISMS OF PATHOGENESIS

The mechanisms of pathogenicity of *E. coli* and similar organisms are slowly being elucidated. Endotoxins probably contribute to the production of disease, as discussed in Chapter 8. The polysaccharide K Ags of some strains are important in inhibiting Ab-C killing and phagocytosis by masking other Ags. Although certain O serotypes are definitely associated with a particular disease, such as UT infections and infantile diarrhea, there is only indirect evidence to indicate that the O Ag is a virulence factor, other than the fact that rough strains lacking O Ag are highly susceptible to Ab-C killing.

Studies on the pathogenesis of *E. coli*-induced diarrhea and colitis have led to the conclusion that *E. coli* can cause intestinal disease by at least 3 to 4 separate mechanisms and that the strains possessing characteristic virulence properties are clustered within distinct geographically related serogroups. The serologic characteristics of *E. coli* strains that cause acute diarrhea are summarized in Table 19–4.

J. MECHANISMS OF IMMUNITY

Immunity to the coliforms is also poorly understood. Natural "barriers" are important in preventing infection. For example, under normal circumstances, *E. coli* seldom infects the UT, in spite of frequent exposure of the urethra to the organism; small numbers undoubtedly enter the tract from time to time by way of the urethra, but sphincters controlling urinary flow and other mechanisms prevent most of them from reaching the bladder and the few that do are either killed or flushed out by the urinary flow before they can become established.

Humoral Abs against H, O and K Ags are commonly produced during *E. coli* infections; however, the roles that they play in protecting against subsequent infection are only partially understood. Recent evidence has shown that H and O Abs offer little or no protection against invasive strains of *E. coli*, whereas anti-K Abs function as effective opsonins that neutralize the antiphagocytic and antibactericidal properties of the capsular Ags.

Persistent or recurring kidney infections with *E. coli* often lead to substantial titers of humoral Abs against somatic O Ags of the infecting strain; in contrast during bladder infections, humoral Ab titers seldom rise above the normal range. Therefore, serum Ab titers against O Ags are a useful diagnostic aid in differentiating pyelonephritis from cystitis, but have no demonstrable role in effective immunity.

When serum Ab titers are high, small amounts of anti-O Ab are demonstrable in the urine of some patients with UT infections involving the kidney; this suggests that humoral proteins, including Abs, are being excreted. On the other hand, patients with cystitis, who lack increased humoral Abs, may nevertheless have low titers of urinary Abs, indicating that such Abs are being produced locally in inflamed areas of the bladder mucosa. Immunoglobulins of the class IgA have been demonstrated in urine, but their significance in protection against cystitis is not known. However, it is probable that both Abs and nonantibody antimicrobial factors of unknown nature function on the bladder mucosa and thus oppose infection. There is evidence from animal studies that bacteria adhere to kidney cells by means of pili, thereby permitting colonization which may be a necessary prerequisite for establishing infection. The presence of Abs in the UT could protect against infection by preventing bacterial adherence.

K. LABORATORY DIAGNOSIS

Laboratory diagnosis of infections with coliform bacteria is relatively simple. Urine, pus, or other specimens are cultured on blood agar, eosin methylene blue agar (EMB), MacConkey, or any of a variety of selective and differential

Table 19–4. Serologic Characteristics of *Escherichia coli* Strains that Cause Acute Diarrhea

	Enteropathogenic	*Enterotoxigenic*	*Enteroinvasive*
Pathogenic mechanism	Unknown	Enterotoxin (LT,ST)	Epithelial cell invasion
Age groups affected	Infants	Infants and adults	Adults and infants
Epidemiology	Sporadic cases and outbreaks	Sporadic cases and outbreaks	Sporadic cases and outbreaks
O groups associated	26, 55, 86, 111, 114, 119, 125, 126, 127, 128, 142	6, 8, 15, 25, 27, 78, 148, 159	28ac, 112ac, 124, 136, 143, 144, 152, 164

From Rowe, B.: *Escherichia coli* in acute diarrhea. Lab-Lore *7*:449, 1977.

media. Colonies become visible within a day or less. A fecal odor is characteristic of *E. coli* and the flat colonies have a distinctive metallic sheen on EMB agar. Lactose fermentation, along with other biochemical tests, helps to differentiate *E. coli* from some of the highly pathogenic Gram-negative rods.

Fluorescent Ab techniques are useful as presumptive tests for screening for enteropathogenic *E. coli,* but agglutination tests with pure cultures are essential for final identification.

The diagnosis of UT infections usually depends on quantitative culture of clean-voided or catheterized urine. It has been established that the presence of more than 10^5 organisms per ml in a clean-voided urine specimen indicates true infection. Fewer than 10^3 bacteria per ml usually denotes contamination during urine collection. However, when clinical symptoms are present, bacterial counts between 10^3 and 10^5 require that the culture be repeated. In addition, Gram-stained smears of one drop of uncentrifuged urine will reveal the presence of bacteria only if more than 10^5 organisms per ml are present; therefore, the direct Gram-stained smear is useful for gaining presumptive evidence of UT infection.

L. THERAPY

Most of the enteric bacteria are susceptible to one or more of a number of antibiotics, including the tetracyclines, sulfonamides, chloramphenicol, ampicillin and cephalexin. Nitrofurantoin is frequently employed for treating UT infections due to coliforms.

It is sometimes necessary to correct a predisposing condition in order to permit successful treatment of infections due to the enterobacteria. For example, a congenital anatomic defect in the urinary tract that results in continual exposure of the tract to heavy contamination or that predisposes to infection by causing retention of urine should be corrected surgically; otherwise antibiotic therapy is merely a temporizing measure.

The treatment of patients with UT infections varies with the category of infection. The *E. coli* strains that cause uncomplicated acute infections are usually sensitive to all of the antibacterial agents generally used for Gram-negative infections. Consequently, the less toxic agents should be employed, and sensitivity testing is not required. Recurring infections or chronic bacteriuria are usually caused by organisms that are resistant to antimicrobial agents, hence sensitivity testing should be carried out in these instances. As a rule nosocomial infections frequently involve drug-resistant strains, whereas infections acquired outside the hospital are caused by drug-sensitive organisms.

M. RESERVOIRS OF INFECTION

The principal reservoirs of *E. coli* are the intestinal tracts of humans and animals. Members of the genus *Enterobacter* and some of the other coliforms can exist as free living saprophytes as well as parasites.

N. CONTROL OF DISEASE TRANSMISSION

Since the coliforms are widespread in nature, the usual measures for preventing bacterial infection are not generally applicable. However, iatrogenic infection due to these organisms can be controlled by careful attention to aseptic techniques. In addition, nursery epidemics due to enteropathogenic *E. coli* and cross-infections in hospitals can be effectively controlled by careful epidemiologic studies and follow-up. Traveler's diarrhea can usually be prevented only by drinking water that is properly treated (boiled) and avoiding potentially contaminated raw foods and vegetables.

KLEBSIELLA

The most common species of this genus, *K. pneumoniae,* is found in the alimentary and respiratory tracts of about 5% of normal persons. Although the organism occasionally causes UT infections and chronic infections in various organs, it is better known as a cause of severe hemorrhagic pneumonia, accounting for about 2 to 3% of all cases of acute bacterial pneumonia. The mortality rate in untreated *Klebsiella* pneumonia is high (50%); therefore, prompt recognition of the infecting organism and early treatment are essential. It is especially important to distinguish between pneumonia due to *K. pneumoniae* and pneumonia due to other organisms, such as *Streptococcus pneumoniae,* because the antibiotics appropriate for treatment vary depending on the infecting agent.

The incidence of Gram-negative bacillary infections of the respiratory tract is increasing as a result of various clinical procedures. A recent study of 213 patients admitted to a medical intensive care unit revealed that the respiratory tract of 45% of these patients became colonized with Gram-negative bacilli on the first hospital day! About 12% of the 213 patients subsequently developed nosocomial infections. *Klebsiella pneumoniae* was isolated most frequently; other isolates commonly included species of *Enterobacter* and *Pseudomonas aeruginosa.*

Klebsiella pneumoniae is nonmotile; it produces a large polysaccharide capsule resulting in the formation of viscous mucoid colonies. The capsule is highly antiphagocytic. There are over 80 K Ags and more than 10 "O" Ags currently recognized in the genus. The organism can be identified directly in sputum by means of the capsular swelling test conducted with specific antisera. It is sometimes impossible to diagnose pneumonias produced by opportunistic Gram-negative bacilli early enough to permit effective therapy. Consequently, methods for rapid diagnosis are being sought, such as those to detect bacterial Ag in serum or other body fluids. For example, soluble capsular Ag of *Klebsiella pneumoniae* can be detected in sputum extracts by counterimmunoelectrophoresis, thus, permitting the diagnosis of *Klebsiella* pneumonia within 1 to 2 hours rather than the 1 to 2 days which is required for isolation and identification of the organism.

Klebsiella infections are difficult to treat because of the marked tissue necrosis and suppuration produced by the organism. The most effective antibiotic for treatment is usually gentamicin or cephalothin; however, this varies with the strain; penicillin is not effective.

On rare occasions, other species of *Klebsiella* cause disease of the upper respiratory tract; they are *K. ozaenae,* which can cause atrophy of the nasal mucous membranes, and *K. rhinoscleromatis,* which causes granulomatous lesions in the nose and throat.

ENTEROBACTER

The genus *Enterobacter* contains 5 species *(E. aerogenes, E. agglomerans, E. gergoviae, E. sakasakii* and *E. cloacae)* that inhabit soil, water and the large intestine of mammals. *Enterobacter* species are motile, except for *E. agglomerans;* all species of *Klebsiella* are nonmotile.

The majority of isolates are involved in UT infections. It appears that *E. cloacae* is the most common isolate from UT infections although the remaining 4 species also have been isolated from clinical material. For example, *E. sakazakii* has been isolated from an infant with meningitis and *E. gergoviae* has been isolated from wounds, blood and sputum.

Species of *Enterobacter* are susceptible to the tetracyclines, aminoglycosides, chloramphenicol, trimethoprim-sulfamethoxazole, nalidixic acid and nitrofurantoin. Most strains are resistant to cephalothin and ampicillin.

HAFNIA

Hafnia alvei, which was formerly classified as *Enterobacter hafniae,* causes infections with a similar pattern to the *Enterobacter* species.

SERRATIA

Serratia species are found in soil, plants, water and animals. The genus *Serratia* produces an extracellular DNAse which differentiates it from the other enteric bacteria. *Serratia marcescens, S. liquifaciens* and *S. rubidaea* represent the potential human pathogens. *Serratia marcescens* is the most frequent human isolate and has been the cause of nosocomial pneumonias, UT infections, wound infections and septicemias. This organism produces a red, water insoluble pigment that can result in a red tinge in sputum from patients with pneumonia.

Serratia marcescens is sensitive to aminoglycosides, chloramphenicol and trimethoprim-sulfamethoxazole.

PROTEUS, PROVIDENCIA AND *MORGANELLA*

Organisms of the tribe *Proteeae* are characterized by the capacity to degrade urea to ammonia and their failure to ferment lactose. The important genera and species are *Proteus vulgaris, Proteus mirabilis, Morganella morganii* and *Providencia rettgeri.* Members of the two species, *P. vulgaris* and *P. mirabilis,* exhibit "swarming motility" on solid agar. For unknown reasons the organisms "swarm" (migrate en masse) in successive directional waves. Swarming overgrowth can be inhibited by adding phenylethyl alcohol or sodium azide to the medium. The characteristic odor of *Proteus* aids in its identification.

Members of the genus *Proteus* are normal intestinal inhabitants, but are also found free-living in soil and water. They are a frequent cause of UT infections, especially *P. mirabilis.* The latter is usually susceptible to penicillin and ampicillin, whereas other species are resistant to these antibiotics. Most isolates of all species are sensitive to the aminoglycoside antibiotics.

Some strains of *P. vulgaris* share Ag determinants with some of the rickettsiae.

Thus, strains OX2, OX19, and OXK are agglutinated by Abs formed during certain rickettsial infections (Chap. 31). Present evidence indicates that there is no genetic relationship between proteus and rickettsiae, but that the sharing of Ag determinants is merely fortuitous.

Morganella morganii, formerly *Proteus morganii,* like *Providencia rettgeri* does not produce H_2S; *P. vulgaris* and *P. mirabilis* produce H_2S. Infections caused by *M. morganii,* as well as treatment, are similar to *Proteus* infections.

Members of the Genus *Providencia* are associated with infections of the urinary tract, blood, lungs and wounds.

EDWARDSIELLA

Only one species, *Edwardsiella tarda,* is in the tribe *Edwardsielleae.* This organism rarely produces human infections. It has been isolated from cases of meningitis and sepsis and from human feces in cases of gastroenteritis. Most isolates are sensitive to the antibiotics used to treat other enteric infections.

CITROBACTER AND *ARIZONA*

These two genera contain opportunistic organisms of the tribe *Salmonelleae* as contrasted to the intestinal pathogens of the Genus *Salmonella* (Chap. 20).

The majority of isolates of the Genus *Citrobacter* come from respiratory and urinary tract infections. Patients with these infections usually have underlying disease or other conditions which impair antimicrobial defenses. *Citrobacter* isolates are usually sensitive to the aminoglycosides and tetracyclines.

Members of the Genus *Arizona* are associated with reptiles and birds. Although human disease is rare, members of this genus can produce gastroenteritis, bacteremia, pyelonephritis, otitis media and osteomyelitis. *Arizona* isolates are sensitive to ampicillin and chloramphenicol.

REFERENCES

Bäck, E., et al.: Enterotoxigenic *Escherichia coli* and other Gram-negative bacteria of infantile diarrhea: Surface antigens, hemagglutinins, colonization factor antigen and loss of enterotoxigenicity. J. Infect. Dis. *142*:318, 1980.

Braun, V., et al.: Functional aspects of Gram-negative cell surfaces. Subcell. Biochem.: *11*:103, 1985.

Clarridge, J.E., et al.: Extraintestinal human infection caused by *Edwardsiella tarda.* J. Clin. Microbiol. *11*:511, 1980.

Cohen, P.S., et al.: *E. coli* colonization of the mammalian colon: Understanding the process. Recomb. DNA Tech. Bull. *8*:51, 1985.

DuPont, H.L., Pickering, L.K.: *Infections of the Gastrointestinal Tract, Microbiology, Pathophysiology, and Clinical Features.* New York, Plenum Press, 1980, pp. 61, 129, 195.

Farmer, J.J., III, et al.: The Salmonella-Arizona group of *Enterobacteriaceae:* Nomenclature, classification and reporting. Clin. Microbiol. Newsletter *6*:63, 1984.

Farmer, J.J., III, et al.: Biochemical identification of new species and biogroups of *Enterobacteriaceae* isolated from clinical specimens. J. Clin. Microbiol. *21*:46, 1985.

Gaastra, W., and DeGraaf, F.K.: Host-specific fimbrial adhesins of noninvasive enterotoxigenic *Escherichia coli* strains. Microbiol. Rev. *46*:129, 1982.

Gorbach, S.L. and Hoskins, D.W.: Travelers' diarrhea. Diagnostic Medium, Oct. 1980.

Graham, D.R. and Band, J.D.: *Citrobacter diversus* brain abscess and meningitis in neonates. JAMA *245*:1923, 1981.

John, J.F., Jr., et al.: *Enterobacter cloacae*: Bacteremia, epidemiology and antibiotic resistance. Rev. Infect. Dis. *4*:13, 1982.
Lerner, A.M.: The gram-negative bacillary pneumonias. Diagnostic Medium, Nov., 1980.
Montgomerie, J.Z. and Ota, J.K.: *Klebsiella* bacteremia. Arch. Intern. Med. *140*:525, 1980.
Olsvik, O., et al.: Characterization of enterotoxigenic *Escherichia coli.* Serotypes, enterotoxins, adhesion fimbriae and the presence of plasmids. Acta. Pathol. Microbiol. Immunol. Scand. (B) *93*:255, 1985.
Porat, R. et al.: Selective pressures and lipopolysaccharide subunits as determinants of resistance of clinical isolates of Gram-negative bacilli to human serum. Infect. Immun. *55*:320, 1987.
Pronk, S.E., et al.: Heat-labile enterotoxin of *Escherichia coli.* Characterization of different crystal forms. J. Biol. Chem. *260*:13580, 1985.
Sack, R.B.: Enterotoxigenic *Escherichia coli*: Identification and characterization. J. Infect. Dis. *142*:279, 1980.
Sack, R.B.: Human diarrheal disease caused by enterotoxigenic *Escherichia coli.* Annu. Rev. Microbiol. *29*:333, 1975.

20

SALMONELLA

Salmonella typhi

Salmonella choleraesuis and *Salmonella enteritidis*

The family *Enterobacteriaceae* contains genera that harbor important pathogens including the genera, *Salmonella (Sal.)* and *Shigella* (Sh). These genera differ markedly in many respects; consequently they will be discussed separately.

Whereas the cause of typhoid fever, *Sal. typhi,* has but one natural host, man, the two remaining species *Sal. choleraesuis* and *Sal. enteritidis,* are harbored by a wide range of animal species as both natural and accidental hosts. Man can serve as an accidental host to many species of salmonellae.

The Kaufmann-White classification scheme introduced in 1934 is still in general use, although the Ewing modified scheme is gaining some support.

Members of the genus *Salmonella* comprise several species and innumerable serotypes, all of which are pathogens. The taxonomy of the salmonellae is exceedingly complex and the present trend is to bring order out of chaos by grouping them into three species, *Sal. typhi, Sal. choleraesuis* and *Sal. enteritidis,* and to further divide them into subspecies and serologic types on the basis of their antigenic makeup.

In physical and chemical structure the salmonellae are similar to *E. coli.* Most salmonellae are motile but unlike *E. coli* most do not ferment lactose. Their heat-stable somatic (0) and heat-labile flagellar (H) Ags resemble those of other members of the family *Enterobacteriaceae.* Some strains of salmonellae have a heat-labile, K-like envelope Ag called "virulence Ag" (Vi Ag) because the strains possessing it are virulent for mice. The tremendous variety and the myriad of combinations of both O and H Ags account for some 2,000 salmonellae serotypes. The complexity of serotypes is further exaggerated by "phase variations" that occur among the H Ags of most strains either as phase 1 Ag(s), which are relatively strain specific, or as phase 2 Ag(s), which are group specific; genetically-controlled changes from one phase to another can occur; strains may be multiphasic, diphasic or monophasic.

Few bacteria are genetically more diverse than the salmonellae. Variants include those that range from rough to smooth colony-producing strains, from motile to nonmotile species and strains, and from Vi Ag-producers to those lacking Vi Ag.

The diseases caused by various salmonellae are similar and fall into three general clinical patterns: (1) the *enteric fevers,* which are characterized by septicemia and enteritis and are most commonly caused by *Sal. typhi* and certain strains of *Sal. enteritidis* (paratyphoid group); (2) *septicemia* with widespread focal lesions, which is most often caused by *Sal. choleraesuis;* and (3) *enterocolitis* or *gastroenteritis,* which can result from any one of some 1,500 serotypes of *Sal. enteritidis.* Serologic typing is so demanding and complex that it can only be done in special laboratories.

Salmonella typhi, the cause of typhoid fever, will be given the most attention and will be discussed first; many of its properties also apply to the other salmonellae, especially the paratyphoid organisms that produce enteric fevers, resembling typhoid fever.

SALMONELLA TYPHI

A. MEDICAL PERSPECTIVES

Typhoid fever is the most serious of the various infections caused by salmonellae and in past centuries was one of the most frequent of the water-borne diseases. Typhoid fever and typhus, which have similar symptoms, were first clearly recognized as distinct diseases by Schoenlein (1839). William Budd (1856–1873) recognized the contagiousness of typhoid fever and noted that the mode of spread was principally by ingestion of food, milk and water contaminated with the feces of typhoid patients. Eberth (1880) discovered the organism in tissues and Gaffky first isolated it in pure culture in 1884. Before the era of modern sanitation, typhoid epidemics were a constant threat and a bane of armies during military campaigns. Typhoid fever is still a disease of major importance in the world at large, especially in underdeveloped countries.

Outbreaks due to carriers acting as food handlers occur even in areas where sanitation is generally good (Sect. M). For example, in 1967, 30 of 72 members of a fraternity at Stanford University contracted the disease. The epidemiology of typhoid fever is often bizarre; for example, a 1959 outbreak involved a water supply contaminated by the feces of seagulls that had fed on sewage effluents discharged by ships at sea.

In the USA, many thousands died from typhoid fever in the year 1900; by contrast, the annual number of cases today does not exceed a few hundred and few deaths occur; 600 typhoid cases were reported in 1981. Nevertheless, because a few long-term carriers persist in the population, small sporadic outbreaks continue to arise.

B. PHYSICAL AND CHEMICAL STRUCTURE

Salmonella typhi possesses a number of Ags in common with other salmonellae, particularly with organisms of the paratyphoid group.

Aside from the general structural properties of the salmonellae described in the introduction to this chapter, the only structural feature of *Sal. typhi* deserving of special comment is the envelope Vi Ag which, when abundant, can cover and mask the O Ag and render the organism inagglutinable by anti-O serum and resistant to phagocytosis. The Vi Ag is a heat-labile polymer of N-acetyl D-galactosamine-uronic acid. Its precise role in virulence of *Sal. typhi* remains to be elucidated.

Endotoxin (LPS) of *Sal. typhi* and related Gram-negative bacteria contains 3 Ags, the highly divergent O *polysaccharide,* which lends serologic specificity, the *core polysaccharide,* which is highly cross-reactive with related bacteria and *lipid A,* the toxic component of LPS. Lipid A is a complex containing a diglucosamine backbone with ester and amide-linked fatty acids. One of its most notable properties is that it serves as a polyclonal B cell activator. Since the Abs to lipid A present in normal serum rise in the face of infection with various salmonellae and related organisms, their potential role in acquired immunity to salmonellosis has been of great interest. However, to date, studies in animal models and man are inconclusive; such Abs can be either protective, ineffective or damaging, depending on the model used.

C. GENETICS

All salmonellae are similar in genetic behavior. Genetic changes in salmonellae are induced by some 80 different phages. For example, certain O Ags are the result of lysogenic conversion, and transduction has been shown to transfer the capacity to synthesize both O and H Ags. Phage typing has been valuable for epidemiologic studies.

Conjugation may occur among different salmonellae or between *E. coli* and salmonellae, resulting in the transfer of genes.

Bacteriocins and multiple drug-resistance factors are frequently transferred among the salmonellae by plasmids.

D. EXTRACELLULAR PRODUCTS

No medically-important extracellular products of *Sal. typhi* are known, with the exception of bacteriocins which may help the organisms to compete in the highly mixed flora of the lower GI tract and thus permit tissue invasion and passage of the organism from host to host.

E. CULTURE

Salmonella typhi is a non-lactose-fermenting, facultative anaerobe with simple nutritional requirements. It grows well on ordinary media under aerobic conditions over a temperature range of 10° to 41°C (optimal 37°C) and a pH range of 6 to 8. In common with all salmonellae, the organism can grow in bile.

The isolation of *Sal. typhi* and other salmonellae from bacteria-rich feces, requires special methods. Therefore, selective or selective-differential media or enrichment culture followed by other media are commonly used for the isolation of these organisms. A sample of the feces or other material to be tested can be cultured in selenite broth or other enrichment media prior to plating, or it can be streaked directly onto a selective-differential medium such as MacConkey's medium, eosin-methylene-blue medium or other selective-differential media. Final identification requires biologic and serologic tests (Sect. K).

F. RESISTANCE TO PHYSICAL AND CHEMICAL AGENTS

Among nonsporulating bacteria, *Sal. typhi* and other salmonellae are average in their susceptibility to most physical and chemical agents. They are readily killed by chlorination and most of the commonly used disinfectants. Although heating at 55°C is lethal, these organisms resist freezing and survive well in chilled and frozen foods. They withstand fairly high concentrations of salts and survive for weeks in sea water as well as fresh water. *Salmonella typhi* is notably resistant to drying and can remain alive in dried sewage for weeks.

G. EXPERIMENTAL MODELS

The mouse has been used for experimental infection with *Sal. typhi* but is unsatisfactory because the organism is not invasive for this animal and the immunologic responses and disease produced do not simulate human typhoid fever. Higher subhuman primates, such as the chimpanzee, provide the closest models for the human disease.

H. INFECTIONS IN MAN

Typhoid fever is the most serious of the enteric fevers due to the salmonellae and is the classical prototype of these infections. It may be mild enough to subside within a week or two after the onset of symptoms, or severe enough to kill the patient within 10 days; however, the usual typhoid case lasts for 4 or 5 weeks. The infection begins as a generalized septicemia characterized by a fluctuating primary bacteremia that is almost constant during the incubation period of 1 to 2 weeks.

Infection commonly results from the ingestion of food or fluids contaminated either directly or indirectly with the excreta of chronic carriers or of typhoid patients. The incubation period varies inversely with the number of organisms ingested and is usually 1 to 2 weeks (range 3 to 30 days). Unusually large infecting doses may provoke early acute gastroenteritis prior to the onset of fever. Histologic study and work with animal models indicate that the organisms invade through the mucosa of the small intestine. They evidently adhere to and colonize on the epithelial surface and are endocytosed by mucosal cells and passaged directly to the submucosa. Apparently they do not grow extensively within or destroy the epithelial cells. They multiply in the submucosa and reach the local lymph nodes where they grow, probably in large part within macrophages, which initially are not "immune" (activated). Organisms soon reach the blood, presumably following liberation from dying macrophages in the lymph nodes. The first bacteria that reach the blood are rapidly cleared by macrophages which accumulate and proliferate in large numbers at bacillary foci. This macrophage hyperplasia causes enlargement of lymph nodes and crowds hematopoietic elements in the bone marrow. Later, when the macrophages of the RES are overwhelmed, a persisting bacteremia sets in and an acute febrile illness of 3 to 5 weeks' duration, attended with many symptoms, gradually develops. The temperature rises in a unique stepwise fashion to a plateau that may sometimes represent extreme pyrexia. Fever is evidently due to endogenous pyrogens liberated by macrophages.

Other symptoms include dull frontal headache, prostration, abdominal discomfort and tenderness, skin rash, nonproductive cough, splenomegaly and leukopenia. Usually the symptoms intensify during the 2nd and 3rd weeks and then gradually decline. At the height of illness, mental dullness and delirium may set in and the characteristic "rose spots" of typhoid fever consisting of areas of petechial hemorrhage due to emboli or bacteria clumped by agglutinins, may appear in the skin. By about the 2nd to 3rd week, substantial metastatic lesions rich in macrophages are also present at various sites, including the bone marrow, lung, lymphoid elements of the lamina propria of the small intestine and colon, and, most singularly, in the biliary tract where the organisms thrive and escape with the periodic discharge of bile to seed the feces with enormous numbers of organisms. Heavy involvement of the solitary lymphoid follicles of the gut, and especially the Peyer's patches of the terminal ileum, often leads to extensive ulceration. Serious complications include hemorrhage due to erosion of blood vessels in intestinal lesions and the rare (1% of cases) but most dangerous of all complications, intestinal perforation, due to deep ulceration; these two complications account for 75% of all deaths. In the event of perforation, surgery should be performed in essentially all cases and should be accompanied by antimicrobial treatment designed to cover all contaminating gut organisms as well as *Sal. typhi* (Sect. L). Other complications include femoral thrombophlebitis, cerebral thrombosis, cholecystitis, pneumonia (often a mixed infection), osteomyelitis, meningitis, endocarditis, toxic mycocarditis, alopecia and abortion. A unique hyaline degeneration of certain voluntary muscles and vascular hyperreactivity to epinephrine and norepinephrine develop.

Strong, although not solid, lifetime immunity to typhoid fever commonly follows recovery. However, in some 10% of the subjects, relapse occurs about 1 to 2 weeks after defervescence. Relapse is most frequent following cure with antibiotics and can recur 2 or 3 times. Relapses are mild and, since Ab titers to H, O and Vi Ag can be high, presumably result from an inadequate cellular immune response. Mortality in untreated subjects varies markedly with different outbreaks but averages about 10%; it is reduced to about 1 to 2% by chemotherapy.

Chronic carriage, often for a lifetime, develops in 3 to 5% of patients (particularly adult females), intestinal carriage due to gallbladder infection being more common than urinary carriage. The gallbladder wall of the carrier is heavily infiltrated with lymphocytes and contains hyperplastic lymphoid follicles. Urinary carriage occurs most often in regions where schistosomiasis is endemic.

I. MECHANISMS OF PATHOGENESIS

The apparent ability of *Sal. typhi* to grow unrestricted within nonactivated macrophages is undoubtedly important in pathogenesis. There is evidence that virulent strains of *Sal. typhi* multiply intracellularly, whereas avirulent species do not. It has been suggested that some unique properties of the cell wall, perhaps the unusual sugars found there, may enable virulent organisms to resist degradation within host macrophages. The O Ag may be necessary for adherence to gut epithelium and may thus serve as a virulence factor.

The mechanisms by which *Sal. typhi* produces tissue injury and disease are not clear, the only known toxin is endotoxin. Since many of the symptoms of typhoid fever can be produced with endotoxin, it has been generally assumed that endotoxin contributes importantly to pathogenesis. During the course of typhoid fever, the bacteremia is cleared long before symptoms subside. Some investigators have attributed the continued symptoms to the slow release of endotoxin from disrupting parasitized macrophages, even in the absence of extracellular bacilli. The alternative suggestion has been made that the synthesis and release of endogenous pyrogen or some other product from activated macrophages may contribute to symptoms of the disease. The endotoxin theory has not been proved conclusively and is challenged by a number of observations. For example, volunteers made highly resistant to endotoxin before experimental challenge infection have been noted to develop typhoid fever equivalent to the disease in nontolerant control volunteers. Anti-lipid A Abs have not been conclusively proved to be protective (Sect. B). Whether Vi Ag conveys virulence to *Sal. typhi* by blocking the access of anti-O Abs to O Ag in the cell wall and/or by its antiphagocytic properties is not known.

J. MECHANISMS OF IMMUNITY

Much remains to be learned about the mechanisms of immunity to *Sal. typhi* and other salmonellae, even after decades of study; however, it is the consensus that cell-mediated immunity is of primary importance. There is no correlation between immunity to typhoid fever and the titer of any known humoral Ab; moreover, passive transfer of a substantial degree of immunity with serum has not been demonstrated conclusively. Humoral Abs are formed early during typhoid fever, usually by the second week. They agglutinate *Sal. typhi* and, together with C, can lyse the organisms. Nevertheless, they do not terminate the disease, which persists for weeks after Abs become demonstrable. Nevertheless, bactericidal Abs may well destroy extracellular organisms and thus inhibit their systemic spread. Secretory IgA Abs may also block adhesion to and endocytosis by host cells. Despite these protective effects the overall ineffectiveness of humoral immunity results from the fact that the organisms can multiply within nonimmune macrophages where they are sequestered from humoral Abs and C.

Cellular immune mechanisms have been intensively studied in the mouse-typhimurium model, and to some extent in man. Evidently effective immunity depends principally on properties of activated macrophages which, in contrast to nonactivated macrophages, can kill or inhibit intracellular salmonellae. Although the development of immune macrophages is specific, resulting from stimuli provided by immune lymphocytes incited to activity by specific Ag, their ability to inhibit and kill salmonellae is nonspecific. Evidence that cellular immunity is of paramount importance in mouse salmonellosis has been provided by passive transfer experiments which show that protection against infection can be transferred to a nonimmune recipient with lymphocytes from an immunized donor, but not with immune sera. Also, immunization is much more effective when living attenuated bacteria or killed bacteria in Freund's complete adjuvant are used (which favor the development of cellular immunity) than when killed saline-suspended organisms are administered. These data, of course, do not deny the possibility that humoral Abs contribute to overall immunity in the mouse model.

Vaccines currently used in man against enteric fevers consist of killed saline suspensions of typhoid and paratyphoid bacteria. The usual combined vaccine, called, *TAB,* (comprised of a pool of *Sal. typhi,* paratyphoid A, and paratyphoid B), offers limited protection. This is in sharp contrast to the marked and long-lasting immunity engendered by the disease itself. If Freund's complete adjuvant could be added to the vaccine, more effective protection could probably be achieved; however, this adjuvant produces severe granulomatous reactions and cannot be used in man. Suitable adjuvants for human use have not been discovered to date.

K. LABORATORY DIAGNOSIS

Because typhoid fever is easily confused with a number of other diseases, laboratory diagnosis is important. It is based on culture of blood, feces and, occasionally abscessed bone marrow, urine or exudates and tests for serum agglutinins. Colonies are readily recognized on selective-differential media and can be picked and identified by biochemical tests and serotyping with known agglutinating antisera. The percentages of positive cultures of blood, stool and urine and of positive agglutination tests are shown in Figures 20–1 and 20–2. A 4-fold increase in serum Abs, especially anti-O agglutinins, with known *Sal. typhi* (Widal test), is indicative of typhoid fever, provided the patient has not recently received typhoid vaccine. A newer more reliable serologic test is for Abs against LPS. It involves determinations of IgG Abs in endemic areas and IgM Abs in nonendemic typhoid areas, using the enzyme-linked-immunoabsorption-assay (ELISA).

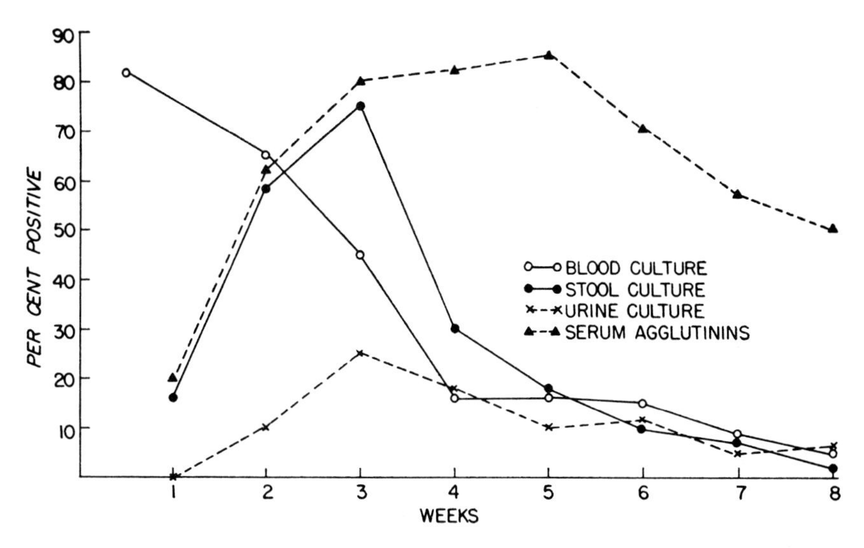

Fig. 20–1. Serum agglutination tests and the incidence of positive blood, stool and urine cultures of patients during the course of typhoid fever. (Adapted from: Dubos, R.J., and Hirsch, J.G.: *Bacterial and Mycotic Infections of Man,* 4th ed. Philadelphia; J.B. Lippincott Co., 1965.)

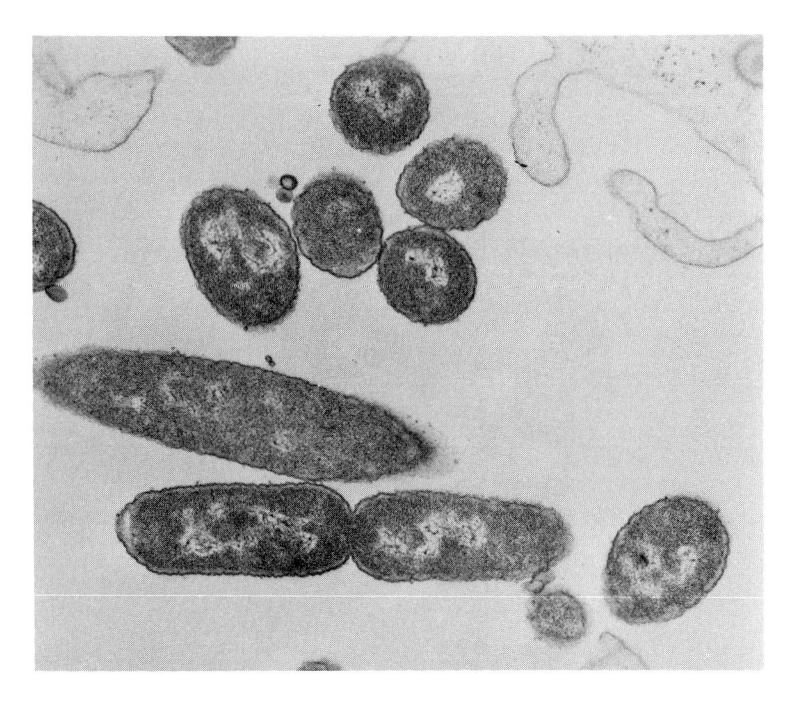

Fig. 20–2. Ultrathin section of extracellular *Salmonella typhi* from a cell culture. Note characteristic thin Gram-negative cell wall. ×25,000.

L. THERAPY

Chloramphenicol is the chemotherapeutic agent of choice. Since a few *Sal. typhi* strains have developed antibiotic resistance, sensitivity testing is necessary. A number of other antibiotics, including oral ampicillin, and trimethoprim-sulfamethoxazole, are useful. Response to therapy occurs slowly over a period of 1 to 5 days; hemorrhage and perforation may develop despite ongoing therapy. Supportive measures include the administration of steroids to combat severe toxemia. Unfortunately antimicrobics suppress the development of agglutinins, promote relapses and do not prevent the acquisition of carriage.

M. RESERVOIRS OF INFECTION

The only reservoirs of infection are human patients, convalescent carriers and chronic carriers. Since the organisms can survive for weeks in water and food, organisms in excreta that reach food and streams may persist for long periods and be carried for long distances without losing their infectivity. Cholecystectomy is useful and often necessary for arresting chronic carriage; antimicrobics are also of some use but sometimes cause typhoid relapses, presumably by altering the normal flora of the intestine.

N. CONTROL OF DISEASE TRANSMISSION

Obviously the principal keys to the control of typhoid fever are: the sanitary regulation of food and water supplies, the elimination of carriage and the regulation of carriers. Except for disasters that disrupt water and food supplies, the major control problem in developed countries relates to typhoid carriers as food handlers. Ostensibly carriers should not be allowed to handle or prepare food eaten by others. Unfortunately, existing laws make it difficult to detect, treat and regulate carriers. This is exemplified by the story of the infamous food handler, "typhoid Mary" who eluded public health control for decades by hopping from job to job; she was directly responsible for many typhoid cases and deaths. The 10-fold decline in typhoid fever in the USA in recent decades has resulted principally from a reduction in the number of chronic carriers due to natural attrition rather than from improved carrier regulation or sanitation. However, further decline in chronic carriers and disease to the point of extinction is improbable so long as carriers from underdeveloped countries are allowed to emigrate to the USA.

Cholecystectomy eliminates carriage in about 85% of subjects and combined ampicillin-probenecid treatment commonly eliminates carriage in those individuals who are not biliary-tract carriers.

Immunization, of adults, is of value because it affords partial protection and may sometimes prevent infection, especially if the exposure dose of organisms is low; in addition, immunization decreases the severity of any subsequent infection that may be contracted. Organisms killed by mild heat treatment are generally used for immunization. When given intramuscularly the vaccine tends to produce slight fever, local inflammation and lymphadenopathy; this can be minimized by administering the vaccine as small, repeated intracutaneous doses. An annual booster is required to maintain the partial immunity afforded by the vaccine. Needless to say, a better vaccine is in the realm of possibility and is sorely needed; a highly promising attenuated oral vaccine is currently being tested.

SALMONELLA CHOLERAESUIS AND *SALMONELLA ENTERITIDIS*

Except for being relatively less virulent for man than *Sal. typhi,* salmonellae of the above species are similar to *Sal. typhi* in most respects. Consequently, discussions of these organisms will be largely limited to the diseases they produce.

These non-typhi salmonellae may be carried by any domestic animal or bird and by some wild warm-blooded and cold-blooded vertebrates. It has been estimated that there are nearly 4 billion domestic animals in the USA at present, and that about 3% of them are infected with salmonellae. Animal salmonelloses are of tremendous economic importance, amounting to an annual loss of about 10 million dollars in the poultry industry alone. Recent outbreaks of salmonellosis in man have been traced to such reservoirs as pet turtles, chicks, ducklings, dogs, chickens, cattle and a variety of other domestic animals. The "infectious dose" for the non-typhi salmonellae is large.

Salmonella choleraesuis infects man as an accidental host. It can cause local or systemic infection and is the most frequent cause of salmonellosis characterized by septicemia; it occurs most often in very old, very young or debilitated individuals. As in typhoid fever, ingested organisms transgress the intestinal epithelium and pass to the lymph nodes and blood stream, from whence they establish widespread foci of infection at various sites, notably bone. However, unlike *Sal. typhi,* they do not infect the gallbladder nor do they produce lesions at the site of entrance, the intestinal epithelium; mortality ranges from 5 to 20%.

Enterocolitis is commonly due to *Sal. enteritidis.* Since the organisms are of low virulence for man and large numbers are required to produce disease, the *usual source of infection is food in which the organisms have multiplied extensively.* Although enterocolitis is the most common of the salmonelloses, it is relatively mild and self-limiting; *Salmonella enteritidis* may sometimes cause asymptomatic infections. Of the 15,000 to 20,000 *Sal. enteritidis* cases reported annually to the USPHS, the relatively few that are fatal occur in the very young, the very old, and debilitated or in immunologically-compromised individuals. As a rule, the symptoms of *"salmonella food-poisoning" begin 12 to 48 hours after eating the contaminated food.* Nausea, vomiting and diarrhea may be accompanied by slight fever and headache. Mild cases may last less than a day, but even the more severe cases usually terminate within 3 to 5 days. Antimicrobial therapy is seldom required and carriage beyond a few days is rare. Fluids and electrolytes should be replaced if extensive diarrhea and vomiting have occurred.

The pathogenesis of enterocolitis probably rests, at least in part, on the activities of endotoxin and/or enterotoxin. The facts that large numbers of bacteria are required to produce disease and that the onset of illness usually follows within hours to a day or two after ingestion of contaminated materials support this concept. The *E. coli*-like enterotoxin produced by many strains of these salmonellae is closely associated with the outer membrane. The enterotoxin is toxic for rabbits when given intravenously and specific Ab is protective.

Solid immunity to Salmonella-induced enterocolitis does not develop and repeated infections can evidently occur with the same serotype. However, these suppositions have not been firmly established as facts. Control of enterocolitis presents major problems. Special attention must be paid to methods of food preparation and preservation. Although, enterocolitis seldom results from food contaminated by human carriers, every attempt should be made to ensure proper personal, as well as plant sanitation in slaughter houses, dairies, food-processing establishments and restaurants. For example, some salmonellae can remain alive at room temperature for 6 to 8 weeks in salami sausages. Also they may survive in dried materials for months, and for several weeks on apparently clean eggshells. Highly acid or alkaline environments (pH < 4.5 or > 9.0) lead to death of salmonellae, but pH effects vary considerably depending on the nature of the suspending medium, temperature and other factors. *Persons who drink raw milk run undue risk of developing enterocolitis.* A number of recent dairy-product outbreaks have occurred in the USA due to a highly virulent strain of *Sal. enteritidis* serotype dublin.

The *incidence* of enterocolitis is impossible to determine accurately because most cases are mild and many are not reported. In the USA, probably only 1 of every 10 to 20 cases is reported; however, the number reported has increased dramatically over the past decade and now reaches some 20,000 to 40,000 annually. This is thought to largely reflect an increase in incidence, as well as improved detection and reporting. The increase may result, in part, from the increasing use of certain processed convenience foods which can harbor salmonellae. The salmonelloses are far from being controlled, and will probably remain a major public health problem for the foreseeable future. When outbreaks occur, extensive epidemiologic studies with phage typing may be required to identify the source of the infection. Serologic typing is of limited epidemiologic value because the same serotypes cause the majority of outbreaks. For example, *Salmonella enteritidis* serotypes A, B and C are responsible for about a fourth of the reported cases of *Salmonella enterocolitis.*

The Riverside, California epidemic of 1969, involving nearly 20,000 cases, reflects some of the unique problems presented in the control of salmonellosis. The source was traced to the municipal water supply, and ultimately to human carriers. The water had been monitored by standard procedures to detect fecal contamination, but was only chlorinated when the tests indicated a certain level of *E. coli* contamination. As a rule, this procedure is adequate because *E. coli* usually occurs in much greater numbers than salmonellae in fecally-contaminated water; however, in this instance the salmonellae were 10 times as numerous as *E. coli*. The preferred procedure, routine chlorination, would have prevented this epidemic.

Salmonellae are commonly present on and in foods such as poultry, eggs and raw milk because they are often present in the gastrointestinal tract of the animals. The organisms are able to survive procedures commonly used to preserve foods, such as freezing (Sect. F). Thus, frozen chickens and turkeys, as well as many other kinds of food, often contain salmonellae that can multiply during food preparation. Unless care is taken, such foods serve as a source of salmonellosis. A common example is the frozen turkey, which takes a long time to thaw. Many cooks have used the dangerous short-cut of allowing the turkey to thaw at room temperature over an extended period. This permits the multiplication of salmonellae on the surface of the bird and in the body cavity. Indeed, if a partially frozen turkey is roasted, the temperature inside the cavity may not rise enough to kill the organisms.

Paratyphoid enteric fever produced by *Sal. enteritidis* serotypes A, B, and C (paratyphoid group) is similar to but milder than typhoid fever. Death is usually due to complications and ranges from 2 to 10%. Chemotherapy is of little value but is used in the very young and the aged. Chronic carriage beyond a few weeks is rare.

Organisms of two other serologically-related genera, Arizona and *Citrobacter* can be confused with salmonellae. *Arizona* species may or may not ferment lactose but produce infections similar to *Sal. enteritidis* infections, i.e. bacteremia, enterocolitis and enteric fever. Their natural hosts include lizards, snakes and gila monsters. *Citrobacter species* do not cause infections in man but can be confused with salmonellae because of their serologic relatedness.

REFERENCES

Bitar, R., and Tarpley, J.: Intestinal perforation in typhoid fever: A historical and state-of-the-art review. Rev. Infect. Dis. *7*:257, 1985.

Branham, S.E.: Classics in infectious diseases: Toxic products of bacterium, enteritidis and of related microorganisms. Rev. Infect. Dis. *6*:579, 1984.

Buchwald, D.S. and Blaser, M.J.: A review of human Salmonellosis: II. Duration of excretion following infection with nontyphi *Salmonella.* Rev. Infect. Dis. *6*:345, 1984.

Collaborative report: A waterborne epidemic of salmonellosis in Riverside, California, 1965: Epidemiologic Aspects. Am. J. Epidemiol. *93*:33, 1971.

Edwards, P.R., and Ewing, W.H.: *Identification of Enterobacteriaceae,* Minneapolis, Burgess Publishing Co., 1972.

Galanos, C., et al.: Symposium V. Immunogenic properties of lipid A. Rev. Infect. Dis. *6*:546, 1984.

Hirschl, A., et al.: Antibody response to somatic antigen of *Salmonella typhi* in areas endemic and non-endemic for typhoid fever. Infect. Control *6*:110, 1985.

Kauffmann, F.: *Serological Diagnosis of Salmonella-Species Kauffman-White-Schema.* Copenhagen, Munksgaard, 1972.

Killion, J.W. and Morrison, D.C.: Protection of C3H/HeJ mice from lethal *Salmonella typhimurium* LT2 infection by immunization with lipopolysaccharide-Lipid A-associated protein complexes. Infect. Immun. *54*:1, 1986.

Le Minor, L.: Taxonomy of Salmonella. In *New Horizons in Microbiology.* Sanna, A. and Morace, G. (eds.) Amsterdam, Elsevier Science Publishers, 1984, p. 45.

Udhayakumar, V. and Muthukkaruppan, U.R.: Protective immunity induced by outer membrane proteins of *Salmonella typhimurium* in mice. Infect. Immun. *55*:816, 1987.

21

SHIGELLA

Members of the genus *Shigella* belong to the family *Enterobacteriaceae.* They are all parasites in the intestinal tracts of their natural hosts, the primates. The virulent shigellae produce bacillary dysentery in man and the higher apes. The organisms are shed in the feces and the disease is transmitted by feces, either directly or indirectly (Sect. H). The genus *Shigella* contains four species, *Sh. dysenteriae, Sh. flexneri, Sh. boydii,* and *Sh. sonnei.* Based on somatic O Ags they fall into the respective serogroups A, B, C, and D. Moreover, since antigenic variations occur within O Ag groups each of the first three species is divided into several O serotypes designated with arabic numbers. *Shigella sonnei* strains belong to some 15 bacteriocin types. Since the shigellae and the diseases they produce are similar, they are discussed as a group with *Sh. dysenteriae* as the prototype.

A. MEDICAL PERSPECTIVES

Bacillary dysentery in man is highly contagious and is likely to occur in epidemic form whenever sanitation and personal hygiene are lax, for example, during military campaigns. It has often played a decisive role in the outcome of battles throughout the centuries. Herodotus reported that the defeat of the Persian Army in 380 B.C. was due to dysentery that swept through the Persian troops. The fabulous military success of Alexander the Great has been attributed in part to his insistence that his armies boil their drinking water, thereby guarding against dysentery and certain other enteric infections. Although bacillary dysentery has been recognized for centuries, its causative agent was not discovered until 1898, when Shiga identified the "dysentery bacillus," later named *Shigella dysenteriae.*

During past centuries shigellosis took a huge toll of lives among infants in the USA and throughout the world. It was at its maximum in the USA during the era of the outdoor nonflushing toilet, and pioneer graveyards abound with its victims. The disease peaked in summer, the season of flies and picnics; among children it was known as "summer complaint."

At present, infection with the highly virulent species, *Sh. dysenteriae,* is rare in the USA; however, infections due to *Sh. flexneri* are relatively common in the southern States and infections due to *Sh. sonnei* are relatively common in the northern States. Outbreaks occur most frequently among migrant farm workers and others who live in inadequate housing. Often fecally-contaminated water supplies are the source of outbreaks. Nonseasonal endemic disease and chronic carriage contribute to the maintenance of the disease in a population and transmission due to poor personal hygiene is the rule. Outbreaks in orphanages and mental institutions often result from person-to-person transmission. The minimum infecting dose is low; experimental studies using human volunteers have shown that as few as 100 ingested *Sh. flexneri* organisms can initiate infection in healthy adult males.

The number of cases of shigellosis reported to the USPHS annually ranges between 15,000 and 20,000; like salmonellosis, many more cases are not reported. Most fatalities occur in young children (peak at 2 years of age).

In the world at large, shigellosis remains an infectious disease of major importance and in certain underdeveloped countries it still takes a staggering toll of lives, especially among infants and young children. Infection due to *Sh.*

dysenteriae is most frequent in the tropics where it is one of the leading causes of infant mortality.

B. PHYSICAL AND CHEMICAL STRUCTURE

The shigellae are nonmotile, nonsporing, nonencapsulated Gram-negative rods possessing characteristic somatic O Ags. Certain smooth-colony strains possess a heat-labile outer envelope Ag which blocks agglutination with anti-O Ab. In Gram-stained smears the shigellae resemble the coliforms, the rods being 1 to 3 μm in length and 0.5 μm in diameter. They are similar to *E. coli* antigenically and biochemically.

C. GENETICS

Like other members of the family *Enterobacteriaceae,* the shigellae show marked variability. The variation of greatest importance to the practicing physician is the readiness with which these organisms acquire multiple drug resistance from organisms of other species and genera by the agency of plasmids. Shigellae are also susceptible to certain bacteriophages and may be altered by lysogenic conversion. Bacteriophage typing has not proven to be of epidemiologic value.

D. EXTRACELLULAR PRODUCTS

The marked severity of *Sh. dysenteriae* infections, which have a fatality rate of about 20% in untreated subjects, results in large part from the production of a potent exotoxin discovered by Shiga (Shiga toxin). Although first designated as a lethal neurotoxin for rabbits, it is now known to produce multiple effects in man. The toxin is a multi-chained proenzyme with a mol. wt. of 68,000; its peptide chains are of two classes. Various *Shigella sp.* produce variable amounts of the toxin, *Sh. dysenteriae* being the highest producer. The toxins of different *Shigella sp.* are probably serologically identical.

E. CULTURE

The shigellae grow as aerobes or facultative anaerobes; most strains do not ferment lactose but a few ferment it slowly. Most strains do not tolerate bile. Their cultural requirements are similar to those of the salmonellae (Chap. 20).

F. RESISTANCE TO PHYSICAL AND CHEMICAL AGENTS

Except for their susceptibility to bile, shigellae are similar to the salmonellae with respect to resistance to environmental factors and antibiotics. Hence, these properties of the salmonellae, described in Chapter 20, are also applicable to the shigellae. For unknown reasons the shigellae do not survive long in fecal specimens even though they can persist in water for 6 months or more.

G. EXPERIMENTAL MODELS

The orally infected subhuman primate is the best experimental model of human shigellosis. The rabbit is commonly used to study the effects of Shiga toxin, which causes paralysis and death when injected intravenously. This effect of the toxin is neutralized by specific antitoxin. Fatal shigellosis can be induced in weanling guinea pigs deficient in folic acid. Other treatments, such as starvation or antibiotic treatment that reduce the natural resistance of the guinea pig permit limited infections.

H. INFECTIONS IN MAN

Unlike the nontyphoid salmonelloses that are commonly transmitted from animal to man, human shigellosis is essentially always transferred from man to man, either directly via the fecal-oral route, or through the agency of mechanical vectors such as the housefly, which gains access to human feces and thence to food. The shigellae are noted for being easily transmitted via fingers, feces, food, flies, and fomites. Unlike the salmonellae, the shigellae do not invade beyond the surface integument of the lower intestine and colon where they enter and destroy epithelial cells.

Asymptomatic infection is rare; it has been reported in the respiratory tract as well as the intestinal tract.

Shigellosis is either a severe form caused by *Sh. dysenteriae* or a less severe form resulting from infection with other *Shigella sp.* In both instances the onset is abrupt and usually follows an incubation period of 2 days (range 1 to 6 days). The first symptoms are fever (50% of cases) and abdominal cramps followed by the onset of painful diarrhea within a few hours. Characteristically, the stools are watery and contain mucus, pus and blood. Nausea is common, but vomiting is usually not a prominent symptom. Within a few hours the diarrhea subsides and tenesmus (dysentery) becomes marked. The lesions that produce diarrhea are confined principally to the terminal ileum, whereas the lesions responsible for tenesmus comprise small ulcerating abscesses in the colon.

In *Sh. dysenteriae* infections the diarrhea is severe and may lead to a cholera-like dehydration. In addition, headache may be intense and sometimes neurologic symptoms develop. The disease is either soon fatal or runs its course within about 10 days. Complications, which are rare, include bacteremia, arthritis, neuritis, conjunctivitis, iritis, vulvovaginitis, and chronic colitis. Fatalities can range as high as 10 to 20%.

Dysentery due to other species of *Shigella* is usually much milder and of shorter duration. The highest incidence occurs in children 1 to 4 years of age. Recovery may be complete within a few days in some subjects, but is usually delayed until about a week.

Bacillary dysentery is notable for the high mortality it causes in infants, and in aged and debilitated individuals. This is true regardless of the *Shigella sp.* causing the infection. Antibodies appear in serum during convalescence and can be demonstrated in the feces as well.

I. MECHANISMS OF PATHOGENESIS

All of the shigellae possess the unique ability to adhere to intestinal and colonic epithelial cells, allegedly because of side-chain receptors on cell wall LPS. The adhering bacilli appear to induce their endocytosis, after which they grow within and kill host cells, evidently by liberating toxin. Following epithelial cell destruction the bacilli persist and grow in the shallow ulcers formed where they attract neutrophils that engulf and destroy at least some of the bacilli. The dying neutrophils probably contribute to the intense inflammation present and liberate endogenous pyrogen to produce fever. The inflammation produced, in turn, leads to destruction of the bacilli. Deep tissue invasion and septicemia only occur in immunocompromised individuals.

Although endotoxin probably contributes to pathogenesis, Shiga exotoxin is evidently the major pathogenetic agent. Prevailing evidence indicates that an alleged separate "enterotoxin" does not exist but instead that the Shiga toxin is cytotoxic and serves both as an "enterotoxin" and a neurotoxin. Toxin is evidently only one factor of virulence; others are concerned with adherence, endocytosis and growth within host cells.

The factors that endow the shigellae with virulence and pathogenicity are multiple, subtle and extremely complex. Studies on pathogenesis and virulence have been based principally on genetic experiments employing tissue cultures and plasmid hybridization between virulent and avirulent strains of shigellae and related organisms such as *E. coli.* These studies have shown that virulence is multifactorial, although some factors remain obscure. However, some of the required factors and related steps in disease production have been defined, namely, that the organism (1) must be able to adhere to the mucus that coats epithelium through the interaction of LPS with a lectin-like substance in the mucus, (2) must be able to digest the glycoprotein of mucus to allow close contact with epithelial cells, (3) must possess plasmid-encoded outer membrane proteins that induce bacillary endocytosis, (4) must be able to grow within their host cells and kill them and, (5) must be able to persist and grow extracellularly in ulcerating lesions.

The Shiga toxin is a virulence factor that exerts cytotoxicity by inhibiting protein synthesis by host cells and in some unknown manner causes diarrhea and fluid loss.

The most singular aspect of hybridization studies to date is the fact that, although plasmid transfer has succeeded in endowing *E. coli* with the capacity to adhere to, become endocytosed by, and grow within tissue culture cells, such organisms are not fully virulent and do not produce disease in the living host. This fortunately implies that the ready generation of "new pathogens" in nature by plasmid hybridization is highly unlikely.

J. MECHANISMS OF IMMUNITY

Solid immunity to shigellosis is of short duration and reinfection with the same serotype can occur. Humoral Abs are formed against toxin and Ags of the cell wall, but the relative roles they may play in acquired immunity are uncertain. Although antitoxin affords some passive protection against the enteric effects of the toxin, it is possible that secretory IgA (sIgA) Abs play the major role in the short-term component of immunity by blocking the adherence of organisms to gut epithelium and possibly their subsequent engulfment by these cells. The observation that, over time, epidemics are caused by a succession of serotypes and that adults are less susceptible to shigellosis than children implies that a component of long-term immunity is generated by infection. Killed parenteral vaccines engender serum Abs but do not afford effective protection against shigellosis. Several attenuated vaccines for oral use are now available that provide protection up to 1 year; however, they are of limited practical use because of the serotype specificity and short duration of the immunity produced. Unfortunately, oral vaccines do not terminate carriage.

K. LABORATORY DIAGNOSIS

Definitive diagnosis of shigellosis is made by isolating the organism from the feces. Bits of bloody mucus in the feces are characteristic; they are examined microscopically and are the most likely components to yield positive cultures. Since the organisms die rapidly in feces, fecal specimens should be cultured promptly following collection using selective and differential media. Microscopic examination of mucus in fecal specimens often aids in differentiating shigellosis from salmonellosis or other diarrheas caused by enteric organisms. Differentiation is based on the numbers of neutrophils present, neutrophils being most abundant in shigellosis. The fact that shigellae are nonmotile is helpful in laboratory diagnosis. Final identification of isolates is based on biochemical tests and serologic analysis. Serum agglutinins appear late after infection and agglutination tests are only of value for retrospective diagnosis.

L. THERAPY

Shigellosis in adults is usually so transient and mild that chemotherapy is unnecessary and fluid replacement can be achieved orally. In children, however, artificial fluid replacement, electrolyte balancing and antibiotic treatment are important therapeutic measures. Unfortunately plasmid-induced drug resistance is common and drug resistance testing is conducted either on isolates and/or the epidemic-strain organism. Effective antimicrobics used for designing chemotherapeutic regimens include trimethoprim-sulfamethoxazole (the drug of choice), ampicillin, tetracycline, nalidixic acid, pivmecillinam and oxalic acid. Chemotherapy, which reduces the severity of disease, is of control value because it shortens the period of convalescent carriage.

M. RESERVOIRS OF INFECTION

Man is almost invariably the reservoir of human shigellosis. On rare occasion, man may contract the disease from subhuman primates. The organisms are usually carried for only a short time (up to 1 month) after recovery from disease; however, it is singularly important that *occasional patients develop the state of chronic carriage with intermittent shedding of organisms, which serves to perpetuate the organisms in human populations!*

N. CONTROL OF DISEASE TRANSMISSION

Isolation of infectious patients and the usual individual and public sanitary measures generally applicable to enteric diseases are the keys for controlling shigellosis. Adequate and prolonged antibiotic therapy during the disease helps to prevent the establishment of the short-term and chronic carrier states. Mass chemoprophylaxis is a useful measure for controlling severe epidemics. Detection, treatment and control of carriers is of utmost importance; food handling by carriers should not be allowed so long as carriage persists. The extent to which carriage resulting from inapparent infection may contribute to epidemics is probably minor. Although, in theory, proper hygiene and sanitation could eradicate shigellosis in the USA, this goal is difficult to achieve, principally because of the occasional chronic carrier and because travelers and immigrants reintroduce the organisms into the population.

REFERENCES

Kelly, Michael T., et al.: Enterobacteriaceae. *In* Manual of Clinical Microbiology, 4th Ed. Lennette, E.H., et al. (eds.) Am. Soc. for Microbiol. Washington, D.C. 1985, p. 263.

Keren, D.F., et al.: Effect of antigen form on local immunoglobulin: A memory response of intestinal secretions to *Shigella flexneri.* Infect. Immun. *47*:123, 1985.

Levine, M.M.: Bacillary dysentery: Mechanisms and treatment. Med. Clin. North Am. *66*:623, 1982.

Olsnes, S., et al.: Subunit structure of *Shigella* cytotoxin. J. Biol. Chem. *256*:8732, 1981.

Sansonetti, P.J., et al.: Genetic analysis of virulence of *Shigellae* and enteroinvasive *Escherichia coli.* In *New Horizons in Microbiology.* Sana, A. and Morace, G. (eds.), Amsterdam, Elsevier Science Publishers, 1984, p. 53.

22

VIBRIO AND CAMPYLOBACTER

Vibrio cholerae

Choleragenic *Vibrio cholerae* 01

Other *Vibrio cholerae* Variants of Medical Importance

Other *Vibrio* species of Medical Importance

Vibrio parahaemolyticus

Vibrio alginolyticus

Vibrio vulnificus

Vibrio fluvialis

Other Genera of *Vibrionaceae* of Medical Importance

Aeromonas

Plesiomonas

Campylobacter (Family: *Spirillaceae*)

The family *Vibrionaceae* contains four genera, *Vibrio, Aeromonas, Plesiomonas,* and *Photobacterium.* Most species of this family are oxidase positive and are capable of both fermentative and respiratory activities, and most are aerobic or facultatively anaerobic. All genera except *Photobacterium* contain species that are pathogenic for humans.

Most *Vibrionaceae* are short, motile, nonsporing, Gram-negative, salt-tolerant, curved rods that occur singly or in S-shaped or spiral-shaped chains; motility is usually effected by a single, thick, sheathed, polar flagellum. Some species possess fimbriae.

Most *Vibrionaceae* are aquatic organisms that naturally inhabit marine or brackish waters; many of them are true halophiles and all grow best in media with added salt. The halophiles require media containing at least 0.5% NaCl and may tolerate concentrations as high as 10% or more. Although many species of the *Vibrionaceae* are aquatic saprophytes, while others are opportunistic pathogens for humans and a wide range of animals as accidental hosts. Whereas some species only produce intestinal diseases, others produce extraintestinal infections or a combination of both.

Although members of the genus *Vibrio* are the most common cause of serious infections in humans, members of the genera *Aeromonas* and *Plesiomonas* are increasingly common causes of intestinal and extraintestinal diseases.

VIBRIO CHOLERAE

The many variants within the species *V. cholerae,* fall into three major groups based on overall characteristics. They belong to 6 serogroups, 01 through 06, based on differences in their somatic O Ags.

The first major group consists of strains belonging to serogroup 01 that produces cholera. Serogroup O type 1 (01) comprises two biotypes, cholerae (classic) and El Tor. *V. cholerae* (classic) is sensitive to 50 IU polymyxin disc, whereas *V. cholerae* biotype El Tor is resistant (Table 22–1). Three Ags, A, B, and C divide 01 into the serotypes, Ogawa, Inaba and Hikojima. The biotypes classic *V. cholerae* and El Tor are the etiologic agents of typical cholera. The other serogroups (02–06) are involved in producing milder forms of the disease. The three serotypes in serogroup 01 can interconvert in natural infection as well as in experimental infections.

The *two other major groups* of *V. cholerae* organisms comprise the nonpathogenic, atypical *V. cholerae* 01 group, which contains the 01 Ag but does not produce enterotoxin and the non-01 *V. cholerae* group, which lacks the 01 Ag and sometimes produces cholera-like enterotoxin and diarrheal disease but never causes classic epidemic cholera (Table 22–2). Since organisms of the 3rd group do not agglutinate in anti-01 serum they are sometimes called "nonagglutinating vibrios" (NAG vibrios).

Table 22–1. Differentiation of *Vibrio cholerae* Biotypes in Serogroup 01

	Biotype	
Test Reactions	*Cholerae*	*El Tor*
Agglutination of chicken RBC	–	+
Sensitivity to polymyxin	+	–
Voges-Proskauer Reaction	–	+

Table 22–2. Some Properties of *Vibrio cholerae* Variants

Name	*Contains 01 Ag*	*Common Group Designation*	*Exotoxin Produced*	*Diseases Commonly Produced*	*Probable Primary Reservoirs*	*Common Vehicles of Secondary Infection*
*V. cholera** biotype cholerae (classic *V. cholerae*)	+	*V. cholerae* 01 (choleragenic *V. cholerae*)	Classic cholera enterotoxin	Cholera (endemic or epidemic)	Brackish marine waters and aquatic life therein	Fecally contaminated drinking water and food
*V. cholerae**, biotype El Tor	+	*V. cholerae* 01 (choleragenic *V. cholerae*)	Classic cholera enterotoxin	Cholera (endemic or epidemic)	Brackish marine waters and aquatic life therein	Fecally contaminated drinking water and food
Atypical *V. cholerae* 01	+	Atypical *V. cholerae* 01	None	None	Brackish marine waters and aquatic life therein	None
Non-01 *V. cholerae*	contains other 0 Ags	Nonagglutinating (NAG) *V. cholerae* (NAG vibrios)	Cholera like enterotoxin, heat-stable toxin or none	Gastroenteritis (often cholera like) and ear and wound infections	Brackish inland and marine waters and aquatic life therein	Raw seafood and exposure to brackish and marine waters

*Strains of the group choleragenic *V. cholerae* 01, can be subdivided into three serologic groups (serotypes), Ogawa, Inaba and Hikojima.

CHOLERAGENIC *VIBRIO CHOLERAE* 01

Major emphasis will be placed on organisms of this group because of the high frequency and seriousness of the infections they produce. Although both the classic and El Tor biotypes can incite epidemics, the El Tor biotype is less prone to cause severe cholera than the classic *V. cholerae* biotype. As evidence of this the ratio of symptomatic to asymptomatic infection in classic cholera is 1:7, whereas the ratio in El Tor infections is 1:20. El Tor organisms also persist longer in nature than the classic *V. cholerae* organisms.

Cholera is a toxin-induced disease that results from a noninvasive local infection on the mucosa of the small intestine. The disease results from a unique exotoxin called cholera enterotoxin or choleragen. Experimental trials in man indicate that, in general, relatively large doses of organisms are required to produce infection and that enormous doses are needed to induce severe disease. These high dose requirements are evidently due in large measure to the bactericidal effects of gastric acidity and serve to explain why cholera is rarely transferred by person-to-person contact. In patients with achlorhydria, the choleragenic dose may be below 10^6 organisms. Individual variation in natural resistance to cholera is striking. However, the factors concerned in such resistance are largely obscure, although debilitated, malnourished and immunosuppressed individuals are known to be highly susceptible. Persons belonging to blood group O are also uniquely susceptible. During epidemics, infection results largely from the ingestion of water or food contaminated with the feces of cholera patients and carriers, and classic epidemic disease can only be maintained under grossly unsanitary conditions of personal and public hygiene.

Until recently it has been generally held that humans are the only natural host and permanent reservoir of classic *V. cholerae,* which tends to die off in fresh waters in a matter of weeks and in most marine waters in a matter of months. However, there is growing evidence that the El Tor biotype and possibly the classic biotype can maintain themselves in the marine life of certain favorable estuaries and salt marshes for extended periods. Indeed it has recently been proposed that the permanent habitats of both biotypes reside in the aquatic life of brackish water in certain geographically favorable locales and, moreover, that these habitats serve as the primary reservoirs of endemic cholera (Sect. M).

A. MEDICAL PERSPECTIVES

In the course of history, epidemic cholera with its swift spread and high fatality rate has caused more public panic than any infectious disease except bubonic plague. Without prompt and adequate treatment 40 to 80% of its victims die of cardiovascular collapse within a few hours or days after onset of massive diarrhea.

Cholera has been known and feared for thousands of years; as early as the 5th century B.C., Thucydides described what was probably a cholera outbreak in Athens, and the disease was reported to be epidemic in the army of Alexander the Great about 320 B.C. Seven well-documented great pandemics swept the world during the 19th and early 20th centuries, several of which spread across the USA on pioneer wagon trails. In the USA, cholera deaths totaled 150,000 during the 2nd pandemic, which began in 1832, and 50,000 during the 4th pandemic of 1866. The 3rd pandemic (1846–1862) caused 140,000 deaths in France alone. Some of the most significant advances in understanding the nature of cholera were made by a British physician, John Snow, who is also famous for being the world's first anesthesiologist. This highly perceptive observer studied the London epidemics of 1849 and 1854. His paper, "On the

Mode of Communication of Cholera," is one of the classics of medical literature. In it he clearly showed that the 1849 outbreak stemmed from water supplied by the Broad Street pump, a public water source in London. Although the causative agent was not known, Dr. Snow deduced that contamination of the water was responsible for the disease.

The precise conditions which may favor certain locales as permanent habitats of cholera vibrios are not known, but probably include mild salinity, abundant organic matter and warm temperatures. Indeed, most areas of endemic cholera are in tropical and subtropical climates, and both endemic and epidemic cases of cholera peak during warm seasons of the year when multiplication of the organisms is greatest.

After Koch's discovery of the cholera vibrio in 1883, many pioneering studies were conducted. Pfeiffer discovered the phenomenon of antibody-plus-complement lysis of Gram-negative bacteria (Pfeiffer phenomenon) when he described vibriocidal Abs in 1894. About the same time, Ag-Ab precipitation reactions were first observed, using Abs and soluble Ags of *V. cholerae.*

The reasons for the sudden rise and fall of cholera epidemics are still far from clear. For example, cholera swept China in 1940 and again in 1946 only to be followed by a lengthy cholera-free period ending in 1961, during which time not a single case was reported. The extent to which herd immunity may influence the epidemiology of cholera is uncertain.

The 7th pandemic of cholera involving the El Tor biotype began in Indonesia in 1961 and spread over Asia and Africa where 100,000 cases were reported in 1971. Since cases of El Tor infection first appeared along the Gulf coast of the USA in 1981 and later appeared on the Mexican coast, it seems evident that this organism has become endemic in the area.

Prevention of cholera in underdeveloped countries is difficult and the disease continues to offer a major public health challenge despite improvements in sanitation and treatment. There is little room for complacency in view of the fact that the World Health Organization (WHO) reported some 86,000 worldwide cases of cholera in 1981. In recent decades research on cholera has received massive support from many governments and the WHO. The isolation and characterization of cholera enterotoxin has led to a better understanding of the pathogenesis of the disease and marked improvements in therapy, which when prompt and adequate can reduce the mortality to less than 1%. Despite these advances the future of cholera is difficult to predict for the obvious reason that it is impossible to accurately forecast what the future in underdeveloped countries holds with respect to the predisposing factors of overpopulation, poverty and famine, namely, unsanitary conditions and lack of medical care. If good sanitation were attainable throughout the world, epidemics of cholera should not occur and only sporadic cases contracted by primary transmission in endemic areas would be expected.

B. PHYSICAL AND CHEMICAL STRUCTURE

The LPS of *V. cholerae* resembles the LPS of other Gram-negative bacteria by possessing lipid A, core-region polysaccharides, and O-specific outer side-chain polysaccharides that determine the variability of somatic O Ags.

Surface receptors for phage are present in cholera vibrios and bacteriophage typing is useful for epidemiologic studies. Bacteriophage treatment of cholera patients, which is of limited therapeutic value, has been supplanted by other more effective therapeutic procedures.

C. GENETICS

As mentioned earlier, the genetic serotype variants of the classic and El Tor biotypes are based on three spontaneously interconvertible Ags, A, B, and C. Cholera vibrios readily develop plasmid-induced resistance to antibiotics. It is also known that enterotoxin production is governed by a chromosomal gene.

D. EXTRACELLULAR PRODUCTS

The classic and El Tor biotypes of *V. cholerae* produce cholera enterotoxin and a variety of extracellular enzymes, including mucinase and neuraminidase. The mucinase digests the surface mucin of the intestinal mucosa and thus exposes receptors that permit adherence of the organisms to the brush border cells.

In vitro studies have shown that classic cholera enterotoxin (choleragen) is toxic for all cells that possess membrane-bound adenyl cyclase. In cholera the brush border epithelium of the intestine is the target of the toxin because of the selective adherence of the organism to these cells and not because they are uniquely susceptible to the toxin.

The enterotoxin is a heat- and acid-labile, diarrheagenic protein exotoxin that causes the severe diarrhea and rapid loss of fluid and electrolytes seen in cholera. The toxin has a molecular weight of about 85,000 and consists of three major subunits: A_1, A_2, and B. Subunit A_1 possesses enzymatic activity and elicits the toxin-specific biologic response by ADP-ribosylating a component of the adenylate cyclase system, the GTP-regulatory protein. Thus, adenylate cyclase activity is stimulated and the intracellular level of cyclic AMP (cAMP) is markedly increased. The increased level of intracellular cAMP leads to the rapid passage of electrolytes and fluid into the lumen of the small intestine; since fluid loss may reach 1 liter per hour and continue for many hours, fatal dehydration often results. Subunit A_2 is thought to be required for internalization of subunit A_1, and subunit B binds to the cholera toxin-specific cell surface receptor (ganglioside G_{M1}) located on the villi of the brush border cells of the small intestine. Various studies have shown that there are five B subunits in each toxin molecule, and that they are arranged in a ring around a central core which contains the enzyme A_1; A_2 connects A_1 to the B ring through a disulfide bond, which must be reduced for expression of maximal enzymatic activity.

E. CULTURE

Cholera vibrios grow readily on a variety of simple media and are unique among Gram-negative enteric bacteria in their preference for an alkaline environment (range 8.0 to 9.5). They grow best under aerobic conditions at pH values near 8.0 and are rapidly killed in an acid medium. The optimal temperature for growth of the organisms is 37°C; however, they grow over a wide range of temperatures (14° to 42°C). Cholera vibrios can tolerate concentrations of bile or tellurite salts that kill most other enteric organisms.

F. RESISTANCE TO PHYSICAL AND CHEMICAL AGENTS

As noted by John Snow more than a century ago, cholera vibrios survive and may even multiply in fecally-contaminated water providing the salt concentration is above physiologic values. They are, however, readily killed by acids, most disinfectants, drying and heat (56°C, 15 minutes). Cholera vibrios have been shown to survive in fresh water for days to weeks and on the surfaces of moist fruits or vegetables for as long as 2 weeks. The organisms are susceptible to certain antibiotics including the tetracyclines (Sect. L). Under favorable conditions *V. cholerae* may sometimes grow in foods and marine waters.

G. EXPERIMENTAL MODELS

A variety of animals can be infected experimentally. Infant rabbits and mice and about 35% of mongrel dogs respond to oral administration of the organisms by developing a disease that resembles human cholera. Adult guinea pigs can also be infected if gastric acidity is neutralized before the organisms are administered.

The canine model has been highly instructive. Some of the dogs that develop severe cholera after oral infection recover without treatment. The experimental canine disease closely mimics cholera in man; intensive, watery diarrhea occurs for 36 to 48 hours. Carriers can be detected among dogs that have recovered from experimental cholera and carriage may last at least 26 months.

The pathogenesis of cholera has been extensively studied using live organisms and enterotoxin in the rabbit intestinal loop model and the canine Thiry-Vella intestinal loop model.

H. INFECTIONS IN MAN

Cholera in man, is strictly limited to the small intestine. The organisms never invade the bloodstream or deep tissues. After an incubation period of a few hours to a few days, an afebrile illness begins abruptly with a feeling of abdominal fullness and loss of appetite, which is soon followed by onset of diarrhea and vomiting. The hallmark of cholera is an odorless but copious, watery diarrhea with as many as 20 to 30 fluid bowel movements in 1 day. Mucus discharged into the watery bowel content gives the stool a characteristic appearance referred to as "rice-water" stools.

The diarrheal fluid is composed largely of water and salts and is low in protein. More than a liter of fluid may be lost each hour, resulting in rapid and devastating dehydration, often with a loss of 5 to 10% of body weight within a few hours. The skin becomes loose and the eyes sunken and other effects of fluid loss rapidly become apparent. The blood becomes so viscous that the circulation is impaired and hypovolemic shock may occur; anuria is common.

Loss of bicarbonate ions in cholera stools and in vomitus is of particular significance, because it can rapidly lead to metabolic acidosis. Consequently, the symptoms and complications of acidosis are often superimposed on the picture of severe dehydration and cardiovascular collapse. The fatality rate in untreated cases ranges from 40 to 80%.

I. MECHANISMS OF PATHOGENESIS

The principal virulence factor directly responsible for the violent diarrhea in cholera is cholera enterotoxin. However, the organisms produce additional virulence factors that contribute to their ability to adhere to and colonize on the intestinal mucosa where toxin is produced and exerts its effects. For example, as stated in Section D, pathogenic strains possess virulence properties that enable them to reach and adhere selectively to the microvilli of target cells of the brush border of intestinal epithelium; these virulence characteristics include motility, mucinase production, chemotactic responsiveness, and a surface adhesin.

A role for LPS endotoxin in pathogenesis is not well established and at present, there is no compelling reason to believe that any directly injurious factor other than enterotoxin contributes to the pathogenesis of cholera. However, since anti-LPS Abs block adherence, the adhesin is presumed to be a component of LPS or closely associated with it.

Despite the severity of diarrhea in cholera little tissue injury occurs. Goblet cells exhibit hyperplasia but inflammation is minimal and the mucosa remains intact.

J. MECHANISMS OF IMMUNITY

Acquired immunity to cholera is weak and short-lived (months to 1 to 3 years at most) regardless of whether it follows natural infection or is induced by vaccines; moreover, the protective mechanisms involved in immunity are ill understood. The weakness of acquired immunity to cholera is attested to by the observation that some individuals in endemic areas of cholera suffer repeated attacks. Complement-dependent vibriocidal anti-O Abs and enterotoxin-neutralizing Abs appear in serum after infection. However, the sites of action, specificities and Ig class or classes of the protective Abs that may be involved have not been identified with certainty. Since *V. cholerae* is a noninvasive organism that adheres superficially to the microvilli or brush border cells of the intestine, it is presumed that the adherent organisms and the enterotoxin they produce can only be reached by free Abs at the mucosal surface. Although digestive enzymes in the intestine are profoundly altered in cholera, it seems probable that the principal if not the only coproantibodies that could act would be enzyme resistant secretory IgA Abs. Theoretically such Abs could act either as anti-adhesins to block the adherence and colonization of organisms on the mucosa or as anti-enterotoxins.

Vaccines, including organisms killed by formalin or heat, engender immunity which is greatest when vibriocidal IgG and IgM Abs against somatic O Ags peak in a matter of a few weeks to months. Cholera vaccines are reported to protect some 50% of subjects. Although vaccines may not induce appreciable levels of secretory IgA in the intestinal mucosa, they may prime gut lymphoid tissue for an IgA anamnestic response at the outset of exposure to the organism. If secretory IgA Abs are the most protective against cholera, oral vaccines would be expected to be the most effective.

K. LABORATORY DIAGNOSIS

During epidemics, the clinical picture of cholera virtually provides the diagnosis, and quick confirmation can be made if examination of the stool by darkfield microscopy is possible. The vibrios exhibit a characteristic, darting motility which is abrogated by adding a drop of specific antiserum. The direct fluorescent Ab technique is also useful for rapid diagnosis since it has a reliability of over 90%. Definitive diagnosis of cholera demands isolation of the organism.

Cultural procedures vary greatly; if delay is involved, a transport medium should be used. Initial culture is commonly conducted using a nonselective medium, such as taurocholate gelatin Agar (TGA) together with a selective medium, such as thiosulfate-citrate-bile salt-sucrose agar (TCBS agar) and alkaline peptone enrichment broth; subculture is sometimes necessary. Fluorescent Ab techniques can be employed to identify the vibrios after 4 to 10 hours of growth in peptone broth cultures.

Speciation of cultured organisms depends on biochemical and/or serologic tests using anti-O sera. Bacteriophage typing, which is useful for epidemiologic studies, is of no aid in diagnosis.

L. THERAPY

Since the pathologic effect of *V. cholerae* is simple severe dehydration without appreciable tissue injury and is readily reversible, the benefits of fluid and electrolyte replacement therapy are prompt, dramatic and nearly always successful. With adequate therapy the death rate from cholera can be reduced to less than 1%. Treatment consists of promptly replacing the volume of fluid lost with a sterile intravenous solution containing appropriate electrolytes, such as Dacca solution; oral replacement therapy may be effective in mild cases.

The especially designed "cholera cot" is used for determining the amount of fluid lost. This bed provides a hole under the buttocks so that the copious diarrheal fluid can be collected in a bucket placed under the bed and then measured.

The recent success in developing oral therapy to substitute for intravenous therapy represents a tremendous achievement. It is based on the finding that the addition of glucose or glycine to the electrolyte solution enhances the absorption of ingested water and ions. Following initial rehydration by intravenous infusion (if necessary), the glucose and salt solution is fed by mouth or through a stomach tube. This simplified method reduces the cost of treatment some 50-fold.

Tetracyclines and certain other antibiotics, such as ampicillin, chloramphenicol and trimethoprim-sulfamethoxazole, are effective in ameliorating and shortening the course of the illness as well as the period during which vibrios are excreted. Antibiotic resistance of epidemic strains must, of course, be checked. Antibiotics should also be used to protect the patient's family members who may be exposed to the same source of infection.

M. RESERVOIRS OF INFECTION

A clear understanding of the endemicity and epidemicity of cholera has been sought for many decades. The early view that man is the sole natural host and primary reservoir of *V. cholerae* is now being seriously questioned.

The currently favored theory is that a permanent or primary reservoir of *V. cholerae* 01 resides in the aquatic life of estuaries and salt marshes that exist in a certain few tropical and subtropical areas in the world, including Bangladesh and several locales in Asia, Indochina, Africa and the Gulf of Mexico. It is alleged that the sporadic cases of endemic cholera that typify these areas result largely from eating raw or inadequately cooked seafoods and, moreover, that these sporadic cases contracted by *primary transmission* commonly, if not invariably, serve as secondary reservoirs to incite the local epidemics that arise in areas of endemicity.

The theory further holds that local epidemics can spread by secondary transmission to far-distant nonendemic areas where environmental conditions are grossly unsanitary. This pattern of secondary spread was probably responsible for the cholera pandemics of past centuries before the sanitary treatment of public sewage and drinking water supplies was introduced.

In any event the observations that the organisms die rapidly in most natural waters and that the infecting dose must be large appear to readily account for the failure of epidemic cholera to spread from endemic to nonendemic areas where good sanitation exists.

Epidemic cholera is transmitted principally by fecally-contaminated water, ice or food, and from patients or convalescent carriers. Long-term carriage of *V. cholerae* is rare, although vibrios may occasionally be carried in the gallbladder for long periods. However, organisms from chronic carriers tend to be relatively avirulent strains. Because it tends to become concentrated in bile, tetracycline is useful for eliminating carriage. However, the tendency for drug-resistant strains to develop questions the wisdom of attempting to halt epidemics by mass prophylaxis with antibiotics.

N. CONTROL OF DISEASE TRANSMISSION

Prevention of cholera depends on adequate public and personal sanitation; the disease is virtually unknown in countries where sanitation is good. The most important control measures are modern sewage treatment, water purification and proper disposal of contaminated materials from patients. Modern-day control of cholera has been greatly aided by improved therapy and the arrest of carriage with antibiotics, both of which limit the dispersal of the vibrios.

Despite the many sociologic, financial and political problems it faces, the WHO is also constantly working to improve sanitation in underdeveloped areas. By international agreement all cases of cholera should be reported to the WHO. Since infection with *V. cholerae* commonly results from the ingestion of fecally-contaminated water and food rather than by other means, such as fomites and person-to-person contact, sources of contamination are a central issue in the control of infection. The principal sources of contamination of drinking water and food are the feces of cholera patients and certain carriers, including convalescing patients and carriers in the incubationary stage of cholera or temporary asymptomatic infection. Whereas chronic carriers are relatively unimportant in transmission, other carriers are important. Most convalescent carriers are over 50 years of age and shed organisms for weeks and sometimes months. Individuals in the incubationary stages of cholera are of particular concern in the spread of disease by air travel.

Contamination of food is worrisome because *V. cholerae* can survive for several days and even proliferate if conditions, such as pH and temperature, are suitable. The failure of venders to protect food against flies and the frequent practice of freshening vegetables with potentially impure water may also contribute to the spread of infection. Persons in endemic areas should avoid eating fresh vegetables and raw seafood, especially if their antibacterial defenses are compromised in any way. Swimmers should, of course, avoid contaminated beaches and many of those who travel to endemic areas of cholera wisely choose to be vaccinated a month or two before departure. Vaccination affords limited but worthwhile protection for over 50% of persons and is still required by some countries but not the USA. Although vaccination of travelers may be of personal benefit, it is not useful for preventing the country-to-country spread of cholera. Effective present-day treatment of cholera has contributed greatly toward controlling its spread.

Recent observations of major concern are that 01 strain organisms from infected persons can spread to and persist, at least temporarily, in cattle, chickens and shellfish. For example, an El Tor outbreak occurred in the Gulf Coast of Louisiana due to inadequately cooked shellfish. Thus the discharge of organisms in carrier-contaminated sewage into brackish waters of coastal estuaries and their prolonged persistence in shellfish may be expected to result in the long-term endemicity of cholera in such areas and to cause sporadic infection among swimmers and the unwary who partake of raw or inadequately cooked shellfish.

Although from a medical and scientific standpoint, cholera should be easy to control on a worldwide basis, it will probably be a long time, if ever, before this goal is achieved.

OTHER *VIBRIO CHOLERAE* VARIANTS OF MEDICAL IMPORTANCE

Atypical *V. cholerae* 01 is present worldwide in natural waters. It is not pathogenic and does not induce cross-immunity against cholera.

The non-01 variants of *V. cholerae* are an ill-understood heterogeneous group of organisms that are ubiquitous in coastal waters, especially estuarine waters, and the animal life present in such waters. Whereas some variants do not produce an exotoxin, others produce either a heat-stable toxin or a cholera-like, heat-labile enterotoxin. Unlike choleragenic *V. cholerae,* at least some organisms of this group can produce extraintestinal infections, including otitis media and sometimes septicemia.

Those non-01 strains of *V. cholerae* that produce either cholera-like enterotoxin or heat-stable toxin cause sporadic local outbreaks of diarrheal diseases of varying nature that are less severe than epidemic cholera and usually do not require parenteral fluid and electrolyte replacement. Recent small outbreaks of cholera-like diarrheal disease have occurred in coastal areas of Southern and Eastern USA probably as the result of eating shellfish. Whether human carriers are an important source of diarrheagenic strains of non-01 *V. cholerae* found in coastal waters and how long such organisms may persist in marine environments is unknown. Some cases of otitis media and septicemia have been reported in immunosuppressed persons and chronic alcoholics.

OTHER *VIBRIO* SPECIES OF MEDICAL IMPORTANCE

Other *Vibrio* species which are opportunistic pathogens for humans comprise mostly species that are halophilic and are found in marine waters and in some uncooked seafoods. They produce either intestinal and/or extraintestinal infections. Three of these *Vibrio* species that have been reported to produce intestinal infections but are not discussed are *V. mimicus, V. hollisae,* and *V. furnissii. V. damsella,* on the other hand, produces wound infections.

VIBRIO PARAHAEMOLYTICUS

Vibrio parahaemolyticus is a worldwide marine halophile that is naturally associated with zooplankton and other marine life in seacoast waters; it is a major cause of gastroenteritis among individuals who partake of seafoods that are raw or inadequately cooked or refrigerated. The disease is frequent in Japan and several outbreaks have occurred in the USA in recent years. The organism grows readily in foods at temperatures of 25° to 44°C, and salt concentrations of 2% and above.

The pathogenesis of the diarrhea produced by *V. parahaemolyticus* is not fully understood. Possible virulence factors include a cytolytic toxin and an enterotoxin. Diarrheal infections may be either dysentery-like or cholera-like but, being less severe than cholera, seldom require fluid and electrolyte replacement or treatment with antibiotics. Diarrhea is sometimes accompanied by fever, headache, vomiting and liver injury.

Extraintestinal infections can result from exposure of the ear and skin wounds or abrasions to sea water or sea products; septicemia occurs on occasion.

Laboratory diagnosis rests on culture procedures similar to those used for *V. cholerae* except that 2% NaCl is added to the media used. The organisms fall into 12 different O Ag serotypes and 59 K Ag serotypes.

VIBRIO ALGINOLYTICUS

Vibrio alginolyticus is a worldwide halophile that occasionally produces extraintestinal infections in the ear, or skin wounds or burns following exposure to marine waters; septicemia may develop in immunosuppressed persons. Treatment consists of surgical drainage and the administration of erythromycin or tetracycline.

VIBRIO VULNIFICUS

Vibrio vulnificus is a widespread halophile which causes severe wound infections that may progress to septicemia. In addition, "primary septicemia" may sometimes follow ingestion of contaminated seafood, such as raw oysters. Most patients who develop "primary septicemia" have hemochromatosis or chronic preexisting liver disease. Obviously predisposed persons should not eat raw or undercooked seafoods. The fatality rate of septicemia is about 50%, even with antibiotic treatment. Patients with "primary septicemia" present with hypotension, fever, chills, and prostration; the majority also develop metastatic necrotic skin lesions. Possible virulence factors include a polysaccharide capsule, siderophores, a cytotolytic toxin and proteases. Therapeutic measures used in *V. vulnificus* infections include wound debridement and the administration of tetracycline.

VIBRIO FLUVIALIS

Vibrio fluvialis is a widespread halophile that has been reported to produce gastroenteritis and sometimes death, especially in young children. The disease usually resembles cholera but may sometimes present with fever and blood and mucus in the stool. Possible virulence factors include an enterotoxin, cytolysin, cytotoxin and proteases.

OTHER GENERA OF *VIBRIONACEAE* OF MEDICAL IMPORTANCE

AEROMONAS

Occasionally, various species of the genus *Aeromonas* produce disease in cold-blooded animals and humans. For example, *A. hydrophilia* which occurs widely in soil, fresh and marine water, sewage and marine animals causes diarrheal disease, wound infections, osteomyelitis, septicemia and meningitis in humans. The infections are usually opportunistic and most of them occur in systemically immunocompromised individuals, such as cancer patients. The bacterium produces many biologically active substances including an enterotoxin.

PLESIOMONAS

Only one species of the genus *Plesiomonas* is an opportunistic pathogen for humans, namely, *P. shigelloides.* It can cause gastroenteritis and occasionally extraintestinal disease.

CAMPYLOBACTER (FAMILY: *SPIRILLACEAE*)

Vibrio fetus, a curved Gram negative rod was originally a member of the family *Vibrionaceae* based largely on its morphology; it was first isolated in 1909 and was associated with abortions in sheep and cattle. Its potential as a human pathogen was not recognized until 1947, when it was cultured from human blood.

In 1963, *V. fetus* was given a new generic classification, *Campylobacter,* and placed in the family *Spirillaceae.* The recognized species of this genus that have been reported to be human pathogens include *C. jejuni, C. fetus, C. coli,* and *C. laridis. C. jejuni* has received the most attention and is now recognized as a major enteric pathogen of humans. Also, currently there is much interest in a recently described bacterium called *C. pylori* or *C. pyloridis,* because of its frequent isolation from gastric biopsy samples from humans with inflammatory gastroduodenal diseases. However, at the present time it is not clear whether the bacterium causes these disorders or simply reflects secondary colonization of damaged tissue.

C. jejuni exhibits a rapid darting motility due to polar or bipolar flagella. The curved rods vary from 1.5 to 3.5 um in length and 0.2 to 0.4 um in width. It grows best in a microaerophilic environment (5 to 10% oxygen and 3 to 10% carbon dioxide) at 42°C. *C. jejuni* is oxidase, catalase, and nitrate-positive and is sensitive to nalidixic acid; *C. fetus,* an opportunistic human pathogen is similar to *C. jejuni* in the above characteristics except that it is resistant to nalidixic acid and grows poorly at 42°C.

C. jejuni, an etiologic agent of enteric infections in all age groups, causes enteritis with fever, diarrhea, abdominal pain, blood stools, headache, and sometimes vomiting. It is estimated that at least 2 million cases of *C. jejuni* infections occur in the USA annually. The source of human infections is usually food of animal origin, especially raw unpasteurized milk. The organisms are excreted in the feces of apparently healthy domesticated animals. It has been reported that 50 to 100% of chickens, turkeys and cattle excrete *C. jejuni.* In addition, organisms can also be isolated readily from surface waters.

Although *C. jejuni* infections in man are mostly gastrointestinal in nature, urinary tract infections, as well as meningitis and cholecystitis, have been reported. Complications may include Reiter's syndrome, reactive arthritis, and Guillain-Barré syndrome.

C. fetus is an opportunistic pathogen that causes infections in debilitated persons, as well as in women during the third trimester of pregnancy. It can cause fatal septicemic infections in newborns and in elderly debilitated men with cirrhosis, malignancies, and cardiovascular disease. *C. coli* causes diseases in humans similar to those caused by *C. jejuni* although the frequency is not known. *C. laridis* is a thermophilic species that has been isolated from humans, birds and mammals; the most frequent isolates have come from seagulls.

C. jejuni produces at least two exotoxins, a heat-labile enterotoxin and a cytotoxin. The respective roles they play in the pathogenesis of disease has not yet been defined. The enterotoxin, which has a mechanism of action similar to that of cholera toxin, induces a secretory diarrhea by stimulating adenylate cyclase activity in the intestinal mucosae.

Immunity to *C. jejuni* probably involves IgA Abs against the enterotoxin as well as against the colonization Ags (adhesins) that mediate adherence and colonization on the surface of the intestinal epithelium. High titers of complement-dependent bactericidal Abs are found in convalescent phase sera.

Campylobacter species are susceptible to erythromycin, tetracycline, chloramphenicol, and aminoglycosides.

REFERENCES

Arai, T.: Survey of *Plesiomonas shigelloides* from aquatic environments, domestic animals, pets and humans. J. Hyg. *84*:203, 1980.

Blaser, M.J., et al.: Extraintestinal *Campylobacter jejuni Campylobacter coli* infections: Host factors and strain characteristics. J. Infect. Dis. *153*:552, 1986.

Bullock, W.: The History of Bacteriology. London, Oxford University Press, 1938.

Burke, V., et al.: The microbiology of childhood gastroenteritis: *Aeromonas* species and other infective agents. J. Infect. Dis. *148*:68, 1983.

Farmer III, J.J., et al: *Vibrio, Manual of Clinical Microbiology* 4th Ed. Edwin H. Lenette (ed.), Amer. Soc. Microbiol. Washington, DC, 1985, p. 282.

Gangarosa, E.J.: The epidemiology of cholera past and present. Bull. N.Y. Acad. Med., *47*:1140, 1971.

Goodwin, C.S.: *Campylobacter pyloridis,* gastritis and peptic ulceration. J. Clin. Pathol. *39*:353, 1986.

Janda, J.M., et al.: *Aeromonas* species in clinical microbiology: significance, epidemiology, and speciation. Diagn. Microbiol. Infect. Dis. *1*:221, 1983.

Kothary, M.H., and Kreger, A.S.: Production and partial characterization of an elastolytic protease of *Vibrio vulnificus.* Infect. Immun. *50*:534, 1985.

Marshall, B.J.: *Campylobacter pyloridis* and gastritis. J. Infect. Dis. *153*:650, 1986.

McSweegan, E., et al.: Intestinal mucus gel and secretory antibody are barriers to *Campylobacter jejuni* adherence to INT 407 cells. Infect. Immun. *55*:1431, 1987.

Middlebrook, J.L., and Dorland, R.B.: Bacterial toxins: cellular mechanisms of action. Microbiol. Rev. *48*:199, 1984.

Morris, J.G., and Block, R.E.: Cholera and other vibrioses in the USA. N. Engl. J. Med. *312*:343, 1985.

Pathak, A., et al.: Neonatal septicemia and meningitis due to *Plesiomonas shigelloides.* Pediatrics *71*:389, 1983.

Tison, D.L., and Kelly, M.T.: *Vibrio* species of medical importance. Diagn. Microbiol. Infect. Dis. *2*:263, 1984.

Walker, R.I.: Pathophysiology of *Campylobacter* enteritis. Microbiol. Rev. *50*:81, 1986.

Wall, V.W., et al.: Production and partial characterization of a *Vibrio fluvialis* cytotoxin. Infect. Immun. *46*:773, 1984.

Wickboldt, L.G., and Sanders, C.V.: *Vibrio vulnificus* infection. Case report and update since 1970. J. Am. Acad. Dermatol. *9*:243, 1983.

23

BRUCELLA

The genus *Brucella* of the family *Brucellaceae* contains three major species, *Br. abortus, Br. melitensis* and *Br. suis,* which cause natural disease, primarily in cattle, goats and swine, respectively. Since these organisms are obligate parasites, the animal hosts serve as the major reservoirs of infection for man. Human beings become infected by contact with diseased animals or by the ingestion of contaminated milk and dairy products.

Three additional species of *Brucella* are recognized. *Br. neotomae* has been isolated from the desert wood rat present in the western USA. *Brucella ovis,* an important pathogen in sheep, is responsible mainly for ram epididymitis. *Brucella canis* causes epidemic abortion in dogs in kennels, especially beagles. Whereas neither *Br. neotomae* nor *Br. ovis* are etiologic agents of human illness, *Br. canis* has been responsible for a few human infections.

A. MEDICAL PERSPECTIVES

Human brucellosis (undulant fever) was first described in 1863 by Marston, who recorded attacks of the disease experienced by others and by himself during his tour of duty as a British Army surgeon on the Island of Malta. The disease was subsequently referred to as Malta fever. In 1887, Bruce isolated *Br. melitensis* at necropsy from British soldiers who had died of Malta fever. The species *Br. abortus* was first isolated by Bang in 1897 in Denmark from cows suffering from infectious abortion (Bang's disease); Traum, an American investigator, isolated *Br. suis* from a premature swine fetus in about 1914. Recently a new species, *Br. ovis,* was isolated from infected sheep following abortion; another species, *Br. canis,* has been described as cause of abortion in dogs.

Brucellosis is a worldwide endemic disease in goats, cattle and swine as well as in many wild cloven-hoofed animals including deer, elk and Alaskan caribou. In addition, mules, horses, sheep, camels, buffaloes, reindeer, dogs, hares and chicks sometimes acquire the disease naturally. In the USA alone, annual livestock losses due to brucellosis are appraised at about 100 million dollars. It has been estimated that, in the USA, 5% of all female cattle, 15% of all cattle herds and about 1 to 3% of swine are infected.

In domestic mammals it is probable that infection commonly results from ingestion of contaminated food or, less frequently, water. The infection spreads to the mammary glands. Spread to the genital organs, and particularly the placenta, is common and is the cause of abortion. The organisms are shed in milk, urine, feces and vaginal secretions. Virgin heifers, which are relatively resistant to brucellosis, become susceptible following pregnancy. In cows, the organisms have a marked viscerotropism for the fetal portion of the placenta, the placental fluids and the chorion. This remarkable localization of the organism is apparently due to high local concentrations of erythritol which stimulate the growth of the organisms, even within phagocytic cells. Erythritol appears to be the main biochemical determinant of infectious abortion due to brucellae (Sect. I). As a rule, females with clinical brucellosis have high titers of agglutinins in their sera as well as in their milk.

Human brucellosis, a classic example of a zoonosis, occurs most often among individuals who are engaged in occupations concerned with livestock and dairy products such as farming, veterinary medicine, and the processing of meats and dairy products. Cattle and swine represent the principal natural reservoirs of human brucellosis in the USA, whereas the goat is the chief source of disease in countries where goat's milk is a common human food. Human brucellosis is still a substantial medical problem in some areas of the world. It is estimated that only about 1 case in 5 is diagnosed and reported. During the period 1947 to 1980 the number of human cases of brucellosis in the USA reported annually to the USPHS dropped from 5600 to about 200. This low incidence of the disease in the USA at the present time represents a remarkable achievement. It is unlikely that brucellosis will again become a significant problem if constant vigilance for controlling the disease among farm animals is maintained and pasteurization of milk is enforced. The problem is further mitigated by the fact that the mortality rate in untreated persons with brucellosis of 2 to 3% can be reduced essentially to zero when appropriate diagnostic and therapeutic procedures are practiced.

B. PHYSICAL AND CHEMICAL STRUCTURE

Organisms of the genus *Brucella* are small, pleomorphic, nonmotile, sometimes capsulated, nonsporing, aerobic, Gram-negative rods that tend to assume a "coccobacillary form" (0.3 to 1.5 μm in length by 0.4 to 0.8 μm in width). *Brucella* species share a common Ag with *Vibrio cholerae* which explains why individuals vaccinated against cholera form agglutinins that react with the brucellae. Cross-agglutination reactions with *Francisella tularensis* have also been reported. The three major species of *Brucella* contain two dominant Ags designated A and M. *Brucella abortus* contains approximately 20 times as much A Ag as M Ag; in contrast, *Br. melitensis* possesses about 20 times as much M Ag as A Ag. *Brucella suis* exhibits an intermediate antigenic pattern although it resembles *Br. abortus* in having a predominance of A Ag. The distribution of these Ags in the various species is depicted in Figure 23–1. Monospecific anti-A and anti-M sera can be prepared by absorption with appropriate brucellae. It has been proposed that the A and M Ags are composed of glucose, glucosamine, galactose, and hexuronic acids. Brucellae also contain cell wall endotoxins that are similar to those of other Gram-negative bacteria. The O-specific side chains of the LPS are lost when smooth (S) strains dissociate to rough (R) strains.

C. GENETICS

Isolates of various species of *Brucella* exhibit the classical smooth-to-rough dissociation pattern. The S→R mutation is associated with a loss of virulence, a tendency toward spontaneous agglutination and a decrease in the Ags which

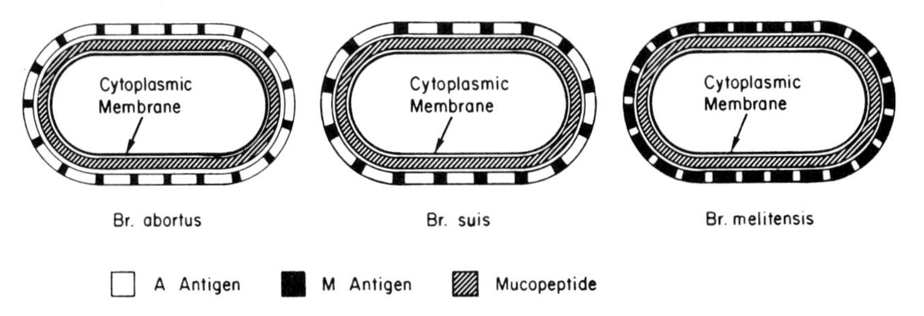

Fig. 23–1. Diagrammatic representation of the ratio of A and M antigens in brucellae.

stimulate agglutinins specific for smooth strains. The attenuated vaccine strain, *Br. abortus* 19, an intermediate mutant, is immunogenic despite its avirulence for cattle and low virulence for man. Taxonomically the *Brucella* species are not closely related to any other genera. However, all species of *Brucella* are homogeneous based on nucleic acid base ratios and DNA homology.

D. EXTRACELLULAR PRODUCTS

The brucellae do not produce any extracellular products of known medical importance. The only apparent toxin produced is an endotoxin.

E. CULTURE

The brucellae are strict aerobes that grow slowly at 37°C in trypticase soy broth or on standard blood agar media pH 6.6 to 6.8. In contrast to the other brucellae that infect man, *Br. abortus* requires 5 to 10% CO_2 for primary isolation from clinical specimens. The major species of the brucellae are differentiated on the basis of their growth patterns in the presence of thionine and basic fuchsin, CO_2 requirement, H_2S production and A and M Ag patterns.

All species produce catalase, urease and oxidase but do not ferment sugars. Whereas *Br. melitensis* usually grows in the presence of 1:100,000 basic fuchsin and 1:100,000 thionine, thionine inhibits *Br. abortus* and basic fuchsin inhibits *Br. suis.* It should be noted that there can be biotype variations in these phenotypic properties as shown in Table 23–1 which can complicate species identification.

F. RESISTANCE TO PHYSICAL AND CHEMICAL AGENTS

Brucellae are readily killed by heat and the usual antiseptics and disinfectants, but are relatively resistant to drying. They can remain viable in contaminated raw milk for as long as 10 days at 4°C and persist in unpasteurized cheese and butter for periods up to 4 months. Brucellae can survive for weeks to months in dust, tissues of dead animals, feces, soil, water and urine.

G. EXPERIMENTAL MODELS

Experimentally infected monkeys develop a nonprogressive disease with undulating temperature patterns; mice and rats are highly resistant. The guinea pig is susceptible to many strains of brucellae, especially *Br. melitensis,* which is often fatal for this animal.

H. INFECTIONS IN MAN

Although man is highly susceptible to infection with virulent brucellae, immunity is readily acquired and most infections remain inapparent. Virulence for man varies widely among strains. Evidently man has little or no natural immunity to virulent brucellae and essentially all of the immunity acquired, although sometimes weak, is developed as a response to primary infection.

In naturally acquired brucellosis the organisms can infect *via* broken skin, the oropharynx, the alimentary tract or the conjunctivae. Thus ingestion, contact and inhalation are the major means of disease transmission. However, in the USA 90% of brucellosis is presently due to contact with infected materials rather than due to ingestion of contaminated milk and milk products. The organisms are seen within macrophages and to a lesser extent within endothe-

Table 23–1. Characteristics of Species and Biotypes in the Genus *Brucella*

				Growth on Dye Media[1]			Agglutination in Monospecific Sera[2]	
				Basic Fuchsin	Thionine			
Species	*Biotypes*	*CO_2 Required*	*H_2S Produced*	*1:100,000*	*1:25,000*	*1:100,000*	*Abortus*	*Melitensis*
B. melitensis	1	–	–	+	–	+	–	+
	2	–	–	+	–	+	+	–
	3	–	–	+	–	+	+	+
B. abortus	1	±	+	+	–	–	+	–
	2	+	+	–	–	–	+	–
	3	±	+	+	+	+	+	–
	4	±	+	+	–	–	–	+
	5	–	–	+	–	+	–	+
	6	–	±	+	–	+	–	+
	7	–	±	+	–	+	+	–
	8	+	–	+	–	+	+	+
	9	±	+	+	–	+	–	+
B. suis	1	–	+	–	+	+	+	–
	2	–	–	–	–	+	+	–
	3	–	–	+	+	+	+	–
	4	–	–	+	+	+	+	+
B. neotomae		–	+	–	–	+	+	–
B. ovis		+	–	+	+	+	–	–
B. canis		–	–	–	+	+	–	–

Adapted from WHO Technical Report Series No. *46*:471, 1971.
[1]Species differentiation is obtained on albimi or tryptose agar with graded concentrations of dyes. Interpretation should be controlled with the reference strains of each species.
[2]Monospecific antisera are produced by absorption with *Br. abortus* or *Br. melitensis*.

lial cells and fibroblasts. After multiplication at the local site of penetration the organisms spread to the lymphatics, lymph nodes and blood stream. Ultimately small granulomatous lesions become established in many organs including the spleen, liver, kidneys and bone marrow. Brucellosis may express itself as an acute, subacute or chronic disease, but seldom takes a fulminating fatal course.

Acute brucellosis usually has an incubation period of 10 to 14 days (range 4 to 45 days). An undulating daily fever pattern is common, although it is not consistently found in patients in the USA; an initial septicemia occurs. Positive blood cultures can usually be obtained during the first 14 to 21 days of the disease; they tend to become negative as the patient develops agglutinins, C-F Abs and opsonins. *Relapses are hallmarks of this disease; when they occur the body temperature is elevated and blood cultures usually become positive again.* Positive, delayed-type skin reactions can usually be elicited with brucellergen, a protein-containing bacillary extract, after the 3rd to the 6th week of the disease. Patients with clinical brucellosis usually exhibit a low-grade fever, which often peaks in the afternoon and early morning (undulant fever), and complain of headache, fatigue, night sweats, weakness, insomnia, anorexia, constipation and generalized aches and pains. Pain on motion of the spine may occur due to a localized spondylitis of one or more of the vertebral bodies. Some 5% of patients exhibit enlarged cervical nodes and about 30% have splenomegaly. Hepatitis is common and jaundice is sometimes seen. Orchitis, if it occurs, is ushered in by a chill followed by high fever. In most patients the primary disease undergoes remission in 3 to 5 months. In a few individuals, brucellosis exhibits a more chronic and extended course.

Chronic brucellosis usually produces weakness, fatigue and vague aches and pains, as well as mental depression, in spite of lack of characteristic abnormal physical findings. Irritability, anxiety and mental depression may become profound and sometimes present major problems in management. The chronic form of brucellosis can last for 1 to 20 years with relapses of varying degrees of intensity; complications involving many tissues and organs may occur. Fortunately the most serious complications, endocarditis and meningoencephalitis, are rare. Abortion due to brucellosis is extremely rare in human beings.

I. MECHANISMS OF PATHOGENESIS

The manner by which the brucellae cause tissue injury is not known but presumably depends, in part at least, on delayed hypersensitivity and possibly endotoxin. The brucellae are primarily intracellular parasites within macrophages and, in accord with expectation, the chronic lesions are granulomas composed of epithelioid cells, giant cells and lymphocytes. *Brucella suis* in particular can produce granulomas exhibiting central necrosis and caseation. Virulence varies greatly among strains of brucellae, but little is known about the factors determining virulence. Since smooth strains are virulent and rough strains are not, cell wall composition is presumed to be related to virulence. The capacity of virulent organisms to metabolize erythritol, a growth-promoting substance, appears to be a virulence factor enabling certain strains (but not others) to infect the ungulate placenta, which is rich in erythritol. However, the human placenta does not contain erythritol which could explain why brucellae seldom cause abortion in humans. For unknown reasons, virulent strains can survive and grow within macrophages better than avirulent strains. In certain strains this property of virulence is enhanced by growing the organisms in media supplemented with placental fluid.

A putative virulence factor of *Brucella* constitutes a surface cell wall carbohydrate that mediates binding to human B lymphocytes. The mechanisms involved, however, are obscure.

J. MECHANISMS OF IMMUNITY

Immunity acquired as a result of natural infection is only partial and, as a consequence, either superinfection, reactivation or reinfection can occur. The low level of acquired immunity is also in accord with the usual relapsing nature and chronicity of the disease. Apparently, bactericidal serum Abs and neutrophils can effectively destroy organisms that are not engulfed and sheltered (at least for a time) by other cells, principally macrophages. In accord with the prediction supported by these circumstances, there is no correlation between serum Abs and acquired immunity in brucellosis, but, instead, there is *strong evidence that the main mechanism of acquired immunity is cell-mediated immunity.* Although it has been observed that immune macrophages can inhibit the intracellular growth of brucellae in vitro more effectively than normal macrophages, there is also evidence that virulent brucellae possess a toxicity for macrophages that can be neutralized by immune serum. *Consequently it may be that the maximum expression of acquired immunity rests on an interplay between cellular and humoral immune forces.* The Ags responsible for acquired immunity are not known; antiendotoxin Abs are not protective.

K. LABORATORY DIAGNOSIS

The clinical diagnosis of brucellosis is often difficult to establish. The disease is commonly categorized among the "fevers of unknown origin." In order to make a definitive diagnosis it is necessary to cultivate the brucellae from samples of blood or aspirates of bone marrow or lymph nodes. It is especially difficult to culture brucellae from patients with subacute or chronic infections. Evidently the intracellular position of the organisms in macrophages is responsible for the frequent failure of blood culture, especially in chronic brucellosis. Although most blood cultures are positive in 2 to 5 days, the cultures should be monitored for 4 to 6 weeks before they are discarded as negative. Other clinical specimens are inoculated into trypticase soy broth incubated in the presence of 10% CO_2, and examined for growth of typical brucellae at 4- to 5-day intervals. *The primary culture should be incubated for 4 to 6 weeks before being discarded as negative!* When growth occurs subcultures are made on a trypticase soy-blood-agar medium and incubated under 10% CO_2. Species identification can be made by employing the tests listed in Table 23–1 as well as with monospecific antisera (anti-A or anti-M). *Cultures of brucellae should be handled with care because of the high risk of laboratory infection.*

Agglutination tests are performed on patients' sera with phenolized suspension of heat-killed smooth organisms. Agglutinin titers above 1:80 to 1:160 are generally interpreted to indicate past or present infection; rising titers in early disease are particularly significant. Agglutinating antibody titers may reach values of 1:640 or greater in acute disease. Titers usually return to a low normal range within a year after successful antibiotic therapy. The agglutinins in early active disease are IgG and IgM, whereas the "incomplete" or non-agglutinating Abs that characterize chronic infection are mainly in the IgG and IgA fractions. *It should be emphasized that patients who have received a brucellergen skin test or cholera vaccine will usually have low titers of serum agglutinins for brucellae.* Because of the value of agglutination tests the brucellergen skin test should not be used routinely as a diagnostic aid. In some patients, brucellae Abs of the "blocking" type which fail to produce agglutination cause the prozone phenomenon. A positive test for blocking Abs, even though agglutinins are lacking, is presumptive evidence of brucellosis. These Abs that appear in chronic disease may be detected by substituting 5% serum albumin for 0.85% NaCl diluent. Antibodies that precipitate with brucella extracts appear late in acute disease; they persist as long as the disease remains active and are the best indicator of an active clinical infection.

A card agglutination test has proven useful for screening purposes because it recognizes more than 90% of patients using the standard tube agglutination test. The card test can be performed and read within 4 to 5 minutes.

Other serologic tests include the enzyme-linked immunoabsorbent assay (ELISA) and complement-fixation (CF) tests. The ELISA which is sensitive, eliminates the prozone problem and can be automated, will probably replace the tube agglutination test.

L. THERAPY

The course of acute brucellosis can be shortened and the complications minimized by the use of a 6-week therapeutic regimen of tetracycline. This schedule must be repeated if a relapse occurs; however, more than 2 courses of tetracycline are seldom administered. Up to 10% of cases can have relapses within the first 3 months after the completion of the first course of therapy. If patients suffer from severe toxemia, depression anorexia and a generalized debilitated condition, adrenocorticoid therapy is recommended as an adjunct to antibiotic therapy. Tetracycline therapy is effective also in chronic forms of the disease. Herxheimer-like reactions are sometimes encountered during chemotherapy.

Many attempts have been made to desensitize patients with repeated injections of increasing doses of brucellae Ags based on the premise that a significant component of the symptom complex is due to a systemic DH reaction. While there may be some merit in this procedure, it is difficult to evaluate and is time consuming. Moreover, it is fraught with the possibility of violent local and systemic reactions and is not generally recommended.

M. RESERVOIRS OF INFECTION

The major reservoirs of human brucellosis in the USA are infected cattle and hogs. Goats are a major reservoir in countries where goat milk is used extensively. The organisms are highly infectious and laboratory infections are easily acquired.

N. CONTROL OF DISEASE TRANSMISSION

Brucellosis in cattle can be controlled moderately well by vaccinating calves with an attenuated live vaccine such as strain 19 vaccine. Another promising vaccine strain is called Rev. 1. In addition, the segregation and slaughter of serologically positive animals provide highly effective control measures. The role that wild animals might play in obstructing efforts to eradicate infection in herds of domestic animals has not been assessed.

Human brucellosis among urban populations in the USA has been essentially eradicated by pasteurization of milk. However, this does not help to minimize the occupational risk of these individuals who are exposed to infected animals and dairy products made from raw milk.

Appropriate human vaccines could be of value for protecting individuals at high occupational risk; nevertheless, public health authorities in the USA have not recommended immunization with the vaccines currently available. However, human brucellosis has allegedly been reduced 10-fold as a result of widespread vaccination in Russia. As one would expect, a search for safer and better vaccines is in progress.

REFERENCES

Kurtz, R.S. and Berman, D.T.: Influence of endotoxin-protein in immunoglobulin G isotype responses of mice to *Brucella abortus* lipopolysaccharide. Infect. Immun., *54*:728, 1986.

Robertson, L., et al.: The isolation of Brucellae from contaminated sources. A review. Br. Vet. J. *133*:193, 1977.

Thimm, B.H.: *Brucellosis Distribution in Man, Domestic and Wild Animals.* New York Springer-Verlag, 1982.

Vella, E.E.: Brucellosis (the Corps Disease). Jr. R. Army Med. Corps, *129*:97, 1983.

Wise, R.I.: Brucellosis in the United States, past, present and future. J.A.M.A., *244*:2318, 1980.

Young, E.J.: Human Brucellosis. Rev. Infect. Dis., *5*:821, 1983.

24

YERSINIA AND PASTEURELLA

Yersinia pestis

Yersinia pseudotuberculosis and *Yersinia enterocolitica*

Pasteurella

The genus *Yersinia,* of the family *Enterobacteriaceae* comprises three species, *Y. pestis, Y. pseudotuberculosis* and *Y. enterocolitica.* Closely related organisms of the genus *Pasteurella* are *P. multocida, P. haemolytica, P. pneumotropica* and *P. ureae.* A summary of differential reactions used to characterize these species is presented in Table 24–1.

YERSINIA PESTIS

Man and the domestic rat are highly susceptible accidental hosts of *Y. pestis* the causative agent of plague; the *natural hosts,* which constitute the permanent reservoir, are relatively resistant wild rodents.

A. MEDICAL PERSPECTIVES

Although plague is a minor health problem in the world today, the ever-present threat of future outbreaks stands as a grim reminder that, *during centuries past, periodic epidemics of this fearsome pestilence extracted the greatest toll of human lives of any epidemic disease.* In fact, plague was one of the major reasons that population control was not needed in past centuries. During the period 600 B.C. to A.D. 1800, more than 196 major epidemics and pandemics occurred causing well over 200 million deaths. In some epidemics, as many as 50% of entire populations were killed. The 50-year Justinian pandemic of the 6th century took 100 million lives and the "black plague" epidemic of the 14th century claimed the lives of one-fourth of the population of Europe (25 million). For some unknown reason the disease was relatively quiescent during most of the 18th and 19th centuries only to reappear with its usual fury near the beginning of the 20th century; mortality during the Hong Kong epidemic of 1894 was 95%. In India alone, 10 million died of plague between 1898 and 1918. Good descriptions of plague epidemics are given in Bocaccio's *Decameron* and Defoe's *Journal of the Plague Year in London.*

The swiftness and destructiveness of the great plague epidemics of the past (most had a major pneumonia component) are difficult to imagine. Petrarch (1348), who wrote about one such epidemic in France, emphasized that future generations would be unable to comprehend ". . . the empty houses, the abandoned towns, the squalid countryside, the field littered with dead, and the dreadful silent solitude which seems to hang over the whole world." *The*

Table 24–1. Differential Reactions of Yersinia and Pasteurella Species

	Motility					
Species	*22°C*	*37°C*	*Indol*	*H_2S Production*	*Urease*	*Ornithine Decarboxylase*
Yersinia pestis	–	–	–	+	–	–
Yersinia pseudotuberculosis	+	–	–	–	+	–
Yersinia enterocolitica	+	–	*	–	+	+
Pasteurella multocida	–	–	+	+	–	+
Pasteurella haemolytica	–	–	–	*	–	*
Pasteurella pneumotropica	–	–	+	+	+	*
Pasteurella ureae	–	–	–	–	+	–

* = usually negative

enormity of the impact of this one disease on the course of human history can never be accurately evaluated. It has caused wars to begin and wars to cease, has brought about major social upheavals and has contributed to the rise and fall of nations. For example, it played a major role in the fall of the Roman Empire and dealt the final blow to the feudal system.

The causative organism was independently discovered by Yersin and Kitasato during the Hong Kong epidemic in 1894. Although it had been noted repeatedly during thousands of years that both rats and human beings sicken and die during plague epidemics, the reasons for the association were not clear until Ogata (1897) suggested and Simond (1900) and the English Plague Commission (1906) proved that the rat flea transmits the causative organism from rat to rat and from rat to man. Plague antiserum was used therapeutically in Yersin's time and a vaccine was introduced by Haffkine in 1896.

During the last 100 years, *Y. pestis* spread widely from Asia, Africa and Europe to other parts of the world, including the USA (San Francisco in 1900). It has spread throughout sylvatic regions in the West, particularly New Mexico, and other limited areas in the USA to establish permanent reservoirs in relatively resistant wild rodent hosts, as well as rabbits (sylvatic plague). Although plague in man commonly results from the bites of insects (usually rat fleas) and involves local lymph nodes (bubonic plague), the organisms may enter the blood stream following the bite and produce a primary septicemia; if a massive lung infection develops, the disease is passed rapidly from man to man via sputum droplets *(pneumonic plague).* In contrast, animal plague almost always results from insect bites and is never pneumonic in type.

The epidemiology of plague is fascinating and stands as a classic model of an insect-transmitted disease in an accidental host. Its complexity is not surprising in view of the large number and variety of rodent hosts (some 200 species and subspecies) and insect vectors involved and the numerous ecologic factors, including meteorologic conditions, that influence their relative numbers and activities. *Yersinia pestis* can survive arctic winters in hibernating rodents with latent disease which reactivates when these animals awaken. The incidence of sylvatic plague among rodents fluctuates with changes in population balances between rodents having varying degrees of susceptibility to the organisms. A high percentage of surviving rodents in areas of sylvatic plague develop immunity to *Y. pestis.* This allows for waves of epidemics among sylvatic hosts as new generations of young susceptible animals arise. *Evidently permanent reservoir areas are only maintained when relatively resistant hosts are among the fauna of the area; in areas where all hosts, such as domestic rats, are highly susceptible the reservoir tends to diminish and disappear.* As expected, permanent sylvatic reservoirs can exist in areas untouched by man or the domestic rat.

The ecologic balances that determine rat epizootics and associated human epidemics are exceedingly delicate and triggering conditions are stringent. Human plague is contracted principally from populations of rodents that have close liaison with man (liaison hosts, commensal hosts). Domestic rats (house rats, barn rats, sewer rats) and semiwild mice, including the multimammate mouse *(Mastomys coucha)* and the spiny mouse *Acomys cahirinus,* which become house guests during inclement weather, constitute the principal sources of human infections. *In general the domestic rat reservoir does not become established unless the population of rats surpasses a certain level and the population of fleas on the rats reaches a certain density (rat-flea index).* Consequently, climates and seasons favoring the propagation of rats and fleas promote epizootics and, in turn, human epidemics which follow in their wake. Since dense populations of domestic rats are found principally in urban environments, epidemics of plague are largely confined to towns and cities.

Human plague is most common in unsanitary surroundings. Epidemics present a *pre-epidemic phase* during which a few cases gradually appear, an *epidemic phase* when cases appear in large numbers and a *declining phase* during which the incidence of new cases declines sharply. Transmission from rat to rat and from rat to man is due to two species of rat fleas which parasitize three species of domestic rats, *Rattus norvegicus* (gray or brown rat), *Rattus rattus* (black rat) and *Rattus alexandrinus* (Egyptian rat). Fleas made hungry by death of their hosts or blockage of their proventriculus by bacilli growing there and in the midgut avidly seek new hosts, including humans, to feed upon. They are especially effective in transmitting the disease because, during their vigorous but futile efforts to feed, they repeatedly regurgitate ingested blood loaded with bacilli into the wound that they inflict. The organisms may grow and persist in the flea for several months, especially at low temperatures. Obviously the flea is both a biologic vector and a reservoir.

Because *Rattus rattus* is more susceptible to *Y. pestis* than *Rattus norvegicus* and lives in closer association with man than its fierce Norwegian cousin, it is the more effective reservoir for inciting epidemics. On the other hand, *Rattus norvegicus* is probably more important than *Rattus rattus* in maintaining a semipermanent reservoir among rats because the latter is too susceptible to the disease. During long interepidemic periods, sylvatic rodents constitute permanent reservoirs for maintaining the organism in nature and serve as sources of reinfection of rat populations.

Under ordinary circumstances, prevention of human plague epidemics is not difficult and can usually be effected by preventing the establishment of plague reservoirs in domestic rat populations. It is accomplished by suppressing local populations of permanent reservoir hosts and rats, together with constant surveillance to prevent the transmission of disease from distant reservoirs to local rat populations. Control has been excellent in recent decades and annual plague deaths in the world at large have been limited to hundreds. Permanent sylvatic reservoirs are difficult to control and probably can never be eliminated, even by extreme measures. This, together with the fact that modern transportation permits rapid transmission of infection for thousands of miles through the agency of rats, fleas and contaminated materials means that plague epidemics will always remain as a public health threat. However, with today's epidemiologic knowledge and tools for controlling rats and insect populations together with vaccines and effective chemoprophylactic agents, all of which our forefathers lacked, *it is difficult to envision why future epidemics of any magnitude should occur in developed countries.* The greatest risks of future epidemics lie in underdeveloped countries where there is danger of over-population with attending social and economic chaos and their sequelae, poverty and poor sanitation. Of course, an unforeseen "Wellesian event" could take place even in a developed country and it is well to remain mindful of the fact that, for unknown reasons, *plague epidemics have had a relentless way of disappearing and reappearing over the centuries. The rise and fall of past epidemics and pandemics may have been related to the composition of rat populations and virulence of organisms.* The decline of plague in Europe during the 18th and 19th centuries may have been due to the fact that the fierce Norwegian rat largely replaced its cousin, the black rat; the black rat is the reservoir par excellence for human plague because of its companionability with man and the intense epizootics that it supports. Rapid passage of the organism in this highly susceptible host permits it to reach the acme of virulence which, in turn, may favor the establishment of the pneumonic form of plague in man. Most of the severe epidemics and pandemics of the past have had a large component of pneumonic plague characterized by "vomiting of blood" and unusually high mortality.

Sporadic cases of plague will probably increase in regions of the USA where sylvatic reservoirs exist; sylvatic infection is transmitted in many ways including bites of insects and rodents, skinning of animals, and consumption of rodent meat. There is special danger when epizootics occur among highly susceptible wild animals such as prairie dogs. Physicians in areas of sporadic plague should be alert to its presence. Moreover, in this day of air travel, plague and many other exotic diseases of distant regions may be encountered by a physician practicing anywhere in the world.

B. PHYSICAL AND CHEMICAL STRUCTURE

Yersinia pestis is a pleomorphic, nonsporing, nonmotile, Gram-negative, enveloped, ovoid rod measuring 0.5 to 0.8 μm × 1.5 to 2 μm. It occurs singly, in pairs and in chains and exhibits bipolar staining with polychrome stains. Pleomorphism is enhanced by growth in unfavorable environments such as on an agar medium containing 3% NaCl.

At least 20 Ags have been detected in *Y. pestis,* including endotoxin as well as other toxins. Fifteen of these Ags have been reported to be shared with *Y. pseudotuberculosis.* However, no serotypes have been described and the precise roles that various bacterial components play in pathogenesis and immunity are not known.

Fraction 1 Ag, referred to as "envelope Ag," is a principal Ag involved in both virulence and immunity. Envelope Ag consists of two serologically identical components (1A and 1B) that differ chemically with regard to the presence of a carbohydrate moiety in 1A but not in 1B.

Virulent *Y. pestis* strains contain two additional Ags V and W. V antigen is a protein Ag with a MW of 90,000. The W Ag is a lipoprotein of 145,000 MW. These Ags appear to confer the ability on *Y. pestis* to establish infection with small infecting doses because they cause the organisms to resist phagocytosis by PMNs. Once infection is established the envelope Ag, coagulase, pesticin and fibrinolysin contribute to pathogenesis. Pestocin I is a bacteriocin produced by *Y. pestis* that can inhibit *Y. pseudotuberculosis, E. coli,* and *Y. entercolitica.* Pestocin acts by converting susceptible bacteria to nonviable stable spheroplasts. Production of coagulase and fibrinolysin appears to be a result of bacteriocinogenic conversion. Strains of *Y. pestis* lacking coagulase and fibrinolysin are infectious for the guinea pig or mouse but the disease is attenuated.

Virulent strains contain a component that results in absorption of hemin to form pigmented colonies. A loss of the ability to carry out synthesis of purine ribotides is also correlated with a loss of virulence. This represents a loss of guanosine monophosphate synthetase. The role of endotoxin in virulence and pathogenesis is not well delineated and is most likely of secondary importance.

Among the several toxins produced by *Y. pestis,* one called plague exotoxin or "murine toxin" is toxic for mice and rats, but not for all animal species. It consists of a cell membrane protein (toxin A) and a cytoplasmic protein (toxin B). It is possible that in their natural state the two components exist as a complex molecule or alternatively that they represent the monomeric and dimeric forms of a single molecule. Murine toxin acts selectively on the mitochondria of certain organs of susceptible animals, but not others. It is not toxic for the mitochondria of resistant animals. It binds to mitochondrial protein, causes mitochondrial swelling and inhibits mitochondrial respiration. Toxicity for mitochondria correlates well with toxicity for the animal species from which the mitochondria are derived. For a discussion of animal toxicity see Section I.

C. GENETICS

A number of colonial types of *Y. pestis,* including smooth and rough, have been described. However, colonial morphology is not meaningful in terms of virulence. For example, virulent isolates from the same patient form both R and S colonies; moreover, these virulent organisms can be rendered avirulent by subculture without change in colony morphology. Mutants resistant to streptomycin have been demonstrated.

D. EXTRACELLULAR PRODUCTS

All of the fully virulent strains produce toxins which are most likely the cause of death from plague. As mentioned previously the soluble heat-labile exotoxin (murine toxin) is composed of two proteins, toxin A (associated with the cytoplasmic membrane) and toxin B (cytoplasmic) with MWs of 240,000 and 120,000, respectively. The LD_{50} of this toxin for mice is less than 1 μg. Other extracellular products of possible medical importance are bacteriocin (pesticin I), coagulase, and a fibrolytic factor, all of which may be related to virulence.

E. CULTURE

Yersinia pestis is a slow-growing organism (generation time 4 hours) with simple nutritional requirements. It grows on common laboratory media as a facultative anaerobe over an extremely wide range of temperature (5° to 45°C) with an optimum range of 25° to 30°C; the range of pH values for growth is about 6.0 and 8.0. Following growth, the pH may rise to 8.8 or 9.0. Surface growth on solid media may require hematin or reducing agents. The organism ferments certain sugars without producing gas. Colonies on 0.1% blood agar are small, round, viscous and granular and have irregular margins; they are pigmented due to accumulation of red cell pigments. The organism can grow in pure bile. Three biotypes *(orientalis, medieavalis* and *antigua*) have been identified on the basis of reduction of nitrates to nitrites and fermentation of glycerol. *Orientalis,* which reduces nitrate but does not ferment glycerol, is the common biotype of Western North America.

F. RESISTANCE TO PHYSICAL AND CHEMICAL AGENTS

Yersinia pestis exhibits the usual resistance to physical and chemical agents possessed by most vegetative bacteria. It is killed within minutes by sunlight, drying, moist heat above 50°C and the common disinfectants. When present in certain animal materials, such as flea feces and dried sputum, *Y. pestis* may survive under natural conditions for weeks to months, especially at low temperatures. The organism remains viable for months in frozen human and animal cadavers. Cultured organisms store well at refrigerator temperatures, and can be preserved for many years by lyophilization.

G. EXPERIMENTAL MODELS

Many species of animals are susceptible to plague, but adult birds are totally resistant. Since bubonic plague is similar in man and animals, various species of animals have been used as models of human disease including mice, guinea pigs, rats, rabbits and monkeys. The favored animal for diagnostic work, the guinea pig, is extremely susceptible; one organism can cause fatal infection. Disease in animals induced with infected fleas usually takes one of two forms depending on host resistance, namely, lymph node (bubonic) infection with septicemia or without septicemia. During the course of septicemia in susceptible animals the blood bacillary load may reach 100 million/ml; in contrast, the load is much lower in man, which reduces the chances of insect transmission of plague from plague patients to healthy human beings or animals. A third form of plague is septicemia without ostensible lymph node involvement.

H. INFECTIONS IN MAN

Although *Y. pestis* infects man by many routes, including the conjunctivae and mucous membranes of the respiratory and digestive tracts, infection usually results from organisms introduced into the skin by infected rat fleas. Under unusual circumstances, bubonic plague may be transmitted from man to man by the human flea *Pulex irritans.*

Bubonic plague is characterized by sudden onset usually marked by low fever, restlessness, prostration, slurred speech, mental confusion, abdominal pain, vomiting and other symptoms. The incubation period is 2 to 5 days. Since fleas commonly bite the legs, the draining lymph nodes involved are usually in the popliteal and/or inguinal regions. Sometimes a discernible primary lesion develops at the site of infection in the form of a small vesicle. If the exposed individual possesses some degree of immunity, a local lesion will not form or will be transitory. In nonimmune individuals the infecting organisms multiply rapidly and soon reach the draining lymph nodes, where they produce extensive hemorrhagic and necrotizing lesions. The resulting exquisitely tender and painful swollen lymph nodes ("buboes") are firm and movable; they present gray areas of coagulation necrosis in the medulla which often suppurate.* Necrosis evidently results from plague toxins. Perilymphatic tissues are intensely edematous. *It is probable that bacilli reach the blood early, at least in small numbers, because they can be cultured from blood in 60 to 80% of*

*The term "bubo" is derived from the Greek work "bubon" meaning groin. It is currently used to designate swollen lymph nodes irrespective of their location.

infected persons. The extent of bacteremia fluctuates markedly during the course of the disease and frequently develops into a fulminating septicemia involving the spleen, meninges, skin and lungs. Lung complications occur in about 5% of patients with bubonic plague and hemorrhages often develop in the skin and mucous membranes. Clotting in small vessels and consumption coagulopathy are presumed to be responsible for hemorrhage. Septicemia leads to hemoconcentration, circulatory failure and death. The mortality rate in persons with untreated bubonic plague without septicemia is 25 to 50% and in those with septicemia near 100%; the overall mortality is 60 to 90%. In the event of natural recovery from bubonic plague, healing of lymph nodes and splenic lesions is slow and organisms may persist for weeks.

Primary *septicemic plague* sometimes develops before lymph nodes become involved. It is presumed to result from the direct introduction of bacilli into the blood stream by feeding fleas and is almost always fatal if not treated.

Pneumonic plague, which is characterized by swift spread and near 100% mortality, is sometimes contracted by persons caring for bubonic plague patients with pneumonic complications. *Under suitable conditions one case can start an epidemic of pneumonic plague even in a nonendemic area.* Passage of pneumonic plague from one person to another by droplet infection appears to be favored by high virulence, high humidity and low atmospheric temperatures. Such conditions contributed to the massive Manchuria epidemic of pneumonic plague in 1910–1911 that took a toll of 60,000 lives. The onset is abrupt and the symptoms include generalized pain and headache, rigor, nausea, respiratory difficulty, fever, and a cough yielding blood-tinged frothy sputum filled with bacilli. At terminus the patient develops extreme cyanosis and death results from suffocation.

I. MECHANISMS OF PATHOGENESIS

The factors of virulence are complex and difficult to quantify. There is compelling evidence that the V and W Ags are of major importance because they permit survival and multiplication inside mononuclear phagocytes. For example, mutants with normal Fraction envelope Ag but lacking V and W Ags are incapable of producing progressive disease. Animal studies have led to the thesis that the determinants of full virulence are linked with or constitute the attributes of toxinogenicity, encapsulation, production of V-W Ags, calcium dependence in culture, ability to synthesize purines, the formation of pigmented colonies on hemin-containing media, resistance to phagocytosis by neutrophils and free macrophages, and especially the capacity to grow within macrophages.

Ability to survive and multiply within macrophages during the early stages of infection seems to be the major virulence mechanism and is probably related to the V-W complex. Plague exotoxin, (murine toxin), and endotoxin undoubtedly contribute to tissue injury and death from plague. The hemorrhagic and degenerative lesions seen in the organs of bacteria-free fetuses carried by plague patients and the "toxic death" which occurs in plague patients subsequent to "bacterial sterilization" effected by chemotherapy are evidently due to toxins. Since species vary in their susceptibility to the toxins, animal studies can only suggest what the modes of action of plague exotoxin may be in man.

The histopathology seen in human plague is similar in many respects to that seen in animals injected with murine toxin and includes edema, hemorrhage, hemoconcentration, serosanguinous effusions in body cavities, and foci of coagulation necrosis in lymph nodes and other organs. Many of these lesions may represent secondary changes and the cells that are the primary targets of toxin remain to be defined.

Current data indicate that Fraction 1 envelope Ag is responsible for resisting phagocytosis, whereas V and W Ags are responsible for survival and multiplication with mononuclear phagocytes. Virulent strains consistently absorb hemin from the medium which results in pigmented colonies. The capacity to synthesize purines is also closely linked to virulence. Lastly, the production of a soluble, heat-labile plague exotoxin as well as an insoluble, heat-stable lipopolysaccharide endotoxin present in the cell wall undoubtedly plays an important role in death from plague.

J. MECHANISMS OF IMMUNITY

The mechanisms involved in immunity to plague differ among different hosts making it difficult to interpret animal results in terms of human immunity. There is no evidence of racial differences in human immunity to plague and all individuals are equally and fully susceptible to primary infection. However, *most plague survivors acquire strong and durable immunity,* and many kinds of Abs are produced. Although antisera convey partial protection against infection, the classes, specificities and activities of the Abs concerned are not known. Antitoxic Abs do not contribute significantly to immunity and serum Abs and C do not lyse plague organisms. Apparently enveloped organisms resist both phagocytosis and killing by free phagocytes, including neutrophils and macrophages. They are more readily engulfed by fixed than by free macrophages. Nonenveloped organisms are engulfed by normal monocytes and grow intracellularly where they synthesize envelope Ag and become more virulent. It is not known whether the macrophages of immune animals engulf and destroy or inhibit organisms more readily than do the macrophages of nonimmune animals; presumably this is the case. Although acquired immunity has been considered to be due largely to the opsonic activity of Abs against the envelope protein and V and, possibly, W Ags, this is uncertain. *Instead, the ability to grow within normal monocytes* suggests that antimicrobial cellular immunity is an important component in *immunity, a possibility that does not appear to have been explored.*

Vaccines in current use consist of live attenuated organisms or formalin-killed virulent bacilli. They afford only partial protection and booster doses every 3 to 6 months are necessary to maintain effective immunity. It will be difficult to devise more effective vaccines until a better understanding of the mechanisms of acquired immunity are at hand.

K. LABORATORY DIAGNOSIS

The organisms are commonly present in abundance in lymph node biopsy material or aspirates, abscesses and blood. In pneumonic plague they are present in the sputum. Direct smears stained with polychrome stain can be highly suggestive of plague. Blood agar, glycerol agar or, if the specimen is contaminated with other organisms, azide-antibiotic-dye selective medium is useful for primary culture. Small delicate colonies develop after 24 to 48 hours at 30°

C. Identification is based on cultural and fluorescent Ab tests together with inoculation of guinea pigs, either subcutaneously or if the specimen is highly contaminated with other organisms, by rubbing the material on an area of freshly shaved skin. Other organisms can be easily mistaken for plague bacilli; hence, the guinea pig test is valuable. In guinea pigs, death occurs in from 2 to 5 days and typical postmortem changes include congested viscera, an enlarged spleen, a granular liver, and pleural effusion. Specific phage is also useful for identifying *Y. pestis.*

L. THERAPY

Early diagnosis and treatment are the keys to therapeutic success; unfortunately diagnosis is often missed or is delayed in sporadic cases because plague is not suspected. *Therapy initiated later than 12 to 15 hours after the appearance of fever is often of little or no value, especially in pneumonic plague.* The drugs of choice are the tetracyclines and chloramphenicol. Sulfonamides should be avoided if possible and should not be used alone. Streptomycin is highly bactericidal but should be used cautiously. If given, it should be combined with tetracycline to obviate the development of streptomycin resistance. Also, in some patients, the administration of large doses of streptomycin has precipitated fatal shock resembling a generalized Shwartzman reaction, presumably because of large amounts of endotoxin and other toxins liberated from the dying bacteria. Drugs are often given by both the intravenous and oral routes and combinations of drugs are sometimes used. Early drug therapy reduces mortality in bubonic disease by about 80%, but is much less effective in pneumonic plague. Serum therapy, which reduces mortality in bubonic disease by about 50%, is no longer used. Supportive treatments to counteract failure of peripheral circulation and pulmonary function are important elements of therapy. Abscesses should not be drained because of danger of massive spread of organisms.

M. RESERVOIRS OF INFECTION

As outlined in previous sections, the principal *reservoirs of human bubonic plague are domestic rats and rat fleas, and sylvatic rodents and the insect vectors that they harbor (principally fleas).* In special situations, plague patients and the human fleas that infest them, and dogs and cats and their fleas can serve as sources of human bubonic plague. *The reservoir of pneumonic plague is the plague patient.*

Chronic carriage in man is not known to exist, but temporary carriage in the throat has been demonstrated during convalescence and in immune contacts; the role of carriers as reservoirs is unknown.

N. CONTROL OF DISEASE TRANSMISSION

As stated in Section A, control of disease transmission in man hinges largely on control of domestic rats and sylvatic reservoirs. Fortunately, this can be accomplished now without resort to the extreme practices of the past, such as the burning of homes. The use of insecticides should precede the destruction of rodents for otherwise the dispossessed fleas seek human hosts and temporarily pose a greater threat to man than if the rats are left unmolested. Mass vaccination and chemoprophylaxis are also useful for preventing or halting epidemics, and vaccination of high-risk individuals in endemic areas is valuable for preventing sporadic plague. Except for the limited instances in which the flea infesting man serves as a vector, measures for preventing man-to-man transmission are largely routine and include quarantine of patients, the use of masks and gloves by attendants of infected patients, and chemoprophylaxis.

YERSINIA PSEUDOTUBERCULOSIS AND *YERSINIA ENTEROCOLITICA*

Yersinia pseudotuberculosis and *Yersinia entercolitica,* unlike *Y. pestis,* are motile when grown at 22° to 25°C. Both of these organisms are characterized and also distinguished from *Y. pestis* by urease production (Table 24–1). There are 6 serotypes of *Y. pseudotuberculosis* which are based on specificity of O and H Ags. The V and W Ags of *Y. pestis* are also present in *Y. pseudotuberculosis.*

Twenty-seven serotypes of *Y. enterocolitica* have been recognized based also on O and H Ag specificity. Serotypes 0:3 and 0:9 are the most common cause of human infection in Europe, Japan and Canada. In the USA serotype 0:8 is responsible for most human infections. Most species of *Brucella* show cross reactions with 0:9 serotype of *Y. enterocolitica.* In the USA *Y. pseudotuberculosis* has been isolated from domestic mammals, deer, rabbits and rodents. Fecal-oral spread of these two species appears to be the major method of transmission. Evidently, both of these organisms are clinically important in the USA as causes of gastrointestinal disease. It has been reported that in certain areas *Yersinia* species are as important as *Shigella.*

Infections of the small intestine can lead to ulcers of the intestinal mucosa in the area of the mesenteric lymph nodes which can result in loss of blood and fluid. Intestinal infections can mimic acute or subacute appendicitis. Diarrhea and lymph node involvement are hallmarks of disease caused by these organisms. Uncomplicated cases of gastroenteritis caused by either *Y. enterocolitica* or *Y. pseudotuberculosis* are clinically indistinguishable from cases of gastroenteritis caused by species of *Shigella* or *Salmonella.*

Definitive diagnosis is dependent on culture of the organisms. Antibiotic sensitivity can be variable so susceptibility testing is necessary.

PASTEURELLA

The genus *Pasteurella* comprises four recognized species; *P. multocida, P. pneumotropica, P. haemolytica* and *P. ureae. Pasteurella* species are small, nonmotile, ovoid rods that exhibit bipolar staining. The organisms are facultative anaerobes that grow on blood agar. Virulent *P. multocida* tends to be capsulated. No exotoxins have been described in any of the *Pasteurella* species, although they contain significant endotoxin activity.

Based on their capsular polysaccharides four major antigenic types (A,B,D,E) of *P. multocida* have been recognized. Types A and D appear to be the most common types among human isolates; type A is the most common cause of respiratory disease. Based on O Ags, *P. multocida* falls into 13 serotypes. In general there is a relationship between serologic type and host range when capsular and somatic Ag types are considered. Based on fermentation patterns of several sugars and sugar alcohols 11 physiologic biotypes of *P. multocida* have been recognized.

The *Pasteurellae,* particularly *P. multocida,* are parasites and pathogens for an unusually wide range of animal hosts. *They cause tremendous losses of livestock but only occasionally infect man as an accidental host.* The organisms are commonly carried in the URT of animals as strains of low virulence. *In animals they are opportunists par excellence and tend to promptly invade mucosae and increase in virulence (with resulting septicemia) whenever host defenses are lowered;* an example is "shipping fever" or "stockyard fever" due to *P. multocida* and *P. haemolytica.* Virulent strains of *P. multocida* also cause epizootics of *hemorrhagic septicemia,* particularly in the tropics and subtropics. In addition, the organisms can act as secondary invaders to cause subacute and chronic infections in animals.

Man infrequently carries *pasteurellae* in the URT, and human *infection, although rare, occurs often enough that physicians should remain alert to its pathogenetic potential. Most infections result from the bites of dogs, cats, rats, other animals, or sometimes man.* The infected bite wounds are painful and tend to suppurate and undergo necrosis. The organisms can spread by contiguity or via lymphatics and the blood stream to produce disseminated disease including meningitis, infectious carditis, cerebellar abscess, osteomyelitis and pneumonia. The second most common form of human infection involves the lungs and usually occurs in patients with preexisting pulmonary disease. The susceptible mouse can be useful in establishing the etiologic diagnosis. Penicillin is the therapeutic drug of choice. No useful vaccine or effective antiserum is available.

REFERENCES

Bottone, E.J., and Sheehan, D.J.: *Yersinia enterocolitica:* Guidelines for serologic diagnosis of human infections. Rev. Infect. Dis., *5*:898, 1983.

Cartwright, F.F.: *Disease and History.* New York, Thomas Y. Crowell Co., 1972.

Chen, T.H. and Elberg, S.S.: Scanning electron microscopic study of virulent *Yersinia pestis* and *Yersinia pseudotuberculosis* type I. Infect. Immun. *15*:972, 1977.

Christie, A.B.: Plague: Review of ecology. Ecol. Dis., *1*:111, 1982.

Foley, J.A., et al.: Reactive arthritis due to *Yersinia enterocolitica.* Clin. Rheumatol., *3*:385, 1984.

Gibbon, E.: *The History of the Decline and Fall of the Roman Empire.* New ed., London, W. Allason, 1816.

Grewal, P., et al.: *Pasteurella ureae* meningitis and septicaemia. J. Infect., *7*:74, 1983.

Hecker, J.F.C.: *The Epidemics of the Middle Ages.* London, Sydenham Society, 1844.

Inoue, M., et al.: Community outbreak of *Yersinia pseudotuberculosis.* Microbiol. Immunol., *28*:883, 1984.

Kadis, S., and Ajl, S.J.: Site and mode of action of murine toxin of *Pasteurella pestis.* In *Microbial Toxins.* Vol. III, *Bacterial Protein Toxins.* T.C. Montie and S. Kadis (eds.). New York, Academic Press, 1970, p. 39.

Kawaoka, Y., et al.: Migratory waterfowl as flying reservoirs of *Yersinia* species. Res. Vet. Sci., *37*:266, 1984.

Oberhofer, T.R.: Characteristics and biotypes of *Pasteurella multocida* isolated from humans. J. Clin. Microbiol., *13*:566, 1981.

Permezel, J.M., et al.: Opportunistic *Pasteurella multocida* meningitis. J. Laryngol. Otol., *98*:939, 1984.

Prpic, J.K.,et al.: In vitro assessment of virulence in *Yersinia enterocolitica* and related species. J. Clin. Microbiol., *22*:105, 1985.

Reed, W.P., et al.: Bubonic plague in the Southwestern United States: A review of Recent experience. Medicine *49*:465, 1970.
Weber, D.J., et al.: *Pasteurella multocida* infections. Report of 34 cases and review of the literature. Medicine, *63*:133, 1984.
Weniger, B.G., et al.: Human bubonic plague transmitted by a domestic cat scratch. J.A.M.A., *251*:927, 1984.
Werner, S.B., et al.: Primary plague pneumonia contracted from a domestic cat at South Lake Tahoe, California, J.A.M.A.,*251*:929, 1984.

25

FRANCISELLA

A single species of the genus *Francisella, Francisella tularensis,* (family: *Brucellaceae*) produces disease in animals and man. This organism (formerly known as *Pasteurella tularensis*) was named after Tulare County, California, where the organism was first isolated by McCoy and Chapin in 1912 from ground squirrels exhibiting a plague-like disease. However, the association of the animal disease with clinical disease in man was not established until Francis started a series of investigations in 1919. The term *tularemia* was coined by Francis to indicate that in man the organism can invade the blood and become generalized. Tularemia is enzootic in all areas of the USA. It is enzootic in most areas of the world that are north of the equator; an exception is the British Isles.

A. MEDICAL PERSPECTIVES

Tularemia, also known as rabbit fever, deer-fly fever and Ohara's disease, is an endemic disease of wild animals that is transmitted to man by direct contact or by insect vectors; man serves as an unnatural or accidental host. The most characteristic manifestations of disease in man include a necrotizing cutaneous or mucous membrane lesion at the site of invasion, coupled with regional lymph node enlargement, septicemia, and a relatively high fever.

Francisella tularensis is highly virulent for man but fortunately man-to-man transmission has never been documented. Wild cottontail rabbits and rodents are the principal reservoirs of *Fr. tularensis* in the USA and accordingly the disease is most prevalent in hunters, linemen, butchers and housewives, who are most likely to come in contact with infected rabbits or their ectoparasites. Human tularemia, a reportable disease, is naturally restricted because of conditions necessary to transmit the disease to man. Accordingly, human cases occur sporadically and usually in small numbers. Nevertheless, since the mortality rate of untreated persons with tularemia is between 6 and 7% early diagnosis and treatment are of utmost importance.

Francisella tularensis has been recovered from over 54 species of arthropods. Approximately one-half of these arthropods are known to transmit this organism to man. Ticks are the most common arthropod vectors which include *Dermacentor andersoni, Dermacentor variabilis* and *Amblyomma americanum.* Other arthropod vectors, which are also involved, include deer flies, mites, black flies, mosquitoes and occasionally lice. Of major importance is that *Fr. tularensis* can be transmitted transovarially in female ticks.

B. PHYSICAL AND CHEMICAL STRUCTURE

Francisella tularensis is a pleomorphic Gram-negative, nonmotile, noncapsulated, nonsporing, bipolar-staining rod ranging in size from 0.3 to 0.9 μm in length and 0.2 to 0.3 μm in width; coccoid and short bacillary forms usually predominate. There is only one serologic type based on the specificity of agglutinins and acquired immunity, although several distinct Ags are present in the cell wall. It appears that these bacteria have one protein Ag in common with members of the genus *Brucella.* Like other Gram-negative bacteria, *Fr. tularensis* cell walls contain lipopolysaccharide.

C. GENETICS

Smooth strains are virulent and as a rule rough strains are avirulent. Cultures of virulent strains usually contain a few mutant cells of low virulence. The live vaccine strain (LVS) of *Fr. tularensis* is a stable attenuated strain that can induce substantial immunity in man.

D. EXTRACELLULAR PRODUCTS

No exotoxins have been identified, although living cells of virulent *Fr. tularensis* produce acute toxicity and death in rabbits when injected intravenously in large numbers ($>10^8$); heat-killed organisms are devoid of this toxicity.

E. CULTURE

Francisella tularensis is an obligate aerobe that grows between 24° and 39°C; it is catalase negative and has an optimum temperature of 37°C; its optimum pH is 6.9. It is a relatively fastidious microbe and requires a rich medium plus blood, glucose and cystine; any organism that grows on plain nutrient agar cannot be *Fr. tularensis.*

Two strains of *Fr. tularensis* can be recognized in the USA. Type A, which exhibits citrulline ureidase activity, accounts for about 80% of human cases. This strain is highly virulent for man and is usually tick borne and associated with rabbits. Type B (negative for citrulline ureidase) is less virulent for humans and is commonly associated with disease of muskrats and beavers.

F. RESISTANCE TO PHYSICAL AND CHEMICAL AGENTS

Francisella tularensis is highly susceptible to heat and the routine antibacterial agents, including the common disinfectants. The organism is particularly sensitive to drying and does not survive well in aging cultures.

G. EXPERIMENTAL MODELS

Domestic rabbits, as well as other common laboratory animals, are extremely susceptible to infection. As few as 1 to 5 organisms of a virulent strain may be lethal.

H. INFECTIONS IN MAN

Early in disease, signs and symptoms include back pain, anorexia, headache, chills, fever, sweating and prostration. Approximately 10% of patients develop a rash which commonly lasts about 1 week. Marked fever is characteristic with temperatures of 104° to 105°F persisting up to 4 weeks in untreated patients.

The most prevalent form of tularemia, accounting for ~80% of human cases, is *ulceroglandular* disease. The primary lesion usually occurs on the fingers or hands as a consequence of contact with the tissues of infected animals. The organism is highly virulent and can even invade skin that appears to be unbroken. After 3 to 4 days a papule appears at the site of exposure and subsequently transforms into an ulcer by the 7th or 8th day. By this time the infection has extended to the regional lymph nodes. Lymphadenopathy is usually of long duration with convalescence lasting from 3 to 6 months.

The *oculoglandular* type of tularemia is the consequence of infecting the eyes with contaminated blood or rubbing the eyes with contaminated fingers. Local ulceration of the conjunctivae occurs with spread to regional lymph nodes. The prognosis is generally favorable if the infection is diagnosed and treatment is instituted promptly. Localized lymph node involvement without a primary skin lesion is termed *glandular* tularemia. Systemic infection that develops without a primary ulcer or localized lymphadenitis is referred to as the *typhoidal* form. The *pulmonary* form can result from inhalation of aerosols generated in the laboratory in the course of handling virulent cultures or may be secondary to hematogenous dissemination. This form of tularemia causes mucoid sputum, hemoptysis, dyspnea, pleuritic pain and cyanosis.

On rare occasions the ingestion of contaminated meat or water can cause primary lesions in the *gastrointestinal* tract. Other rare forms of disease include endocarditis, pericarditis, peritonitis, osteomyelitis, meningitis and appendicitis.

I. MECHANISMS OF PATHOGENESIS

The hallmark of tularemia is an infectious granulomatous lesion that undergoes necrosis, often with attending suppuration of draining lymph nodes. The granulomas resemble tubercles when present in liver, spleen, lung or kidney. Although the organism is not known to secrete any toxins, it is highly toxic for macrophages, which probably explains the marked necrosis that occurs in the granulomas produced. Delayed sensitivity may also contribute to tissue damage.

J. MECHANISMS OF IMMUNITY

Specific acquired immunity to tularemia appears to be primarily due to immune T4 lymphocytes and activated macrophages, although there may well be an interplay with some form of humoral immunity directed against toxic components of the organism. It is of interest that BCG-vaccinated mice express significant nonspecific resistance against many infectious agents, e.g., *Listeria monocytogenes* and *Salmonella typhimurium,* but not against virulent *Fr. tularensis.* This suggests that acquired immunity in tularemia is complex and depends on more than simply the presence of activated macrophages. Immunity following infection is relatively solid, although second infections have been reported.

It is of special interest that sheep and other animals that have been immunized against ticks have a markedly reduced death rate when exposed to ticks carrying *Fr. tularensis* compared to animals not immunized against ticks. The immunologic response to tick Ags markedly discourages a tick from obtaining a blood meal and as a consequence the chance of transmitting *Fr. tularensis* is reduced.

K. LABORATORY DIAGNOSIS

Francisella tularensis usually can be cultured from the mucocutaneous ulcer or regional lymph nodes; blood cultures are commonly negative. It can also be cultured from the sputum of patients with the pneumonic form of disease. The organism is highly infectious and *clinical laboratories should not attempt to culture it unless they have appropriate laboratory facilities for handling dangerous infectious material.*

Serum agglutinins are usually present after the second week of illness and are particularly useful in diagnosis if a 4-fold rise in titer is observed.

Infected individuals develop delayed-type sensitivity; therefore, a skin test with an extract of the organisms results in a delayed skin reaction. The test is specific and may be of diagnostic value in some instances.

L. THERAPY

Streptomycin is bactericidal for *Fr. tularensis* and is the therapeutic drug of choice; chloramphenicol and tetracyclines are also effective although relapses have been reported, presumably because these drugs fail to eradicate subcutaneous foci of bacteria.

M. RESERVOIRS OF INFECTION

Man usually contracts tularemia by contact with the tissues of infected animals. Sources of infection include wild rabbits, carnivores, ungulates, birds, squirrels, woodchucks, muskrats, skunks, foxes, opossums, mice, snakes, ticks and fleas. The Rocky Mountain tick *(Dermacentor andersoni),* the western wood tick *(D. variabilis),* the eastern dog tick *(D. occidentalis)* and the Lone Star tick *(Amblyomma americanum)* function as constant reservoirs of infection because the organism can be transmitted transovarially. One species of deer fly *(Chrysops discalis)* and a species of mosquito present in Sweden *(Aedes cinereus)* can transmit tularemia to man and contaminated water sometimes causes infection.

N. CONTROL OF DISEASE TRANSMISSION

The attenuated live vaccine (LVS vaccine) is highly effective in man and should be employed for immunizing individuals at risk, especially laboratory workers who handle the organisms.

Since wild rabbits are the major source of human tularemia in the USA, hunters should be especially wary of lethargic rabbits. Care should be taken in cleaning game, because skinning infected rabbits is the way that tularemia is contracted most often. It is important that laws preventing the sale of wild rabbits by meat markets be instituted and enforced.

Antibiotic prophylaxis with streptomycin may be indicated following known exposure to *Fr. tularensis.*

REFERENCES

Bell, J.F., et al.: Resistance to tick-borne *Francisella tularensis* by tick-sensitized rabbits: Allergic klendusity. Am. J. Trop. Med. Hyg. *28*:876, 1979.

Buchanan, T.M., et al.: The tularemia skin test. 325 skin tests in 210 persons: serologic correlation and review of the literature. Ann. Intern. Med. *74*:336, 1971.

Eigelsbach, H.T., et al.: Murine model for study of cell-mediated immunity: protection against death from fully virulent *Francisella tularensis* infection. Infect. Immun. *12*:999, 1975.

Evans, M.E., et al.: Tularemia and the tomcat. J.A.M.A. *246*:1343, 1981.

Gordon, J.R., et al.: Tularaemia transmitted by ticks *(Dermacentor andersoni)* in Saskatchewan. Can. J. Comp. Med. *47*:408, 1983.

Halperin, S.A., et al.: Oculoglandular syndrome caused by *Francisella tularensis.* Clin. Pediatr. *24*:520, 1985.

Hutton, J.P., and Everett, E.D.: Response of tularemic meningitis to antimicrobial therapy. South Med. J. *78*:189, 1985.

Jacobs, R.F., and Narain, J.P.: Tularemia in children. Pediatr. Inf. Dis. *2*:487, 1983.
Koskela, P., and Herva, E.: Cell-mediated immunity against *Francisella tularensis* after natural infection. Scand. J. Infect. Dis. *12*:281, 1980.
Koskela, P., and Herva, E.: Cell-mediated and humoral immunity induced by a live *Francisella tularensis* vaccine. Infect. Immun. *36*:983, 1982.
Miller, R.P. and Bates, J.H.: Pleuropulmonary tularemia. A review of 29 patients. Am. Rev. Resp. Dis. *99*:31, 1969.
Schmid, G.P., et al.: Clinically mild tularemia associated with tick-borne *Francisella tularensis.* J. Infect. Dis. *148*:63, 1983.
Van Metre, T.E. and Kadull, P.J.: Laboratory-acquired tularemia in vaccinated individuals: a report of 62 cases. Ann. Intern. Med. *50*:621, 1959.
Young, L.S., et al.: Tularema epidemia: Vermont, 1968. Forty-seven cases linked to contact with muskrats. N. Engl. J. Med. *280*:1253, 1969.

26

PSEUDOMONAS

Pseudomonas aeruginosa

Infections Caused by Other *Pseudomonas* Species

PSEUDOMONAS AERUGINOSA

Members of the genus *Pseudomonas* of the family *Pseudomonadaceae* are ubiquitous in nature. These organisms, commonly called "pseudomonads," have extensive oxidative capabilities and play important roles in degradative cycles in nature. Although most pseudomonads exist solely as free-living saprophytes in soil and water, others serve to parasitize a wide range of plants and both warm-blooded and cold-blooded animals including insects. One species, *Ps. aeruginosa,* is sometimes found among the normal microbial flora of man, especially in the intestinal tract. Although a few species are obligate parasites and pathogens for plants and animals, most species that parasitize animals represent facultative parasites, and none is an obligate parasite of man. Since *Pseudomonas aeruginosa* is the species that most commonly assumes the role of a facultative human parasite, this chapter is centered primarily on a discussion of this organism.

Like many other Gram-negative rods, *Ps. aeruginosa* acts almost invariably as an opportunistic pathogen, most often in the hospital environment. It causes 10 to 20% of all nosocomial infections and is the most feared of the opportunists because the infections that it causes are difficult to treat and are often fatal.

In common with other bacteria that inhabit aquatic environments in nature, *Ps. aeruginosa* can adhere to and colonize on various inert solid surfaces. After the motile "swarmer cell" attaches, it proliferates to form a microcolony of cells embedded in a matrix or glycocalyx (Sect. B). Each cell is encased in an envelope formed by stranded exopolysaccharides. As colonization proceeds an excess of exopolysaccharides accumulates to form the protective glycocalyx in which the microcolony of growing organisms is embedded.* The glycocalyx is composed of an enzyme-resistant polymer alginate comprised of guluronic and mannuronic acids that facilitates adsorption of nutrients and at the same time, protects against hostile elements in the environment, including predatory amoebae and bacteriophages. The glycocalyx constantly liberates motile "swarmer cells" that serve to initiate new colonies; however, many of them are destroyed. It is probable that these unique properties of *Ps. aeruginosa,* together with a marked genetic adaptability, contribute to the versatility of these organisms as opportunists and to the difficulties encountered in treating the infections they produce (Sects. I, J, and L).

A. MEDICAL PERSPECTIVES

Pseudomonas aeruginosa was recognized many years ago as the cause of "blue-pus infections" (Gessard 1882); recently these infections have increased greatly in incidence of increasing numbers of highly susceptible patients, such as the elderly and persons with underlying diseases, especially malignancies, chronic uremia, cystic fibrosis and liver disease. Individuals treated with immunosuppressive drugs, patients on anticancer drugs and patients with serious burns also contribute importantly to the numbers at risk of opportunistic infections. In addition, the use of certain devices has provided new avenues of entrance through integuments; for example, intravenous catheters allow direct entry, and intermittent positive-pressure machines for respiratory support

---

*Strains that produce a great excess of exopolysaccharide and produce mucoid colonies are called *mucoid strains.*

sometimes become contaminated with large numbers of organisms which are consequently introduced into the patient's lungs. The high resistance of many pseudomonads to some of the widely used disinfectants, as well as to many antibiotics, compounds the problem. *Pseudomonas aeruginosa* is responsible for the vast majority of *Pseudomonas* infections in man. The organism produces a wide range of infections in various organs including the lung, eye, kidney, ear, intestine, and in burned skin.

Some clinicians and epidemiologists feel that the future outlook for controlling infections due to *Ps. aeruginosa* has discouraging overtones. They warn that the indiscriminate therapeutic use of the third-generation cephalosporins and new β-lactam drugs may extend the range of drug-resistant strains of *Ps. aeruginosa* in the hospital environment to include essentially all β-lactam drugs. There is every indication that *Pseudomonas* infections will remain a serious problem, at least until more effective immunoprophylactic, immunotherapeutic or chemotherapeutic agents are developed to combat them.

B. PHYSICAL AND CHEMICAL STRUCTURE

Organisms of the genus *Pseudomonas* are piliated, nonsporing, Gram-negative rods of varying size averaging 1.5 to 4.0 × 0.5 to 1.0 μm; in stained smears they closely resemble the coliforms. *Pseudomonas aeruginosa* is motile by means of 1 to 3 polar flagella. The cell walls of the organism contain a weak coliform-like LPS endotoxin and a capsule-like envelope (mucoid or slime layer) comprised of polyuronic acids arranged linearly to form strands. The envelope is a protective immunogen (Sect. J).

The cell wall LPS complex consists of inner core polysaccharides common to all strains and outer serospecific side chain polysaccharides that determine serologic specificity and susceptibility of the organisms to pyocins (bactericins) and bacteriophages.* The outer LPS polysaccharides are sometimes referred to as the O Ags.

The serotyping of outer polysaccharide Ags of *Ps. aeruginosa* is more useful for epidemiologic purposes than pyocin or bacteriophage typing. The scheme of the International Association of Microbiological Societies recognizes 17 serotypes.

C. GENETICS

Genetic-based changes that create difficulties in identifying *Ps. aeruginosa* are the loss of flagella and the ability to produce the characteristic pigment, pyocyanin. More importantly genetic changes leading to antibiotic resistance are frequent; they involve mutation, conjugation, and transduction (Chap. 4).

D. EXTRACELLULAR PRODUCTS

Pseudomonads are notorious for their ability to synthesize and secrete a large number of extracellular products. Among the enzymes secreted by *Ps. aeruginosa* are fibrinolysin, collagenase, elastase, lecithinase, lipase, proteases and hemolysins. A leukocidin, an enterotoxin and 2 exotoxins, designated A and S are also produced (Sect. I).

---

*Pyocins are rod-shaped protein particles that resemble phage tails; they attach to specific receptors on the cell walls of sensitive strains of host bacteria.

Table 26–1. Pseudomonas Species Associated with Human Infection

	Bacteriologic Features				
Species	*Pyocyanin*	*Other Pigments*	*Oxidase*	*Susceptible to Polymyxin*	*Disease*
Ps. aeruginosa	+	+	+	+	Opportunistic
Ps. fluorescens	–	+	+	+	Opportunistic
Ps. cepacia	–	–	+	–	Opportunistic
Ps. maltophilia	–	+	–	+	Opportunistic
Ps. mallei	–	–	+	–	Glanders
Ps. pseudomallei	–	–	+	–	Melioidosis

Both water-soluble and chloroform-soluble pigments are produced by *Ps. aeruginosa,* the most important being the deep-blue, chloroform-soluble pigment, pyocyanin, and the yellowish-green, water-soluble pigment, fluorescein (Table 26–1). Certain of the pigments have strong antibacterial activities against a wide range of Gram-positive and Gram-negative organisms, and against some fungi. The early name given the organism, *Bacillus pyocyaneus* or *Ps. pyocyaneus,* was coined to designate the blue color lent to pus by pigments produced by the organism.

Exotoxin A is a highly lethal diphtheria-like toxin. It is a proenzyme of polypeptide composition (M.W. 66,000) comprised of 2 fragments, fragment A (M.W. 26,000) which carries the enzyme site and fragment B which carries the cell-binding site. The activated enzyme has ADP-ribose-tranferase activity capable of halting protein synthesis by hepatic and other target cells. Cleavage and activation of the native molecule into fragments A and B results from either limited proteolysis or from unfolding of the molecule following splitting of disulfide bonds.

Exoenzyme S (exotoxin S) which is produced by about one-third of clinical isolates has been studied in the mouse model, but its possible participation in human infections has not been determined. Currently there is no certainty that exoenzyme S is toxic for man or animals.

E. CULTURE

Pseudomonads are classified as nonfermentative aerobes that are oxidase positive; however, they can grow anaerobically by using arginine and nitrate as electron acceptors. The organisms are readily cultured because they are not nutritionally fastidious and grow over a wide range of temperature (5° to 43°C). Their ability to use a wide variety of nutrients allows them to grow in materials that support few or no other bacteria. They can survive or even grow in unlikely media such as disinfectant solutions, and often can be isolated from water faucets, sinks, thermometers and similar locales, where they multiply, using the meager nutrients present. They can also tolerate alkaline conditions.

Some species of the genus *Pseudomonas* are psychrophilic and can grow at refrigerator temperatures. Hence, they can multiply in refrigerated blood, saline solutions and foods. Transfusion of stored blood or intravenous fluids contaminated with *Pseudomonas sp.* has caused lethal shock.

F. RESISTANCE TO PHYSICAL AND CHEMICAL AGENTS

Pseudomonads are relatively resistant to quaternary ammonium compounds, benzalkonium chloride, hexachlorophene and many other disinfectants. However, they are susceptible to ethylene oxide or heat (55°C for 1 hour). The organisms are only partially susceptible to drying and can often be found in hospital dusts. They are notably resistant to many antibiotics and readily acquire resistance during antibiotic treatment (Sect. L). Overall, they are the most antibiotic resistant of the important pathogens of man.

G. EXPERIMENTAL MODELS

Experimental *Pseudomonas* infections have been studied in mice and various other laboratory animals. The pathogenicity of different strains of *Ps. aeruginosa* for guinea pigs and mice varies greatly; some strains kill in 24 to 48 hours, whereas others may kill belatedly or not at all. Mutant strains that fail to produce exotoxin A have reduced virulence for experimental animals and man as compared to toxin producers. It has been reported that antitoxin protects mice against otherwise lethal doses of exotoxin(s). Laboratory animals have been used to study the effects of active and passive immunization on *Pseudomonas* infections in burn wounds; variable results have been obtained.

H. INFECTIONS IN MAN

As stated in section A, *Ps. aeruginosa* can cause infection at many sites in the body and in most circumstances in which local and/or systemic antibacterial defenses are low. For example, in burn patients both local and systemic defenses are depressed. Infections seldom occur in persons whose systemic immune defenses are intact unless local defenses are seriously compromised as in cases of severe trauma. Severe local infections such as pneumonia often lead to systemic spread via lymphatics and blood; systemic disease carries a mortality of some 80%.

Certain conditions are strongly predisposing; for example 30% of all infections in burn patients and the majority of infections in individuals with cystic fibrosis are due to *Ps. aeruginosa;* neutropenia is also highly predisposing to infections. Corneal infections are especially dangerous since the cornea can be destroyed within 24 to 48 hours.

Pseudomonas aeruginosa produces a grave toxic pneumonia with symptoms including bradycardia, azotemia, reversal diurnal temperature curves, toxemia, confusion and progressive cyanosis. Infections often result from organisms introduced via urinary catheters, tracheostomy apparatus, nebulizers, or respiratory support equipment. Alternatively, they may be introduced into the blood directly by means of intravenous catheters and injected materials. Meningitis caused by *Ps. aeruginosa* usually is the result of introducing the organism during lumbar puncture.

I. MECHANISMS OF PATHOGENESIS AND IMMUNITY

Although the pathogenesis and immunology of infections due to *Ps. aeruginosa* is unclear, the formidable array of potentially harmful products synthesized makes it virtually certain that host-parasite relationships are multifactorial and highly complex.

Prevailing evidence indicates that *Ps. aeruginosa* has little or no ability to colonize on normal mucosae and that its appearance among the normal flora of the mouth and intestine of some 2 to 10% of healthy individuals is temporary at best. The ability of *Ps. aeruginosa* to adhere to host cells rests principally, if not exclusively, on pili; anti-pilus Abs are protective in animal models.

The failure to adhere and colonize on normal mucosae is evidently due to a protective fibronectin coating on mucosal epithelium that opposes bacterial adherence. Consequently, adherence and colonization only takes place when the fibronectin coat is interrupted or destroyed as the result of trauma or infection, etc. For example, the fibronectin coat is destroyed by the abnormally high local levels of fibronectin-destroying proteases that accompany chronic lung disease and trauma of the respiratory tract. Such "opportunistic adherence" to epithelium exposed by injury also occurs in the urinary tract and eye. Adherence and colonization by *Ps. aeruginosa* is not confined to mucosal epithelium but involves deep tissue cells as well, including cells exposed in immunocompromised patients with skin burns. *Pseudomonas* elastase is alleged to destroy a protease inhibitor produced by bronchial mucosa. Colonization of mucosae opens the way to invasion of deep tissues but does not assure that invasion will occur so long as systemic host defenses remain intact. Indeed, when organisms reach deep tissues as the result of trauma, they do not survive and multiply unless local injury continues and/or systemic immunity is depressed. Although the slime envelope may afford organisms some protection against phagocytic destruction, the major contribution to virulence made by exocellular polysaccharides appears to rest with the glycocalyx which limits the access of phagocytes and humoral factors, to the embedded organisms.

The neutrophil is the principal agent of defense at the mucosal barrier and when lacking, as in patients with neutropenia, invasion of deep tissues by colonizing organisms is a virtual certainty. The sera of most normal persons contain low levels of IgM serotype opsonins which probably contribute substantially to phagocytic defense.

At least three virulence-associated pathogenetic proteases are produced by *Ps. aeruginosa,* including elastase. Although these proteases have limited systemic toxicity and cytotoxic activity, they contribute to invasion and spread of the organisms by destroying fibrin, elastin and the ground substance between cells. They also inactivate C and cleave the IgG molecule. Destruction of the elastin of blood vessels by elastase results in lesions characterized by hemorrhage and necrosis; for example the skin lesion "ecthyma gangrenosum" is seen in some patients with systemic infection.

Pseudomonas aeruginosa produces a heat-stable hemolysin and a heat-labile hemolysin (phospholipase C), which act synergistically with a phosphatase to break down lipids and lecithin to produce necrosis. These hemolysins also degrade pulmonary surfactant with resulting atelectasis.

Two extracellular ADP ribosyltransferases with toxic activity are produced by *Ps. aeruginosa,* namely, the major well-studied heat-labile necrotizing exotoxin A produced by most clinical isolates and exotoxin S produced by about 30% of isolates (Sect. D).

Exotoxin A is highly toxic for blood monocytes and for the bone marrow; this exotoxin may account for the neutropenia seen in fulminating infections. Exotoxin A produces lesions in the skin and eye of experimental animals; it serves as a protective immunogen in animals and is a major virulence factor. It is highly lethal for animals and appears to be the major cause of toxicity seen in systemic *Ps. aeruginosa* infections in man; toxigenic strains are more virulent than nontoxigenic strains in both animals and man. Patients with high titers of Abs against exotoxin A at the onset of septicemia have lower fatality rates than those with low titers.

Although the LPS of Ps. aeruginosa may have some antiphagocytic activity and is a weaker endotoxin than coliform endotoxin, it contributes substantially to systemic toxicity and many of the diverse symptoms of infection, including leukopenia, disseminated intravascular coagulation, activation of the C and fibrinolytic systems. *The serotype oligosaccharide* portion of LPS is a protective immunogen as is the lipid A moiety of LPS.

The enterotoxin of *Ps. aeruginosa* is probably responsible for the diarrhea that accompanies intestinal infections. Phagocytosis, especially by neutrophils, is the principal mechanism of defense against *Ps. aeruginosa.* Although neutrophils can phagocytize and destroy unopsonized organisms, phagocytosis is greatly enhanced by the opsonic activities of Ab and C and maximum immunity requires that both the alternative and classical pathways of C activation are intact. The relative opsonic effectiveness of Abs of different Ig classes is not known. *Pseudomonas aeruginosa* does not appear to be susceptible to the bactericidal action of Ab and C.

Progress toward elucidating the nature of immunity to *Ps. aeruginosa* has been tediously slow despite its importance to the future of immunoprophylaxis and immunotherapy. The problem is especially difficult because of the multiplicity of pathogenetic factors involved and perforce, the multiplicity of specific Abs that may be needed to provide full protection against infection. For example, the observation that mice immunized with exotoxin A are fully protected against multiple MLDs of the toxin but succumb to challenge with live organisms does not mean that Abs against the toxin afford no protection or that full protection is not attainable with a polyvalent vaccine. This illustrates the principle that protective immunogens do not necessarily serve as virulence factors or vice versa (Chap. 8).

Although immunoprophylactic and immunotherapeutic trials with various vaccines and antisera have had only limited success in man, it appears that the most protective immunogens are the serotype oligopolysaccharides (O Ags) of LPS and that the protective Abs generated act as opsonins.

Since the future outlook for chemotherapy is uncertain, there is urgent need to determine the various factors concerned in the pathogenesis and immunology of *Ps. aeruginosa* infections, in order to provide background for improving immunotherapy and immunoprophylaxis.

J. LABORATORY DIAGNOSIS

Although *Pseudomonas aeruginosa* grows at 20° to 42°C on most ordinary media; blood agar is commonly used for bacteriologic diagnosis. The colonies have delicately fimbriated edges and may vary from smooth to rough to mucoid; they are usually apparent because of diffusion of the blue-green pigments into the surrounding medium. Pure cultures have a characteristic grape-like odor of aminoacetophenone, the metabolic product of trimethylamine. The fluorescent pigments produced occasionally also aid in identification of the colonies and in direct diagnosis of pseudomonas infection. For example, examination of a burn area with a Wood's lamp may reveal fluorescence produced by infecting *Ps. aeruginosa* before the infection becomes clinically evident. The ability to grow at 42°C, together with the production of the blue pyocyanin pigment distinguishes *Ps. aeruginosa* from other pseudomonads.

Difficulties in identification arise with the few strains that lack pigments or flagella. In these instances, biochemical tests are especially important. Whereas the *Enterobacteriaceae* ferment many sugars, obligately oxidative pseudomonads cannot ferment carbohydrates and are oxidase positive.

Agglutination tests with serotype (O) antisera or, alternatively, lysis by *Ps. aeruginosa* phages or the production of pyocins are useful tools for detecting non-pigmented strains.

K. THERAPY

Managing the chemotherapy of infections due to *Ps. aeruginosa* is formidable for a number of reasons. The organism is natively resistant to most antimicrobics and has a marked ability to acquire resistance during treatment. Additionally, the in vivo efficacy of antimicrobics is difficult to evaluate or to predict on the basis of in vitro studies. Infections are wide in scope and essentially all are based on marked deficiencies in antimicrobial defense of various kinds and degrees. Moreover, *Ps. aeruginosa* has the unique ability to further lower antimicrobial defenses by cleaving Igs, inactivating C and suppressing the bone marrow progenitors of neutrophils and macrophages; indeed when leukopenia approaches as a result of immunodepression by infecting organisms there is virtually no chance that antibiotics alone can save the patient.

Only those antibiotics that are determined to be active against the infecting organism should be selected for therapeutic use. The aminoglycosides, gentamicin, tobramicin, amikacin and colistin, are active against many strains of *Ps. aeruginosa.* Ticarcillin and carbenicillin are useful even though about 50% of strains are resistant. Third generation cephalosporins, such as cefoperazone, moxalactam and cefotaxime are also used. Extensive trials are being conducted with a promising new cephalosporin, cefsulodin; most strains are susceptible; and resistance to the drug does not develop readily. The drug penetrates the cell wall of *Ps. aeruginosa* better than most antibiotics and acts synergistically with aminoglycosides against some 20 to 80% of isolates. When given alone, cefsulodin compares favorably with a number of other drugs commonly used for a variety of *Ps. aeruginosa* infections.

Topical application of various chemical agents and antibiotics, especially polymyxins and mafenide, to burn areas and mucosae of immunosuppressed patients has been reported to be of some use for preventing colonization.

Because of the limitations of drug therapy in *Ps. aeruginosa* infections, the addition of active and passive immunotherapy to therapeutic management has become increasingly attractive. Although animal trials have usually been favorable, it is no surprise that human trials have yielded irregular and often equivocal results in view of the variable nature of human infections.

The usual approach to active immunization (which actually constitutes both prophylaxis and therapy) has been to initiate a series of closely-spaced vaccine injections as soon as the patient enters the hospital.* The various vaccines used have included polyvalent mixtures of either whole killed cells or serotype polysaccharides. Although the protection generated in most of these trials appears to have resulted principally from the opsonic activity of serotype Abs, it is probable that Abs against other Ags can also contribute substantially to immunity. Consequently, it would appear that the ultimate goal should be to prepare a vaccine containing all of the important protective immunogens of *Ps. aeruginosa.*

Passive immunotherapy in *Ps. aeruginosa* infections is commonly employed in combination with antibiotic therapy. Substantial protection has been achieved in burn patients with pooled immunoglobulins from volunteers immunized with polyvalent vaccine. Pooled human γ globulin and whole fresh blood from normal adults has also been reported to be of some therapeutic benefit. Leukocyte transfusion has been reported to be of benefit in infections attended by leukopenia. Immunotherapy trials in human infections have been limited and obviously should be continued and extended.

Among patients with *Pseudomonas* infections, those with cystic fibrosis (CF) are by far the most difficult to treat. The observation that the incidence of CF is the highest of any of the hereditary diseases of Caucasians (some 30,000 in the USA alone) emphasizes the enormity of the problem of treating *Pseudomonas* infections in this group.

Cystic fibrosis is characterized by a severe depression of local pulmonary but not systemic antimicrobial defense. The genetic defect in persons with CF leads to the production of excessive amounts of mucus by the respiratory mucosa; in addition, the fibronectin coat on respiratory epithelium is reduced and proteases in the respiratory tract are increased, both of which predispose to bacterial adherence and colonization. Antimicrobial defenses provided by alveolar macrophages and the mucociliary escalator are severely depressed. As a consequence of lowered defense the patient with CF commonly experiences a succession of opportunistic lung infections by various organisms, including staphylococci, pneumococci, *H. influenzae* and *Ps. aeruginosa.* Many of the infections in children with CF can be controlled with antibiotics. However, with advancing age the lungs of an increasing number of patients become colonized by mucoid strains of *Ps. aeruginosa,* such that, by adulthood colonization reaches about 90%. Some mucoid strains probably represent transformed strains, whereas others are newly-colonized strains. Because the enormous glycocalyx formed by mucoid strains severely limits the access of leukocytes, opsonins and antibiotics to the organisms, they are virtually impossible to eradicate and continue their relentless destruction of lung tissue. Most CF patients do not survive beyond 20 to 30 years of age; a large majority of fatalities result from *Pseudomonas* pneumonia. Since high levels of anti-*Pseudomonas* Abs and immune complexes exist in the sera of chronically-infected patients, there is no reason to expect that either active or passive immunotherapy would be beneficial, but instead, that such therapy may be harmful. The only group of CF patients who may logically benefit from prophylactic vaccination would be young patients whose lungs have not been

*The objective is to induce the early production of Abs which will prevent and/or suppress colonization by *Ps. aeruginosa.*

colonized by pseudomonads and whose sera lack high levels of anti-*Pseudomonas*-Abs.

L. RESERVOIRS OF INFECTIONS

Pseudomonas aeruginosa occurs literally everywhere in the hospital environment, especially in moist microenvironments. Intestinal colonization among hospital patients is 30 to 60% versus some 2 to 6% in the general population. Patient to patient spread via hospital personnel is a major mode of spread within hospitals. The patient himself or any other person may serve as a source of infection. *Pseudomonas aeruginosa* thrives in or on medical and surgical equipment, solutions, food, water and a variety of other materials. Examples of such reservoirs include contact lens solutions and aerosols from humidifiers, nebulizers and respirators; use of the latter often causes pulmonary infections in predisposed individuals.

M. PREVENTION OF INFECTION

Prevention is the best method for coping with *Pseudomonas* infections. *Patients at risk should be managed with great attention to aseptic techniques.* Sterilization, rather than "disinfection," of equipment should be used whenever possible. All sources of infection in hospitals should be carefully controlled; fresh vegetables are one source that has often been overlooked. A promising method for preventing *Ps. aeruginosa* infection in burn patients is being evaluated, namely, early active immunization with pooled vaccine (Sect. K). Whether immunoprophylaxis will prove to be useful in other high risk patients, such as prospective organ transplant recipients remains to be explored.

INFECTIONS CAUSED BY OTHER *PSEUDOMONAS* SPECIES

A limited number of *Pseudomonas* infections are produced by various species other than *Ps. aeruginosa.* The majority of these are nosocomial infections produced by two species, namely, *Ps. cepacia* and *Ps. maltophilia* (Table 26–1). These organisms produce a wide range of infections, the most frequent being wound and urinary tract infections. Most infections occur in debilitated and immunosuppressed individuals and some result from medical instrumentation; *Ps. cepacia* endocarditis has been encountered in heroin addicts, and the feet of persons who work in subtropical swamps. These two species are resistant to most antibiotics, but the majority of strains of both species are susceptible to chloramphenicol.

Two nonopportunistic *Pseudomonas* species that infect man as an accidental host are *Ps. mallei* and *Ps. pseudomallei. Pseudomonas mallei causes a disease called glanders* in its natural hosts, horses and related animals. Man can be infected accidentally by contact with diseased animals, but this seldom occurs at present because glanders is rare among horse and donkey populations in the USA. Without antibiotic treatment human infections are often fatal.

Melioidosis, a disease similar to glanders, is caused by *Ps. pseudomallei* which resembles *Ps. aeruginosa* in many respects (Table 26–1). Human melioidosis is a tuberculosis-like disease that occurs most often in Southeast Asia, where the organism is found in wild rats and other animals, and in soil and water. The presence of specific Abs in the serum of Malaysians (8%) suggests that asymptomatic infections are frequent. The disease is usually transmitted from rats to man through abraded skin or by the bites of fleas or mosquitoes, but can also be contracted from contaminated food and water and inhalation of aerosols or dust. In recent years, cases of melioidosis have appeared in a number of military service personnel returning from Southeast Asia. The disease frequently involves the lungs and takes various forms, ranging from an acute septicemia that may be rapidly fatal, to chronic infections with widespread granulomatous lesions in lymph nodes and other tissues. Clinical specimens used for diagnosis include sputum, urine, pus, and blood. Trimethoprim-sulfamethoxazole is the therapeutic drug of choice with chloramphenicol, tetracycline and sulfadiazine as alternatives.

REFERENCES

Fong, I.W., and Tomkins, K.B.: Review of *Pseudomonas aeruginosa* meningitis with special emphasis on treatment with ceftazidime. Rev. Infect. Dis. *7*:604, 1985.

Gilardi, G.L.: Pseudomonas, Chap. 30, *Manual of Clinical Microbiology,* 4th Ed. E.H. Lennette (ed). Wash., DC, Am. Soc. Microbiol. 1985.

Lerner, A.M.: The Gram-Negative Bacillary Pneumonias. Weinstein, L. and Fields, B.N. (eds.). *Seminars in Infectious Disease 5*:159. New York, Thieme-Stratton Inc., 1983.

Middlebrook, J.L. and Dorland, R.B.: Bacterial toxins: cellular mechanisms of action. Microbiol. Rev. *48*:199, 1984.

Pennington, J.E.: *Pseudomonas aeruginosa:* Pneumonia the Potential for Immune Intervention. Weinstein, L. and Fields, B.N. (ed.), *Seminars in Infectious Disease 5*:71, New York, Thieme-Stratton Inc., 1983.

Ramphal, R., et al.: *Pseudomonas aeruginosa* adhesins for tracheobronchial mucin. Infect. Immun. *55*:600, 1987.

Young, L.S. (guest editor): *Pseudomonas aeruginosa*—Biology, Immunology, and Therapy: A Cefsulodin symposium. Rev. Infect. Dis. *6*:S603, 1984.

27

MYCOBACTERIUM

Mycobacterium tuberculosis

Mycobacterium bovis

Mycobacterium avium

Mycobacteria That Act as Facultative Pathogens

Mycobacterium leprae

The genus *Mycobacterium* of the family *Mycobacteriaceae* contains a large number of species, both saprophytes and parasites. Organisms of this genus are notable for their lipid-rich cell walls, and "acid-fast" staining properties. A unique property of the mycobacteria is that their cell walls contain N-glycolylmuramic acid rather than N-actylmuramic acid. They are widely distributed in nature and most species have extraordinarily simple nutritional requirements. Pathogenic species attack a wide range of natural hosts throughout the animal kingdom. The slow growth of the mycobacteria may account, in part at least, for the slow development and progression of the diseases they produce.

Man is the sole natural host of *M. tuberculosis* and *M. leprae,* the two most important species pathogenic for man. The cow is the natural host for the closely related species *M. bovis* which, when transmitted to man as an accidental host, is essentially as pathogenic as *M. tuberculosis.* Disease in man occasionally occurs due to certain other species that can adapt to man as facultative pathogens; they are often referred to by the inappropriate terms, "unclassified," "anonymous" or, more commonly, "atypical" mycobacteria. As tuberculosis has waned in developed countries, mycobacteriosis due to these facultative pathogens has increased in relative importance. Mycobacteria that never serve as pathogens for man are of no special medical interest aside from the fact that some of them are members of the normal flora and may be mistaken for pathogens. *Mycobacterium tuberculosis,* which is often referred to as the "tubercle bacillus" because it is rod-shaped, stands as the prototype of the pathogenic mycobacteria and will be discussed in the greatest detail. Mycobacterial pathogens for natural hosts other than man include *M. bovis* (cattle), *M. paratuberculosis* (ruminants), *M. microti* (voles), *M. lepraemurium* (rats), *M. avium* (birds) and *M. marinum* (fish).

MYCOBACTERIUM TUBERCULOSIS

A. MEDICAL PERSPECTIVES

Tuberculosis is favored by overpopulation, poverty, ignorance and hereditary predisposition. The most common form of infection, pulmonary tuberculosis (phthisis, consumption), is a chronic wasting disease, which in its early stages is often mistaken for chronic bronchitis. The Greek word "phthisis" means "consumption of the flesh." Symptoms of tuberculosis commonly include fatigue, weight loss, night sweats, coughing, and spitting of blood.

In the world as a whole, tuberculosis has always ranked as the leading killer among the infectious diseases of man. However, in developed countries the death rate has declined steadily since about the middle of the 19th century, and precipitously since the advent of highly effective drug therapy in 1952. The reasons for this decline in death rate, which began even before the infectious nature of the disease was fully accepted, are not clearly apparent but probably reflect evolutionary changes in innate immunity of host populations and in virulence of the parasite, as well as improved standards of living. Unfortunately, the case rate has not declined as markedly as the death rate.

The origin of tuberculosis antedates the recorded history of man; evidence of the disease, such as the distortions of the spine described by Pott, have been found in the statuary and bones of Egyptians dating as far back as 7,000 years. Asia, Africa and the Middle East were evidently early spawning grounds of many infectious diseases, including tuberculosis.

Tuberculosis probably waxed and waned in Europe during the middle ages, but no records exist to indicate that it reached epidemic proportions until the mid-17th century, after which its incidence declined and rose again to reach a major peak in the mid-19th century. The disease was so prevalent in Europe during the 18th and 19th centuries that "paleness" became fashionable and Alexander Dumas wrote that since everyone has consumption, especially the poets, "It is good form to spit blood from sheer emotion." At that time tuberculosis was thought to be hereditary and the notion grew that it was linked with genius, probably because genius seldom comes to light except by sacrifice and overwork. Among those who have died of tuberculosis are Thoreau, Stevenson, Schiller, Keats, Chopin and the dean of tuberculosis physicians himself, Laennec. In 1815 Thomas Young estimated that tuberculosis caused the premature death of one-fourth of the inhabitants of Europe and in 1882 Robert Koch, the discoverer of *M. tuberculosis,* wrote that, in Europe, tuberculosis killed one-third of all persons of middle age!

The manifestations of tuberculosis are many and varied; consequently, they were not easily recognized to reflect a single causation. Until late in the 19th century the concept, proposed by Laennec in 1819, that lung and miliary lesions were manifestations of one and the same disease continued to be challenged by men of great competence, such as the famous pathologist, Virchow. It is of singular interest that Laennec invented the stethoscope for the express purpose of listening to lung sounds of patients with tuberculosis.

Although Galen, in the 2nd century, proposed that pulmonary tuberculosis is transmitted by the "breath," this concept was abandoned by the Romans. That tuberculosis was not a communicable disease remained the consensus of medical men until the mid-19th century. However, in 1865 the French army surgeon, Villemin, firmly established the infectious nature of both human and bovine tuberculosis by infecting rabbits and guinea pigs. Robert Koch cultivated and identified *M. tuberculosis* as the causative agent of the disease in 1882, and received the Nobel prize in 1905.

The number of cases of active clinical tuberculosis in the world today probably exceeds 50 million, at least 20 million of whom are infectious. Each year the number of new cases approximates 10 million and the number of deaths may be as high as 3 million. Currently about half of the people in the world are infected with *M. tuberculosis.*

In advanced countries, tuberculosis is no longer the sinister and dreaded "white plague" of centuries past and sanatoria that once had long waiting lists are now virtually empty. However, in underdeveloped areas the tuberculosis problem is still immense; in some countries 25% of all deaths among young adults are due to the disease and the number infected often exceeds 90%.

Currently in the USA there are at least 25 million infected individuals, as indicated by positive reactions to tuberculin tests; of these, about 250,000 have active disease. Whereas only about 3 to 5% of young adults are tuberculin positive, the large majority of the elderly are tuberculin reactors. The incidence of infection is highest in urban populations, but is markedly lower today than in past decades when in large cities the number of infected individuals sometimes exceeded 95%; the current annual infection rate is 14 per 100,000. Alcoholism and drug addiction with associated malnutrition strongly predispose to the development of active disease, consequently the highest incidence of active cases and mortality rates occur among single elderly males residing in slum areas in large cities. Most children in the USA are tuberculin negative; conversion to positivity is usually due to a single identifiable source such as a family member or teacher. Prior to the advent of modern treatment, economic losses due to tuberculosis were staggering because of the prolonged course of the disease and the high cost of care and treatment. The tragic social and economic consequences of the disease have been especially profound in past years. Home life was commonly disrupted by the absence of a parent in the household and most families became financially impoverished. This state has been greatly mollified in recent years because ambulatory treatment with drugs is usually possible.

The future short-term outlook for the control of tuberculosis in developed countries is good food despite the problem of emerging drug-resistant strains. However, since many factors serve to determine infection rates, case rates, and the effectiveness of therapy on mortality rates, the long-term outlook for tuberculosis, even in a single country such as the USA, is difficult to predict. If, as seems reasonably certain, the long-continued control of tuberculosis within a region will eventually lead to the development of a more susceptible population, the difficulty of controlling the disease may increase with passing generations.

B. PHYSICAL AND CHEMICAL STRUCTURE

Mycobacterium tuberculosis is a nonmotile, nonsporing, capsulated, slender, pleomorphic rod measuring about 0.4 μm × 3 μm. The organisms are sometimes curved and in stained preparations often have a beaded appearance due to volutin bodies and unstained vacuoles. They stain with great difficulty, but, once stained, retain stains even when such drastic destaining methods as treatment with acid-alcohol are employed, hence the term "acid-fast." The acid-fast staining procedure commonly employed is the Ziehl-Neelsen method. Although the properties determining acid-fastness are unknown, physical integrity of the cell is essential. *Mycobacterium tuberculosis* tends to stain weakly Gram-positive; however, its chemical composition more closely resembles the Gram-negative than the Gram-positive bacteria.

The cell wall of *M. tuberculosis* is a complex 3-layered structure containing enormous quantities of lipids and waxes (up to 60% of the dry weight of the cell), which evidently account, in part at least, for many of the organism's unique properties among pathogens. These include its high resistance to physical and chemical agents and to intracellular killing by phagocytes, its hydrophobic character, its slow growth, its acid-fastness, and its marked adjuvant and granulomagenic activities. The cell's hydrophobic surface promotes phagocytoses by neutrophils and by macrophages, which are attracted chemotactically to the organism. Cell wall complexes are rich in antigenic proteins and polysaccharides.

The roles that lipids and waxes play in tissue responses and sensitivity reactions in tuberculosis remain uncertain despite extensive investigation. However, it is of interest that the so-called "wax D" of the cell wall, which is composed of peptidoglycolipids, has adjuvant and directive effects on the immune response similar to those of whole bacilli. When wax D is mixed with tuberculoproteins, it directs the immune response to the proteins largely along the path leading to delayed hypersensitivity (DH). An old but interesting phenomenon observed over 40 years ago is as follows: when a foreign protein, such as egg albumin, is mixed with tubercle bacilli, especially when Freund's incomplete adjuvant mixture is also included, the immune response to the protein is strongly directed toward the development of DH. On the other hand, injection of free uncomplexed proteins alone, including tuberculoproteins, leads to Arthus and immediate-type sensitivities.

The adjuvant activity of wax D appears to be due to a cell wall peptidoglycolipid existing on the cell surface as a network of fine branching filaments 130 Å in diameter; the filaments are unique to organisms of the genus *Mycobacterium.*

The tubercle bacillus also contains the so-called cord factor which has some adjuvant activity; however, its alleged contribution to virulence is doubtful (Sect. D). Although little is known about the Ags of different mycobacteria, many species share common Ags, both surface and cytoplasmic. This is evidenced by cross-reactions displayed in Ouchterlony gel diffusion tests and in skin sensitivity tests conducted with proteins (tuberculins) derived from various mycobacteria. Human and bovine tubercle bacilli are closely related and their respective tuberculins are essentially of equal potency for eliciting cutaneous tuberculin reactions in individuals infected with either of the two organisms. Other mycobacteria, such as the "atypical mycobacteria," are so distantly related to *M. tuberculosis* that individuals infected with most of these organisms commonly give only weak or negative reactions to standard tuberculins derived from *M. tuberculosis.*

C. GENETICS

Little is known about the mechanisms responsible for genetic changes in *M. tuberculosis,* especially in the important areas of medical interest, virulence and antibiotic resistance. Bacteriophages for *M. tuberculosis* have been described, but their possible role in genetic changes is unknown; they have been of some epidemiologic value. Freshly isolated strains of *M. tuberculosis* from untreated patients with pulmonary tuberculosis have not been found to vary significantly in their virulence for experimental animals. However, isolates from persons having a form of skin tuberculosis (lupus vulgaris) and from patients treated with isonicotinic acid hydrazide (INH or isoniazid) are often of low virulence for experimental animals. Organisms of reduced virulence have also been obtained by long subculture under suboptimal growth conditions that evidently provide growth advantage to low-virulence variants, e.g., the bovine bacillus of Calmette and Guerin (BCG) and the human avirulent strain, H37Ra.

D. EXTRACELLULAR PRODUCTS, VIRULENCE FACTORS AND PROTECTIVE IMMUNOGENS

Little is known about extracellular and cell surface products of *M. tuberculosis* or the possible role that such substances may play in pathogenesis. Being a facultative intracellular parasite, the shed substances, surface components, or metabolic activities of *M. tuberculosis* could exert toxic effects on parasitized macrophages; in particular, large numbers of growing bacilli within macrophages appear to be cytotoxic. However, no soluble external component of potential cytotoxicity has been described. Some surface components could have low toxicity.

Lipid-rich surface complexes in virulent strains of *M. tuberculosis* have long been thought to contain a virulence component. One such complex was originally called "cord factor" because its presence on virulent but not avirulent-strain organisms was associated with growth to form long pallisade-like bacillary aggregates resembling serpentine cords. Cord factor, a glycolipid, was assigned the chemical composition, trehalose-6-6′-dimycolate. Recent evidence strongly supports the concept that virulence factors reside in associated mycosides, certain sulfolipids and sulfatides.

Recent in vitro studies with nonactivated alveolar macrophages from normal rabbits have shown that when live virulent human-strain H37Rv organisms are engulfed by these phagocytes, they display the unique ability to adhere strongly to phagosomal membranes and to fragment them within hours, thus liberating essentially 100% of the bacilli into the cytoplasm (Fig. 27–1). Heating these virulent organisms at 65° C for 30 minutes or exposure to the chemotherapeutic agent, isonicotinic acid hydrazide (isoniazid or INH), abolished their ability to disrupt the phagosomal membranes of nonactivated macrophages.

Whereas organisms of the avirulent strain H37Ra did not adhere to and disrupt phagosomal membranes, the lowly-virulent organisms, BCG, adhered to and caused limited membrane disruption with the liberation of some 20 to 25% of the ingested bacilli into the cytoplasm.

In contrast to nonactivated macrophages, lysosome-rich activated macrophages from BCG-immunized rabbits and an occasional "normal rabbit" were found to possess phagosomes that resisted disruption by H37Rv-strain organisms (Fig. 27–2). These results are consistent with the accepted view that, whereas the antimicrobial effector functions of activated macrophages are nonspecific, their activation need not be specific. The additional observation that the cholesterol-protein ratio of the cytoplasmic membranes of alveolar mac-

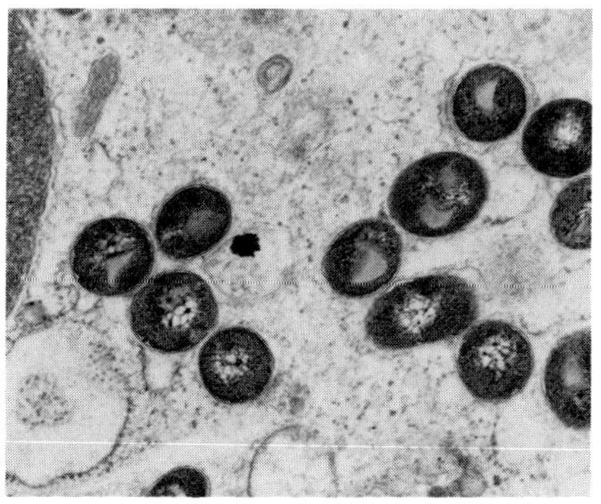
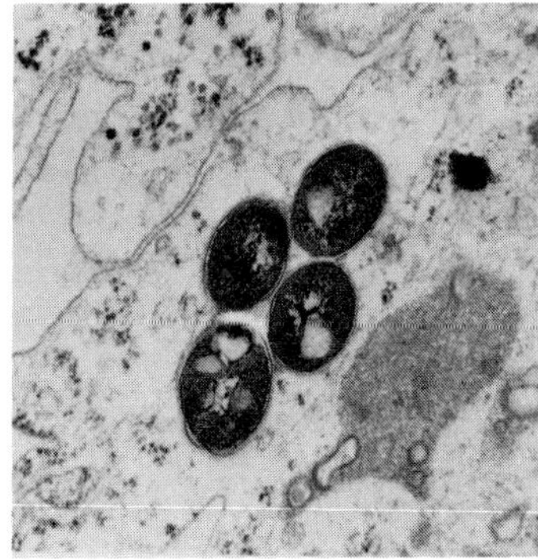

Fig. 27–1. Intracellular virulent *M. tuberculosis* (H37Rv strain) in normal rabbit alveolar macrophages. The absence of a phagosomal membrane is a common finding. Ultrathin section. ×20,300. (Courtesy of E.S. Leake.)

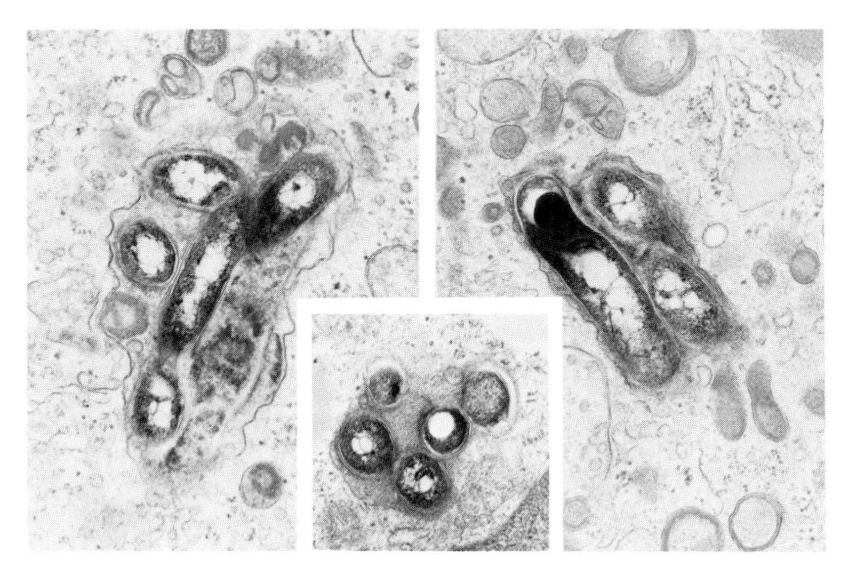

Fig. 27–2. Well-defined phagosomal membranes surrounding intracellular virulent *M. tuberculosis* (H37Rv strain) are commonly seen, as well as accumulation of electron dense material within the phagosome. Ultrathin sections. ×20,300. (Courtesy of E.S. Leake.)

rophages is increased 5-fold by BCG-immunization indicates that resistance of activated macrophages to the disruptive effects of virulent bacilli rests on a marked change in phagosomal membrane composition. The results make it reasonably certain that the virulence factor(s) of *M. tuberculosis* resides in its surface structure, although its precise chemical nature is not known. One proposal is that virulence may rest with moieties presented by one or more of the mycosides, sulfolipids or sulfatides. Since virulence factors of microbes do not necessarily serve as immunogens, there is no compelling reason to believe that any virulence factor will prove to be a protective immunogen. Instead it is probable that the most important immunogen(s) of *M. tuberculosis* comprise one or more of its constituent proteins and that they elicit cell-mediated immunity (CMI) by acting in combination with adjuvant components of the organism.

In addition to their disruptive effects on the phagosomal membrane, virulent bacilli have been reported to inhibit the fusion of lysosomes with phagosomes.

Beginning with the elegant experiments of Max Lurie several decades ago it has been repeatedly shown that, whereas virulent tubercle bacilli are inhibited and killed by "immune" (activated) macrophages within days to weeks, they grow freely in nonactivated macrophages and eventually destroy them.

Immune lymphocytes but not Abs have been shown to passively protect against tuberculosis. However, effective, long-lasting active immunity has only been produced by live mycobacteria, such as BCG vaccine, presumably because a protective level of CMI is only fully maintained over time by the continuing stimulus provided by the protective immunogen(s) unique to living bacilli.

E. CULTURE

Although *M. tuberculosis* can grow on simple synthetic media, it is common practice in diagnostic work to use a complex medium, such as Lowenstein-Jensen medium, because small inocula are sensitive to trace amounts of various toxic components present in synthetic media. Since trace amounts of free fatty acid, such as oleic acid, stimulate growth, known amounts of fatty acid and serum albumin are sometimes added to synthetic media; the albumin binds the fatty acid with an affinity that is proper for maintaining a level of dissociated free fatty acid suitable for growth. Synthetic media containing the wetting agent, Tween 80, plus albumin are useful for quantitative experimental work since they promote rapid dispersed growth of these slow-growing hydrophobic organisms.

Mycobacterium tuberculosis is a strict aerobe. Its shortest generation time is about 12 to 18 hours and visible growth under the best of conditions is commonly not achieved before 1 to 2 weeks or more; usually at least 3 weeks are required before growth from clinical specimens becomes apparent. Increased CO_2 tensions of 5 to 10% enhance growth, which occurs over a pH range of 6.0 to 8.0. The temperature growth range is 30° to 42°C (optimum 37°C). Colonies are characteristically buff-colored, wrinkled and dry.

F. RESISTANCE TO PHYSICAL AND CHEMICAL AGENTS

Mycobacterium tuberculosis is highly resistant to drying in the presence of organic matter and survives in dried sputum for days to weeks. It is also highly resistant to acids, alkalies and dye-stuffs. Phenols are among the best disinfectants and moist-heat treatment equal to that of pasteurization is required to destroy the organism. This unique resistance to chemical agents is of advantage for destroying contaminating microbes in clinical specimens before culture. Chemicals are also used for preparing selective media (Sect. J).

G. EXPERIMENTAL MODELS

Tuberculosis is largely a disease of domesticated animals and man. Its rarity in wild animals probably rests primarily on lack of adequate patterns of transmission and/or lack of close contact between diseased and healthy individuals. The range of pathogenicity of the medically important mycobacteria that produce tuberculosis and tuberculosis-like disease in man and certain animals is given in Table 27–1. Some subhuman primates are highly susceptible to *M. tuberculosis* and when in close contact with man can readily contract and transmit the disease to their kind and to man. However, tuberculosis is not a natural disease of these animals in the wild.

Guinea pigs and mice have been used most extensively in experimental tuberculosis. Whereas guinea pigs have the advantage of developing DH closely simulating that developed by man, virulence can be more easily quantitated in the more resistant mouse.

H. INFECTIONS IN MAN

Certain infectious diseases, characterized by chronicity and granuloma formation, are produced by facultative intracellular parasites that are not effectively destroyed by humoral factors in the extracellular environment or by

Table 27–1. Pathogenicity of *M. Tuberculosis* and Related Mycobacteria for Man and Certain Animals

Animal	*M. tuberculosis*	*M. bovis*	*M. avium*	*Atypical Mycobacteria*
Man	+ + +*	+ + +	rare	– to + + +
Guinea pig	+ + + +	+ + + +	–	– to +
Rabbit	+	+ + + +	+ +	–
Mouse	+ + +	+ + +	–	– to + + +
Cow	–	+ + +	–	?
Chicken	–	–	+ + +	?
Swine	+	+ +	+	?
Cat	+	+ +	–	?
Dog	+ +	+		

*The degree of pathogenicity is designated by plus signs ranging from + to + + + + and in animals is based on experimental infections.

neutrophils, but, instead, are phagocytized and grow within macrophages; tuberculosis is the prototype of such diseases.

Man is highly susceptible to invasion by tubercle bacilli but is remarkably resistant to progression of the infection and the development of clinical disease. Although *M. tuberculosis* may infect many organs and tissues by many routes, the lung is by far the most common site of infection (about 85%), evidently because of the high frequency of exposure, and the high oxygen tensions that it provides, particularly anteriorly and in the lower lobes. The usual routes of invasion are the respiratory and intestinal tracts. For unknown reasons the parenchyma of a few organs, such as the pancreas, heart and thyroid gland, are extraordinarily resistant. The only infection produced by *M. tuberculosis* that will be discussed in detail is *pulmonary tuberculosis.*

Because of the great complexity and controversial nature of host-parasite relationships in tuberculosis, the pathogenesis and immunology of the disease cannot be suitably dealt with as separate subjects. They will be considered together in the discussion of the natural history of the disease, after which mechanisms of immunity will be discussed separately.

It is notable that, whereas early primary infection is characterized by essentially unrestricted growth of bacilli at the infection site and spread of organisms to local lymph nodes and blood, late primary, reactivation and reinfection tuberculosis are characterized by restricted growth of bacilli and a tendency for the organisms to remain localized at the infection site or to spread solely by contiguity unless the tubercle ruptures into a bronchus or a vein. Another major point worthy of emphasis is that the progress of the individual lesion appears to be related to the load of bacilli it carries, and to the level of DH and CMI that develops. When hypersensitivity and cellular immune activities are strong and the load of bacilli is low, proliferative healing lesions dominate; when the load of bacilli is high, progressing exudative lesions dominate.

1. Primary Infection

Primary infection is usually acquired by inhalation of droplet nuclei generated by individuals with active cavitary disease who cough and thus discharge organisms into the air.* The droplet nuclei must have a diameter of < 10 μm in order to reach the alveoli where infection is initiated. Inhaled droplets of larger size are arrested short of the alveoli and are usually removed by the mucociliary blanket and swallowed. The larger airways of the respiratory tract are highly resistant to infection. Studies on human necropsy material and experimental tuberculosis in animals suggest that the following events take place. Most of the inhaled bacilli reaching the alveoli are engulfed promptly by *resident alveolar macrophages* (largely nonactivated) and grow readily in such cells for a week or more. Many of the macrophages initially parasitized are eventually killed, presumably as the result of the growth and metabolic activities of the engulfed bacilli. Most of the organisms released by dying macrophages are promptly engulfed by increasing numbers of infiltrating neutrophils and monocytes, which either remain in the area or migrate to the local lymph nodes.† Whether free bacilli enter afferent lymphatics is uncertain. Neutrophils cannot destroy bacilli, which are soon released on death of these short-lived phagocytes. Alternatively, neutrophils laden with bacilli may be engulfed by macrophages. A few bacilli usually escape from the lymph nodes to the blood and lodge at distant sites. However, these few blood-borne organisms are usually destroyed by "immune macrophages" without leaving permanent evidence of a lesion. Alternatively, blood-borne organisms may persist in tissues for many years and on occasion may serve to cause reactivation disease. Thus, during the initial phase of infection, essentially unrestricted local growth of organisms occurs both extracellularly and within nonactivated (nonimmune) macrophages, and lymph node and systemic blood-borne spread takes place before acquired immunity, with its localizing influences, develops. After about 2 to 3 weeks, when the first evidence of the immune response appears, all events become increasingly influenced by the developing force of DH and CMI. As tubercles form the overall intracellular growth of the organisms gradually slows, evidently in substantial measure because the infiltrating macrophages become activated and capable of destroying bacilli. However, immunity does not become fully developed until host cells assemble in large numbers and form mature tubercles. If the developing immune forces are too weak or the numbers of infecting microorganisms are too large, they are not restrained by tubercle formation and the disease may progress and spread. Indeed, if the infecting dose of bacilli is large enough and/or the immune response is weak, the primary infection may progress and kill the patient rather than undergo spontaneous arrest, as occurs in the large majority of cases. For example, primary infections often progress in the immunologically immature infant and the immunologically compromised adult.

The tubercle, a typical allergic granuloma, frequently presents centrally located giant cells (usually containing bacilli) immediately surrounded by epithelioid cells and a peripheral mantle of macrophages, lymphocytes and a few plasma cells (Figs. 27–3 and 27–4); whether plasma cells contribute to events within the tubercle is not known. The tubercle has long been recognized as the sine qua non of resistance to tuberculosis, but it is only recently that the relation between the composition and function of this "adaptive structure of

---

*Bacilli survive for weeks in particles of dried sputum, which also constitute a potential source of infection.

†Monocytes are young nonactivated macrophages.

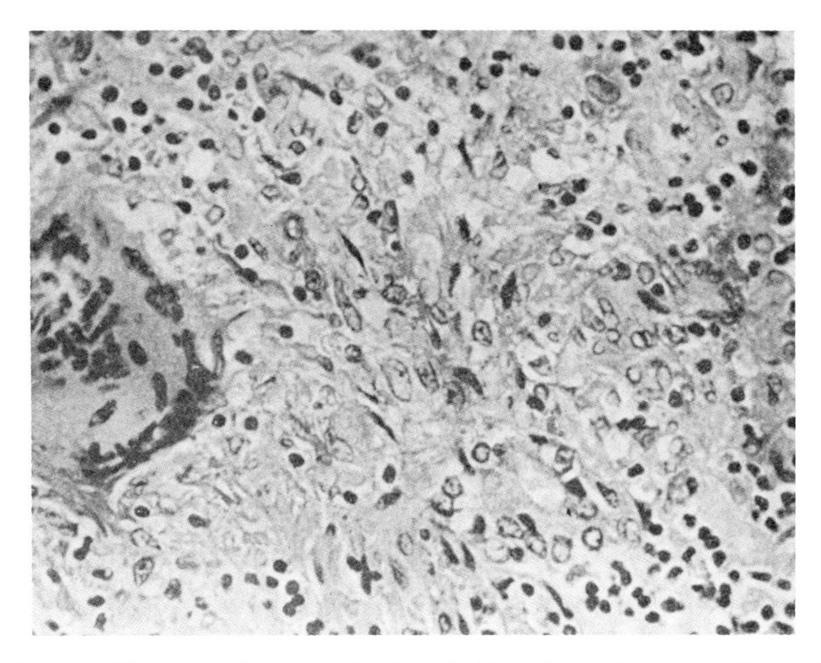

Fig. 27–3. The center of a young tubercle in the liver of man, showing a well-marked epithelioid cell reaction, a giant cell and a peripheral zone of lymphocytes. The vesicular nuclei of the epithelioid cells and the cytoplasmic extensions of the giant cell are well shown. (From Florey, H.: *General Pathology.* Philadelphia, W. B. Saunders Co., 1970.)

immunologic defense" in immunity has become meaningful. Although the precise ways in which the various cells of the tubercle may contribute to immunity are unknown, it is virtually certain that specific T effector cells (Te cells) generated by Ags of the bacillus play a major role by reacting with locally released Ags to yield macrophage-activating lymphokines.

Epithelioid cells are elongated, nonmotile, derivatives of macrophages. Epithelioid cells develop from macrophages present in the tubercle, but the factors that incite their development are unknown. They have little or no capacity to phagocytize but appear to have a high capability to inhibit and kill engulfed microbes; they usually contain few if any bacilli. It has been shown that monocytes attracted to the periphery of the tubercle by chemotaxis become activated under the stimulus of lymphokines as they pass through the peripheral mantle of lymphocytes. It is also probable that, as macrophages move centrally, lymphokines exert additional effects on them, namely, immobilization, transformation into epithelioid cells and fusion to form giant cells. If a heavily parasitized activated macrophage is to survive in the tubercle, transformation to an epithelioid cell with cessation of phagocytosis may be of advantage since otherwise the heavy load of engulfed organisms might become so great that the cell would be killed. By interdigitation of their processes epithelioid cells literally form a solid multilayered wall around the center of the developing tubercle and may thus limit the spread of bacilli.

The Langhans-type giant cell of the tubercle, with its multiple nuclei dis-

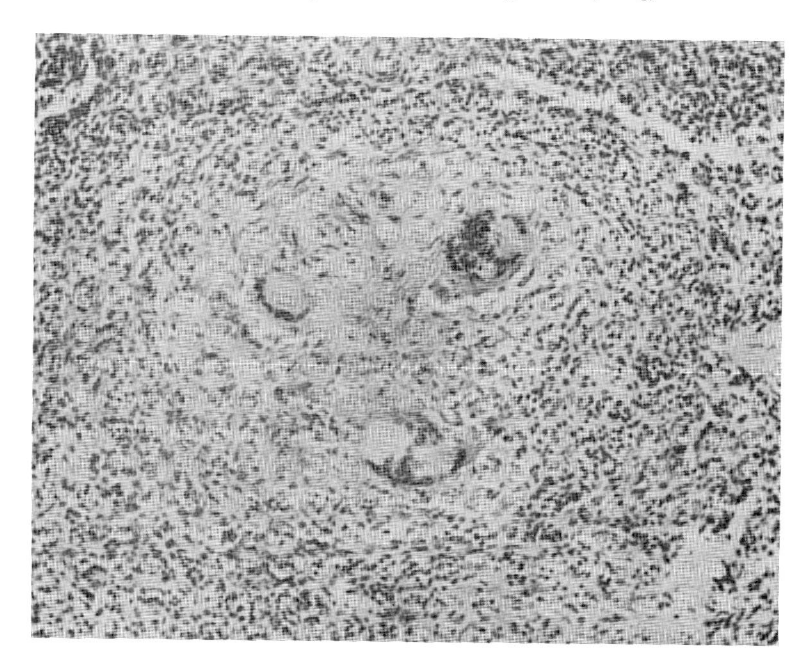

Fig. 27–4. Microscopical tubercle in the spleen of man. At the center of the palely staining mass of epithelioid cells the earliest signs of necrosis can be seen. (From Florey, H.: *General Pathology.* Philadelphia, W. B. Saunders Co., 1970.)

tributed at the periphery of its cytoplasm, is the result of fusion of numerous macrophages. Giant cells occur in association with epithelioid cells; they often contain viable bacilli, and often occupy the centers of small tubercles undergoing caseation necrosis. Recent in vitro experiments have shown that lymphokines cause macrophages to aggregate and fuse to form giant cells. Giant cells may function as fail-safe structures for handling bacillary aggregates that are too large for handling by single macrophages.

In those primary tubercles that do not heal readily, central caseation necrosis usually occurs, apparently as the result of death of epithelioid cells (Fig. 27–4). Whether death of other types of cells alone would provide the conditions necessary for the development of caseation necrosis is doubtful. The necrotic material consists of the partially digested remains of dead cells which lend to it a cheese-like consistency. Although the precise reason for caseation necrosis remains an enigma, the probability that the DH reaction makes a major contribution to this event suggests that the lymphokines generated in DH reactions are the principal cause of this type of necrosis. There is good evidence that lack of nutrition, due to the unique avascularity of the tubercle, does not serve as the sole basis of necrosis. Whereas fibroblasts proliferate vigorously under the stimulus of a lymphokine, vessels do not grow into the tubercle.

The growth of tubercle bacilli is markedly inhibited in areas of caseation necrosis, presumably because of lack of oxygen, the presence of toxic fatty acids, and possibly additional factors, such as lysozyme and other cationic cell components. Growth of bacilli is most favored in the outermost zone of necrosis near the periphery of the tubercle, where the organisms multiply extracellularly under favorable oxygen tensions and beyond the reach of macrophages. It is notable that, when softening *(liquefaction)* of caseous material occurs, explosive growth of bacilli takes place, often leading to catastrophic spread of the organisms and sometimes death. Whatever the factors are that cause central necrosis of epithelioid cells, they evidently affect other cells as well, for fibroblasts at the periphery of such tubercles do not fare as well as before and healing and vascularizaiton of the necrotic area does not occur until the bacilli are destroyed by invading macrophages. Calcium deposits are sometimes laid down following caseation and are often present in healed primary lesions. Residual walled-off caseous materials, scarred areas and calcium deposits are of particular interest since bacilli remain alive in them for years and serve as a source of organisms leading to reactivation disease.

The phenomenon of softening of the caseous material presents another enigma. Although softening invariably occurs when secondary infection with pyogenic bacteria takes place, which theoretically might result from enzymes liberated by dying neutrophils, it can also occur spontaneously in certain tubercles but not others. The softening of large caseous lesions is usually accompanied by erosion of bronchi and communication of such lesions with the airways of the lung. The liquefied caseous material, together with increased availability of oxygen from the airways of the lung provides an environment favorable for an explosive multiplication of bacilli. Unfortunately, these events prompt extensive coughing and the highly-infectious caseous material is disseminated in the form of an aerosol. Discharge of caseous material leads to the formation of a cavity harboring residual exudate and necrotic debris teeming with bacilli. The fibrous capsule lining the cavity provides a barrier to the normal cellular defense mechanisms as well as to chemotherapy. The organisms can persist for an indefinite period in the residual caseous material lining a cavity, particularly if the cavity does not close. The above events also provide a mechanism for disseminating large numbers of bacilli into the bloodstream, as occurs when progressing tubercles erode blood vessels. *Cavities are the greatest deterrent to healing and present the most difficult problem in clinical management.* Because of diffusion problems in cavities chemotherapeutic agents may lack effectiveness; obliteration by collapse therapy may fail and surgical excision is sometimes the only recourse. Swallowed sputum sometimes leads to tuberculosis of the intestine.

The key to softening of caseous material is one key to tuberculosis, for if softening could be averted tubular and vascular spread would be restricted and healing would be the rule; moreover bacilli would not be discharged to infect others and the chain of infection would be broken. When softening occurs the symptoms of tuberculosis, fever, night sweats and wasting, intensify. Fever in tuberculosis results from liberation of pyrogens, presumably, from dying leukocytes.

The tubercle is a hotbed for cell-interactions and cell activities governed largely by lymphokines, monokines and interleukins, some of which probably provide negative feedback to counterbalance excessive stimulatory effects of others. Information concerning these agents has been gained largely from in vitro trials in which lymphocytes from sensitive hosts were incubated with tuberculin and the culture fluids tested, or in which extracts and cell concentrates from diseased organs and tissue lesions were examined. (Myrvik & Weiser, *Fundamentals of Immunology*, 2nd Ed., Lea & Febiger, 1983, pp. 169, 173).

Delayed hypersensitivity is commonly measured by injecting tuberculoproteins (tuberculin) from heated bacillary cultures into the skin, (Sect. J and *Fundamentals of Immunology* pp. 142, 170). Positive DH reactions are triggered within a few hours following contact of tuberculin with sensitive T effector cells that migrate from the circulation to the test site. The lymphokines generated produce a cascading influx of neutrophils, lymphocytes (both sensitive and nonsensitive), and monocytes (activated macrophages do not circulate). Many of the infiltrating monocytes become activated and multiply. A grossly visible reaction appears after a delay of 6 to 12 hours, peaks at 24 to 48 hours and then fades. Epithelioid cells are not formed and no structure resembling a tubercle develops.

By contrast, when either the insoluble complex, tuberculin-glutaraldehyde, or a suspension of killed bacilli is injected into tuberculin-positive animals, cell infiltration and macrophage activation, aggregation and transformation proceed slowly (peak 7 to 10 days) to finally present a structure resembling a noncaseating tubercle. These different responses generated by tuberculin vs bacilli or complexed tuberculin are best explained by differences in the levels of free Ag at the test site over time. *The results are consistent with the concept that in tuberculosis the development of noncaseating tubercles rests on the slow release of Ag from a limited bacillary load, whereas caseating lesions result from the rapid release of Ag from a heavy bacillary load.*

About 1 to 2 weeks after a primary infection is initiated appreciable numbers of sensitized Te cells begin to appear in the circulation and in developing lesions. At this time the "nonsensitive primary lesion" is converted to a "sensitive primary lesion" characterized by marked changes in cell composition and dynamics of development. At the outset of sensitization the mobilization and activation of lymphocytes and macrophages accelerates and some macrophages either transform to epithelioid cells or aggregate and fuse to form giant cells; when the bacillary load is high caseation necrosis and lesion progression are prone to occur.

Two general types of lesions, *exudative* and *productive* or proliferative occur after the development of DH and CMI. Exudative lesions are associated with disease progression, whereas productive lesions are associated with regression and healing. In exudative lesions the acute inflammatory components, neutrophils and exudate are prominent. Productive lesions comprise tubercles which, although they may caseate, eventually heal without the escape of large numbers of bacilli to trigger DH reactions in surrounding tissues.

Primary lesions of the lung usually heal without leaving radiologically detectable evidence of their existence. When found, radiologically detectable healed lesions are usually solitary despite the probability that primary seeding is multifocal. Such a solitary lesion is usually located randomly at any site deep in the lung. As it grows beyond microscopic size, caseation necrosis occurs and satellite tubercles sometimes develop that fuse to form caseating conglomerate tubercles. Caseating lesions also form in the draining lymph nodes. The radiologically detectable primary lesion and its affected draining lymph nodes is called the *"primary complex"* or *"Ghon complex"*; it contrasts sharply with typical lesions of subsequent disease (reactivation or reinfection tuberculosis) in which the draining lymph nodes are not substantially involved. *Lesions of the primary complex usually escape clinical notice, and commonly heal and remain quiescent throughout life. Alternatively, primary lesions may sometimes progress and produce fatal disease,* especially in infants, debilitated individuals and individuals belonging to races with low genetic resistance. The balance between host resistance and virulence of organisms is evidently a delicate one and is easily shifted by the size of the inoculum, the physiologic state of the host, or the fortuitous rupture of a tubercle into the bloodstream, a tubular structure or a body cavity. Small numbers of living bacilli commonly persist in healed primary lesions for many years, indeed often for the lifetime of the individual, e.g., "once infected always infected" is an old adage. Although such bacilli are virtually impossible to detect microscopically, they have been demonstrated by inoculating animals with necropsy material. Further evidence that bacilli persist in appreciable numbers in healed lesions is the observation that tuberculin sensitivity commonly persists for the lifetime of the individual. The lesions in which bacilli can persist are often minute areas of scar and can sometimes be found after tuberculin sensitivity has become negative. *The precise circumstances that permit organisms to survive in healed lesions are unknown.* However, bacilli can remain alive for years on laboratory media stored under semianaerobic conditions.

2. Reactivation and Reinfection

In regions where the incidence of disease is low, most cases of clinical tuberculosis are the result of reactivation arising from the residual bacilli of primary infection (reactivation tuberculosis) rather than from exogenous bacilli (reinfection tuberculosis). Hence, in contrast to previous decades of high incidence of disease, when it was of advantage to bear healed primary lesions and to be tuberculin positive and immune in order to meet the high risk of exogenous reinfection, it is currently disadvantageous to have experienced a primary infection because the risk of reactivation now outweighs the low risk of reinfection.

Obvious but unanswered questions are: Why do old healed, primary lesions sometimes reactivate in adulthood and why does reactivation disease in adults progress to severe clinical disease more often than does primary infection itself? Although the arguments on this key question of reactivation are many and varied, the explanation probably rests with the fact that most primary infections occur in children of ages 3 to 12 years, a period in life when resistance is highest and progressive disease rarely eventuates. There is strong evidence that reactivation during puberty and adulthood often occurs for physiologic reasons such as hormone changes, malnutrition (especially protein deficiency), overcrowding and stress. For example, after the release of inmates of Nazi concentration camps, the speed with which many of those with far-advanced tuberculosis recovered was astounding and indicated clearly that the conditions of imprisonment severely depressed resistance.

The influence of age, sex and hormones on resistance to tuberculosis is well known, but the mechanisms involved are obscure. Whereas children between 3 and 12 years of age are highly resistant and seldom develop progressive primary disease, infants are highly susceptible, presumably because of immunologic immaturity, and often die a few months after infection. Susceptibility decreases substantially during adolescence and reactivation tuberculosis among young adults is more frequent in females than males, perhaps because of hormonal influences. Resistance falls again in the aged as immunocompetence declines.

Following primary infection, immunity commonly wanes over the years, allegedly because of declining numbers of bacilli. Waning immunity plays a role in reactivation tuberculosis as evidenced by the observation that individuals in whom tuberculin sensitivity declines to low levels are more prone to develop reactivation disease than those whose sensitivity persists at moderate levels. Healthy persons whose tuberculin sensitivity rises to high levels in the course of time are also prone to develop reactivation disease; indeed their high DH may reflect early reactivation. Reactivation tuberculosis commonly develops in the apices of the upper lobes of the lung, presumably from residual organism carried to the site via the circulation during primary infection. Growth of bacilli in the apices may be favored because the uniquely low pulmonary arterial inflow of blood at these sites when in the erect position may impair defenses.

I. MECHANISMS OF IMMUNITY

Experiments with outbred animals have shown that essentially all members of a species become infected when challenged with a uniform dose of virulent bacilli, but that development of the resulting lesions varies greatly among individuals, thus indicating that genetic variation lies primarily in variations in the capability to acquire immunity rather than in some nonspecific resistance mechanism existing at the time of infection. This concept is supported by studies in inbred rabbits and species of animals of differing susceptibilities in which it has been found that resistant animals respond most rapidly with regard to tuberculin sensitization, lesion development and growth inhibition and destruction of bacilli. The slower pace at which susceptible animals mount these activities is alleged to result in a bacillary overload which is subsequently difficult to overcome.

Studies involving hundreds of sets of twins indicate that marked genetic variation in susceptibility to tuberculosis also exists in human populations. This is exemplified by the Lubeck disaster of 1926 in which 251 newborn infants were mistakenly fed a large dosage of virulent bacilli instead of BCG vaccine; by 4 years later 72 had died of progressive tuberculosis, 5 had died of other causes, 127 presented radiologically detectable lesions and 47 did not present detectable lesions.

There is abundant evidence that, in the course of evolution, exposure to infection has played an important role in selecting those individuals who most readily acquire effective immunity to *M. tuberculosis.* Certain American Indians and Eskimos, who first experienced tuberculosis when contacted by Europeans, displayed greater susceptibility to the disease than people of European stocks. The primary disease in primitive people tends to simulate disease in infants, often being progressive and fatal.

The tubercle, a prototype of the allergic granuloma, is a remarkably effective structure of immunologic defense against agents that resist humoral defenses (*Fundamentals of Immunology,* Chap. 11). Consequently elucidation of the mechanisms of acquired immunity to tuberculosis may be best achieved by studying antibacterial events and forces within the tubercle. Tubercle bacilli seldom exist free of a tubercle for long periods because satellite tubercles form readily as a localizing response to freed organisms, especially after Te cells responsible for CMI and DH have developed.

Formation of the primary tubercle is heralded by infiltration of the infection site by nonactivated macrophages (monocytes) and a few neutrophils that are chemotactically attracted to the organisms and ingest them. Essentially all bacilli soon reside within macrophages since bacilli-laden neutrophils are either engulfed by macrophages or soon die and release bacilli, which in turn are ingested by macrophages. Within hours some phagocytes carrying bacilli reach the draining lymph nodes. During this time it must be envisioned that Ags released from liberated bacilli reach and sensitize immunocompetent lymphocytes both at the infection site and in the draining lymph nodes. Evidently sensitized Te cells in the circulation migrate to the infection site in increasing numbers and accumulate around the clustered macrophages; on meeting local Ag they proliferate and release various lymphokines, including chemotaxins that attract additional nonactivated macrophages and lymphocytes to the site. Because of their many effects on macrophages and lymphocytes, lymphokines and interleukins evidently bring about a cascade of reactions terminating in the mature tubercle with all of its structural and functional attributes, often including caseation necrosis. Since the key cell in acquired immunity is the sensitized Te cell, the continued maintenance of a high level of systemic CMI demands a persisting high level of these cells in the circulation, lymph nodes and submucosal sites, a state which, in turn, requires a continuing stimulation by Ag that is only supplied by the bacilli of chronic infection. Local CMI in the tubercle could exceed systemic CMI provided by sensitized circulating Te cells.

The controversy as to whether DH (allergy) to *M. tuberculosis,* as determined by the intracutaneous test with tuberculin, is harmful or beneficial in tuberculosis has raged for decades. The original champions of the view that DH is harmful rather than beneficial claimed that the abolition of DH to tuberculin (desensitization) does not compromise immunity but, instead, assists it. However, this claim and the many reports that immunity can be induced without the concomitant induction of cutaneous DH and vice versa does not necessarily prove that DH is either harmful or beneficial under any and all circumstances, because local lesion DH reactivity doubtlessly exceeds cutaneous DH reactivity and because the heat-denatured proteins in tuberculin could not conceivably contain all of the Ag epitopes released from the bacilli. Additionally it is possible that the Te cells responsible for CMI and DH belong to 2 respective subsets. Hence, although there is uncertainty as to whether DH may bolster CMI in the face of a low bacillary load in the tubercle, there is little doubt that in the face of a high bacillary load the DH reaction to locally released tuberculoprotein is the principal, if not the sole, cause of caseation necrosis which, in turn, predisposes to lesion progression. However, in most individuals, the lesions of a primary infection, including caseating lesions, readily heal unless softening occurs.

The factors concerned in bacillary overloading and its consequences deserve emphasis. Although in natural primary infections heavy exposure to bacilli predisposes to bacillary overloading, *the most evident factor of importance in resistance to overloading is the speed of the CMI response;* a rapid response deters overloading because it leads to the rapid accumulation of Te cells and early predominance of activated over nonactivated macrophages in the developing tubercle. This circumstance is often duplicated in tubercles resulting from endogenous superinfection, such as miliary tubercles that develop rapidly in the face of a high level of CMI; such lesions often heal even though the bacillary-laden "mother tubercle" continues to progress and expand peripherally. The frequent coexistence of healing and progressing lesions in the lungs of tuberculous patients is notorious. Indeed an experimental analogy to the healing tubercle of endogenous superinfection was developed a century ago when Robert Koch showed that, whereas a primary subcutaneous lesion induced in a guinea pig will progress slowly and allow a spread of infection that eventually kills the animal, a similar super-infecting dose of bacilli injected at a distant site 4 to 6 weeks after the first injection will incite a rapidly developing lesion that heals quickly without attending lymph node spread (Koch phenomenon).

Although, as stated earlier, the circumstances that lead to reinfection and reactivation tuberculosis remain an enigma, they are doubtlessly multifactorial, waning specific immunity, natural constitutional immunodeficiency, and artificial immunosupression being the most likely causes.

From the above discussion it must be envisioned that the *early developing tubercle* of a primary infection is largely populated by nonactivated macrophages and that they continue to serve as "bacillary incubators" until the surrounding mantle of sensitized Te cells thickens to the point where essentially all of the infiltrating monocytes become activated before they make contact with bacilli.

Thus far, the numbers and distribution of the various subsets of T cells and activated and nonactivated macrophages in developing tubercles has not been adequately studied.

J. LABORATORY DIAGNOSIS

The diagnosis of pulmonary infections due to *M. tuberculosis* and related organisms is based largely on clinical, x-ray and bacteriologic findings. The tuberculin skin test is helpful in certain cases.

Bacteriologic evidence is necessary to firmly establish a diagnosis. The procedures used include culture, smears for detection of acid-fast bacilli, and animal inoculation. The materials examined are either fresh unaltered specimens or concentrates of specimens prepared by various physical and chemical procedures designed to selectively kill or inhibit contaminating bacteria and to concentrate tubercle bacilli. Specimens include such materials as sputum, gastric washings, spinal fluid, joint fluid, feces, lung aspirates and biopsied tissue.

The finding of typical acid-fast rods, even with strong clinical evidence of tuberculosis, constitutes only presumptive evidence that the disease is tuberculosis and further confirmatory tests should be performed. For example, on occasion, the nonpathogen *M. smegmatis,* a normal inhibitant of the genitalia, has been mistaken for *M. tuberculosis.* In years past, numerous individuals with other diseases have been mistakenly sent to tuberculosis sanatoria because of faulty bacteriologic diagnosis based on smears alone!

Cultural tests are usually sufficient to identify *M. tuberculosis* with reasonable certainty and animal inoculation is only resorted to under special circumstances. Niacin is produced by cultured *M. tuberculosis* and *M. microti* but not by other mycobacteria, and thus serves as a useful diagnostic aid. In some instances, finding the organism by any method may be difficult; repeated and intense search is always recommended in suspected cases since these bacilli are often shed intermittently and in small numbers.

The frequent occurrence of drug-resistant variants possessing altered pathogenicity for animals and the fact that atypical mycobacteria can produce lesions indistinguishable from those produced by *M. tuberculosis* have introduced complex and difficult problems in the diagnosis and therapy of mycobacterial diseases.

The preferred intracutaneous tuberculin test (Mantoux test) with various tuberculins (protein derivatives of bacilli) is becoming increasingly useful in the diagnosis of the mycobacterioses. Tuberculin elicits a delayed skin reaction (tuberculin reaction) in sensitive animals and man. "Old Tuberculin" (OT), a heated culture filtrate of *M. tuberculosis,* was first prepared by Koch in 1880. The tuberculin preparation "Purified Protein Derivative" (PPD) which was first isolated and purified from OT by Florence Seibert is commonly used for skin testing; it contains derivatives of low molecular weight (less than 10,000), of various proteins of the tubercle bacillus. Even though the proteins of various mycobacteria are related and cross-hypersensitivity occurs among the mycobacterioses, skin tests, with proper concentrations of various tuberculins made from different species of mycobacteria, reveal marked differences in the size and intensity of reactions and can be of value in differential diagnosis. In infections due to *M. tuberculosis* (except for rare infections with *M. bovis* or "atypicals") a reaction of 10 mm or more in diameter to 5 tuberculin units (TUs) of standard PPD is generally indicative of infection with *M. tuberculosis.* Moreover, comparative skin tests with the various tuberculins are commonly confirmatory with respect to ruling out the possibility of infection due to atypical mycobacteria.

In developed countries the tuberculin test is of great value in the control of tuberculosis because of its diagnostic and case-finding usefulness. In interpreting the results of tuberculin testing, it must be kept clearly in mind that essentially all infected individuals give positive reactions regardless of whether their infection is clinically manifest or not. Moreover, whereas a negative reaction has a high exclusion value in diagnosis, a positive reaction only indicates the possibility of clinical tuberculosis. As the incidence of infection and clinical disease in a population decreases to low levels, the value of the tuberculin test as a screening device for case finding and as a diagnostic tool rises. Since the reaction to the test converts from negative to positive soon after infection, the test has special value among groups of individuals who may experience frequent and heavy exposure, such as laboratory workers, doctors and nurses. The test is also useful for detecting new cases among those who have had limited contacts with tuberculous patients, especially children. Today, in most communities in the USA, a positive reaction in a primary school child signifies that the source of infection is not far away and that it may involve other contacts as well. Cutaneous tuberculin sensitivity may vary in intensity and temporarily may decrease or disappear during the course of high fever, in exanthematous disease, miliary tuberculosis, sarcoidosis, the terminal stages of pulmonary tuberculosis and during steroid treatment. Negative reactions in such cases may be specific for tuberculosis or may represent *total anergy* without response to any of the common skin test Ags, such as mumps and histoplasmin. The intensity of the tuberculin reaction probably depends on the number of circulating T-lymphocytes sensitive to PPD. For reasons that are not always clear, an occasional individual may fail to react to tuberculin even after a natural infection or after BCG vaccination.

Although no reliable relationship exists between the level of tuberculin sensitivity and extent and severity of tuberculous infection, the intensity of the tuberculin reaction is not without significance. As a general rule, relatively high levels of sensitivity are found in persons with recently acquired infection, in those with caseous nonpulmonary tuberculosis (e.g., in lymph nodes and bones), and in those who are in continuous contact with open tuberculosis but who show no signs of active disease. Fluctuating levels of sensitivity to tuberculin are seen in patients with serous membrane tuberculosis, e.g., tuberculous pleuritis; these apparently reflect varying degrees of exudation and resorption of fluid. In persons with far advanced tuberculosis and acute forms of the disease, sensitivity is usually low or absent, and it appears to be a general truth that low levels of sensitivity in persons with active, progressive tuberculosis signal a poor prognosis.

The persistence of tuberculin sensitivity requires a certain low level of infection below which the tuberculin test becomes negative. Hence, with time the DH of a primary infection can sometimes fade to complete negativity even though a few live bacilli may remain. Such negativity does not mean that CMI fades equally or that the anamnestic response will not provide protection against reinfection or reactivation exceeding the protection that arises during a primary response. Although cutaneous DH and CMI appear to commonly coexist at similar levels, wide divergence sometimes occurs.

K. THERAPY

Older forms of therapy such as bedrest, lung collapse and surgical excision of infected tissue are limited in their effectiveness and have become, in large measure, unnecessary because of the marked success achieved with drug therapy. First-line drugs include streptomycin (SM), rifampin (RM), isoniazid (INH), ethambutol (EMB), thioacetazone and pyrazinamide. Second-line drugs include viomycin, cycloserine, ethionamide (ETA), para-aminosalicylic acid (PAS), kanamycin and capreomycin. The therapeutic regimens afforded by these numerous drugs are many and complex; short term and long term regimens are possible in which drug resistance is usually avoided. In accord with experience with other diseases, drug treatment of tuberculosis is not a perfect therapeutic measure; its limitations result principally because drug-resistant strains of organisms develop, because the drugs fail to reach organisms within necrotic tissue, and because maximal action of drugs is often blocked in these environments. Owing to the months to years of therapy involved and attending opportunity for multiplication of organisms, drug-resistant mutants commonly arise unless a second or third drug is administered simultaneously. Ethambutol has largely supplanted PAS as a companion of INH because it is more active than PAS and causes fewer adverse effects. One of the most effective combinations is INH and EMB; another is SM plus ETA. However, the use of SM has been limited because of its toxicity. A more recent and unusually effective agent is RM. A combination of INH and RM is among the least toxic and most highly effective of all. The drawbacks to the use of RM are that it has immunosuppressive activity and is hepatotoxic for some patients.

Although INH-resistant mutants retain their virulence for man and are transmitted from man to man, they have low virulence for the guinea pig, lack catalase and peroxidase activity, and grow more slowly than wild-type organisms. Laboratory tests for drug resistance are important guides to chemotherapy; the tests are performed with drugs added singly and in combination. The impact that drug-resistant mutants may have on the future pattern of tuberculosis in human populations is unpredictable.

Based on his belief that DH is protective, Koch introduced prolonged immunotherapy for selected patients using small spaced s.c. doses of tuberculin carefully gauged to give minimal systemic DH reactions. The systemic DH reaction reflects the reactivity of Te cells of the tubercle and is more sensitive than the cutaneous reaction, which only reflects the numbers of circulating sensitive Te cells. The systemic DH reaction, as evidenced by intensification of the symptoms of tuberculosis, i.e. fever, malaise etc., represents the cumulative effects of focal inflammatory reactions around tubercles, which may sometimes exhibit hemorrhage and necrosis. Ostensibly the symptoms of tuberculosis must reflect, in part at least, a low-grade systemic DH reaction. The systemic effects are due to liberated products of inflammation including endogenous pyrogens from leukocytes around tubercles. Severe reactions involve a generalized increase in vascular permeability, hemoconcentration, circulatory shock and often death.

Unfortunately, in their enthusiasm some early clinicians became so uncautious that they mistakenly gave excessive doses of tuberculin that were sometimes fatal. Although the use of Koch's therapy waxed and waned over the decades, a few physicians continued its use for certain selected cases of tuberculosis until the mid-century advent of effective drug therapy.

L. RESERVOIRS OF INFECTION

Man is the only important reservoir of *M. tuberculosis.* The disease is often contracted from tuberculous individuals who are unmindful or unaware of the nature of their disease, as well as from patients with recognized disease. On rare occasion the disease is contracted from experimental animals or from subhuman primates in close association with man (pets, animals in zoos, etc.). Bacteriophage typing has promise for epidemiologic studies.

M. CONTROL OF DISEASE TRANSMISSION

Present-day efforts to prevent tuberculosis in the USA are largely directed at case finding and regulation of "shedders." Tuberculin screening tests, x-ray of contacts and high morbidity groups, and search for sources of infection, especially in families and among school children, have been the most effective means of discovering new cases. *Periodic tuberculin surveys are important for determining the effectiveness of community control of tuberculosis.* Infection with tubercle bacilli and tuberculin sensitivity commonly persist throughout life. Consequently the percentage of reactors is cumulative and increases with age. Tuberculin testing carried out in sufficiently large test groups that are proportionately representative of the population with respect to age, sex, race and economic level *furnishes the best available measure of the prevalence of tuberculous infection and has predictive value. For example, it has been found that when the infection rate in the age group under 14 years drops below 1%, tuberculosis in the population is on the way to virtual extinction.*

Prevention of tuberculosis by BCG vaccination has definite but limited value. Whereas it has been widely used in various countries of Europe and other areas in the world, principally for vaccinating infants, its use in the USA has been primarily experimental and has been largely restricted to "high exposure" groups. For such groups, chemoprophylaxis promises to replace BCG vaccination completely. In developed countries the vaccine has the disadvantage of rendering the person tuberculin positive, thus destroying the value of the tuberculin test for mass screening, diagnosis and surveillance for "tuberculin converters" among contacts and in high-risk groups. The tuberculin test for surveillance studies is highly valuable and its use for this purpose should be encouraged. The value of BCG vaccination in underdeveloped countries remains controversial. Trials using BCG as a dual vaccine against tuberculosis and leprosy have shown no benefit in some geographic areas, possibly because of subject contact with cross-reacting saprophytic mycobacteria in the environment.

Prophylactic treatment with INH has been used with good success in high risk groups, including recent "tuberculin test converters," close "contacts" with "bacillary shedders," persons at high risk of reactivation tuberculosis, and high-mortality tuberculin-positive individuals such as infants, diabetics, silicotics and immunosuppressed individuals. In rare instances INH prophylaxis has caused hepatitis. Prophylaxis with the combination INH-RM or other combinations should prove to be more successful than INH alone, although more work with combinations is needed.

In view of the late development of highly effective therapeutic and prophylactic drug regimens, it is probable that tuberculosis could be almost totally eliminated from a country such as the USA were it not for the high incidence of active tuberculosis among new immigrants from underdeveloped countries, which in some instances may exceed 50 times the incidence of the disease in the USA.

MYCOBACTERIUM BOVIS

Mycobacterium bovis closely resembles *M. tuberculosis* and is highly pathogenic for man as well as for the cow, its natural host.

Human infections with *M. bovis* tend to involve lymph nodes and bone and are commonly contracted in childhood as the result of drinking milk from tuberculous cows. The principal means of shedding from cows is via the feces; on rare occasion, herdsmen acquire pulmonary infection by inhaling the dried feces and possibly pulmonary droplets of tuberculous cows. Ingested organisms are carried by macrophages either through the mucosa of the oropharynx to infect the cervical lymph (scrofula) or to the mucosa of the intestine to infect the mesenteric lymph nodes. Organisms disseminating from the lymph nodes via the bloodstream tend to localize selectively in joints and bones, particularly the vertebrae (Pott's disease). *The control of bovine tuberculosis stands among the greatest achievements in public health.* In the USA, tuberculosis due to *M. bovis* has been largely eliminated in both man and animals as the result of widespread pasteurization of milk and the legislated slaughter of tuberculin-positive animals. In man, tuberculosis due to *M. bovis* stands as a classic example of how subtle the properties of an organism must be that determine its capacity to maintain itself in a natural host. Why does *M. bovis,* which is essentially identical with *M. tuberculosis* in its morphologic and biochemical properties and which is highly virulent for man as an accidental host, fail to adopt man as a natural host when the infection becomes pulmonary? Obviously, and for unknown reasons, the organism is incapable of establishing a permanent chain of communicability from one human being to another. A parallel observation is that sheep are not natural hosts for *M. bovis* despite their high susceptibility to experimental infection.

MYCOBACTERIUM AVIUM

Mycobacterium avium, a natural pathogen for birds, sometimes infects mammals, particularly swine, as accidental hosts. Avium infections rarely, if ever, occur in man and most of the alleged cases probably occurred as the result of severe immunodeficiency or alternatively, represented infections due to the closely related organism, *M. intracellulare.* It is unlikely that *M. avium* and *M. intracellulare* are identical species since *M. avium* is distinct on the basis of DNA relatedness and serology, and is uniquely adapted to grow best at 44°C.

MYCOBACTERIA THAT ACT AS FACULTATIVE PATHOGENS

This group of diverse mycobacteria is comprised of numerous species, commonly designated by such unsuitable terms as *"atypical"* and *"anonymous"* (Table 27–2). Most of the group produce tuberculosis-like lesions. The present discussion will be limited to a few species because full coverage is more appropriate for texts devoted to clinical microbiology and medicine.

Severe infections produced by the group are few, although they are increasing in number and relative importance among mycobacterial infections because of the widening use of immunosuppressants and the declining incidence of tuberculosis.

In common with all mycobacteria, members of the group are nonsporulating, nonmotile, Gram-positive, aerobic, acid-fast, pleomorphic rods, which exhibit a surface network of peptidoglycolipid filaments. Slow-growing species predominate over fast-growing species, and most species are chromogenic, i.e., produce yellow to orange pigments; some grow at 30°C but not at 37°C. Little is known about the natural habitats and geographic distribution of these microbes; with few possible exceptions, they are saprophytes that can become facultative parasites for man adapted to exist as either nonpathogens or pathogens. Since knowledge on reservoirs of infection is meager, prevention of infection is difficult. Severe infections in man are most frequent in immunodeficient and debilitated individuals, especially elderly males. On occasion, certain species have been isolated from the respiratory tract of healthy human

Table 27–2. Some Properties of Representative "Atypical" Mycobacteria

Group	Representative Species	Growth		Pigment Produced	Sites of Human Disease			
		At 25°C	*At 37°C*		*Lung*	*Lymph Nodes*	*Bones & Joints*	*Disseminated*
I. (Photochromogens) Pigment only produced on exposure to light	*M. kansasii*	Slow*	Slow	Yellow	+	+	+	+
II. (Scotochromogens) Pigment production not light dependent	*M. scrofulaceum*	None	Slow	Yellow to orange	+	+	+	+
III. (Nonchromogens) Little or no pigment	*M. intracellulare* (Battey bacillus)	None	Slow	None or faint yellow	+	+	+	+
	M. xenopei	None	Slow§	Faint buff	+	–	–	–
IV.	*M. fortuitum*	Rapid	Rapid		+	–	–	–

*Rapid growers produce visible growth in a week or less. Slow growers produce visible growth only after 1 to 3 weeks or longer.
§Also grows slowly at 45°C.

subjects where they may possibly exist as temporary members of the normal flora. Perhaps special circumstances, including suitable exposure, genetic susceptibility and, more importantly, compromise of host defenses predispose to infection. In this context these mycobacteria appear to be similar to many of the opportunistic fungi.

Whether latent infection with species that are cross-reactive with *M. tuberculosis* serve to create a significant level of immunity to the latter or vice versa is an unresolved but distinct possibility.

Because of the frequency of the high drug resistance of this group a combined chemotherapeutic-surgical approach is favored for treatment. Lung resection is commonly resorted to in patients with infection due to *M. intracellulare* (Battey bacillus).

Person-to-person transmission of group species does not occur and the organisms are usually not virulent for experimental animals; isolates from patients are often difficult to identify. Division of the group into 4 subgroups, the Runyon subgroups I through IV, is currently of only limited value for identification; speciation is now possible and needed for case management, serology and thin-layer chromatography of lipids.

Mycobacterium intracellulare is a natural saprophyte that sometimes acts as a facultative pathogen to produce disease in diverse species of birds, mammals, and man, particularly in immunosuppressed persons such as AIDS victims. Human infections most frequently involve the skin, kidney, lymph nodes, bones or lung where cavitary disease results. The organism is highly resistant to chemotherapeutic agents and even prolonged treatment with a combination of as many as 6 agents is sometimes unsuccessful; relapses are frequent. Sensitivity to a species-specific tuberculin develops.

Mycobacterium fortuitum is a fast-growing species that often infects birds and poikilothermic animals but rarely infects man. Local lesions are occasionally seen among drug addicts and subjects who receive implants of foreign materials, such as artificial heart valves; spread to the lung sometimes occurs, especially in immunodeficient individuals. Combined chemotherapy is usually effective and surgery is seldom necessary.

Mycobacterium kansasii infections occur in the Midwest USA; lesions are most frequent in skin, cervical lymph nodes and the lung where they produce cavitary disease. The tuberculin test becomes positive and chemotherapy is usually effective.

Infections due to M. ulcerans are largely confined to tropical and subtropical climates. The organism is of singular interest because infection is limited to low-temperature areas of the body, principally the skin, where ulcerating granulomatous lesions develop. This unusual temperature dependency evidently results from the inability of the organism to grow at temperatures above 33°C, as can be demonstrated in culture. The lesions produced experimentally in mice occur selectively in cool areas, especially the feet, ears, and tail. The disease is highly resistant to chemotherapy, consequently wide excision and skin grafting is often necessary.

Infections with *M. marinum (M. balnei)* were first reported in swimmers and the disease was termed "swimming-pool disease." The organism is a saprophyte that is facultatively parasitic and pathogenic for man as well as for frogs, fishes and other animals. It sometimes causes lesions on the hands and arms of fish fanciers as the result of contact with organisms shed into aquaria by infected fish. The ulcerous granulomas produced resemble the lesions of sporotrichosis. The infection tends to extend through superficial lymphatics to draining lymph nodes, but tubercle-like granulomas are not formed. Primary lesions usually heal spontaneously and chronic lesions respond well to chemotherapy.

MYCOBACTERIUM LEPRAE

Leprosy (Hansen's disease) is a chronic debilitating disease that cripples, disfigures and blinds. However, it usually kills only after many years, principally as the result of infection with secondary agents. Leprosy has a long incubation time that may range from months to many years (average 3 to 5 years). Disfigurement of face and limbs is marked because nerves are destroyed leading to atrophy of muscle and bone; the regional anesthesia produced results in inadvertent injury due to lack of sensation to temperature and trauma. A similar natural disease in rats, rat leprosy, is caused by *M. lepraemurium.*

A. MEDICAL PERSPECTIVES

The origin of leprosy is shrouded in antiquity. Early evidence of the disease has been found in Asia dating back to 1400 B.C.

It is a singular observation that among early civilizations, especially the civilization of the early Christian era, fear of the disease was so intense that almost every means of exterminating its victims was used; this was done with the excuse that blame for the disease lay with the victim, who was being punished for some sin by a spiritual being. Early Christian beliefs fostered cruel treatment of lepers because suffering was generally accepted as divine punishment for sins both present and original. Isolation, torture and extermination of lepers were common practices for centuries. The Romans left lepers in the mountains without food or clothing. Lepraphobia became so intense in the 12th century that it was an easy matter to have one's creditor or enemy banished by simply initiating the gossip that "he has leprosy." Harsh treatment of lepers is by no means extinct today, and fear of the disease severely limits treatment and control. It is difficult to recruit doctors for research and patient care. In Nigeria, in 1967, only 3 physicians were available to treat half a million leprosy patients! *Only when leprosy is universally recognized simply as an unfortunate disease and not as a disgrace will injustices be terminated and treatment improved.*

The history of the spread of leprosy is of great interest. During the 12th and 13th century lepers became so numerous in Europe that strict isolation became necessary and some 19,000 "Lazar houses" (named after Lazarus of biblical times) were established for this purpose. Apparently leprosy was spread from the Middle East to Greece in the centuries before the Christian era, thence to Rome with the returning armies of Gnaeus Pompey in 62 B.C. and finally to most of Europe by the Romans. The second massive spread of leprosy in Europe occurred during the 6th and 7th centuries with the return of Christian Crusaders from the Middle East. The disease assumed epidemic proportions beginning with the 11th century; its incidence peaked in the 15th century and declined sharply to near zero by the mid-16th century when the few remaining Lazar houses, which were really houses of isolation, neglect and death, were replaced by leprosy hospitals. Leprosy spread with the slave trade from Africa to South America, where it remains today. The few foci established in the USA by immigrants are now under control.

The reasons for the sudden decline of leprosy in Europe in the 16th century remain an unsolved mystery. Does this decline in leprosy stand as an example of host-parasite evolution? Did loss of virulence of *M. leprae* or increase in immunity of the population play a part? To what extent did isolation practices, better living conditions or mortality due to other diseases play a part? One attractive hypothesis is that the waves of other highly lethal diseases, such as bubonic plague (which in the first epidemic of the 14th century killed one-fourth of the total population of Europe) and tuberculosis (which killed 1 of every 3 Europeans between 15 and 40 years of age), so completely decimated the highly susceptible leper population that the chain of communicability was not strong enough to maintain the disease. The possibilities for breaking the chain of communicability in leprosy are good since, in most populations, relatively few individuals (probably about 3 to 10%) appear to have high innate susceptibility to the disease.

Currently, leprosy is largely restricted to underdeveloped tropical and subtropical regions and is rare in developed countries. For example, there are only about 4,000 cases of leprosy in the USA, most of whom contracted the disease elsewhere. The number has increased in recent years due to increasing numbers of immigrants from countries where leprosy is prevalent.

In the world at large the disease is still of major importance. Although the number of persons with active disease has been estimated to range between 10 and 15 million, only a small fraction of these are under treatment. In some limited areas, 10% of the population are diseased.

Although it is often stated that *M. leprae* is an obligate intracellular parasite, it is by no means certain that the organism cannot grow extracellularly. Its usual intracellular position may be due to the readiness with which it is phagocytized.

Mycobacterium leprae is a typical acid-fast rod resembling *M. tuberculosis.* It occurs in enormous numbers in the lesions of patients with the lepromatous form of leprosy; *the total estimated mass of organisms in such patients exceeds that of any other microbial disease.* Certain strains of *M. leprae* grow more rapidly in the mouse footpad than others, but the meaning of this observation in terms of virulence for man is not evident.

Whether the causative agent of a natural leprosy-like disease in the 9-banded armadillo is identical to *M. leprae* and whether it is naturally transmissible to man is not known. However, the isolates from armadillos show 100% DNA homology with human isolates of *M. leprae* which strongly suggests common identity. One theory states that the armadillo got infected from contact with human lepers.

B. PHYSICAL AND CHEMICAL STRUCTURE

Hansen's discovery of *M. leprae* in the lesions of leprosy patients in 1878 provided the first reliable description of a bacterial agent as the cause of human disease. Studies on organisms isolated directly from tissues indicate that *M. leprae* is similar to *M. tuberculosis* in its chemical composition, being especially rich in lipids and waxes.

Mycobacterium leprae contains many Ags both surface and cytoplasmic, some of which cross-react with Ags of other myocbacteria. Certain polysaccharides stimulate high titers of Abs that cross-react with Ags of various other mycobacteria. Proteins elicit a DH reaction as evidenced by the 48-hour Fernandez skin reaction, which resembles a tuberculin reaction. Since many proteins of *M. leprae* and other medically important mycobacteria are cross-reactive, skin test preparations for determining DH may be only partially specific; the same is true of serologic tests conducted with protein preparations. One specific Ag for serologic tests in leprosy is glycolipid I; a specific cell wall-associated protein may also prove to be serologically useful.

C. GENETICS

Little is known about the genetics and virulence of different strains of *M. leprae.*

D. EXTRACELLULAR PRODUCTS

No biologically important extracellular products of *M. leprae* are known.

E. CULTURE

Mycobacterium leprae has never been cultured in vitro on nonliving media; this has greatly hampered leprosy research. Reports that *M. leprae* can be grown in cultured Schwann cells remain to be confirmed. Its doubling time in the mouse footpad is about 12 days. The apparent low temperature dependency of *M. leprae* accounts for its preferential growth in low-temperature areas of the body and its infectiousness for certain poikilothermic and hibernating animals.

F. RESISTANCE TO PHYSICAL AND CHEMICAL AGENTS

Mycobacterium leprae is probably similar to *M. tuberculosis* with respect to resistance to most physical and chemical agents; however, lack of culture methods has made it difficult to test this directly.

G. EXPERIMENTAL MODELS

Most animals are highly resistant to *M. leprae,* the infections produced being mild and self-limiting. To date, infection has been produced in mice, rats, hibernating ground squirrels, hedgehogs, certain monkeys, hamsters, and the 9-banded armadillo. Mice and armadillos are the most suitable experimental animals; the mouse can be infected in the footpad with as few as 1 to 10 bacilli. The lesions in Shepard's mouse model occur principally in low-temperature areas such as the feet and ears, a characteristic remarkably like *M. ulcerans* infection and leprosy in man. The armadillo model has been extremely useful for propagating large numbers of *M. leprae.* Armadillo grown organisms are the main source for preparing the skin test preparations lepromin (suspension of killed *M. leprae*) and leprolin (a water soluble extract of disrupted *M. leprae*).

H. INFECTIONS IN MAN

Leprosy is a chronic systemic disease that involves primarily low-temperature areas of the body including the nose, ears, extremities and low-temperature areas of the skin. The incubation period is long, averaging 3 to 5 years, but may be as long as 10 to 20 years.

Most cases of leprosy belong to one of two polar forms of the disease, lepromatous leprosy (LL) and tuberculoid type leprosy (TT); other cases comprise borderline forms of the disease that lie between the two polar forms. The borderline forms include, borderline tuberculoid (BT), borderline lepromatous (BL) and a form between the two, (BB). Although the reason for such diverse forms of the disease is an enigma, they are thought to be in large measure the outcome of variations in the immunologic response to the organism. *Essentially all of the organisms are found within macrophages and the Schwann cells of nerves;* a few may invade muscle cells, particularly the erector-pili muscles. *The organism has a unique predilection for nerves which aids in diagnosis.*

Lepromatous leprosy is presumed to reflect high genetic susceptibility to the organism. The lesions of this form of the disease consist of masses of macrophages loaded with enormous numbers of bacilli (multibacillary disease) and lipid-containing vacuoles; a continuous bacteremia exists which is often accompanied by phlebitis. It is notable that the macrophages in the lesions of LL do not mature to form organized granulomas containing epithelioid cells.

Tuberculoid leprosy presents mature granulomas composed of cell collections resembling the tubercles of tuberculosis but without caseation; bacilli are sparse (paucibacillary lesions) and difficult to find and lymphocytes are abundant. Whereas the BT and BL forms of disease may transform to their respective polar forms or to BB, the BB form may transform in either direction. The polar forms of disease seldom change.

When heat-killed test organisms (lepromin) isolated from biopsied tissue are injected into the skin a local granuloma forms after 3 to 4 weeks in the majority of normal children and adults; such individuals seldom, if ever, develop LL. Except for infants, 90% or more of normal individuals will respond to initial or repeated testing. This reaction, the "Mitsuda reaction," is presumed to represent an "allergic granuloma," resulting from the development of DH to the injected organisms.

Tuberculoid patients commonly develop a granulomatous Mitsuda reaction but, in addition, show an earlier 48-hour reaction to the injected organisms, "the Fernandez reaction," which is presumed to represent a delayed-type reaction due to sensitivity existing at the time of the test.

The Fernandez reaction can also be elicited by leprolin, a protein-containing extract of *M. leprae.*

Lepromatous patients never show Fernandez or Mitsuda reactions. The reason for this is not known, but presumably is in some way related to the high susceptibility of the lepromatous patient to *M. leprae.* The immunologic deficiencies in LL are clearcut and appear to be of two kinds: *first, mild to moderate generalized deficiencies in lymphocyte function,* demonstrable by such evidence as a partially depressed capacity of lymphocytes to support the mixed-lymphocyte reaction and to reject skin allografts; and *second, a complete depression of the specific immunologic capacity of Te cells* as demonstrated by failure in vitro to yield positive macrophage-migration inhibition reactions in the presence of leprolin. The nonspecific lymphocyte deficiencies in TT are of a much lesser degree than those of LL.

A notable feature of the histopathology of LL is that bacillus-laden macrophages accumulate in the paracortical areas of lymph nodes with attending depletion of lymphocytes in these areas. This may contribute to the nonspecific immunologic deficiency seen in LL.

I. MECHANISMS OF PATHOGENESIS

Essentially nothing is known abut the pathogenesis of leprosy. Bacilli are easily phagocytized by both neutrophils and macrophages but do not appear to be notably toxic for phagocytes, at least when numbers of bacilli are not excessive. The organisms seem to be remarkably resistant to growth inhibition and destruction by lepromatous macrophages in which they multiply to reach enormous numbers. Since the bacilli have not been clearly shown to be toxic, the ultimate destruction of lepromatous macrophages may result from the physical effects of the large accumulation of dividing bacilli and the lipids they produce. There is some evidence indicating that organisms may accumulate to "rupture" the phagosomes and exist as masses of bacilli free in the cytoplasm. The lipids that accumulate within phagosomes to produce "foamy macrophages" are largely of bacterial origin.

Nerve damage is probably secondary to cell infiltration and allergic inflammation. In fact, essentially all forms of tissue injury may be on an immunologic basis. Immunologic injury results from humoral as well as DH reactions. For example, the local lesions of "erythema nodosum leprosum" (ENL) that are sometimes seen in LL, especially in the face of drug treatment, result from circulating soluble immune complexes formed as a consequence of the sudden liberation of large amounts of Ag from dying bacilli. A generalized acute form of the disease is Lucio's phenomenon, characterized by widespread ulcers due to local deposits of immune complexes in dermal blood vessels. When BT disease shifts toward TT disease a flareup of lesions may occur (reversal reaction) which leads to severe nerve damage; reversal is sometimes triggered by chemotherapy. Nerve damage may be due to demyelinating enzymes liberated by activated macrophages.

J. MECHANISMS OF IMMUNITY

As in tuberculosis, acquired immunity is cellular rather than humoral. Even though some 80% of most human populations are highly resistant to *M. leprae,* it is probable that most, if not all, individuals become infected following heavy prolonged exposure, albeit few develop disease.

When normal adults experience a primary infection with *M. leprae,* the large majority show no signs of disease but express infection by mounting immune responses, as indicated by DH skin reactions to soluble protein Ags of *M. leprae* and positive tests for serum Ab using a specific glycolipid from *M. leprae.* Within the infected group a few develop lesions that heal spontaneously and a few develop chronic paucibacillary disease with persisting low levels of serum Abs and DH; a lesser number (3 to 6%) develop multibacillary disease with high levels of serum Abs, but no DH.

Leprologists have generally held the view that uninfected Mitsuda-positive adults seldom, if ever, develop multibacillary disease upon becoming infected and conversely, that essentially all Mitsuda-negative adults belong to a small genetically susceptible group who upon becoming infected almost invariably develop multibacillary disease. It has been alleged that the Mitsuda-negativity commonly displayed by infants is due to immunologic immaturity, especially as it may relate to the characteristic high T suppressor cell (Ts cell) activity of infancy.

The long-held view that the degree of susceptibility to leprosy is markedly influenced by genetic factors is currently under careful scrutiny using advanced genetic methodology. Although there is strong evidence that susceptibility to familial TT rests on a chromosome-6 immune-response gene closely linked to the HLA-D locus, no such evidence exists for susceptibility to LL.

An alternative view has recently been advanced, namely that in man specific acquired immunologic tolerance may account for cases of multibacillary leprosy, and that tolerance induction may occur as the result of an accidental early "feeding" of Ag (presumably bacilli) into the blood from some unspecified nonimmunogenic locus of primary infection. It is also possible that the entrance of bacilli into the blood could result from the bite of an insect vector or some similar event. This view is based on the observation that leprosy-resistant mice that would be Mitsuda-positive if tested, become Mitsuda-negative (tolerant) and develop leprosy following the injection of *M. leprae* intravenously but not subcutaneously or intradermally. These observations on tolerance induction with *M. leprae* are in perfect accord with the principle that DH is most effectively induced by injecting Ag intradermally where it first meets an abundance of Langerhans' dendritic cells (Langerhans' cells) and that tolerance is most effectively induced by administering Ag i.v. so as to bypass Langerhans' cells, (*Fundamentals of Immunology,* pp. 145, 146) and reach and preferentially stimulate the precursors of Ts cells in the spleen.

It is possible that some heretofore unsuspected perturbation of the immune system, either acquired or genetically determined, is the key to susceptibility to multibacillary leprosy. In any event, it is highly likely that the LL patient with total specific anergy was predestined to develop this type of disease from the moment of infection and moreover, that the macrophages of the lepromatous patient fail to cope with *M. leprae*, be it because of a primary failure of macrophage activation and/or for some other reason; indeed the fault leading to high susceptibility must perforce rest either with the failure of the precursors of T effector cells to become sensitized and, in turn, activate macrophages, or within the macrophages themselves. The observation that the average incubation period for LL is about twice as long as for TT suggests that, as in tuberculosis, highly susceptible individuals may be "slow immunologic responders."

The question as to whether potential T cell and Langerhans' cell abnormalities may be responsible for susceptibility to LL has prompted studies on the T cell subsets in blood and lesions, especially those that are concerned in CMI (*Fundamentals of Immunology,* pp. 127–131, 190–191). The results have varied widely and the exact patterns of various abnormalities remain to be determined.

Claims have been made that, whereas the predominant cell in TT lesions is the T helper cell, the predominant cell in LL lesions is the Ts cell; the T helper cell (OKT4), a source of interferon gamma (IFNγ), is sparse in LL lesions. Other findings that may relate to high susceptibility suggest that bacillary overloading enhances Ts cell activity, that certain mycobacterial components, particularly a unique phenolic glycolipid, may preferentially activate Ts cells, that interleukin 2 production is lacking and that both specific and nonspecific Ts cells may contribute to immunologic suppression. The significance of abnormalities in the distribution of T cell subsets in the blood and lesions of leprosy patients, as they may pertain to susceptibility to LL, can be questioned because such abnormalities are probably in large measure the result rather than the cause of the disease. It seems likely that they are principally a reflection of profound disturbances caused by the enormous bacillary load attained, namely disturbances such as the crowding out of lymphocytes by macrophages in the paracortical areas of lymph nodes and the high levels of hydrocortisone observed in LL. If it were somehow possible to study uninfected Mitsuda-negative adults and LL subjects during the initial phases of disease development, the results should be more meaningful than those of past studies on patients with fully developed LL.

Despite the fact that in vitro studies on the lymphocytes of patients with LL have the above shortcomings, they have shown that when their circulating lymphocytes are exposed to Ags of *M. leprae* they fall into two equal-sized groups with respect to proliferation and the production of lymphokines, including IFNγ, *nonresponders* and *low responders.* Freeing the test preparation of OKT8$^+$ T suppressor cells and monocytes did not reverse nonresponsiveness. These results, of course, do not deny that Ts cells and/or monocytes may oppose the differentiation and sensitization of lymphocytes in the course of LL and thus account for the anergy and lack of immunity in nonresponders. The responsiveness of low-responder lymphocytes could be enhanced by monocyte depletion or the addition of interleukin-2.

The macrophages of LL patients appear to remain nonactivated and to have little, if any, capacity to digest *M. leprae* since organisms killed by drug treatment persist in the lesions for years. It has been reported that whereas in Mitsuda-positive individuals injected killed bacilli are digested within 3 to 4 weeks, they persist for at least several months in Mitsuda-negative individuals. It has been proposed that the defect in LL may rest with the failure of macrophages to process and properly present Ag to T inducer cells, perhaps because macrophage membranes are injured by bacilli.

The frequent coexistence of leprosy and tuberculosis presents the possibility that cross-relationships exist with respect to serology, DH and CMI. To date there is no convincing evidence of substantial cross-immunity in humans despite the finding that BCG vaccination protects against *M. leprae* in the mouse model. The relatively high morbidity and mortality of tuberculosis in LL is probably due to nonspecific factors. There is likewise no convincing evidence of substantial cross-reactivity with respect to cutaneous DH. The lowered tuberculin reactivity reported in some LL patients may have been due to nonspecific effects of lepromatous disease in the skin test sites or elsewhere. The observation that the majority of healthy persons who are negative to both tuberculin and lepromin can be rendered lepromin positive with BCG vaccine does not necessarily signify cross-reactivity since the second lepromin test was a repeat test. Neither does the finding that an occasional LL patient is rendered lepromin positive by BCG vaccination indicate that cross-reactivity exists. Moreover, the reverse is not true, i.e., leprosy patients do not give false-positive reactions to tuberculin. Serologic cross-reactivity is more certain and about a dozen cross-reactive Ags have been found.

K. LABORATORY DIAGNOSIS

Laboratory diagnosis of leprosy is based on the finding of typical acid-fast rods in smears of scrapings from skin lesions and the nose, or sections made of affected skin or nerves. The presence of acid-fast bacilli in nasal scrapings is not a uniformly reliable indicator of leprosy, since other acid-fast organisms are sometimes present in nasal scrapings from normal persons. Bacilli can often be found in slit-skin smears before disease becomes clinically manifest. The organisms are usually sparse in patients with TT, but may be found in sectioned tissues after careful search. Typical nerve lesions, together with characteristic clinical findings, usually permit accurate diagnosis.

Two diagnostically important attributes unique to leprosy bacilli are: (1) they produce the phenolase (dopase) that converts dihydroxyphenylalanine to a pigmented compound and (2) their acid-fastness can be abolished by extraction with pyridine.

L. THERAPY

Chemotherapy of leprosy is effective for arresting and reversing the disease. However, treatment over many months to years is usually required and drug reactions and drug resistance are frequent obstructions to treatment. Fortunately chemotherapy often renders the patient noninfectious within 1 to 2 weeks. Combined chemotherapy, which minimizes the development of drug reactions, is recommended by the WHO for all types of leprosy; whereas treatment for 6 months with dapsone (DDS) and RM is advised for paucibacillary cases, multibacillary cases should be treated for a minimum of 2 years or until slit-skin smears are negative using DDS, RM and clofazimine or under some circumstances, ethionamide or prothionamide instead of clofazimine. Unfortunately, RM, the most effective drug, entails a treatment cost of some $500.00 per year as compared to a few dollars for the sulfones.

M. RESERVOIRS OF INFECTION

Patients with leprosy are the only known natural reservoirs responsible for human infections. The status of the 9-banded armadillo as a potential reservoir for man is uncertain.

N. CONTROL OF DISEASE TRANSMISSION

The nasal mucosae of patients with LL or borderline leprosy is the chief "portal of exit" of the bacilli and the usual route of infection is via the respiratory tract. Infection also occurs through skin abrasions and possibly the bites of insect vectors. The offspring of mothers with LL probably become infected in utero on occasion. The spread of infection in nearly all cases of leprosy is dependent on close "contact".

Methods for the control of disease transmission include proper sanitary practices between patients and their "contacts", case finding and drug treatment of infectious patients and their close contacts. Infectious patients should be isolated during early chemotherapy as long as they remain infectious. Other potential methods for control of transmission in areas of endemic leprosy include chemoprophylaxis of Mitsuda-negative individuals, mass chemoprophylaxis and mass vaccination; to date BCG vaccine trials have been of limited and questionable value.

Devising a suitable vaccine against leprosy has been difficult because culture and selection of attenuated variants of *M. leprae* is not possible and because protective Ags have not been identified. However, the observation that a com bined vaccine of live BCG and killed *M. leprae* can provide better protection in animal models than BCG alone has led to therapeutic trials in borderline and LL patients using the combined vaccine. The mixture was more effective than either agent alone; in some cases it was reported to awaken skin test reactivity to lepromin and to cause degradation of bacilli in skin. These promising results suggest that it may be possible to develop a suitably effective prophylactic vaccine composed of BCG and/or other mycobacteria mixed with killed *M. leprae* grown in armadillos. This would be a major step toward the worldwide control of leprosy and is a high-priority aim of the WHO Committee on the Immunology of Leprosy (IMMLEP).

Considering the low percentage of infectious individuals among leprosy patients, the high resistance of the majority of the uninfected and the marked effectiveness of chemotherapy for inducing noninfectiousness in the infected, it seems safe to predict that leprosy could ultimately be abolished from the world were it not for the problems of overpopulation, poverty, ignorance, superstition, and the difficulties of recruiting trained personnel. However, since the chain of infection is easily broken by drug usage, it is virtually certain that leprosy could be eradicated from most local closed populations.

REFERENCES

Beiguelman, B.: Editorials. Macrophages versus lymphocytes in leprosy. Int. J. Lepr. *50*:501, 1982.

Bloom, B.R.: Rationales for vaccines against leprosy. Int. J. Lepr. *51*:505, 1983.

Bloom,, B.R., Mehra, V., Grosskinsky, C., and Brosnan, C.: *Immunology of Leprosy—Progress in Immunology.* Yamamura, Y. and Tada, T., (eds.) Academic Press, New York, 1984.

Convit, J., et al.: Specificity of the 48-hour reaction to Mitsuda antigen. Use of a soluble antigen from human and armadillo lepromin. Bull. WHO *52*:187, 1975.

Duncan, M.E., et al.: A clinical and immunological study of four babies of mothers with lepromatous leprosy, two of whom developed leprosy in infancy. Int. J. Lepr. *51*:7, 1983.

Florey, H.: *General Pathology,* 4th Ed., Philadelphia, W.B. Saunders Co., 1970.

Fukunishi, Y., et al.: Ultrastructural features of macrophages of armadillos infected with actively multiplying *Mycobacterium leprae.* Int. J. Lepr. *52*:198, 1984.

Gillis, T.P., et al.: Immunochemical characterization of a protein associated with *Mycobacterium leprae* cell wall. Infect. Immun. *49*:371, 1985.

Good, R.C., Opportunistic pathogens in the genus *Mycobacterium.* Annu. Rev. Microbiol., *39*:347, 1985.

Gordon, J., and White, R.G.: Surface peptido-glycolipid filaments on *Mycobacterium leprae.* Clin. Exp. Immunol. *9*:539, 1971.

Humber, D.P.: Immunological aspects of Leprosy, Tuberculosis and Leishmaniasis. Proceedings of Meeting at Addis Ababa, Ethiopia. Excerpta Medica, Amsterdam, 1980.

Joklik, W.K., Willett, H.P., and Amos, D.B.: *Zinssers Microbiology,* 18th Ed. Norwalk, Connecticut, Appleton-Century-Crofts, 1984.

Kaplan, G., et al.: An analysis of T cell responsiveness in lepromatous leprosy. J. Exp. Med. *162*:917, 1985.

Leake, E.S., et al.: Phagosomal membranes of *Mycobacterium bovis.* BCG-immune alveolar macrophages are resistant to disruption by *Mycobacterium tuberculosis* H37Rv. Infect. Immun. *45*:443, 1984.

Lurie, M.B.: *Resistance to Tuberculosis: Experimental Studies in Native and Acquired Defensive Mechanisms.* Cambridge, Harvard University Press, 1964.

Myrvik, Q.N., Leake, E.S., and Wright, M.J.: Disruption of phagosomal membranes of normal alveolar macrophages by the H37Rv strain of *Mycobacterium tuberculosis.* Am. Rev. Respir. Dis. *129*:322, 1984.

Rea, T.H., et al.: Peripheral blood T lymphocyte subsets in leprosy. Int. J. Lepr. *52*:311, 1984.
Rich, A.R.: *The Pathogenesis of Tuberculosis,* 2nd Ed. Springfield, Charles C Thomas, 1951.
Shepard, C.C.: Immunity to leprosy and the Mitsuda reaction. Int. J. Lepr. *52*:74, 1984.
Sherris, J.C., et al.: *Medical Microbiology.* New York, Elsevier Publishing Co., 1984.
Stead, W.W., and Bates, J.H.: Evidence of a "silent" bacillemia in primary tuberculosis. Ann. Intern. Med. *74*:559, 1971.
Strober, A.: Natural suppressor (NS) cells, neonatal tolerance, and total lymphoid irradiation. Annu. Rev. Immunol. *2*:219, 1984.
Wallach, D., et al.: The cellular content of dermal leprous granulomas: An Immunohistological approach. Int. J. Lepr. *52*:318, 1984.
Walsh, G.P., et al.: Naturally acquired leprosy in the nine-banded armadillo: A decade of experience 1975–1985. J. Leukocyte Biol. *40*:645, 1986.

28

MYCOPLASMA AND UREAPLASMA

The family *Mycoplasmataceae* comprises two genera, *Mycoplasma* and the *Ureaplasma.* These microorganisms are unique in two respects, first because they lack classical cell walls and second because they are the smallest of the free-living organisms. In fact, a smaller cell volume could not conceivably accommodate all of the cytoplasmic inclusions necessary for independent growth; studies on DNA homology indicate that most species are not closely interrelated. Various species of the *Mycoplasmataceae* can be readily differentiated on the basis of metabolic differences and by specific antisera. Most species that occur worldwide include saprophytes and parasites for a wide range of plants, animals, and man, some of the latter being pathogenic. The *Mycoplasmataceae* that parasitize birds, animals and man adhere to mucosae to produce surface infections; they seldom invade the bloodstream.

Early studies on the *Mycoplasmataceae* were greatly delayed because they differ markedly from most bacteria and require special conditions for growth on cell-free media. Although a number of species of *Mycoplasma* and *Ureaplasma* parasitize man as part of the normal flora, only three species are known to be pathogenic for man, *M. pneumoniae, M. hominis* and *U. urealyticum.* Of these the latter two usually cause infection only as "opportunists." Although the diseases caused by these three organisms are rarely life threatening, they are sufficiently frequent and debilitating to be of public health concern.

MYCOPLASMA PNEUMONIAE

A. MEDICAL PERSPECTIVES

Near the end of the 19th century two French microbiologists, Nocard and Roux, discovered an unusual microbe as the causative agent of *pleuro-pneumonia* of cattle. Unlike other bacteria the organism was found to lack a cell wall and to pass filters that retain most bacteria. Over the years other "pleuro-pneumonia-like" organisms were isolated from a variety of sources, including a human case of primary atypical pneumonia. The causative agent, *M. pneumoniae,* was generally thought to be a virus and was called Eton's agent.

With respect to the future outlook for controlling infections, caused by the *Mycoplasmataceae,* it is reasonable to expect that continuing studies with monoclonal Abs and other modern immunologic techniques will soon lead to the development of better methods for early diagnosis and more effective vaccines. Such vaccines could be of value for preventing and/or ameliorating infections due to *M. pneumoniae,* especially among military personnel. It should also be possible to determine whether a person is susceptible or immune to mycoplasmal disease by measuring protective Abs in serum.

B. PHYSICAL AND CHEMICAL STRUCTURE

Because of the lack of a classical bacterial cell wall *M. pneumoniae* is extremely pleomorphic and stains poorly or not at all. The organism varies in diameter from about 0.2 to 0.3 μm and in shape from coccoid to stellate to branching filamentous, multinucleated forms. The cells are rich in lipids, most of which are steroids derived from the external environment and incorporated in the triple-layered cytoplasmic membrane. These membrane steroids are unique to all species of the *Mycoplasmataceae*; they serve to strengthen and stabilize the cell membrane and enable the cell to resist differences between external and internal osmotic pressures. It has been reported that the cell membrane of *M. pneumoniae* adheres to specific host cell receptors presented by cell membrane proteins. The cell genome comprises double-stranded, circular DNA (mol. wt. 5×10^8) contained within a single nuclear body. Ribosomes are present but mitochondria are lacking. In common with other species of the *Mycoplasmataceae, M. pneumoniae* has limited biosynthetic capabilities, including an inability to synthesize nucleic acid precursors.

Cell membrane Ags include species specific and strain specific protein Ags together with Ags that cross-react with other *Mycoplasmataceae,* distantly related microbes and even host Ags.

C. GENETICS

Antibiotic-resistant mutants have been observed among the mycoplasmas, but genetic mapping has not been successful because suitable genetic recombination systems for mycoplasmas have not been found. The organisms have not been observed to undergo conjugation, transduction, or transformation to any significant extent. However, this may reflect a lack of adequate experimental methods, rather than failure of the mycoplasmas to recombine genetically. The recent discovery of viruses that parasitize mycoplasmas opens the possibility that transfer of genetic material may occur by transduction of other viral-
associated transfer mechanisms.

The percent of the bases guanine and cytosine in the DNA of *Mycoplasma* species ranges from 23 to 41, whereas the guanine and cytosine content of the DNA in bacteria with cell walls ranges from 25 to 75%. *M.;pneumoniae* contains about 40% guanine and cytosine in its DNA.

D. EXTRACELLULAR PRODUCTS

The mycoplasmas are catalase-negative and some strains generate hydrogen peroxide, which produces beta hemolysis. Other strains produce alpha hemolysis. None of the human pathogens, including *M. pneumoniae,* is known to produce toxins; in contrast, some murine mycoplasmas elaborate neurotoxins and are lethal for mice.

E. CULTURE

Most mycoplasmas grow best aerobically but can adapt to grow as facultative anaerobes. For example, *M. pneumoniae* can adapt to grow anaerobically, with or without 10% CO_2, in rich cell-free medium supplemented with 30% serum or ascitic fluid. After 2 or 3 days at 37°C in broth cultures, mycoplasmal growth is usually not visible to the naked eye, but the organisms can be seen in stained, centrifuged sediments. On agar-solidified media in sealed Petri dishes, small colonies form after 2 to 6 days incubation at 37°C. The colonies are less than 0.5 mm in diameter and a lens must be used to see them. Typically, they have an inverted "fried egg" appearance, resulting from a dense, central embedded core surrounded by a less dense surface growth. Growth can be inhibited by adding specific antiserum to the culture medium.

F. RESISTANCE TO PHYSICAL AND CHEMICAL AGENTS

Mycoplasmas, including *M. pneumoniae,* resist concentrations of thallium acetate that inhibit most other bacteria. Incorporation of 1:10,000 thallium acetate will prevent the growth of most contaminating bacteria in a specimen such as sputum, but permits growth of most mycoplasmas.

All strains of mycoplasma are resistant to penicillin, cephalosporins and other antibiotics that act on bacterial cell walls. However, most strains are sensitive to tetracyclines, erythromycin, kanamycin and gold salts; they are rapidly destroyed by detergents.

G. EXPERIMENTAL MODELS

The widespread occurrence of mycoplasmas in the normal flora of animals and as animal pathogens present a variety of ready-made models for general studies on the mycoplasmas; the hamster is susceptible to *M. pneumoniae.*

H. INFECTIONS IN MAN

Mycoplasma pneumoniae most commonly produces pharyngitis, tracheobronchitis and pneumonia.

Mycoplasma pneumoniae also produces allergic reactions, including a bullbous eruption of mucous membrane (Stevens-Johnson syndrome) and erythema multiforme.

Pneumonia due to *M. pneumoniae* is transmitted by pulmonary aerosol droplets. The disease is rare before the age of 6 months, probably because of protective maternal Abs, and is most common between the ages of 5 and 15 years; it constitutes 20% of all pneumonias and has a long incubation period of some 2 to 4 weeks. Since pneumonia due to *M. pneumoniae* is less severe than most bacterial pneumonias, it has been called "walking pneumonia." The disease is characterized by gradual onset, headache, malaise, a hacking relatively nonproductive cough and a pulmonary infiltrate composed of neutrophils, macrophages, lymphocytes, and plasma cells. The infection continues for about 10 days to 2 weeks and presents a late elevation of the white blood cell count. Cold agglutinins usually appear at 2 to 3 weeks. Complications and fatalities are rare and recovery is usually uneventful. Convalescent carriage may extend for months to a year or more. Rare complications include nonpurulent otitis media with bullous myringitis, meningoencephalitis, myocarditis, pericarditis, polyradiculitis and pancreatitis.

Milder respiratory infections characterized by sore throat and pharyngitis with minimal pulmonary involvement also occur, particularly in the 1 to 5 year age group. These patients are usually treated symptomatically.

I. MECHANISMS OF PATHOGENESIS

The pathogenesis of disease produced by *M. pneumoniae* is poorly understood. The organism has not been shown to produce a toxin or to directly invade host cells. However, in common with other pathogenic mycoplasmas, *M. pneumoniae* attaches to epithelial cells of the respiratory tract where it grows and reproduces. Attachment evidently depends on both nonspecific and specific factors, the latter being specific protein receptors on the host cell membrane. There appears to be a tendency for end-attachment of filamentous forms and a continuation of their flexing and gliding movements after attachment. Attachment interferes with ciliary action of the engaged host cells and causes them to die and desquamate with resultant inflammation. It has been alleged that host cell damage, as first reflected by alterations in nucleic acid metabolism, may result from local accumulation of metabolites such as H_2O_2. The potential role of autoimmunity in association with cross-reactive anti-tissue Abs has been emphasized by some investigators.

J. MECHANISMS OF IMMUNITY

Immunity to *M. pneumoniae* is not fully understood; nevertheless it is substantial and since most adults have been exposed, many effectively resist infection. Uncertainty exists concerning the duration of acquired immunity after infection and the extent to which later exposure to the organism may boost such immunity. In any event solid immunity is not permanent and reinfection sometimes occurs.

Nonspecific Abs generated by glycolipids of the outer membrane are probably nonprotective; they include cross-reactive IgM cold hemagglutinins, reagins of syphilis, and Abs that react with streptococcus MG.

The specific Abs that are evidently responsible for acquired immunity comprise secretory IgA Abs of temporary nature and mycoplasmacidal IgG Abs that persist for many months after infection. The major role that secretory IgA Abs may play in protection by blocking the adherence of organisms to host cells deserves greater attention.

K. LABORATORY DIAGNOSIS

With respect to diagnosis and treatment it should be kept in mind that several nonpathogenic species of mycoplasma are among the normal flora of the mouth and moreover, that mycoplasmal pneumonia resembles several other pneumonias, including legionellosis, Q fever, psittacosis, and the pneumonias due to adenovirus, syncytial virus, and influenza virus.

Diagnosis of *M. pneumoniae* infections depends on the nature of the infection. For lack of other suitable methods *early diagnosis* of pneumonia must rest on clinical symptoms. A method for identifying the organism in sputum smears using the fluorescent Ab technique has not been perfected to date; however, this should be possible with a specific monoclonal Ab. *Late diagnosis* is based on positive culture of the organism from clinical specimens, together with exclusion of other possible etiologic agents. Pulmonary secretions are cultured on special enriched selective media containing the antimicrobial agents, thallium acetate and penicillin; detectable colonies do not arise before 2 to 3 weeks of incubation. Since *M. pneumonia* is hemolytic, final overlaying of the incubated plates with RBC-containing agar results in zones of hemolysis around morphologically typical partially submerged colonies shaped like an inverted fried egg.

Specific Complement (C)-fixing Abs can be detected late in the disease using *M. pneumoniae* Ag. Specific Abs that suppress in vitro growth of the organism can also be detected late in the disease. Several other tests for specific Abs include C-mediated mycoplasmacidal and radioimmunoprecipitation procedures. Tests for specific C-fixing Abs are commonly used as an aid to diagnosis; a 4-fold rise in titer during disease development indicates that the disease is due to *M. pneumoniae.* Although cold hemagglutinins are nonspecific and are not always present in pneumonia cases, the test is simple and can sometimes aid in diagnosis.

L. THERAPY

Early administration of tetracycline is the usual therapy; in pneumonia, however, some clinicians prefer erythromycin because it is also effective against *Legionella pneumophila.* If the infecting organism is resistant to tetracycline, erythromycin is the second drug of choice.

M. RESERVOIRS OF INFECTION

The sole reservoir of infection with *M. pneumoniae* is the infected human being, often suffering from minor illness with cough. The organism persists in the respiratory tract for at least a month after infection, even when titers of humoral Abs are present. The duration of the carrier state has not been adequately studied but in some instances carriage has been shown to persist for several months to a year or more.

N. CONTROL OF DISEASE TRANSMISSION

Controlling the transmission of mycoplasmal pneumonia is difficult. The disease is endemic in world populations, particularly in temperate climates, but can be epidemic especially in crowded environments. Widespread epidemics occur at intervals of about 4 to 6 years; they comprise small outbreaks that occur randomly in families, schools and barracks; the majority of individuals within each unit are usually affected. Judging from serologic studies only a small percentage of infected individuals develop pneumonia; some exhibit mild symptoms and others have no symptoms. Since a degree of immunity is attained even following mild infections, immunization with a vaccine should be possible. However, if an effective vaccine were to be developed, it is unlikely that it would be widely used because of the low case rate as compared to the infection rate and because death rates are low even among the most severely affected. The greatest benefit of a vaccine would be to protect military personnel against epidemic disease.

Since avoiding contact with infected individuals and potential carriers is relatively ineffective, the only remaining control measure is prophylactic chemotherapy with tetracycline which can greatly reduce the severity of disease but not its incidence.

MYCOPLASMA HOMINIS

Mycoplasma hominis has many properties in common with *M. pneumoniae.* Together with *U. urealyticum, M. hominis* is a frequent inhabitant of the genitourinary tract, especially in the female; the tract commonly becomes seeded with *M. hominis* and *U. urealyticum* during birth and again in later years at the onset of sexual activity. The organism is also found together with several other pathogens in certain obstetric and gynecologic infections, such as the post abortion and postpartum fevers. *Mycoplasma hominis* has been isolated from blood in a small percentage of such cases. Although *M. hominis* probably participates in the pathogenesis of the "mixed infections," its contributions to them remain to be determined. However, it is generally accepted that *M. hominis* is fully responsible for certain infections in immunodeficient persons, such as infection in individuals with agammaglobulinemia and occasional cases of meningitis and pneumonia in neonates. These rare infections in neonates raise the question as to whether they reflect failure of the mother to provide protective maternal Ab. In cases of meningitis it is found that *M. hominis* is the only organism that can be detected and that therapy with tetracycline or lincomycin rapidly clears the organisms from the spinal fluid. The fact that mycoplasmas cause arthritis in several species of animals led to the proposal that rheumatoid arthritis may be due to mycoplasmas including *M. hominis.* However, the support for this proposal, namely, purported isolation of the etiologic agent in tissue culture was flawed because mycoplasmas were long overlooked as frequent contaminants of tissue culture cell lines. Claims have been made that *M. hominis* may be responsible for some cases of arthritis in patients with hypogammaglobulinemia and in postpartum mothers. *The membrane Ags* of *M. hominis* are principally proteins, several of which are strain specific. Some of the specific Abs produced are mycoplasmacidal, but it is not certain whether they can act in other protective ways i.e., as opsonins or to block adherence to host cells. There is some evidence that the level of bactericidal serum Ab correlates with a decreased risk of postpartum fever. A reasonable assumption is that the local secretory IgA Abs play an important role by blocking adherence of organisms to host cells.

Diagnosis depends on culture of specimens and identification of the organism by cultural and biochemical utilization tests, together with the exclusion of other possible agents.

Therapy is commonly carried out with tetracycline, the drug of choice. Clindamycin promises to be an excellent alternative drug. Unlike most other mycoplasmas, *M. hominis* is resistant to erythromycin.

UREAPLASMA UREALYTICUM

Ureaplasma urealyticum, a frequent inhabitant of the urogenital tract is unique among *Mycoplasmataceae* because of its ability to attack urea with the production of ammonia, its relative resistance to thallium and the extremely small partially submerged colonies it produces. Some 14 serotypes are known to exist. Despite some controversy it is generally held that *U. urealyticum* is the sole cause of about half of the cases of urethritis (so-called nongonococcal urethritis) the others being due largely to *Neisseria gonorrhoeae* and *Chlamydia trachomatis.*

Diagnosis of "pure" *U. urealyticum* infections rests with positive culture of specimens to the exclusion of other potential contributors. Since many urogenital tract infections are probably mixed infections, evaluating the possible contributions of the various organisms involved becomes exceedingly difficult. Although the potential role of *U. urealyticum* and *M. hominis* in chorioamnionitis and spontaneous abortion is controversial, a few cases have been reported. The amniotic fluids removed by transabdominal amniocentesis early in pregnancy were discolored and yielded pure cultures of one of these organisms.

Chemotherapy is best carried out with tetracycline, since it is also effective against *Chlamydia.*

REFERENCES

Archer, D.B.: The structure and functions of the *Mycoplasma* membrane. Int. Rev. Cytol. *69*:1, 1981.

Holmes, K.K.: *Mycoplasma hominis*—A Human Pathogen. Sex. Trans. Dis. *11*:159, 1984.

Hu, P.C., et al.: *Mycoplasma pneumoniae* infection: Role of a surface protein in the attachment organelle. Science *216*:313, 1982.

International Symposium on *Mycoplasma hominis*—A human pathogen. Sex. Trans. Dis. *10*:225, 1983.

Kotani, H., and McGarrity, G.J.: Ureoplasm infection of cell cultures. Infect. Immun. *52*:437, 1986.

Woese, C.R., et al.: Phylogenetic analysis of the mycoplasmas. Proc. Natl. Acad. Sci. USA *77*:494, 1980.

29

LISTERIA

Listeria monocytogenes, a facultative parasite and pathogen of the genus *Listeria,* previously placed in the family *Corynebacteriaceae,* but now assigned to the family *Lactobacillaceae,* infects many species of mammals and birds as well as crustaceans and ticks. It is the only member of the genus that is pathogenic for man. The organism is a widely distributed saprophyte in nature and has been isolated from streams, sewage, mud, silage, and vegetation as well as from the animals that it infects.

A. MEDICAL PERSPECTIVES

Listeria monocytogenes was first isolated and characterized in 1926 by Murray, Webb and Swann during an epizootic among guinea pigs and rabbits. The propensity of this organism to induce a monocytosis led to its species name. The first case of human infection with *L. monocytogenes* was described and identified by Nyfeldt in 1929. Subsequently, this microbe has been shown to cause disease in such diverse animals as goats, lemmings, foxes, horses, cows, domestic fowl and wild birds. In animals, genital tract infection of the pregnant female with subsequent perinatal infection of the offspring is a characteristic syndrome associated with listeriosis; this syndrome also occurs in human beings.

In man, listeria acts as an opportunistic pathogen; meningitis is the most common form of human listeriosis in the USA. Other forms of human disease include febrile pharyngitis with cervical and generalized lymphadenopathy. As in many opportunistic infections in man, listeriosis is being recognized with increasing frequency. The case rate and mortality rate are highest among infants under 4 weeks of age (attack rate 0.9/100,000) and in adults over 40 years of age (attack rate 0.1/100,000). Because of the relatively high incidence of animal infections, the animal reservoir is extensive and control of human infection rests on methods that prevent transmission of the disease from animals to man. Better means for diagnosing human listeriosis are needed in order to advance our understanding of its epidemiology and to institute more effective measures for its control.

B. PHYSICAL AND CHEMICAL STRUCTURE

Listeria monocytogenes is a pleomorphic, nonsporing, Gram-positive rod, ranging in size from 0.2 to 0.4 μm × 0.5 to 2 μm. It possesses peritrichous flagella and exhibits a characteristic tumbling motility when grown in broth culture at room temperature. The organism possesses a heat-labile flagellar (H) Ag and a heat-stable somatic (O) Ag. Four main serologic groups with one to several serotypes (serovars) in each of the groups are recognized (Table 29–1). Serotype 4b is the predominant strain in United States and Canada. Serotypes 1/2a, 1/2b and 4b cause more than 90% of all clinical *Listeria* infections worldwide.

A chloroform-soluble lipid that can be extracted from *L. monocytogenes* has been termed "monocytosis-producing agent"; when injected into mice this lipid induces a marked monocytosis.

C. GENETICS

Nonmotile mutants readily develop and may mutate back to motile forms. As a rule the organism grows as a short rod in smooth colonies and as a long

Table 29–1. Serovars of *Listeria monocytogenes*, *Listeria innocua* and *Murraya grayi*

Designation		*O-Antigens*															*H-Antigens*				
Paterson	*Seelinger-Donker-Voet*																				
1	1/2 a	I	II	(III)													A	B			
	1/2 b	I	II	(III)													A	B	C		
2	1/2 c	I	II	(III)														B		D	
3	3 a		II	(III)	IV												A	B			
	3 b		II	(III)	IV												A	B	C		
	3 c		II	(III)	IV													B		D	
4	4 a			(III)		(V)		VII		IX							A	B	C		
	4 ab			(III)		V	VI	VII		IX							A	B	C		
	4 b			(III)		V	VI										A	B	C		
	4 c			(III)		V		VII									A	B	C		
	4 d			(III)		(V)	VI		VIII								A	B	C		
	4 e			(III)		V	VI		(VIII)	(IX)							A	B	C		
	5			(III)		(V)	VI		VIII		X						A	B	C		
	7?			(III)									XII	XIII			A	B	C		
*Listeria innocua**	6a (4f)			(III)		V	VI	VII		IX						XV	A	B	C		
	6b (4g)			(III)		V	VI	VII		IX	X	XI					A	B	C		
Murraya grayi (ssp. grayi)				(III)									XII		XIV						E
(ssp. murrayi)				(III)									XII		XIV						E

*nonhemolytic strains
From: *Medical Microbiology and Infectious Diseases,* Braude, A.I. (ed), Philadelphia, W.B. Saunders Co., 1981.

filamentous forms in rough colonies; it exhibits marked pleomorphism in transitional strains. Aside from variation in antigenic types and colonial forms, little is known about genetic variability of the organism. After several subcultures it usually becomes avirulent for mice. Passage through mice can select for virulent mutants; virulence of laboratory strains is usually maintained by this method.

D. EXTRACELLULAR PRODUCTS

The virulent strains produce an oxygen-labile hemolysin that has cardiotoxic activity. A toxic protein is produced that is lethal for mouse macrophages, presumably because of its lipolytic action.

E. CULTURE

Listeria monocytogenes is facultatively anaerobic. It grows on blood agar media over a wide range of temperature (4° to 37°C); colonies may exhibit a narrow zone of β-hemolysis. *Listeria monocytogenes* is difficult to distinguish from diphtheroids on morphologic grounds; however, a test for motility is helpful because diphtheroids are never motile. At 20°C the organisms exhibit swarming characteristics in soft agar. When grown at room temperature, they are peritrichously flagellated but usually possess a single polar flagellum when grown at 37°C. Avirulent *Listeria* isolates have been classified as *L. innocua.* A new genus *Murraya* has been proposed for closely related non-pathogens. Mannitol is fermented by *Murrayi grayi* which distinguishes it from *L. innocua* and *L. monocytogenes* (Table 29–2).

F. RESISTANCE TO PHYSICAL AND CHEMICAL AGENTS

Although *L. monocytogenes* can be considered to be "average" in terms of susceptibility to routine sterilization and disinfectants, it has remarkable survival properties in soil and organic materials at ambient temperatures.

G. EXPERIMENTAL MODELS

Mice, rabbits and guinea pigs are susceptible to listeriosis. About 10^4 organisms cause a lethal infection in a mouse. The organism has been used extensively in studies on the mechanisms of cellular immunity, employing the mouse model.

H. INFECTIONS IN MAN

Culturally proven listeriosis is preponderantly a disease of immunologically immature or immunocompromised individuals. The most common infection

Table 29–2. Properties of *Listeria* Species and Closely Related Organisms

	Beta Hemolysis	*Mannitol Fermentation*	*Pathogenicity*
L. monocytogenes	+	–	+
*L. innocua**	–	–	–
Murrayi grayi†	–	+	–

*New species to designate nonhemolytic, nonpathogenic strains of *L. monocytogenes.*
†Proposed new genus and species to designate nonpathogenic *Listeria* species, *L. grayi* and *L. murrayi.*

in adults is a leptomeningitis that may or may not be associated with bacteremia. *Listeria monocytogenes* is a leading cause of meningitis in cancer patients and renal transplant recipients which indicates the opportunistic nature of listeriosis. Normal adults may develop the disease upon sufficient exposure; for example, cutaneous listeriosis has been reported in veterinarians after carrying out bovine obstetrical procedures. A *marked monocytosis is characteristic of listeriosis in man as well as animals.*

Infections of the newborn range from meningitis that becomes apparent within 4 weeks postpartum to disseminated disease in aborted, premature and stillborn infants; some infected neonates die within a few days after birth. Infants born with disseminated listeriosis are usually critically ill with cardiorespiratory distress, vomiting and diarrhea; hepatosplenomegaly may be evident. Disseminated listeriosis, which has been given names such as "granulomatosis infantiseptica," "miliary granulomatosis" and "pseudotuberculosis," was described by Henle in 1893. Necropsy of infants with this form of listeriosis reveals widely disseminated abscesses and granulomas of varying sizes involving the liver, spleen, adrenal glands, lungs, pharynx, CNS, GI tract and skin. Whereas the meconium of normal newborn infants does not contain bacteria, the meconium of a neonate with listeriosis contains numerous organisms; hence a Gram-stained smear and culture of the meconium should be performed whenever there is gross contamination of the amniotic fluid with meconium, or particularly if a mother has an unexplained fever at the onset of labor.

Listeriosis in a pregnant woman may be asymptomatic or symptomatic; symptomatic infections are characterized by malaise, chills, mild diarrhea, back pain and itching of skin. Infections that are asymptomatic prior to delivery can become symptomatic a week to a month after delivery. However, the disease in the mother is usually benign and self-limiting.

I. MECHANISMS OF PATHOGENESIS

Infection with *L. monocytogenes* is characterized by monocytosis (~30% of total white cell count). This is caused by a lipid from the organism that seems to correlate with virulence. The oxygen-labile hemolysin as well as the lipolytic moiety are thought to contribute to pathogenesis.

Listeria monocytogenes also contains an antiphagocytic surface component which is endotoxin-like. The listerial endotoxin is thought to be responsible for the transient cold hemagglutinin syndrome observed in some patients. The Abs produced in this case react with the host's own RBC which results in C-mediated lysis.

The basic disease process characteristically involves the RES, and is marked by septicemia leading to granulomas, necrosis and suppuration. It is probable that DH contributes to the tissue damage observed. The fact that most infections are inapparent offers further evidence that *L. monocytogenes* is primarily an opportunistic pathogen.

J. MECHANISMS OF IMMUNITY

Extensive experimental data indicate that *cell-mediated immunity is the main mechanism of acquired immunity in listeriosis.* Some nonspecific agents such as lysozyme and β-lysin probably also contribute to resistance; however, there is abundant evidence from animal experiments that functioning thymus-derived lymphocytes and activated macrophages are the sine qua non of resistance to *L. monocytogenes.*

K. LABORATORY DIAGNOSIS

A definitive diagnosis depends on isolation and identification of *L. monocytogenes;* no useful serologic tests are available. Listeriosis should always be considered when cultures from patients with meningitis, conjunctivitis, endocarditis, bacteremia or polyserositis lead to such clinical laboratory reports, such as "diphtheroids" or "nonpathogens." The best chances of success in isolating *L. monocytogenes* can be achieved by holding specimens in glucose broth at 4°C and subculturing at weekly intervals.

L. THERAPY

The overall mortality rate in untreated persons with listeriosis is 70%. Useful chemotherapeutic drugs include sulfonamides, penicillin, tetracyclines and erythromycin; however, ampicillin and erythromycin are the drugs of choice. In meningitis, intravenous infusions of penicillin G plus oral erythromycin are commonly given; their use is generally continued for 5 to 7 days after clinical signs have abated. Listeriosis in the pregnant female is usually treated by administration of tetracycline and erythromycin. An equally successful therapeutic regimen for fetal listeriosis has not been devised.

M. RESERVOIRS OF INFECTION

The organisms are widespread in nature and cause disease in many species of animals. Contact with fluids of infected animals, ingestion of contaminated milk and inhalation of contaminated dust probably represent the major sources of human infections. In 1971, the United States Department of Agriculture reported 231 cases of bovine listeriosis in 154 herds and 125 cases in sheep and goats from 48 herds. It is a puzzling observation that cases of disease in animals occur most frequently in the spring, the period when human listeriosis is infrequent; the numbers of human cases peak between June and September.

N. CONTROL OF DISEASE TRANSMISSION

Prevention of human listeriosis requires pasteurization of milk and elimination of diseased animals by slaughter. Effective control is hampered by difficulties in recognizing the infection, both in animals and in man.

Efforts should be made to prevent nosocomial spread of the organisms in neonatal units and among patients receiving immunosuppressive drugs.

REFERENCES

Canfield, M.A., et al.: An epidemic of perinatal listerosis serotype 1/b in hispanics in a Houston hospital. Pediatr. Infect. Dis. *4*:106, 1985.

Fleming, D.W., et al.: Pasteurized Milk as a Vehicle of Infection in an Outbreak of Listeriosis. N. Engl. J. Med. *312*:404, 1985.

Gervais, F. et al.: Analysis of macrophage bactericidal function in genetically resistant and susceptible mice using the temperature-sensitive mutant of *Listeria monocytogenes.* Infect. Immun. *54*:315, 1986.

Hof, H.: Virulence of different strains of *Listeria monocytogenes* Serovar 1/2a. Med. Microbiol. Immunol. (Berl). *173*:207, 1984.

Listeriosis Outbreak Associated with Mexican-Style Cheese-California, MMWR, *21*:357, 1985.

Nieman, R.E., and Lorber, B.: Listeriosis in adults: A changing pattern. Report of eight cases and review of the literature, 1968–1978. Rev. Infect. Dis. *2*:207, 1980.

Paterson, J.S.: Antigenic structure of organisms of genus *Listerella.* J. Path. Bact. *51*:427, 1940.

Pollock, S.S., et al.: Infection of the central nervous system by *listeria monocytogenes:* A review of 54 adult and juvenile cases. Q.J. Med. *53*:311, 1984.

Relier, J.P.: Perinatal and neonatal infections: Listeriosis. J. Antimicrob. Chemother., *5*:51, 1979.

Samra, Y., et al.: Adult listeriosis—A review of 18 cases. Postgrad. Med. J. *60*:267, 1984.

30

TREPONEMA, BORRELIA, AND LEPTOSPIRA

Members of the family *Spirochaetaceae,* order Spirochaetales, are classified under three genera: *Treponema, Borrelia* and *Leptospira.* The spirochaete that causes the newly discovered disease, Lyme disease, remains to be classified. The *Spirochaetaceae* are helical-shaped and have the unique attribute of being highly flexible. They are characterized by sinuous flexing and rotary movements brought about by an axial filament that consists of a bundle of intertwined flagella-like contractile fibrils coiled in spiral fashion around the cell. Many species are parasitic for a wide range of animal hosts; some are pathogenic.

TREPONEMA

The genus *Treponema* contains a number of pathogenic and nonpathogenic species; most if not all are microaerophiles or anaerobes. Many of the nonpathogenic species belong to the normal flora of the alimentary tract and only appear to be able to exert pathogenetic effects by participating as opportunists in mixed infections. *None of the pathogenic treponemes has been cultivated in vitro and the only known distinguishing characteristics of these organisms are the range of natural hosts that they attack, the experimental animals that they can infect and the nature of the diseases that they produce. The prototype species of the human pathogens of this genus is* Treponema pallidum, the cause of human syphilis. Man is its only known natural host. Other human pathogens are *T. pertenue,* the cause of *yaws, T. carateum,* the cause of *pinta,* and a subspecies variant designated, *T. pallidum epidemicum.*

On a worldwide basis the treponematoses continue to be among the major afflictions of mankind despite the availability of highly effective therapeutic agents including penicillin, the "queen of antitreponemal drugs."

TREPONEMA PALLIDUM

A. MEDICAL PERSPECTIVES

Syphilis (lues) is a serious, widespread, and often fatal disease encompassing all levels of society.* It is spawned by sexual promiscuity and thrives on ignorance and the social stigma that it carries.

The origin of malignant syphilis is highly debated. *One theory* is that it was first introduced into Europe in 1493 by Columbus' sailors, who were alleged to have contracted the disease in the West Indies. Malignant syphilis was mentioned in an edict of the Diet of Worms issued in 1495 and treatment with mercury was introduced in 1497. Dr. Ruy Diaz de Isla treated several of Columbus' sailors, including the pilot, Pinzon of Palos, and later (1539) wrote a book about the "new disease," which stands as the first clear description of malignant syphilis. *An alternative theory* on origin is that the malignant form of the disease, first reported in 1494, resulted from a large-step mutation of some progenitor organism, then existent in the Old World, comparable to one of the present-day treponemes that cause the nonvenereal treponematoses of milder nature. In any event, some of the sailors of Columbus participated in

*The name "syphilis" is derived from the principal character of the poem "syphilis sive Morbus Gallicus," written by Fracastor in 1530, about an imaginary swineherd with the disease. Syphilis is sometimes called "lues," meaning plague or pestilence.

the seige of Naples in 1495 where the first recorded epidemic of malignant syphilis ("the great pox") appeared and spread over Europe with the disbanding of troops. *The high malignancy of the disease at that time and the sweeping nature of the epidemic are expressions characteristic of a new infectious agent in a host population.* Many deaths (about 25%) occurred during the secondary stage of the disease, which is now less severe (deaths from secondary syphilis no longer occur). All attempts to control the original epidemic were unavailing and, in desperation, the Parliament of Paris of 1496 decreed that all persons with overt syphilis must leave the city within 24 hours. The disease spread rapidly to other parts of the world and persisted in Europe to become one of the great scourges of the 16th century.

Bell's observation (1793) that syphilis and gonorrhea are distinct entities was confirmed by Ricord in 1831, and Haensell (1881) infected the eyes of rabbits with exudates of human lesions. Discovery of the causative organism by Schaudinn and Hoffman in 1905 was quickly followed in 1906 by Wassermann's diagnostic serologic test. Ehrlich's discovery of arsphenamine (606) and neoarsphenamine was made a few years later. These therapeutic arsenicals, which replaced the highly toxic mercurials used earlier, marked the birth of modern antimicrobial chemotherapy; Wagner-Jauregg received the Nobel prize for his work on malarial fever treatment of neurosyphilis in 1927 and Kettering's electronic fever cabinet was introduced soon thereafter. *The sad addendum to the remarkably rapid advances made during the opening decades of this century is that little has been added to our understanding of T. pallidum or the pathogenesis of syphilis during subsequent years.*

In the Western World, syphilis was a major cause of death (mortality rate about 12% in untreated subjects) prior to the first widespread use of penicillin in about 1948. It is still a major killer on a worldwide basis and will always stand as a threat to public health whenever vigilance is relaxed. *In developed countries the case rate for syphilis showed a precipitous drop of some 85% after 1948;* the death rates due to late cardiovascular, CNS and congenital syphilis dropped even more dramatically than the overall case incidence.

The control of congenital syphilis by penicillin treatment has been especially successful because women seeking prenatal care are commonly required by law to undergo serologic tests for syphilis and, consequently, the diagnosis is made and treatment given if the reactions to the tests are positive. Because of common failure to report the disease, the incidence of syphilis in pregnant women remains as the best available indicator for estimating the incidence of syphilis in the public at large.

In the USA the incidence of all forms of syphilis dropped from over 5% in 1943 to less than 1% in 1957. This sharp drop was probably due in substantial degree to the widespread general use of penicillin during that period, which undoubtedly effected cures in many cases of syphilis, both diagnosable and undiagnosed. *In the years since the low ebb in 1957 when only 6,500 new cases were reported, the incidence of reported cases has fluctuated upward.* In 1980 27,000 cases were reported with a peak incidence in persons between 20 to 24 years of age; about half of whom were either homosexual or bisexual. The present infectious pool is probably about a half million. The actual number of new cases can be presumed to have been 5 to 10 times the number of cases reported.

Because of the great therapeutic effectiveness of penicillin, high hopes were initially held that syphilis might be essentially eradicated from large populations by mass penicillin treatment. Officers of WHO initially expressed the view that, with proper cooperation, yaws, the sister disease of syphilis, might be eradicated from large areas within 5 to 10 years by mass penicillin treatment. Since then, mass antitreponemal programs of the WHO have involved the treatment of 50 million persons with various treponematoses. Although these programs have greatly reduced the incidence of these diseases, they have not served to eradicate any of the treponematoses in any large area.

The most disappointing aspect of the history of syphilis is that, despite the discovery of a highly effective therapeutic tool, which was so fervently hoped for in the past, a shocking rise in the incidence of new cases has occurred in many countries in recent years.

Needless to say, the future outlook for the control of syphilis will only be good so long as effective therapy persists, and major efforts are sustained in the areas of research and medical education, and full cooperation is achieved between the public, the physician and public health authorities. *Under present circumstances, eradication of syphilis from any large population will only be possible providing physicians report all cases of the disease that they encounter to public health authorities.*

B. PHYSICAL AND CHEMICAL STRUCTURE

The morphology of *T. pallidum* is singularly deserving of attention because of its importance in identification of the organism and its bearing on invasiveness, pathogenicity and immunity. *Treponema pallidum* is a slender, pleomorphic, helical-shaped organism with tapered ends, each terminating in a nosepiece; it averages about 0.18 μm in diameter and varies in length from about 5 to 20 μm (Fig. 30–1). The organism possesses a cell wall, which is delicate and lacks rigidity; muramic acid is present, and lysozyme enhances lysis by Ab and C. The organism is encased in a special semi-permeable outer envelope (periplast) surrounded by a mucoid coat (slime layer). The envelope, which is 70 to 90 Å thick, is presumed to be of lipoprotein nature and to protect the organism against osmotic stress. Three contractile fibrils (Nichol's strain) arise from each end of the cell, each attached to a basal cytoplasmic granule in the protoplasmic cylinder. The fibrils lie between the cytoplasmic membrane and the cell wall. Each fibril is attached to one or the other end of the cell and has a free end. The fibrils, which may exceed the length of the cell, overlap and entwine to form the axial filament; on contraction they force the cell to assume its helical shape (Fig. 30–2).

The coils of the body of the organism are about 0.3 μm deep and are spaced at intervals of about 1.0 μm. The organism is highly motile and exhibits rapid, rotary, corkscrew-like movements, interrupted by marked flexing, around its long axis. Since it has no anteroposterior polarity, movement may be either forward or backward or alternating. The unique motility, flexibility and slender shape of *T. pallidum* probably contribute to its great power to pass bacterial filters and to invade tissues and spread rapidly throughout the body.

Contrary to general belief, *T. pallidum* stains readily with basic dyes and is regarded to be Gram-negative. However, because of its small diameter, the organism cannot be observed readily by ordinary light microscopy unless

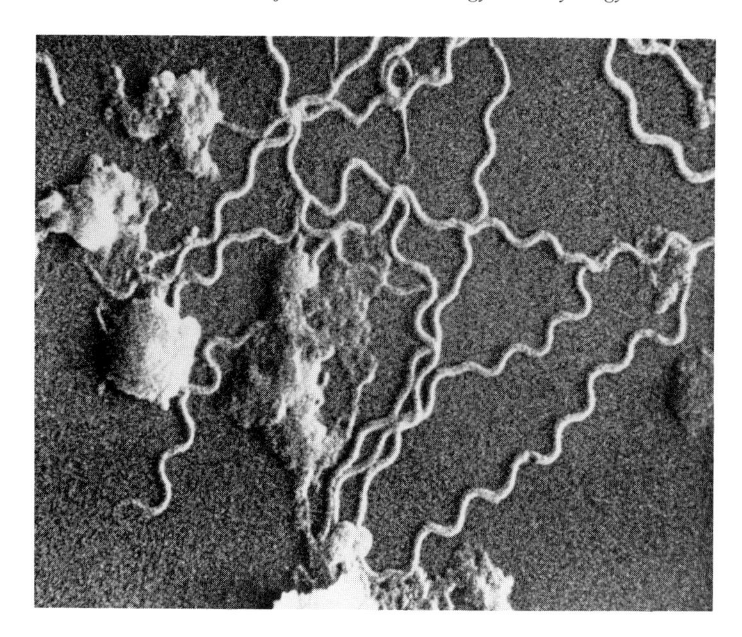

Fig. 30–1. Scanning electron micrograph of *Treponema pallidum.* (From: *Atlas of Scanning Electron Microscopy in Microbiology.* Yoshi, Z. et al. Williams & Wilkins, Baltimore, 1976.)

stained by some special method such as "silver staining" which leads to the deposition of silver on the cell surface. The organism can be observed by darkfield illumination, phase contrast microscopy and negative staining.

Little is known about Ags of *T. pallidum* because the organism has not been grown in vitro in continuous culture; obtaining pure suspensions of organisms from infected tissues is essentially impossible. Although group-specific proteins and polysaccharides have been reported, no individual treponemal Ags have been identified as being responsible for pathogenicity, virulence, immunity or the induction of the Abs measured in serologic tests (Sects. I and J). The Ags responsible for the C-activating Ab (*Treponema pallidum* immobilizing Ab or TPI Ab) is presumed to be highly labile. Freshly-isolated organisms,

---

Fig. 30–2. A, The cytoplasmic membrane (CM) is seen as a borderline between zones 1 and 2 of the organism. Beyond CM is cell-wall material only (CW). CW is seen in close contact with CM except in the regions of lower electron density, blebs (B), and the part of the cell where the fibrillar bundle (F) passes straight along the cytoplasmic body (arrow). Formalin fixation; negative staining; 1% ammonium molybdate. Magnification 90,000 ×. B, The three zones at the tip of the organism are clearly illustrated. Three fibrils (F) with insertion points (IP) are seen in zone 3, whereas 5 fibrils are seen more centrally. Two thin fibrils are present (arrow) and dense inclusions (I) are seen. Formalin fixation; negative staining; 1% phosphotungstic acid. Magnification 90,000 ×. (From Jepsen, O. B., Hougen, K. H., and Birch-Andersen, A.: Acta Pathol. Microbiol. Scand. *74*:247, 1968.)

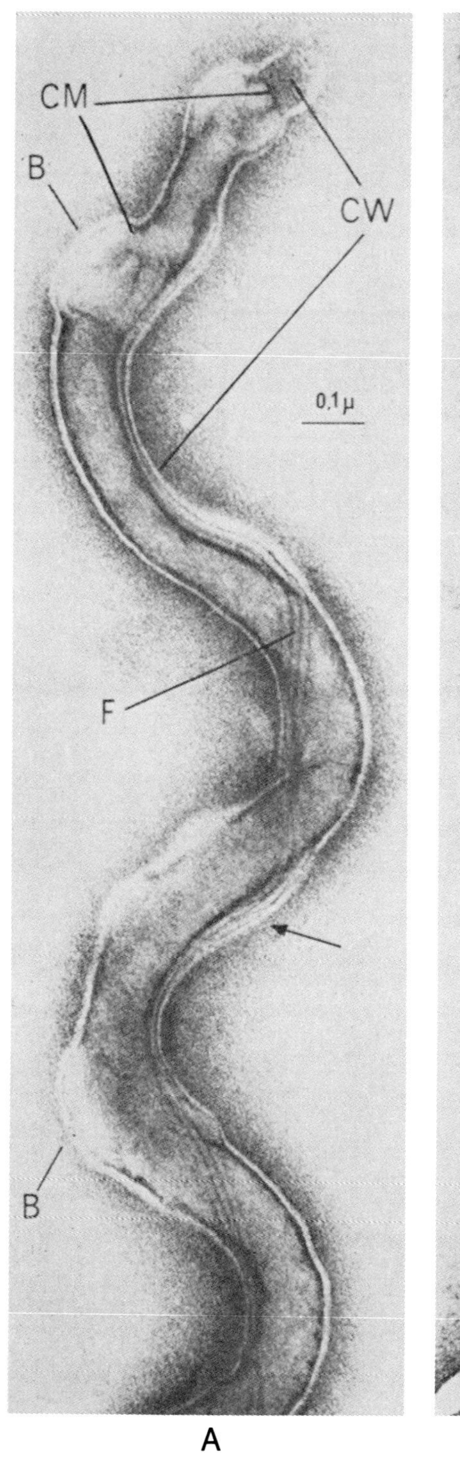

A

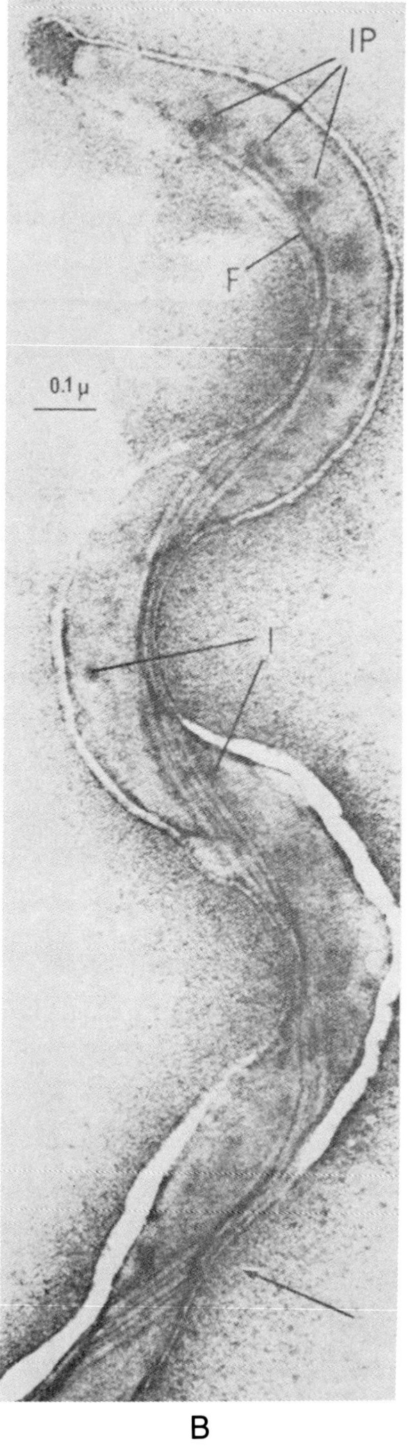

B

which are initially resistant to TPI Ab and C, become immobilized by this system after a few hours of incubation in vitro, which presumably leads to loss of integrity of the surface coat material and unmasking of Ag. The envelope has been shown to contain a family of proteins of varying molecular weights. When incubated in vitro the organism also synthesizes and liberates several unique low molecular weight proteins. It has been reported that certain surface proteins contribute to attachment of treponemes to host cells (Sect. I). The surface mucoid or slime layer comprises principally mucopolysaccharide, probably of host origin, together with other host components, including proteins ceruloplasmin and transferrin. The mucoid coat is alleged by some investigators to be antiphagocytic and to impede the access of Ab to underlying Ags (Sect. I) The envelope surface binds the above host materials, especially mucopolysaccharide.

C. GENETICS

Organisms from different patients have varying degrees of immunogenicity and virulence for the rabbit, but whether rabbit virulence parallels human virulence is not known. Isolates that are mildly virulent for the rabbit can be made more virulent by rabbit passage. The observations that *T. pallidum* harbors plasmid DNA and that fragments of treponemal DNA can be cloned in *E. coli* portend near-term advances in our knowledge of syphilis.

D. EXTRACELLULAR PRODUCTS

Treponema pallidum has been reported to secrete the enzyme, mucopolysaccharidase, and to liberate several low molecular weight proteins as noted above.

E. CULTURE

Treponema pallidum and related pathogenic treponemes have probably never been cultured continuously in vitro even in the chick embryo or tissue culture. Many nonpathogenic species have been cultured anaerobically. *Treponema pallidum* is a microaerophile that requires a low redox environment; the organism utilizes O_2 and possesses a cytochrome oxidase system. Maintenance media can support its continued motility for 10 days or longer in an atmosphere of 95% N_2 and 5% CO_2. Needless to say, failure of culture has greatly hampered research on *T. pallidum*.

The shortest generation time of *T. pallidum* in the rabbit is between 24 and 33 hours. This slow growth rate probably contributes to the long incubation time of 2 to 10 weeks observed in human syphilis and the slow therapeutic action of penicillin.

F. RESISTANCE TO PHYSICAL AND CHEMICAL AGENTS

Treponema pallidum is extremely sensitive to heat, drying, osmotic stress and aerobic conditions. *Because of this, transmission of syphilis is limited largely to direct contact between moist integuments of infected and uninfected individuals rather than agents such as fomites and aerosol droplets.*

The organism remains alive for 1 to 3 days in blood stored at 4°C; consequently, on rare occasions the transmission of syphilis has followed transfusion of relatively fresh blood. *Treponema pallidum* can be stored in the virulent state for many years at the temperature of solid CO_2, or preferably liquid N_2, especially if suspending media containing 15% glycerol are used. However, the organism does not withstand lyophilization, presumably because of the drying involved. Maintenance of motility outside the body does not ensure retention of rabbit infectivity, which, for unknown reasons, is usually lost within hours.

The in vivo temperature growth range of *T. pallidum* in the rabbit is 30° to 39°C (optimum 34°C); at temperatures of 40°C or above the organism does not survive for extended periods. In in vitro environments it is killed by 1 hour of exposure at 41.5°C. This temperature susceptibility evidently accounts for the occasional benefits of fever therapy formerly employed in man and for the selective distribution of experimental lesions in the testes and cool regions of the skin and extremities of rabbits (internal body temperature 38° to 39°C).

The organism is sensitive to many chemcal agents, including iodides, bismuth, mercury salts, arsenicals, and certain antibiotics.

G. EXPERIMENTAL MODELS

Although natural infection with *T. pallidum* is limited to man, many animals, including guinea pigs, mice, rats, rabbits, hamsters and subhuman primates, are mildly susceptible to experimental infection. In most of these animals, tissue responses to the presence of the organism are mild or absent and in many the growth of the organism is severely limited; however, infection can persist for months to years. *Only primates regularly develop the late lesions of tertiary syphilis.* Certain Abs, called "reagins," that are produced in man do not develop in small laboratory animals, presumably because of lack of tissue destruction necessary to their production (Sect. J).*

In the commonly employed experimental animal, the rabbit, a mild chronic infection can be produced by inoculating the cool skin, extremities or the favored site, the testes. The standard laboratory organism is the Nichols' strain of *T. pallidum* and the site used to produce organisms for laboratory use is the testicle; the animal room temperature is maintained at 16° to 18°C.

Only one or a few organisms are required to infect the testicle of the rabbit, and it is probable that man is as susceptible to infection as the rabbit. Although the organisms spread readily from the injection site and remain alive, their growth is apparently restricted at most sites, and metastatic lesions only occur in abundance in the nose and cool areas of the skin and bones of the extremities where the temperature is 1 to 3 degrees lower than the internal body temperature. Immunity to superinfection is characterized by a strong tendency of the organisms to remain *localized* at the injection site and to decline rapidly to very low numbers.

Apparently, passage of *T. pallidum* in the rabbit does not reduce its virulence for man since both accidental and intentional human infections have resulted from organisms maintained for some 40 years by rabbit passage.

H. INFECTIONS IN MAN

Syphilis, "the great imitator", is a systemic perivascular and vascular disease in which essentially every organ may become infected. So many and varied

*The term "reagin" is unfortunate and confusing because it is also used to designate the IgE Ab of the immediate sensitivities, such as hay fever.

are its manifestations that Osler wrote, "to know syphilis is to know medicine." *In man, one of the most notable characteristics of syphilis, and other treponematoses as well, is that the disease commonly progresses in stages with intervening periods of quiescence, each stage presenting lesions of different morphologic characteristics.* If untreated the outcome in most populations is: ⅓ recover spontaneously, ⅓ remain latent and ⅓ develop late destructive lesions.

Infection commonly results from contact of mucous membranes, and some 90% of cases of noncongenital syphilis in man are contracted during sexual intercourse. Most of the remaining 10%, including those in infants and children, are acquired by kissing, with resulting primary lesions (chancres) on the lips or in the oral cavity (Colle in 1631 was the first to prove that syphilis could be transmitted by common drinking cups and kissing). "Local epidemics" due to unusual modes of nonvenereal transmission have been reported, including wet nursing, common use of razors or public cigar cutters, carriage by means of surgical or musical instruments, and tattooing and glassblowing.

The organisms readily penetrate breaks in the skin or mucosa, no matter how miniscule, and possibly transgress intact mucosa as well, approximately 50% of those exposed become infected.

After breaching mucosa or skin the organisms quickly invade the local perivascular lymph spaces, pass to the local lymph nodes and reach the blood stream within a matter of minutes to a few hours at most. Bacteremia, which can be demonstrated readily by injecting defibrinated blood of the patient into the rabbit testicle, is a constant feature in the large majority of persons having primary syphilis and can occur sporadically in any stage of the disease.

Within 3 months (usually 2 to 4 weeks) after exposure the treponemes multiply and reach large numbers in the local lymphatics and the draining lymph nodes, as well as at the site of entrance where the local lesion, the chancre, develops. The chancre is often solitary and is commonly the size of a pea. The time of its appearance probably bears an inverse relation to the size of the infecting dose of organisms. Since the *chancre is commonly painless,* unless it is extragenital or becomes secondarily infected, and is frequently in a concealed position, the patient is often unaware or unmindful of its presence. *Forty to 60% of patients (especially women) pass through the primary and secondary stages of syphilis without knowing that they have the disease.* The chancre originates in the corium; early infiltrating PMNs are soon followed by other leukocytes. The fully developed chancre is indurated (hard) and *contains large numbers of lymphocytes, macrophages and plasma cells; in fact, plasma cells are singularly prominent in the lesions of all stages of syphilis.* Endothelial proliferation and obstructive endarteritis of infected blood vessels cause necrosis and attending ulceration of the overlying epidermis. The ulcerated lesion yields a serous exudate rich in treponemes and the draining lymph nodes, which are swollen, hard and *painless,* yield treponemes on aspiration. Healing of the chancre, which occurs within a few weeks, is accompanied by the local disappearance of most of the organisms, although a few may persist for months to years.

The secondary stage is a generalized infection; it usually appears after a quiescent period of about 6 to 12 weeks, but can occur before the chancre heals. During this period the organisms slowly multiply and invade widely, often involving the skin, mucous membranes, eyes, lymph nodes, bones, the CNS and the walls of major vessels. *The lesions and symptoms of secondary syphilis are legion, the most prominent being a macropapular rash. Although the secondary lesions are histologically similar to primary lesions and contain enormous numbers of treponemes, they usually do not lead to extensive and life-threatening tissue destruction and commonly heal spontaneously, without scarring, within a few weeks or months, a situation quite in contrast to the highly destructive lesions of tertiary syphilis.* Immune complex nephritis and arthritis sometimes occur. In some 25% of cases several alternating relapses and remissions of the secondary stage may occur within 1 to 2 years, apparently because of a weak and vacillating immune response. Relapses are more common than is generally believed; they range from 15 to 20% and include late latent as well as early syphilis. The term "early syphilis" includes primary, secondary, early relapsing and early latent forms of the disease as contrasted with later stages.* During the secondary stage, the saliva, semen and the uterine secretions (particularly at the time of menstruation) contain enormous numbers of treponemes. In contrast, patients with late syphilis are usually noninfectious except during pregnancy. Among patients with secondary syphilis who are not treated, the *majority progress* to the *tertiary* stage either insensibly or after a latent period that may be brief or may extend for 2 to 20 years.

The severity of tertiary syphilis bears an inverse relation to the severity of secondary syphilis (law of inverse relation of Brown and Pearce). In about 30 to 50% of persons with early untreated syphilis the disease progresses steadily and positive serologic tests for *specific Abs* are usually continually obtained; about 20 to 30% of this group eventually die of the disease. The overall mortality due directly to syphilis is about 12%; however, the disease also contributes to death attributed to other causes. In another 25% of untreated subjects, infection remains latent for long periods *(late latent syphilis)*; the specific serologic reactions remain positive and reactivation occasionally occurs. In the remaining 25% of those untreated the lesions and symptoms subside completely and reactions to serologic tests become negative. *(However, despite spontaneous clinical cures it can be seriously questioned whether spontaneous bacteriologic cures take place simultaneously.)*

The *late stages of syphilis progress slowly, but are nevertheless the most destructive;* they are characterized by the *focal lesion, the gumma,* a painless granuloma with central coagulation necrosis, which is of gum-like consistency, and the *diffuse lesion,* involving principally the walls of blood vessels and the CNS. *Patients in whom diffuse lesions prevail present the most tragic cases, for during long years of apparent good health the treponemes continue with their tardy but terrible acts of destruction that are only too familiar to the pathologist.*

The development of the gumma of syphilis is sometimes precipitated by trauma. Gummas produce symptoms readily; they are often solitary and usually develop in skin, bones, testes, liver and in the brain where they cause the symptoms of a tumor. They vary greatly in size and are composed of epithelioid cells and a few foreign-body giant cells and Langhans' giant cells surrounded

*Latent disease of less than 4 years duration is defined as *early latent* syphilis and more than 4 years' duration as *late latent syphilis.*

by *macrophages and enormous numbers of lymphocytes and plasma cells.* The tissue destruction involved is presumed to be largely on a hypersensitivity basis with vascular occlusion accounting for some of the necrosis. Gummas contain relatively few organisms, but reactions to serologic tests are usually strongly positive. Large gummas tend to heal centrally and progress peripherally. They are readily arrested by penicillin therapy and heal rapidly. Individuals who develop gummas often fail to develop late CNS and cardiovascular syphilis due to diffuse lesions, presumably because immunologic mechanisms concerned in the formation of gummas are associated with strong immunity. Syphilis involving gummas is sometimes termed "benign syphilis." This is unfortunate because if the gumma is in a vital spot, such as the heart or brain, it can be highly lethal.

In contrast to *gummas, diffuse lesions* are characterized by slow, insidious destruction of blood vessels and other tissues, such as those of the CNS. Unfortunately they often progress asymptomatically for many years until irreparable damage of the entire vessel wall has occurred. In diffuse lesions, as in most lesions of the disease, the organisms show a predilection for perivascular lymph spaces. *Perivascular infiltration with lymphocytes, plasma cells and macrophages is the hallmark of the diffuse lesion.* Destruction of the valvular endocardium and walls of small vessels, such as the supporting vasa vasorum of the walls of large vessels, often occurs. Small and medium-sized arteries, such as cerebral and coronary arteries, may suffer obliterative endoarteritis, which restricts the circulation and leads to severe disturbances in the function of many tissues and organs. Cerebral hemorrhage, aortic valvular disease and aneurysms of the thoracic aorta are common results of such lesions.

Late CNS syphilis commonly becomes symptomatic earlier than syphilis of the aorta; it involves the meninges *(syphilitic meningitis)* and parenchymal tissues of the brain *(general paresis)* and the spinal cord *(tabes dorsalis).* In some cases, destruction of the optic nerve occurs and leads to blindness.

Congenital syphilis results from the passage of the highly invasive treponeme from mother to fetus in utero, an uncommon occurrence among bacterial diseases. *Pregnancy* markedly *suppresses the symptoms of syphilis* in the mother, probably, in part at least, as a result of estrogenic hormone effects. *However, a mobilization of treponemes in the blood is favored by pregnancy and bacteremia often occurs in the pregnant female with late syphilis; thus the opportunities for infection of the fetus are great during pregnancy.* Infection of the fetus is seldom manifest before about the 5th month of gestation, allegedly because a partial barrier against treponemal invasion is afforded by Langhans' cell layer of the early placenta and/or because immune responses are lacking in the fetus.

Maternal anti-treponemal IgG Abs afford little or no protection to the fetus; the treponemes may reach tremendous numbers and cause great damage, especially in the viscera. The infected fetus may be born dead, either at term or prematurely, born alive with clinical evidence of syphilis or born in apparent good health and subsequently develop clinical disease.

The hallmarks of congenital syphilis are based on arrests of embryonic development. The lesions, which are evident both early and late after birth, are numerous and may resemble those of the secondary and tertiary stages of noncongenital syphilis in adults. *The stigmas of late congenital syphilis,* e.g., the telltale indicators that accompany late lesions in the maturing individual, usually include at least two of "Hutchinson's triad," namely notched permanent incisors, and interstitial keratitis of the cornea leading to corneal opacity and blindness. Since these late development conditions result from injury of embryonic tissues, they are not helped by penicillin treatment.

Prevention of congenital syphilis can be readily accomplished by penicillin treatment initiated early in pregnancy to suppress maternal bacteremia. In such cases the newborn infant, although uninfected, may show a positive reaction to serologic tests (due to maternal IgG Abs); this reactivity steadily lessens in early postnatal life. When syphilis is acquired by the mother late in pregnancy, the infected newborn can be apparently healthy and serologically negative. Whereas congenital syphilis is rare in the USA, it remains an enormous public health problem in many underdeveloped countries where it may account for some 20% of infant mortality.

I. MECHANISMS OF PATHOGENESIS

The pathogenesis of syphilis has remained an enigma for almost a century. There is no evidence that *T. pallidum* forms a toxin that could account for the lesions seen in syphilis. However, late advances in syphilology have lead to new well-supported hypotheses that are deserving of mention despite lingering controversy. Advances concerned with three major properties of the organism serve as the principal bases for an understanding of pathogenesis, namely the surface mucoid coat, mucopolysaccharide, treponemal mucopolysaccharidase and the nature of attachment of treponemes to host cells.

One of the most impressive features of treponemes is that they display avid end-attachment to host cells (only about 5 to 10% attach at both ends). This occurs both in vivo and in vitro but has only been well-studied in vitro. When freshly-prepared virulent organisms are added to live cultured host cells the majority attach randomly to the host cell surface where they remain as motile, fully-capsulated, dividing organisms over observation periods of a week or more. There is some evidence that engaged host cells liberate factors that enhance the viability of treponemes and somehow enable them to evade the immune forces of the host. The percentage of treponemes that attach to host cells varies directly with the virulence of various strains of the organism. The host cell is not injured until the number of attached microbes becomes excessive (some 200 or more), at which time the cell disintegrates. Treponemes that do not attach to host cells lose motility after about 9 hours of culture. The treponeme receptors and corresponding cell receptors that are responsible for attachment have not been identified with certainty; however, they are probably envelope proteins. Attachment to host cells is not unique to *T. pallidum*; cultivable nonpathogenic spirochaetes similarly attach to and colonize the mucosa of the monkey and human colon.

Whether unattached treponemes in infected tissues fare as well as attached organisms with respect to viability and division is not known.

Whereas tissue treponemes are principally extracellular, they are seen within various cell types on rare occasion, including cells that are either incapable or capable of nonprofessional endocytosis. Attachment and subsequent direct penetration of the cell membrane of these cells is the probable mode of entrance.

Another impressive feature of *T. pallidum* is that the surface mucoid coat appears to be necessary for the continued viability of the organism in tissue culture. Some investigators believe that the coat material is solely of host origin, whereas others believe that *T. pallidum* can synthesize mucopolysaccharide capsular material as well and that this attribute accounts for the enormous amounts of mucopolysaccharide seen in syphilitic lesions. Since the surface mucopolysaccharide is extremely soluble and dissipates and deteriorates readily the treponeme must constantly resynthesize or reabsorb mucopolysaccharide to replenish its mucoid coat and maintain its viability in culture.

Since materials in the mucoid coat lack antigenicity for the host, they may impede the access of certain Abs to underlying Ags, as evidenced by the old observation that freshly isolated treponemes only become fully susceptible to killing by TPI Ab after several hours of in vitro incubation in a cell-free medium; apparently incubation allows the dissipation of coat material. The enhancing action of lysozyme on the TPI reaction may likewise rest, in part at least, on its precipitating action on mucopolysaccharide in the coat; the mucopolysaccharide is also oxygen sensitive and can readily undergo oxidative-reductive depolymerization. The claims that the surface mucopolysaccharide is antiphagocytic and that it exerts nonspecific immunosuppression remain controversial (Sect. J). Whether integrity of the mucoid coat is as important to viability in vivo as in vitro is obscure.

Although, the mucoid coat may afford some resistance to phagocytosis, coated treponemes are engulfed by activated macrophages, especially in the presence of opsonizing Abs that gain access to surface Ags, perhaps at points of discontinuity in the coat. Even though monocytes (nonactivated macrophages) fail to engulf unopsonized treponemes in vitro, they have not been tested in the presence of opsonins. Treponemes engulfed by activated macrophages are readily digested, but their fate in cells that are not professional phagocytes is unknown; perhaps in such positions that can survive by being sheltered from immune forces during periods of latency.

Although it has been proposed that mucopolysaccharidase may effect treponemal attachment by reacting with its substrate on the host cell surface, it is more likely that attachment is due to envelope proteins. In any event the mucoid coat, treponemal attachment, and the activities of mucopolysaccharidase may account for or contribute to many events observed in syphilis including; (1) attachment to and entrance into lymphatics and thence to blood, (2) attachment to and penetration of blood vessel endothelium to reach perivascular sites, (3) entrance and destruction of various cells and tissues, such as vascular components, with resulting endoarteritis and aneurysms, (4) blockade of certain effector functions of immunity, such as opsonization, phagocytosis and Ab-C killing, and (5) immunosuppression that results in a slowly developing, vascillating immune response.

The roles that immediate and DH reactions may play in tissue injury remain controversial. Immune complexes are known to cause tissue injury in syphilis, e.g., that immune complex nephritis of secondary syphilis, and DH reactions doubtlessly cause injury in tertiary disease, including the necrosis seen in the gumma.

J. MECHANISMS OF IMMUNITY

Less is known about the nature of immunity to syphilis than of immunity to most any other bacterial disease. Since no single protective immunogen is known with certainty, it has not been possible to explore the bases for the lack of complete cross-immunity between treponemal isolates.

Levels of total serum globulins are usually high in syphilis and the Abs produced are of two kinds; (1) so-called *reagins,* which are autoantibodies against tissue Ags and (2) treponemal Abs specific for treponemal Ags. Reagins are chiefly of the 19S class, especially early in the disease, and functionally may be either "complete" or "incomplete." They are measured by either complement fixation (Wassermann type) tests or flocculation (Kahn type) tests using tissue-derived Ags. They are not specific for syphilis and appear sporadically in certain unrelated infectious and noninfectious diseases.

Tissue Ags, which are presumed to represent structural cellular Ags released from injured host cells, evidently engender *reagins* of two classes. Reagins of one class react with tissue Ags only, whereas those of the other class are cross-reactive, i.e., react with both Ags of *T. pallidum* and tissue Ags. *The level of reagins in the serum of syphilitics correlates well with the rate of tissue destruction and is the best guide for assessing the effectiveness of therapy.* Tests for reagins frequently yield either *false-positive* or *false-negative* reactions, the latter being especially frequent in early primary syphilis, latent syphilis or in tertiary syphilis in which tissue destruction is limited. Cold autoantibodies specific for the Tj^a antigen of human red cells often arise in late and congenital syphilis; following chilling of the extremities they act with C to destroy RBCs and cause paroxysmal cold hemoglobinuria.

The second general category of Abs, the *treponemal Abs,* consists, on the one hand, of *"specific Abs" that react only with T. pallidum and closely related pathogenic treponemes* and, on the other hand, of *"group Abs" directed against cross-reacting Ags present in a wide variety of distantly related treponemes* as well as *T. pallidum.* Treponemal Abs are principally of the class IgG (7S), but include IgM (19S) and occasionally IgA Abs as well. Specific Abs cause agglutination and some belonging to the class IgG are cytotoxic for the organism in the presence of C; they cause early loss of motility detectable by the TPI test. Specific treponemal Abs can also be detected by a number of other tests, including the fluorescent treponemal Ab test (FTA test), an indirect fluorescent Ab test. This test is conducted by applying suspected serum to known dried *T. pallidum* on a slide, followed by washing, adding fluorescein-tagged antihuman globulin serum and examining with the fluorescence microscope. The specificity of the FTA test can be improved by properly absorbing the suspected serum with Ags of the nonpathogenic Reiter strain treponeme to remove many of the cross-reactive "group Abs" (FTA-ABS test). *Of the current routine tests for syphilis and closely related treponematoses, the FTA-ABS test is the most specific, reliable and sensitive.* Treponemes of the normal flora, especially when they participate in opportunistic infections, may incite "treponemal Abs" that are not absorbable by Reiter Ag and thus abrogate the specificity of the FTA-ABS test. The highly specific TPI test is used occasionally to detect false positive reactions to FTA-ABS tests. Despite their shortcomings, the tests for treponemal Abs have been of great value in diagnosis, positivity being largely limited to syphilis and closely related treponematoses. They have been especially useful for detecting the false-positive reactions encountered in reagin screening tests for syphilis.

Treponemal Abs tend to appear somewhat earlier in the course of the disease than reagins and often persist indefinitely after spontaneous or therapy-induced clinical cure. Consequently, a *positive* FTA-ABS *test provides no certain evidence of clinical activity or necessity of treatment. It simply indicates that the subject has or has had one of the treponemal diseases.* In a minority of untreated patients with late CNS or cardiovascular syphilis, especially those with tabes dorsalis or long-standing congenital disease, the reaction may be negative.

The formation and distribution of Abs in CNS *syphilis and congenital syphilis deserve special consideration. In* CNS *syphilis, Abs* are produced locally by infiltrating cells. Since Abs do not readily pass the blood-brain barrier in either direction unless the barrier is severely damaged, syphilis limited to the CNS may be accompanied by Abs in spinal fluid alone, or, if not so limited, by Abs in both serum and spinal fluid. For example, in 40% of persons with tabes dorsalis, positive reactions are limited to tests on the spinal fluid.

Since 7S as well as 19S Abs are represented among reagin and treponemal Abs, the newborn infant of a syphilitic mother with positive serologic reactions may show positive reactions for both reagin and treponemal Abs as the result of maternal transfer of 7S Abs on the one hand and the formation of 19S Abs by the fetus on the other.* Demonstration of treponemal IgM Abs in fetal blood by use of anti-H chain serum constitutes presumptive evidence of congenital syphilis. In some instances, maternal IgM Abs may reach the fetus through a "leaky" placenta. Such leakage can be detected by comparing the ceruloplasmin concentration in the serum of the mother and child. If the infant has active disease, the serologic reaction will remain positive, but, if free of infection, it will become negative in a matter of weeks to 3 to 4 months owing to metabolic decay of maternally-derived Abs. If the mother contracts the disease late in pregnancy, the serologic reaction of both the mother and infant may be negative at birth and become positive later.

Apparently man possesses little or no innate immunity to syphilis and all individuals are susceptible (although not necessarily equally so) to primary infection. *A substantial level of acquired immunity develops in most infected individuals;* however, it is insufficient in the majority to accomplish early arrest of the disease. Acquired immunity is manifested by marked resistance to natural or experimental superinfection, occasional spontaneous permanent arrest of disease following the secondary stage (some 25% or more of untreated individuals), a paucity of tissue treponemes in late lesions, slowly progressing disease, and evidence of an anamnestic response on reinfection. *Acquired immunity is not absolute* and heavy inocula of organisms into the skin of syphilitic patients will often produce lesions characteristic of the stage of existing disease. Also *immunity sufficient to prevent the development of clinically evident lesions on natural reexposure does not necessarily protect against tissue invasion by superinfecting organisms.*

A strong stimulus by the treponemes appears to be necessary for the initiation and maintenance of immunity, as indicated by the following observations: (1) immunity to superinfection does not arise until after the full development of the primary lesion, (2) an inverse relation exists between the severity of the secondary lesion and the extent of tertiary lesions and (3) immunity to rein-

*Antibodies responsible for false-positive maternal serologic reactions can also be transferred to the fetus.

fection wanes following cure with therapeutic agents, the decline being especially rapid following cure effected in the early stages of the disease. *The alleged need for living organisms to maintain high immunity has led to use of the term "infection immunity."* It is notable that maximum immunity to either reinfection or superinfection is not gained earlier than about 3 months in the rabbit and some 2 or more years in man. In rabbits, measurable immunity can persist for substantial periods after the organisms have been eliminated with drugs. *Total immunity probably could result from a combination of 2 or more of 3 general mechanisms,* namely, humoral Abs acting in various ways including: immobilization, lysis, opsonization and blockade of attachment to host cells, cellular immunity (possibly of an unusual kind) and nonantibody humoral factors of immunity, such as lysozyme.

Except for its greater strength, more rapid development and lack of a tertiary phase, experimental syphilis in the rabbit resembles human syphilis. Within hours and for a few days after intradermal injection of 10^6 treponemes they enter lymphatics and reach the draining lymph nodes and the blood stream from which they disperse widely to reach many sites in the body. Within a week lymphocytes, plasma cells and macrophages accumulate in appreciable numbers at the primary site of infection to initiate a visible chancre. Simultaneously germinal centers develop in the cortical areas of the draining lymph nodes. At about 10 days serum Abs appear and, as evidence of developing DH, lymph node lymphocytes respond in vitro to treponemal sonicates with blast transformation. Although the dermal lesions heal in about 4 to 8 weeks, the animal may still be at least partially susceptible to superinfection. The appearance of serum TPI Abs does not necessarily correlate with the onset of chancre regression. When treponemes are injected intratesticularly, infection progresses more rapidly and the numbers of treponemes in the testes drop sharply after about 2 weeks, at which time electron microscopy reveals extensive phagocytosis by activated macrophages. Antibodies may contribute to treponeme elimination by acting alone as, bactericidins, or anti-attachment agents or alternatively, as opsonins to promote phagocytosis by activated macrophages, the latter being an expression of collaboration between humoral and cellular immune forces.

As evidence of the role of humoral immunity in syphilis the injection of large doses of immune serum can markedly depress the development of intradermal lesions in the rabbit.

A role for CMI in the rabbit is supported by the observation that immune T cells can transfer passive protection and that selective inhibition of the CMI response by cyclophosphamide treatment increases the severity of experimental lesions. Whether immune T effector cells can act directly against treponemes or indirectly by lymphokine effects is not known. Continuing susceptibility to superinfection at a time when the chancre is healing or has healed may simply mean that the strength of local humoral and cellular forces within the immune-cell rich chancre are inordinately high as compared to systemic forces that can be mobilized from the circulation to a site of rechallenge.

Secondary stage disease in the rabbit is rare and mild, presumably because immunity is stronger than in man and/or because immunosuppression is less apt to occur.

There is strong and increasing evidence that the singularly slow development and vascillating nature of systemic immunity in man, as exemplified by frequent relapses, reflect transitory periods of excessive immunosuppression. The immunosuppressants that have been indicted include, immune complexes, especially during secondary disease, and capsular mucopolysaccharide. Immunosuppression is the greatest during primary and secondary disease.

Depression of the cellular immune response in man is indicated by depressed in vitro reactivity of T lymphocytes to treponemal Ags, a paucity of lymphocytes and histocytic infiltration in the paracortical areas of lymph nodes and a depressed or negative DH response to skin tests with treponemal extracts (luetin test of Noguchi). Indeed "depression" of DH and CMI may represent desensitization due to excess Ag, which causes sensitized T cells in the circulation to sequester in lymphoid organs.

Depression of humoral responses in man appear to involve certain Abs but not others, e.g., TPI Abs fail to appear in the serum of two-thirds of patients with primary disease and one-third of patients with secondary disease. Suppression of nonspecific as well as specific immune responses occurs in early syphilis. Whether the C-dependent cold-reacting anti-lymphocyte autoantibodies present in sera during primary and secondary syphilis contribute to immunosuppression is uncertain.

Hypersensitivity reactions may play a role in acquired immunity to syphilis. Positive skin test reactions to killed *T. pallidum* or its extracts, characteristic of immediate-type hypersensitivity have been reported in early preserologic syphilis; in late syphilis they are of the delayed type.

A clear understanding of acquired immunity to syphilis will probably only be gained by extensive studies on lymphocyte subset activities and use of the monoclonal Ab technique for identifying the various Ag-Ab systems involved.

K. LABORATORY DIAGNOSIS

Since *T. pertenue* and other pathogenic treponemes are strongly cross-reactive with *T. pallidum,* none of the tests for syphilis is specific unless other treponemes are ruled out.

The laboratory diagnosis of syphilis is commonly based on the direct demonstration of *T. pallidum* in material from the lesions or draining lymph nodes and/or serologic tests for Abs in serum and cerebrospinal fluid. Since *Abs do not arise until some 1 to 3 or more weeks after the chancre first appears, the only available test for judging early primary lesions involves detection of the organism.* Under darkfield illumination, *T. pallidum* is easily confused with nonpathogenic treponemes, and the examination is only reliable when it is conducted by experts and is repeated if necessary. A highly-promising modified darkfield test has recently been developed in which a known absorbed fluorescein-tagged syphilitic serum specific for the organism is used (FADF test).* *The physician should be mindful of the fact that atypical lesions due to mixed infections sometimes occur and that the frequent dominance of organisms other than T. pallidum can be misleading.* Failure to find treponemes by darkfield examination is not certain evidence that the disease is not syphilis.

Since treponemes are usually scarce in the lesions of late syphilis and can only be demonstrated directly by painstaking examination of biopsy material

*Fluorescent antibody darkfield (FADF) test.

which is usually unavailable, *the only useful routine methods for laboratory diagnosis of tertiary syphilis are serologic tests.* Serologic tests, which are legion (over 200 have been proposed), are concerned (1) with the detection of reagins by testing the patient's serum and spinal fluid with tissue Ag preparations** and (2) with the detection of *treponemal Abs* by tests with a laboratory strain of *T. pallidum.* Improved variations of these tests are constantly being made. The use of different tests by different laboratories has led to much confusion in reporting and interpretation of results. Consequently, it is advisable for the physician to make a concerted effort to become thoroughly acquainted with the tests used in local laboratories. Tests in current use in most laboratories include the Venereal Disease Research Laboratory test (VDRL test), a flocculation type of slide test for reagin, and the FTA-ABS and *microhemagglutination treponeme* (MHA-TP) tests for treponemal Abs. *Serologic tests for syphilis are not as reliable as has been generally believed and proper interpretation of results demands keen judgment by experienced personnel,* taking into consideration the clinical aspects of the case as well as the limitations of the tests themselves. Since tests for reagin are nonspecific, biologic false-positive reactions associated with treponemal disease must be ruled out by tests for treponemal Abs. Rabbit inoculation is resorted to in special circumstances. Detailed descriptions of late modifications of tests for syphilis can be found in books on laboratory diagnosis and USPHS publications.

L. THERAPY

Although the development of arsenicals for the treatment of syphilis was one of the greatest achievements in the history of medicine, penicillin and certain other antibiotics have much greater treponemicidal activity and much lower toxicity than arsenicals.

Penicillin kills T. pallidum slowly because the organism grows slowly. Treatment is usually effective within about 2 to 6 weeks. Late syphilis is the least responsive, presumably because the organisms are multiplying less rapidly than during early disease and because tissues destroyed prior to treatment do not regenerate. If the patient is sensitive to penicillin, other antibiotics, including tetracyclines, cephaloridine, carbomycin, synnematin, and especially erythromycin, may prove useful. Antibiotic resistance tests are done in animals. Fortunately, resistance to penicillin does not tend to develop readily.

Serologic tests for reagin are highly useful in therapeutic management because they correlate with the rate and extent of tissue destruction. The VDRL test should be run at 3-month intervals for 1 to 2 years or longer as a measure of the adequacy of treatment. A persisting or rising VDRL titer implies continued disease or possibly reinfection.

The life-threatening systemic "Herxheimer" or "Jarisch-Herxheimer" reaction with accompanying exacerbation of local lesions and associated destruction of liver, which was a troublesome problem in the days of arsenical treatment, particularly in the therapy of paretics and patients with secondary syphilis or congenital syphilis, is an even greater problem with penicillin treat-

**Cerebrospinal fluid tests should be included in the management of all cases of syphilis because of their prognostic and diagnostic value in CNS disease.

ment. For example, penicillin treatment of infants with congenital syphilis is attended by the Herxheimer reaction in 30% of these patients, evidently because of the enormous numbers of organisms in the lesions. It is presumed to represent an allergic or, possibly, a toxic reaction caused by the sudden release of treponemal components following initiation of treatment. Reactions usually begin about 4 hours after initiation of treatment and persist for 10 to 15 hours. They are accompanied by fever and heavy infiltration of neutrophils into local lesions and can be fatal in patients with late cardiovascular or CNS syphilis. Since early penicillin treatment of primary infections destroys the organisms before immunity can develop, the patient often remains fully susceptible to exogenous reinfection following cure with penicillin.

Although it has been the consensus that permanent clinical and serologic cures can often be effected with arsenicals and antibiotics, especially in early syphilis, the concept that, following such cures, the body is completely ridded of the organisms has always been seriously challenged. There is now substantial evidence that an occasional patient cured clinically and serologically with penicillin continues to harbor live *T. pallidum* in the inner ear, the CNS, and, most notably, in the anterior chamber of the eye. These are sheltered sites where penetration of immune forces and penicillin is limited, especially in the absence of inflammation. Although the status of *T. pallidum* in possible sites of sequestration will only be settled by further investigation, present evidence suggests that great care should be exercised to ensure that the treatment of late syphilis is rigorous and adequate.

M. RESERVOIRS OF INFECTION

Infected human beings are the only natural reservoirs of infection. Experimental laboratory animals have been a source of accidental human infection on rare occasions.

N. CONTROL OF DISEASE TRANSMISSION

Inasmuch as syphilis is acquired chiefly by sexual intercourse, effective measures of control consist of education, early detection and adequate treatment of cases and contacts. Since early syphilis is readily cured with penicillin and late syphilis is seldom infectious, it should be possible to eradicate the disease from a closed population. *However, this can never be achieved unless essentially all persons with early disease and contacts of individuals with infectious syphilis are detected and treated with a sense of immediacy.* Unfortunately, in recent years indifference on the part of patients, the public and even the medical profession has made control of syphilis far less effective than it could be. Abandonment of the hospital practice of conducting routine serologic tests for syphilis on all incoming patients has aggravated the problem. *The weakest links in control are case reporting and detection of contacts. It is a sad commentary that for every case reported to public authorities at least 5 to 10 additional cases known to the physician go unreported.* Also, because of the social stigmas involved, the detection of cases and contacts is exceedingly difficult, especially among teenagers and sexually perverted individuals; the hunted are like a will-o'-the wisp. The detection and management of contacts are best conducted by public health experts; the trails are long and may even extend to foreign countries. Moreover, the fact that many individuals, especially females, remain unaware that they have contracted the disease is a great deterrent to case finding and frequently a great tragedy since the disease may reach the late destructive stages before it is recognized. Another deterrent to case and contact detection is secretive self-treatment which, being frequently inadequate, often leads to false cures and the development of late disease.

Military experience has shown that no local prophylactic measure has more than limited value, in part because of faulty use of such measures and in part because the measures are inadequate. For example, the infecting treponemes often invade tissues beyond the reach of such agents as topical ointments within minutes to a few hours after exposure. Inadequate prophylactic measures often mask infection by suppressing the early lesions, leaving the individual unaware of any need for further treatment. Moreover, inadequate suppression of early infection depresses the specific immune response and predisposes to the development of severe tertiary lesions. Chemoprophylaxis is only effective when it meets the minimum requirements for curative treatment. In instances of known exposure, especially of the pregnant female, chemoprophylaxis is justified but should always be supervised by a physician.

No suitable vaccine for syphilis has been developed to date; if such were available, it should be highly useful for immunizing patients after completing drug treatment and especially for preventing infection among prostitutes. This is supported by the observation that venereal syphilis is rare in areas of endemic yaws which generates treponemal cross-immunity. Attempts to identify and isolate protective immunogens of *T. pallidum* by the use of monoclonal Abs and to develop attenuated strains of treponemes are promising approaches for producing a vaccine.

TREPONEMA PERTENUE

Although yaws can be distinguished from syphilis on epidemiologic and clinical grounds and by infection in rabbits, the causative organism, *T. pertenue,* is genetically, serologically and morphologically indistinguishable from *T. pallidum.* Since organisms closely resembling *T. pertenue* have been isolated from African baboons, man may not be the only natural host.

Unlike syphilis, yaws is not a venereal disease and is probably contracted largely by skin contact and by insect bites. Infection tends to occur early in life and the majority of a population often become infected. Yaws is most prevalent in moist tropical regions where little clothing is worn. Indeed it is thought that clothing greatly limits infection and breaks the chain of transmission. High humidity favors the persistence of open skin lesions and disease transmission.

The course of yaws is similar to the course of syphilis. During the first few years, early relapses involving skin may occur; late latent disease with relapses is common. Many of the late lesions of yaws closely resemble those of syphilis but are largely confined to skin and bones. Mutilating deformities of the face and feet are common. Immunity is weak and superinfection often occurs. Singular characteristics are that the cardiovascular and nervous systems are seldom involved and congenital disease is rare. Penicillin is highly effective and has been used widely in WHO campaigns to control yaws. However, eradication of yaws has posed a new problem, namely, that of creating a population lacking the protection against syphilis naturally conferred by yaws.

In 1954 the number of cases of yaws in the world was estimated to be 50 million. The numbers today are fewer but are beginning to rise because of lapses in control measures.

TREPONEMA CARATEUM

Pinta due to *T. carateum* is another nonvenereal treponemal disease. It is endemic in various areas including Mexico, Cuba and Central and South America. The organism is serologically and morphologically indistinguishable from *T. pallidum* and *T pertenue*; it is infective for the chimpanzee but not other laboratory animals. In the skin, a lower temperature area, the only apparent site of attack, the organism is almost exclusively in the lower malpighian layers of the epidermis. It produces chronic, scaly, pigmented lesions that often persist for decades and sometimes result in atrophy, scarring, and depigmentation. In common with yaws, infection usually occurs in childhood, presumably as the result of skin contact. Mass chemotherapy with penicillin has been effective, and at present it is probable that fewer than one million cases exist.

TREPONEMA PALLIDUM EPIDEMICUM

Diseases that probably represented endemic nonvenereal syphilis have been recorded from time to time from the 17th century onward and continue to persist in many areas of the world, especially in Africa and Asia.

Bejel, a modern example of nonvenereal syphilis is endemic among the Arabs of the Upper Euphrates Valley. Another example of nonvenereal endemic syphilis occurred in Bosnia, Yugoslavia but has been completely eradicated by mass treatment with penicillin. Owing largely to unsanitary conditions, the majority of individuals with bejel become infected in childhood, probably in such ways as by the use of common drinking utensils; essentially all individuals in endemic areas are infected by the time they are adults. Skin, bones, and especially mucous membranes of the URT are often involved; gummas are common but congenital disease and lesions of the cardiovascular and nervous systems seldom occur. Multilating deformities of the face and limbs often develop. Reactions to serologic tests for syphilis are positive and penicillin treatment is effective. Immunity is weak and superinfection can occur. The causative organism is infectious for the rabbit.

INTERRELATIONS OF THE TREPONEMATOSES

Interrelationships between the treponematoses is a subject of great interest and speculation. Although cross-immunity exists, especially between syphilis and yaws, its nature is not clearly understood.

It has been suggested that differences in disease patterns and organ tropisms of species of treponemes may have resulted from evolution involving both ecologic and geographic separation of groups of human beings. For example, whereas yaws, which is characterized by skin-to-skin transmission, is favored by warm, humid climates where little clothing is worn, the disease can also exist in temperate climates where personal hygiene is poor and numerous members of the family sleep together because of overcrowding. It has been proposed that in the evolution of civilized man the advent of improved clothing and living conditions restricted the opportunity for skin-to-skin transmission and introduced selective pressures leading to the emergence of new mutant strains of treponemes best fitted to accomplish mucosa-to-mucosa transmission, especially by venery, for their maintenance in the host population. Selection may subsequently have led to changes that endowed the organism with increased capabilities for invading internal organs, the CNS and the fetus.

BORRELIA

Members of the genus *Borrelia* have a coarser spiral structure than the treponemes and are large and thick enough to be seen in stained smears by phase and darkfield microscopy. The prototype species of known pathogenetic significance for man is *Borr. recurrentis,* the principal species that causes relapsing fever. Other less pathogenic species such as *Borr. vincenti* are occasionally present among the flora of mixed infections, such as ulcerative lesions of the mouth and skin and lung abscesses, but their possible contributions to these infections are not known.

During relapsing fever, *Borr. recurrentis* is abundant in blood; it was first observed by Obermeier (1873), in blood from a patient, appearing as a highly motile, thread-like spiral rod with dimensions averaging about 8 to 30 μm $\times$ 0.3 to 0.5 μm. A wide range of animals, principally rodents and their arthropod parasites of the genus *Ornithodoros,* serve as natural hosts, and reservoirs of the various *Borrelia sp.* that cause relapsing fever; man and certain animals are usually accidental hosts. The bite of this nocturnal tick usually goes unnoticed.

Relapsing fever in the USA (principally in the West and Southwest) is of limited medical importance; only 382 tick-borne cases have been reported in the last 25 years. However, from the time of Hippocrates it has been documented as a disease of major medical importance in Europe, Asia and Africa. It continues to be a human disease of great importance in Northern Africa and Central Asia. Certain *Borrelia* sp. cause enormous losses in the poultry industry.

Borrelia recurrentis is closely related to *T. pallidum,* but unlike the latter is not sensitive to O_2; it has an outer slime-like coat and exhibits marked instability in antigenic composition in the face of host immunity. Because of this, residence in different mammalian and insect hosts has led to great heterogeneity of strains and difficulty in classification. *Some investigators have given the variants species status (usually named after the region where isolated), whereas others regard the variants to be strains within a species.* The species name *Borr. recurrentis* is sometimes used to designate the organism responsible for *louse-borne disease* but not *tick-borne disease.* Louse-borne strains probably do not adapt to become tick-borne strains or vice-versa.

Relapsing fever in man consists of (1) tick-borne disease contracted from numerous species of ticks as biologic vectors that parasitize rodents and (2) louse-borne disease passed from person to person by human body and head lice, and sometimes the bedbug. *Borrelia recurrentis* grows on egg embryos; culture has recently been achieved on nonliving media.

Tick-borne relapsing fever is an endemic worldwide disease. The natural reservoirs for man are wild rodents and other small animals and their insect parasites, principally ticks of the genus *Ornithodoros.* The causative organism is passed transovarially in the tick. The ticks infect man primarily by shedding contaminated coxal gland fluid into the bite wound while feeding; tick-borne disease is seldom fatal.

Louse-borne relapsing fever is an epidemic disease that is often mistaken for typhus. The causative organism apparently has man and certain other primates as its only natural hosts. Organisms released from feeding lice crushed by scratching invade through the bite wound. Many *large epidemics* of *louse-borne* disease have occurred in the past with mortalities ranging from 5 to 70%. During the 1942 to 1944 epidemic in the Eastern Mediterranean region, 1 million cases and 50,000 deaths occurred. A pandemic in the early 1900s involved 50 million cases and 5 million deaths, usually from myocarditis.

The disease begins abruptly, usually within a few days after exposure. The symptoms are many; they often include high fever, nausea, photophobia, prostration, splenic enlargement, cough, thirst, epistaxis, delirium, jaundice and leukocytosis; sometimes petechiae appear in the skin. The organisms invade widely and grow freely in the blood stream. The organisms produce a heat-labile pyrogen which lacks the properties of classical endotoxin. Unopsonized live organisms are not readily phagocytized by neutrophils; on rare occasion they have been seen within endothelial cells.

The most singular aspect of relapsing fever, either tick-borne or louse-borne, is the relapsing nature of the fever. The initial fever, which lasts for 3 to 10 days, is followed by a remission of a few days before the second attack occurs. This cycle of recurrent fever and relapses of decreasing intensity may be repeated on as many as 4 or more occasions, each relapse being due to *a new and antigenically-distinct variant.* During each remission the organisms disappear from the blood, allegedly because of newly formed opsonins and bactericidal Abs specific for the Ag unique to the new variant that caused the attack.

In animal work it has been shown that serum taken after each attack protects against the organism(s) responsible for the preceding attack(s) but not succeeding attacks. Apparently any given strain of *Borr. recurrentis* is genetically capable of expressing certain antigenic phase variations that determine the number of relapses that can occur in the animal it infects.

Recent work with monoclonal Abs and Western blot analysis indicate that antigenic variations responsible for relapses are exceedingly complex and that a single serotype can suddenly convert to multiple mixtures of serotypes independent of Abs being produced. However, only one major protein unique to each serotype predominates.

Little is known about immunity to the organism except that cross-immunity between strains is limited, and that specific immunity has been passively transferred in the rabbit and in man on an experimental basis. The protective Abs are probably directed at a number of surface Ags. One of the 3 Abs of greatest importance in immunity achieves immobilization of organisms without C participation; a second achieves lysis with C participation and the third promotes phagocytosis of living organisms. To what extent these various Ab activities contribute to immunity is not known.

A formalized vaccine has been shown to be effective in the rabbit. However, the development of a useful vaccine for man is difficult because of the bewildering number of antigenically different strains present in any area of endemic disease.

Relapsing fever activates latent kala-azar, but the mechanism involved is not known.

Diagnosis of relapsing fever is often not made until a relapse(s) occurs, even though the organism is easily identified by observing typical borelliae in routine or thick blood smears specially stained with Giemsa stain, or alternatively in darkfield preparations of fresh blood; culture is possible and infection with some pathogenic strains can also be estabished in mice by injection of blood. To date no useful serologic tests have been developed.

Therapy with penicillin, tetracyclines, streptomycin or chloramphenicol is effective but is sometimes complicated by the Jarish-Hexheimer type reaction.

Preventive measures of importance are rodent and insect control and avoidance of insect bites. Practical vaccines and antisera have not been developed.

LEPTOSPIRA

A new classification has been suggested for organisms of the genus *Leptospira* in which the basic taxon is the serotype. Two species would be recognized, *L. interrogans,* to represent the pathogenic strains, and *L. biflexa,* to represent the nonpathogens (largely saprophytes). According to the new classification the prototype member of the pathogens (over 180 serotypes or serovars and 18 sero groups have been described), formerly called *L. icterohemorrhagiae,* would be designed *L. interrogans* serotype *icterohemorrhagiae.* The leptospira have a wide range of natural hosts, principally wild rodents and other small animals, in which they commonly produce mild or asymptomatic infection. Each of the 18 serogroups tends to have only one or a few natural hosts. Man and some domesticated animals serve as accidental and terminal hosts. In these hosts the disease can be serious and often fatal. Infections, known by many different names, have been described in numerous species of animals.

Leptospirosis is a broad term applied to any of the similar infections in different hosts due to the various leptospires. *The organisms produce systemic disease and reach the kidney tubules where they produce urease and colonize* in large numbers. Since such colonization occurs in unnatural as well as natural hosts, members of either of these host groups can serve as chronic urinary carriers. For example, the Norway rat, a common natural host of serotype *icterohemorrhagiae,* can remain a carrier for its lifetime.

Although many thousands of cases of human leptospirosis occur annually throughout the world, the disease is not of great importance in the USA where the annual number of hospitalized leptospirosis patients seldom exceeds 100. However, because the disease is often mild, the total number of cases far exceeds this number.

Members of the genus, Leptospira, are thin, enveloped, flexible, thread-like organisms 7 to 20 μm × 0.1 to 0.2 μm. They possess some 12 to 18 tightly coiled spirals and in most strains the terminal third of one end of the cell body is bent in the form of a hook (Fig. 30–3). Motility depends on an axial structure called the axistyle. The leptospires possess strain-specific surface Ags and a genus-specific somatic Ag. However, their role in determining serologic inter-

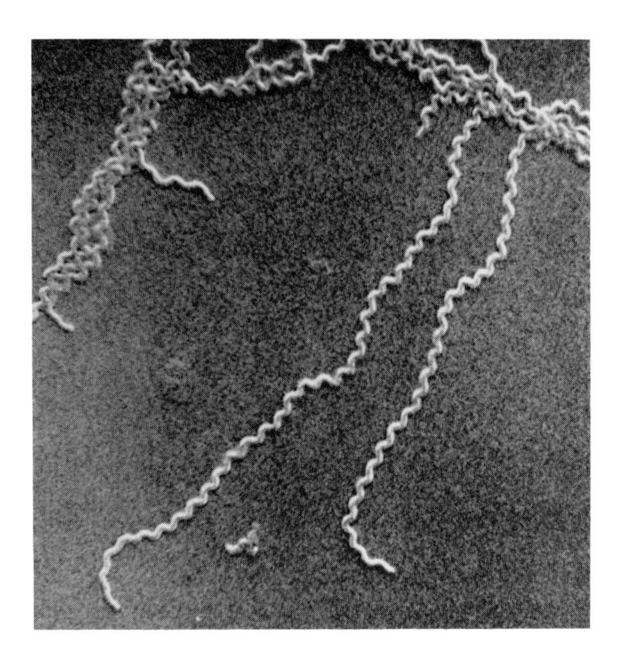

Fig. 30–3. *Leptospira interrogans* serotype *icterohemorrhagiae.* (From: *Atlas of Scanning Electron Microscopy in Microbiology.* Yoshi, Z., et al. Williams & Wilkins, Baltimore, 1976.)

relationships between strains and immunity has not been elucidated. The organism is highly mutable and mutants can be selected with immune serum.

The organism grows slowly under *aerobic conditions* at 30°C on serum-containing media (pH 7.2), on synthetic media and in the chick embryo, which it kills within a few days. Pathogens do not grow below about 15°C and do not multiply in natural waters. They survive freezing and are destroyed at 50° to 55°C.

Among laboratory animals, weanling rabbits, young guinea pigs, hamsters and chicks are the most susceptible; jaundice and hemorrhage are prominent features of experimental disease.

Human leptospirosis is commonly contracted by contact with organisms present in the urine of infected animals. *The organisms can remain alive for months in contaminated neutral or alkaline waters and wet soils to infect man through skin abrasions or mucous membranes, particularly of the mouth and conjunctiva.* Infection can also result from direct contact with urine or tissues of infected animals. For example, herdsmen in close contact with infected swine often contract the disease, hence the alternative term "swineherd's disease." *The disease in man is largely occupational,* being common among individuals working in water or damp surroundings where rats exist, such as in rice fields, around docks, in mines and in abattoirs. Epidemics have resulted from swimming in waters polluted with the urine of infected livestock. In the USA, dogs are a more frequent source of human infection than has been generally realized. Human infections are often subclinical. *Clinical leptospirosis is a biphasic systemic disease* with an incubation period of 1 to 2 weeks; its acute phases are characterized by bacteremia and organisms in the cerebrospinal fluid. The disease appears abruptly, after an incubation period of 3 to 30 days, with numerous symptoms, often including severe headache, chills, fever, muscle pain, cough, nausea and vomiting, conjunctival suffusion, cutaneous hyperesthesia and stiff neck. After 4 to 9 days the organisms disappear from the blood and CSF and a remission of 1 to 3 days ensues before the CNS symptoms and Abs characteristic of the second "immune phase" of 1 to 3 days duration appears. The second phase is often diagnosed as "aseptic meningitis." In rare instances a second relapse occurs during convalescence.

A clinically severe form of the disease with jaundice was described by Weil in 1886 *(Weil's disease)* and later identified as leptospirosis by Inada and Ido (1915), who discovered the etiologic agent. Jaundice is seen in about 25% of leptospirosis patients. It appears on about the 3rd to the 6th day of illness and does not peak until the 2nd stage is reached. Hepatic involvement, together with necrosis of renal tubules and serious renal dysfunction, occurs. Widespread hemorrhage from capillaries is observed in various sites including the skin, GI tract, adrenals and lung. Endocarditis and iridocyclitis are sometimes present. *Pretibial fever* (Fort Bragg fever), due to serotype *autumnalis,* was described in 1942. Its distinctive features are splenomegaly and a pretibial rash on about the 4th day of illness. Serum Abs appear after about 1 to 2 weeks of illness and persist for many years. They can be measured by several procedures, including agglutination. The immune adherence *phenomenon of Rieckenberg* is of particular interest; when leptospires are sensitized with Ab and C they adhere to RBCs for some unknown reason.

Mortality in leptospirosis, which usually results from renal failure, varies widely with the subject's age, being some 10% in patients under 50 years of age and about 55% in older patients. In jaundiced patients the mortality ranges from 15 to 40%.

Pathogenetic mechanisms in leptospirosis are largely obscure. Although endotoxin production is limited, a cytotoxic protein is produced; this has been reported to be present in greater abundance in culture fluids from virulent than from avirulent strains. A leptospiral *hemolysin* is evidently responsible for intravascular hemolysis and severe anemia. The agents and events responsible for jaundice, azotemia, thrombocytopenia, hypoprothrombinemia and hemorrhage have not been defined. Saprophytic leptospires adhere reversibly to solid substrates but whether or not pathogens adhere to host cells has not been studied.

The mechanisms of acquired immunity to leptospirosis are poorly understood. However, immunity to infection is largely serotype specific and little cross immunity between serotypes occurs. Convalescent serum containing IgM and IgG agglutinins protects guinea pigs against experimental infection even though the organisms are not readily killed in vitro by specific Ab and C. However, there is evidence that neutralizing Abs for the hemolysins of leptospires are formed; these may protect against invasion as well as against other pathologic effects of the organisms, including excessive red cell destruction and anemia. Although phagocytosis has been observed, its possible role in immunity has not been elucidated nor has cellular immunity been evaluated.

Laboratory diagnosis during the first week is based principally on daily blood culture and, if indicated, spinal fluid culture; thereafter urine should be cultured on selective media. Leptospires tend to be present in the blood in small numbers during the first week of disease and subsequently in the urine. Multiplication in vitro is slow and several days are required before growth appears. All leptospires utilize long chain fatty acids as a major carbon and energy source and require fatty acids for lipid synthesis. Because of the small numbers of leptospires in blood and their frequent confusion with blood debris, neither the direct smear nor the darkfield procedure is recommended. Providing suitable formalized serotype antigens are available for agglutination tests, a demonstrated rise in specific Ab titer during the course of the disease and during convalescence is of diagnostic significance; the development and evaluation of other serologic tests is being continued. Animal inoculation is a necessary and useful adjunct or substitute for culture in some instances.

Therapy includes administration of antimicrobial drugs as early as possible in the course of the disease. Unfortunately, early diagnosis is seldom made and treatment after the 5th day is of doubtful value. The agents used include penicillin G, tetracycline, streptomycin and the macrolide antibiotics. Supportive treatment is highly important. Peritoneal dialysis has been suggested when azotemia develops. No effective antiserum is available.

Prevention of disease rests largely on sanitary measures. Protective vaccination has been proposed for individuals with a high risk of infection.

LYME DISEASE SPIROCHAETE

The lyme disease spirochaete is probably a distinct genus with a world-wide distribution. It is known to be carried by at least 3 ticks of the genus Ixodes, *Ixodes damminis, Ixodes pacificus* and *Ixodes ricinus,* and completes its life cycle in the white-footed mouse and the white-tailed deer. The organism grows readily on *Borrelia*-media and has some properties in common with borreliae and treponemes. One proposal is that it be designated, *Borrelia burgdorferi.* Lyme disease in man comprises the clinical syndrome *erythema chronicum migrans* which may be attended by arthritis, meningitis and myocarditis. Antibodies of the IgM and IgG classes appear in 3 to 6 weeks. Treatment with penicillin and tetracycline is useful.

REFERENCES

Burke, J.P., et al. (eds.): International symposium on yaws and other endemic Treponematoses. Rev. Infect. Dis. *7*, Supplement 2, 1985.

Cockburn, T.A.: The origin of the treponematoses. WHO Bull. *24*:221, 1961.

Fitzgerald, T.J.: Pathogenesis and immunology of *Treponema pallidum.* Annu. Rev. Microbiol. *35*:29, 1981.

Harwood, C.S. and Canale-Parola, E.: Ecology of spirochetes. Annu. Rev. Microbiol. *38*:161, 1984.

Johnson, R.C., et al.: Active immunization of hamsters against experimental infection with *Borrelia burgdorferi.* Infect. Immun. *54*:897, 1986.

Marchitto, K.S., et al.: Monoclonal antibody analysis of specific antigenic similarities among pathogenic *Treponema pallidum* subspecies. Infect. Immun. *45*:660, 1984.

Norgard, M.V., et al.: Cloning and expression of the major 47-kilodalton surface immunogen of *Treponema pallidum* in *Escherichia coli.* Infect. Immun. *54*:500, 1986.

Rice, N.S., et al.: Demonstration of treponeme-like forms in cases of treated and untreated late syphilis and of treated early syphilis. Br. J. Vener. Dis. *46*:1, 1970.

Schell, R.F. and Musher, D.M.: Pathogenesis and Immunology of treponemal infection. In *Immunology Series.* N.B. Rose (ed.), New York, Marcel Dekker, 1983.

Sell, S. and Norris, S.J.: The biology, pathology and immunology of syphilis. Int. Rev. Exp. Pathol. *24*:203, 1983.

Stamm, L.V. and Bassford, P.J. Jr.: Cellular and extracellular protein antigens of *Treponema pallidum* synthesized during in vitro incubation of freshly extracted organisms. Infect. Immun. *47*:799, 1985.

Steere, A.C., et al.: The spirochetal etiology of Lyme disease. N. Engl. J. Med. *308*:733, 1983.

Titus, R.G. and Weiser, R.S.: Experimental syphilis in the rabbit: Passive transfer of immunity with immunoglobulin G from immune serum. J. Infect. Dis. *140*:904, 1979.

Willcox, R.R.: The treponemal evolution. Trans. St. Johns Hosp. Dermatol. Soc. *58*:21, 1972.

31

RICKETTSIA AND COXIELLA

Members of the genera *Rickettsia, Coxiella,* and *Rochalimaea* of the family *Rickettsiaceae* (order *Rickettsiales*), are obligate intracellular parasites that usually have arthropods as their natural hosts and participants in their life cycles. Some of them are highly pathogenic for man, an accidental host. The major rickettsial diseases of man and their etiologic agents are listed in Table 31–1. In contrast to the other rickettsiae, *Cox. burnetii* can survive extracellularly for long periods and hence is usually transmitted via inhalation of contaminated dusts and not by an arthropod vector.

It is apparent from Table 31–1 that the pathogenic members of the genus *Rickettsia* share many characteristics; for example, each of them is transmitted to man by an arthropod vector (lice, fleas, or mites); each spreads systemically to cause a febrile disease attended with a rash and headache. However, the rickettsioses differ in their severity and in both the type and the distribution of the rash. Rashes are of two sorts *typhus-like* and *spotted fever-like* or *pox-like.* The cycle of transmission of some of the pathogenic members of the genus *Rickettsia* is also indicated in Table 31–1.

In this chapter, *R. prowazekii* will be discussed in greatest detail. Many of its properties apply to the other rickettsiae and to *Cox. burnetii.*

RICKETTSIA PROWAZEKII AND *RICKETTSIA TYPHI*

A. MEDICAL PERSPECTIVES

Rickettsia prowazekii causes epidemic louse-borne typhus and *R. typhi* causes rat flea-borne endemic typhus (murine typhus). In common with certain other infectious diseases, typhus has played a larger part in shaping the history of the world than has any single ruler or even any nation. Epidemics of typhus have been responsible for deciding the outcome of many battles and campaigns including Napoleon's retreat from Moscow. It is probable that typhus fever was introduced into Spain around 1490. Ferdinand and Isabella, while battling the Moors for possession of Granada, fought a far greater battle with typhus; in one campaign during that war, 3,000 soldiers were killed by the Moors, whereas 17,000 died of typhus fever! The disease followed armies of that time, spreading across Europe to the East, and gradually to the North.

One of the most violent epidemics of typhus in all history occurred in Serbia near the start of World War I. This epidemic may well have been a crucial factor in deciding the ultimate outcome of the war. In less than 6 months, over 150,000 people died of typhus, including about a third of all the physicians of Serbia. Although in an excellent position to sweep through Serbia, the attacking Austrians knew better than to enter the area and were delayed for 6 crucial months until the epidemic subsided.

Russia bore the brunt of epidemics of typhus and a number of other infectious diseases during the turbulent years of World War I and the Russian revolution. At least 25 million cases of typhus fever occurred in Russia during the years 1917 to 1921, resulting in 2.5 to 3 million deaths. Although subsequent outbreaks of typhus have occurred, they have not been of such magnitude.

The striking coincidence of wars and typhus epidemics is not accidental, but is simply due to the fact that the rickettsiae of epidemic typhus are transmitted via the bite of the body louse, the bane of military personnel in the

Table 31–1. Some Characteristics of the Major Rickettsioses of Man

Disease	*Etiologic Agent*	*Commonly Transmitted to Man Via:*	*Principal Reservoir(s)*	*Principal Geographic Sites*	*Symptoms*	*Case Mortality Rate*	
						Untreated	*Treated*
Epidemic typhus	*Rickettsia prowazekii*	Body lice Head lice	Man with active or latent infection	Asia, Africa, South America	High fever (104°F or more); rash beginning on trunk and spreading peripherally	10% to 40%	5%
Endemic (murine) typhus	*R. typhi (R. mooseri)*	Rat to rat fleas to man	Rat Rat flea	Worldwide	Fever and rash as above, usually less severe than epidemic typhus	<5%	<5%
Rocky Mountain spotted fever (RMSF)	*R. rickettsii*	Wood ticks or dog ticks to man	Rodents, Dogs, Dog tick, Wood tick	The Americas	High fever; rash beginning peripherally and spreading to the trunk, palms of hands, soles of feet	5% to 80% average 20%	6%
Rickettsial pox	*R. akari*	House mice to mites to man	House Mouse	USA, Russia, Korea, Africa	Vaccinia-like lesion at site of infection by mite; fever, vesicular rash resembling chickenpox	No deaths reported	No deaths reported
Scrub typhus	*R. tsutsugamushi (R. orientalis)*	Wild rodents to mites to man	Wild rodents, Mites	Asiatic Pacific Region	Indurated lesion at site of infection by mite; fever, rash, and pneumonitis	1% to 50%	< than untreated
Q Fever	*Coxiella burnetii*	Inhalation of barnyard dust	Marsupial rats, ticks, cattle, sheep, milk	Worldwide	Pneumonitis, no rash, endocarditis and pneumonia rare	Rarely fatal <1%	Rarely fatal <1%

field. Whenever conditions of crowding or poor hygiene (lack of opportunity to bathe and launder clothing) permit body lice to proliferate on man, typhus is likely to appear. Outbreaks of epidemic typhus occur frequently in certain populations of the world, due to poor personal hygiene and the persistence of latent infection in some individuals; however, epidemic typhus occurs only rarely in the USA.

Louse-borne trench fever caused by *Rochalimaea quintana* was prevalent in the armies of World War I and in the Eastern Front armies of Wold War II.

B. PHYSICAL AND CHEMICAL STRUCTURE

Rickettsiae are tiny nonmotile (about 0.3 to 0.6 μm), Gram-negative coccobacilli that occur singly, in pairs, or short chains. They are pleomorphic and may also take the form of rods or filaments. Giemsa or Macchiavello stains are commonly used to demonstrate rickettsiae; they appear blue with Giemsa and red with the Macchiavello strain. Their typical Gram-negative bacterial cell walls contain peptidoglycan and an endotoxin-like lipopolysaccharide; *R. prowazekii* presents an amorphous capsule. Both group-specific and type-specific Ags are demonstrable by the CF test. The Weil-Felix test involving Abs that cross-react with *Proteus* polysaccharides is useful in the diagnosis of most rickettsial diseases.

C. GENETICS

Although rickettsiae share antigenic determinants with some strains of the genus *Proteus,* there is no evidence to indicate that the two organisms are genetically related.

D. EXTRACELLULAR PRODUCTS

Cytolytic toxins are produced by some rickettsiae; they lyse sheep and rabbit but not human erythrocytes. Although the chemical nature of the toxins has not been defined, one component is evidently endotoxin. Anti-endotoxin Abs can protect mice against toxic death induced with large doses of organisms.

E. CULTURE

Rickettsiae grow readily in the yolk sac of embryonated chicken eggs. They can also be propagated in rapidly-growing cultured cells. Although members of the genus *Rickettsia* can only grow within host cells, one member of the *Rickettsiaceae, Rochalimaea quintana* (the cause of trench fever), can be grown on blood agar in the absence of host cells. Under optimal conditions the generation time of rickettsiae is about 18 hours.

Although rickettsiae have lost some of their metabolic capabilities, they have retained the capacity to carry out certain energy-yielding reactions, e.g., oxidative phosphorylation, and to synthesize proteins from amino acids. It has been postulated that they possess an extremely permeable cytoplasmic membrane that permits them to utilize essential preformed nutrients from the host cell; however, it is possible that this apparent "leakiness" may be an artifact caused by laboratory manipulation of the organisms.

F. RESISTANCE TO PHYSICAL AND CHEMICAL AGENTS

Extracellular rickettsiae cease to metabolize and gradually lose infectivity; they are moderately susceptible to heat and most of the common disinfectants but can survive in dried arthropod feces for months.

G. EXPERIMENTAL MODELS

Guinea pigs and mice can be infected with rickettsiae, but the present trend is toward utilizing cell cultures or infected embryonated chicken eggs to study the interaction of rickettsiae with host cells. Arthropod hosts have also been used for experimental studies.

H. INFECTIONS IN MAN

Man is the only natural host of *R. prowazekii*, the etiologic agent of epidemic typhus fever, a worldwide disease. The body louse, *Pediculus corporis*, is the usual vector, although other species of *Pediculus*, especially *P. capitis*, the head louse, can transmit the infection from man to man. The body louse lives in clothing, kept warm by body heat. Several times a day it takes a meal of blood from its human host. While feeding, the louse defecates; because the bite is irritating, the person usually scratches and rubs louse feces into the open wound. If the feeding louse is infected with *R. prowazekii*, the human host readily becomes infected. Conversely, if the human host is infected, the uninfected feeding louse becomes infected. The rickettsiae multiply within cells of the intestinal tract of the louse and huge numbers are shed into the feces. As Zinsser has pointed out in his erudite account of typhus fever, "Rats, Lice and History," the louse fares worse than man. In fact, *R. prowazekii* infections are invariably fatal for lice within 1 to 2 weeks; this is in contrast to most rickettsial infections of arthropods, which do not lead to disease of the vector.

A mild recrudescent form of *R. prowazekii* infection is called Brill's disease. This mild form of typhus fever is of short duration and does not produce the characteristic rash, but leads to a specific IgG Ab response of the anamnestic type. Brill's disease represents the reactivation of an inapparent or latent infection with *R. prowazekii* that often remains silent for many years. Individuals with latent infection and patients with Brill's disease can initiate new epidemics of typhus fever if they become louse-infested. Brill's disease is now being observed among immigrants entering the USA. Also flying squirrels in Southestern USA have recently been found to carry *R. prowazekii* and a few cases of epidemic typhus have apparently resulted from contact with flying squirrels. This potential source of infection has become a matter of public health concern.

Epidemic typhus fever is often fatal (10 to 40%), and lice feeding on patients who die of typhus become heavily infected. Soon after death, as the body temperature decreases, these lice leave the corpse and seek a new host to provide warmth and nourishment. Conditions of crowding favor the spread of typhus, because crowding provides ideal opportunities for new hosts to become infested, and subsequently infected, before the lice succumb to the rickettsiae. The symptoms of epidemic typhus fever appear 5 days to 3 weeks after exposure to infected lice. For several days the disease resembles influenza, with high fever (103° to 104° F or higher), chills, severe headache, and depression. On the 4th or 5th day after onset, the characteristic rash appears on the trunk and gradually extends peripherally. The original pink spots become purplish and finally brown as the rash fades. The face is usually free of rash, and no spots are seen on the mucous membranes. Cough and CNS manifestations are not uncommon. The name "typhus" is derived from a Greek word "hazy," because of the frequent mental haziness and delirium that occurs.

Endemic typhus or murine typhus of man has many of the characteristics of epidemic typhus, but is generally much less severe. The causative organism, *R. typhi*, produces only inapparent latent infections in the natural rat host and in the fleas that are vectors for its transmission from rat to rat, and from rat to man. The incidence of endemic typhus in the USA has dropped markedly and fewer than 100 cases are reported annually. Biotypes of *R. typhi* occur: one of them, *R. canada* has caused outbreaks in Canada. The few deaths that result from endemic typhus are usually in elderly or debilitated persons.

I. MECHANISMS OF PATHOGENESIS

The rickettsiae contain endotoxin-like lipopolysaccharides that, in large doses, kill laboratory animals within a few hours. However, rickettsial toxins differ from most endotoxins in that they act strongly to incite the production of protective Abs. The fever, rash and intravascular coagulation produced during the rickettsioses are probably due, in part at least, to rickettsial endotoxins.

The principal sites of proliferation of rickettsiae in infected hosts are within phagocytes and endothelial cells of small blood vessels. Parasitized endothelial-cells proliferate and at least some are killed. Inflammation develops and thrombosis of the vessels occurs.

The steps in parasitization are adherence followed promptly by endocytosis and phagosome destruction. Apparently adhesins on the rickettsial surface combine with cholesterol-containing receptors of surface components of the host cell. Both cells must be viable and engulfment is rapid. After engulfment most rickettsiae promptly destroy the phagosomal membrane by phospholipase action and then multiply within the cytoplasm; at least some host cells are killed with attending release of organisms. However, organisms can also be released from living host cells by exocytosis. The regulation of phospholipase A activity with relation to the destruction of phagosomal and cytoplasmic membranes is not understood. In vitro observations with RBCs and L cells have shown that mere adherence can lead to cell lysis as the result of destruction of the cytoplasmic membrane by phospholipase. Rickettsiae of the spotted fever group are alleged to grow in the nucleus as well as the cytoplasm of cells.

J. MECHANISMS OF IMMUNITY

The results of both clinical experience and animal studies indicate that IgG Abs protect against rickettsial infection. Some IgG Abs probably block the entrance of organisms into host endothelial cells, whereas others appear to act as anti-endotoxins. Specific Abs of the class IgG that protect guinea pigs against challenge with *R. prowazekii* are, at least in part, directed against soluble Ags associated with cell-wall endotoxins.

Peritoneal macrophages of immunized animals phagocytize the specific organisms more rapidly than do normal macrophages, especially in the presence of immune serum. It is also probable that Ab-mediated inhibition of adherence to endothelial cells represents a protective mechanism of acquired immunity.

A clear understanding of the pathogenesis and immunology of the rickettsioses is sorely needed and remains a major concern of the USPHS.

In typhus, as in most infectious diseases, early Abs are predominantly of the IgM class and late Abs are of the class IgG. Although it is probable that latent infection seldom follows cases of epidemic typhus, antigenic stimulation is well maintained during latency, as shown by the prompt production of 7S but not 19S Abs during recrudescent disease (Brill's disease). The location of rickettsiae during latent disease and the means by which they persist and evade host defenses have not been determined. The observation that physiologic resting forms remain unchanged within cultured cells over a period of days to months suggests that similar inactive, resting organisms may persist within host cells in vivo.

K. LABORATORY DIAGNOSIS

Rickettsiae can often be cultured from clinical specimens and animals or from vectors, although laboratory personnel run high risk of becoming infected in the course of carrying out the procedures involved. The hazards of working with rickettsiae in the laboratory are emphasized by their history; the organism, *Rickettsia prowazekii,* was so named for Howard Ricketts and Stanislaus von Prowazek, two investigators who died from infections contracted while working with these organisms. Because of the high risks involved in laboratory culture, serologic testing of the patient's serum with commercial Ags is commonly used as an alternative. Methods used for serologic diagnosis are similar in most of the rickettsial diseases.

A specific CF or agglutination test with suspensions of Ags prepared from yolk-sac cultures of rickettsiae is the most reliable method for diagnosis. A 4-fold rise in the titers of Abs against cross-reactive Ags of various strains of *Proteus vulgaris* in the Weil-Felix reaction also serves to indicate the nature of the infecting agent (Table 31–2). A rise in titer of specific CF Abs in the patient's serum during the course of the disease serves to identify the infecting species of rickettsiae with reasonable certainty. Other useful serologic tests conducted in special laboratories include the ELISA test, latex agglutination test, hemagglutination test, IF test and the mouse toxin neutralization test.

Table 31–2. Weil-Felix Reactions in Rickettsial Diseases

	Antigen		
Serum From Patient With:	*OX-19*	*OX-2*	*OX-K*
Epidemic typhus	+	+	–
Brill-Zinser disease	+/–	+/–	–
Endemic typhus	+	+	–
Scrub typhus	–	–	+
Rocky Mountain spotted fever	+	+	–
Q fever	–	–	–
Rickettsial pox	–	–	–

L. THERAPY

When administered early after infection, tetracyclines and chloramphenicol are therapeutically effective against all of the rickettsiae. Since the organisms are inhibited but not killed, treatment should be continued for 4 to 5 days after the fever subsides.

Sulfonamides should not be used for treating the rickettsioses because they enhance growth of the organisms and promote the disease process.

M. RESERVOIRS OF INFECTION

The reservoirs of rickettsial disease are indicated by Table 31–1.

N. CONTROL OF DISEASE TRANSMISSION

Rickettsial diseases are best prevented by interrupting the cycle of infection. Epidemic typhus depends on transmission by lice; therefore, elimination of lice prevents its spread. Endemic typhus requires rats and rat fleas for its transmission from rat to rat and rat to man; hence the elimination of either the rat or the flea, or both, will prevent transmission of the disease to man. In areas where infected rats are known to exist, insecticides should first be used to get rid of the flea vectors in the environment before using rat poisons. If rats are killed first, their infected fleas will quickly seek man as a source of nourishment and will thus intensify the spread of infection.

Vaccines consisting of formalin-killed rickettsiae prepared from cultures in yolk sacs of chick embryos are available for immunization against typhus and certain other rickettsioses. The typhus vaccine contains both *R. prowazekii* and *R. typhi*; it lessens the severity and mortality of typhus fevers, but does not decrease their incidence. Rickettsial vaccines are recommended only for persons at high risk of contracting the rickettsioses.

RICKETTSIA RICKETTSII

Rocky mountain spotted fever (RMSF), which is sometimes called rickettsial spotted fever or tick-borne typhus, is caused by *R. rickettsii*; it is not uncommon in the USA. About 1000 cases are reported annually to the USPHS. The disease may occur wherever the species of *Dermacentor* ticks that serve as the vectors for *R. rickettsii* are found and is the prototype of the spotted fevers. Although numerous cases are reported from the Rocky Mountain area, the disease is most prevalent in the Eastern USA (Fig. 31–1). Organisms closely related to *R. rickettsii* occur which might be variants; whether they deserve species status is uncertain.

Reservoirs of RMSF comprise various rodents, dogs, and other animals. Ticks feeding on contaminated blood become systemically infected; their ovaries permit rickettsial growth and infected ova transmit the organisms from one generation of ticks to another. Growth of rickettsiae in the salivary glands of the tick and the presence of organisms in their saliva permits the introduction of large doses of rickettsiae into the skin of the human host when the tick feeds.

The spotted fevers of man due to *R. rickettsii* have a relatively short incubation period of 3 to 12 days, and an onset similar to that of typhus fever or influenza, with high fever and chills. The rash appears earlier than in typhus,

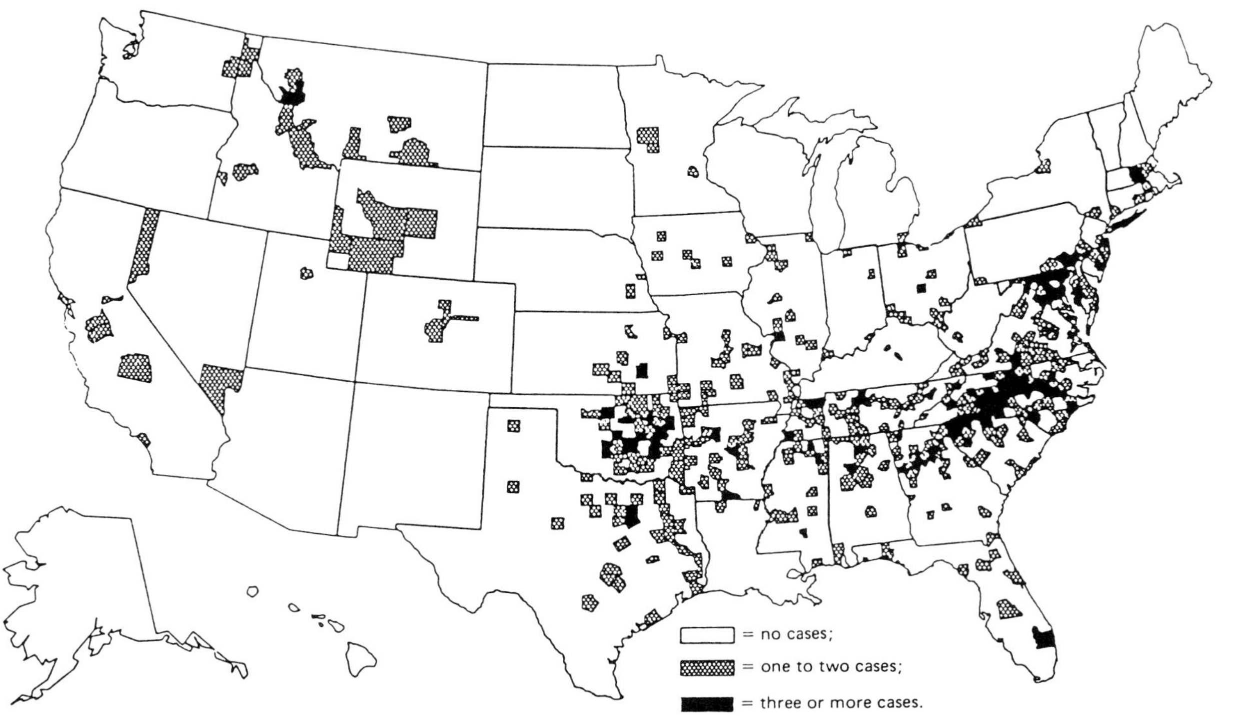

Fig. 31–1. Rocky Mountain spotted fever. Number of cases reported by county in the United States, 1978. *(Reprinted from Rickettsial Disease Surveillance Summary 1978. Centers for Disease Control. U.S. Department of Health and Human Services.)*

becoming apparent on the 2nd to 5th day of illness. The eruption begins peripherally, on the feet, hands, or forehead, and spreads to the trunk. The fever usually lasts for 2 to 3 weeks and subsides gradually. Most cases occur in individuals under 15 years of age. In fatal cases, death most often occurs during the 2nd week of illness. Major symptoms of the spotted fevers and the related rickettsial pox due to *R. akari* are briefly described in Table 31–1. Depending on the geographic area the mortality in untreated cases ranges from 5% to 80% possibly due to virulence differences among strains.

It has been shown that avirulent strains of R. rickettsii fail to attach to human host cells, and cannot infect them. Studies on the virulence of *R. rickettsii* in guinea pigs have shown that organisms growing in ticks are avirulent until after a blood meal is taken. The change from avirulence to virulence has also been studied in vitro. It depends on utilization of factors from host RBCs, and occurs at 35°C but not at low temperatures.

The principal Abs responsible for acquired immunity to RMSF are probably anti-adhesins, although other Abs may act as intracellular "antitoxins"; whereas rickettsiae exposed to specific Abs are engulfed and killed by peritoneal macrophages of the guinea pig, unexposed rickettsiae are phagocytized and grow.

Diagnosis of RMSF is based on clinical and serologic findings including the Weil-Felix reaction, the microagglutination test, the FA test and the type-specific CF test conducted on sera taken at intervals to detect rises in Ab levels. A 4-fold rise in titer occurring during the course of infection, a Weil-Felix titer above 1:160 in early disease, and characteristic clinical symptoms constitute a presumptive diagnosis.

Chemotherapy is usually conducted with chloramphenicol, the drug of choice, with tetracycline as an alternate drug.

Control of infection by these organisms includes clearing brush where ticks and mites propagate; in addition, the use of insecticides and rodent control are excllent measures for suppressing tick and mite populations and preventing the transmission of some of these rickettsioses. Cattle dips to keep livestock free of ticks is also valuable. In endemic areas ticks should be watched for and promptly removed from the human body. A period of several hours is required after attachment of the tick before virulent rickettsiae are passed to the host; thus, prompt removal of ticks may prevent transmission of infection.

RICKETTSIA AKARI

Rickettsial pox caused by *R. akari* is a mild disease characterized by a vesicular rash that resembles chicken pox. The disease occurs chiefly in urban buildings heavily infested with mice. Man acquires the disease through the bite of the mouse mite. Mite bites characteristically give rise to an eschar (black, hemorrhagic lesion) and may be used as a diagnostic aid. Rickettsial pox is found world-wide and several hundred cases are diagnosed in the USA each year. The disease is not fatal but may predispose immunocompromised patients to more serious secondary infections.

RICKETTSIA TSUTSUGAMUSHI

Rickettsia tsutsugamushi (orientalis) is the cause of a typhus-like disease known either as tsutsugamushi fever or scrub typhus or chigger-borne typhus.

The disease is transmitted by the bite of a mite larva (chigger). It differs from epidemic typhus in that a punched-out eschar is commonly present at the location of the site of the bite. Transovarial transmission of the rickettsiae occurs in certain mites and allows them to serve as reservoirs of scrub typhus. Tsutsugamushi fever occurs over a wide area in the Asiatic Pacific region, including Japan, Korea, Formosa, Malaya, Burma, New Guinea and the Philippines. The fatality rate varies from near 0 to 50% and chemotherapy is highly successful.

Preparing a suitable vaccine has been difficult because of strain variations with respect to protective immunogens.

COXIELLA BURNETII

Coxiella burnetii is the cause of Q fever, a worldwide disease that differs from the rickettsial diseases in several ways. Originally described in Australia, human Q fever was contracted by man from the marsupial rat as a natural host and reservoir. The disease is most common in Australia, the Mediterranean area and Western USA. Ticks, which can serve as both vectors and reservoirs for *Cox. burnetii,* are responsible for transmitting the organisms among cattle and sheep. Infected domestic animals disseminate huge numbers of the organisms in secretions and excreta. *Coxiella burnetii* is much more resistant to adverse conditions than are most of the rickettsiae; it not only survives drying, but also is more resistant to heat than most pathogens. Therefore, pasteurization of milk at 71.5°C (161° F) for 15 seconds is necessary to assure killing of coxiellae.

Human Q fever is usually contracted via inhaled contaminated dusts, often from barns or animal sheds; improperly pasteurized milk is sometimes a source of infection. The disease is a febrile pneumonitis, but, in contrast to most rickettsial diseases, a rash is not usually seen and the disease is rarely fatal unless severe pneumonia or endocarditis develops. Serum Abs may persist for as long as 7 years after an attack of Q fever, suggesting that *Cox. burnetii* may become latent within immune hosts.

Coxiella burnetii is morphologically identical to the rickettsiae but forms endospore-like bodies. In nature, the organism possesses a surface polysaccharide coat (Phase I) that is antiphagocytic. Following adaptation to growth in the chick embryo, this surface Ag is lost; these organisms are known as Phase II strains. It appears that as infection progresses, parasites and host so accommodate to each other that both proceed to propagate freely.

Following entrance into cells by pinocytosis or phagocytosis, *Cox. burnetii* does not disrupt the phagosomal membrane, but instead, multiplies within the phagosome as do most intracellular bacteria. Growth occurs within the phagosomes of both professional and nonprofessional phagocytes. There is evidence that coxiellae interfere with the regulation of energy metabolism by host cells, but have low cytotoxicity.

The Weil-Felix test is negative but CF tests are useful for diagnosis. Treatment is sometimes indicated, especially when endocarditis is suspected. Tetracycline alone or combined with trimethoprim-sulfamethoxazole is used for treating cases of endocarditis.

REFERENCES

Burgdorfer, W, and Anacker, R.L.: *Rickettsia and Rickettsial Diseases.* New York, Academic Press, 1981.
Font-Creus, B., et al.: Mediterranean spotted fever: a cooperative study of 227 cases. Rev. Infect. Dis. *7*:635, 1985.
Hase, T.: Developmental sequence and surface membrane assembly of Rickettsiae. Annu. Rev. Microbiol. *39*:69, 1985.
Hattwick, M.A., et al.: Rocky Mountain spotted fever: epidemiology of an increasing problem. Ann. Intern. Med. *84*:732, 1976.
Kaufman, R.S.: Rickettsial pneumonias. Seminars in Infect. Dis. *5*:268, 1983.
McDade, J.E. and Newhouse, V.F.: Natural history of *Rickettsia rickettsii.* Annu. Rev. Microbiol. *40:* 287, 1986.
Moulder, J.W.: Comparative biology of intracellular parasitism. Microbiol. Rev. *49*:298, 1985.
Ormsbee, R.A.: Rickettsiae, Chapter 84. *Manual of Clinical Microbiology,* 4th Ed. Editor-in-Chief, E.H. Lennette. Wash. DC, Am. Soc. Microbiol. 1985.
Walker, D.H.: Rickettsial Diseases: an update. *Current Topics in Inflammation and Infection.* Baltimore, Williams & Wilkins, 1982, p. 188.
Zinsser, H.: *Rats, Lice, and History.* Boston, Little Brown & Co., 1935.

32

CHLAMYDIA

Organisms of the genus *Chlamydia* belong to the family *chlamydiaceae* and the order *Chlamydiales.* The *Chlamydia* are tiny, coccoid, nonmotile Gram-negative, obligate, intracellular bacteria; they resemble the rickettsiae in many ways but are not transmitted by arthropods. The *Chlamydia* differ from all other bacteria by having a unique 2-stage intracellular developmental cycle, but nevertheless possess all of the usual characteristics of bacteria except that they lack some of the metabolic activities needed for extracellular reproduction. The *Chlamydia* infect a wide range of vertebrate hosts.

At present only two species of *Chlamydia* are recognized, *Chlamydia trachomatis* and *Chlamydia psittaci.* The two species can be differentiated by two simple criteria; *Chl. trachomatis* produces glycogen rich iodine-staining cytoplasmic inclusions and is inhibited by sulfonamides, whereas *Chl. psittaci* lacks both of these characteristics.

CHLAMYDIA TRACHOMATIS

Chlamydia trachomatis variants belong to three distinct biologic groups with respect to the disease syndromes they produce, namely, classical ocular trachoma, the inclusion conjunctivitis-genitourinary complex of diseases and lymphogranuloma venereum (LGV) (Table 32–1). These three biologic groups are sometimes referred to as "agents" of the different infections and spectrums of disease they produce.

A. MEDICAL PERSPECTIVES

Trachoma was the first chlamydial disease to be recognized as a distinct clinical entity. Trachoma is a Greek-derived term meaning "rough"; its original clinical usage was to designate pebble-like lesions of the eyelid.

Reference to trachoma as a "contagious eye disease" appeared in Greek writings of the 4th century B.C. and Egyptian writings of the 2nd century B.C. Aristophanes alluded to individuals who resorted to self-induced "contagious ophthalmia" to avoid marine service and Xenophon wrote about a contagious eye disease among Greek troops in "The Retreat of the Ten Thousand." Hippocrates and Celus mentioned the use of copper ointments for treating the disease. In common with other highly contagious diseases, trachoma becomes epidemic in armies in the field where soldiers live close together under poor hygienic conditions. For example "military ophthalmia" plagued the armies of Napoleon during his Egyptian campaign.

Trachoma is seen only occasionally in the USA, principally among the Indians of the Southwest. On a worldwide scale, however, the WHO estimates that 500 million to 1 billion people currently have trachoma, and that 6 million of them are totally blind.

The observation that *Chl. trachomatis* is an agent of oculogenital disease and that the agent can be transmitted sexually as well as by the ocular route was made in the early 1900s but it was not until 1963 that the relationship between genital and ocular disease was fully recognized. Indeed, it is only recently that emphasis is being placed on *Chl. trachomatis* as a cause of sexually-transmitted infections as well as infections of the eye. Man is the principal natural host of *Chl. trachomatis*; most infections remain localized to their target cells, the

Table 32–1. Some Characteristics of *Chlamydia trachomatis* Biogroups

Biogroup	*Infecting Serotypes*	*Infections Produced*	*Mode of Transmission*	*Principal Regions of Occurrence*
Classical trachoma agent	A,B,C	Classical trachoma	Direct contact, fomites, flies	Arid and subarid
TRIC agent	D,E,F,H, I,J,K	Inclusion conjunctivitis, urogenital tract complex of diseases, pneumonia	Direct contact, fomites, sex acts	Worldwide
Lymphogranuloma venereum agent	L_1,L_2,L_3	Lymphogranuloma venereum	Sex acts	Tropical regions

The serotypes that cause Lymphogranuloma venereum are sufficiently unique that they will eventually probably be designated a separate species. Public health officials tend to lump the serotypes of LGV and the urethral/cervical group together as organisms that cause STD (i.e., sexually transmitted disease).

With respect to the mode of transmission and type of disease, the boundaries of the various serotypes of *C. trachomatis* are not rigid. All types can cause inclusion conjunctivitis and, potentially, trachoma. Furthermore, STD organisms can be transmitted nonsexually (although that is not their normal means of perpetuation in nature), while nonsexual types can probably be transmitted sexually (although that is not their normal means of transmission).

squamoepithelium of mucosae. The exception is the LGV group of organisms that infect lymphoid cells and hence spread via draining lymph nodes.

Lymphogranuloma venereum was described as a distinct clinical entity of unknown infectious nature during the 18th century; currently, some 500 cases are reported in the USA each year.

Tremendous research efforts are being made to learn more about *Chl. trachomatis* infections and high priority has been given to the development of effective vaccines. Until such vaccines become available, or until sanitation and personal hygiene become universal, the incidence of infections due to *Chl. trachomatis* will probably not change appreciably and trachoma will remain the most common cause of blindness.

B. PHYSICAL AND CHEMICAL STRUCTURE

Two forms of *Chlamydia* occur during the 2-phase developmental cycle of intracellular growth (Sect. E), the *elementary body (EB) or infectious particle,* and the reproductive form, the reticulate body (RB). After a phase of intracellular multiplication the RB gives rise to the EB to complete the cycle (Sect. E, Fig. 32–1).

Since the genus *Chlamydia* is comprised of closely-related but distinct organisms, it is not surprising to find that its members share a common group-specific Ag as an expression of relatedness and that differing Ags occur as an expression of individual strain differences. A monoclonal Ab preparation has been produced against an outer membrane protein of *Chl. trachomatis* that distinguishes it from *Chl. psittaci.* Monoclonal Abs will undoubtedly soon be produced that will identify most of the various strains of *Chlamydia.*

The group-specific Ag that characterizes the genus *Chlamydia* is a heat-stable cell wall carbohydrate-lipoprotein complex that carries a 2-keto-3 deoxyoctanoic acid moiety that serves as the immunodominant epitope; it can be dem-

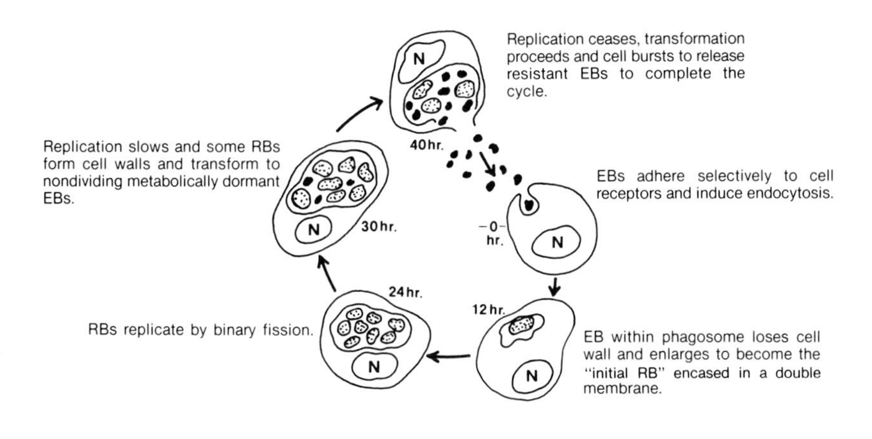

Fig. 32–1. The replicative cycle of *Chlamydia.*

onstrated by the CF test. The Ags that differentiate strains of chlamydiae are toxic heat-labile outer membrane proteins of EBs that are demonstrable by tests using specific antisera, namely the microimmunofluorescence test (Micro-IF test) and/or the neutralization tests for mouse toxicity and tissue culture infectivity.

Antisera against species-specific Ags can be used to determine whether a given strain belongs to one or the other of the two species; for example a monoclonal Ab specific for *Chl. trachomatis* has been produced. Each of the various strains of *Chl. trachomatis* has been found to contain a *major outer membrane protein* (MOMP) that comprises about 60% of the total cell wall and surface proteins. Each MOMP molecule has a constant region and a variable region and each of the identical MOMP molecules of a given strain of *Chl. trachomatis* contains both strain-specific epitopes and species-specific epitopes i.e., each molecule presents a mosaic of different epitopes.

By the use of antisera against various strains of *Chl. trachomatis* it has been possible to separate them into at least fifteen different groups or serotypes (serovars) designated A through M including L_1, L_2, and L_3. Moreover, as mentioned above, the serotypes fall into three groups (biogroups or biotypes) with respect to the diseases they produce (Table 32–1), namely, the classical trachoma agent, the TRIC agent and the lymphogranuloma venereum agent groups.

C. GENETICS

Antibiotic-resistant mutant strains of *Chl. trachomatis* have been selected in the laboratory.

D. EXTRACELLULAR PRODUCTS

Little is known about extracellular products of *Chl. trachomatis.* However, group specific hemagglutinins are shed into the media of infected cell cultures; they agglutinate the RBCs of mice, hamsters and chickens and appear to be complexes composed of DNA, lecithin and protein. Hemagglutinating activity is neutralized by specific Abs.

E. CULTURE

Chlamydia trachomatis shares with all chlamydiae a unique developmental cycle which can be demonstrated in infected cell cultures (Fig. 32–1). The tiny elementary bodies (infectious particles) adhere selectively to receptors on susceptible squamocolumnar epithelial cells and induce their endocytosis (phagocytosis).* Elementary bodies (EBs) inhibit phagosome-lysosome fusion and persist in the phagosome where they transform to the larger RNA-rich reproductive forms called *reticulate bodies* (RBs) that stain blue with the Giemsa stain. The RBs divide by binary fission and eventually transform to EBs to complete the developmental cycle.

Although events during the developmental cycle are not fully understood,

---

*Endocytosis is a broad term for designating both engulfment of particulates (phagocytosis) and engulfment of fluid droplets (pinocytosis): The term phagocytosis, is often used to designate engulfment of particulates by "professional phagocytes" especially large particles. The vesicle formed by the cytoplasmic membrane as the result of endocytosis is designated by various synonyms, including phagocytic vacuole, endocytic vesicle, endocytic vacuole, cytoplasmic vesicle and phagosome.

the population of organisms at the end of the logarithmic phase is a mixture of RBs and EBs. The organism-packed phagosomes at this stage constitute the *inclusion bodies* that stain purple with the Giemsa stain. Transformation of RBs to EBs accelerates during the stationary phase of the growth cycle.

In the course of reproduction of RBs some of the host cells may burst and die, especially host cells in which *high* multiplication of RBs occurs. Those cells in which *low* multiplicity occurs live and reproduce. Thus within a culture some cells may burst to provide EBs to infect new host cells, whereas other host cells may reproduce and transmit infection to daughter cells. Whether organisms may pass directly from infected to uninfected cells without extracellular exposure is not known. The EBs survive well in the extracellular environment, whereas RBs do not.

It is of singular interest that antisera against *Chl. trachomatis* block attachment of EBs to host epithelial cells and, moreover, that if endocytosis of antiserum-treated EBs is induced by centrifugation the organisms fail to transform and grow.

Evidently in the course of evolutionary adaptation to obligate intracellular parasitism, chlamydiae have retained little or no ability to generate energy for their own needs and hence must depend on the host cell for energy. The high permeability of the cytoplasmic membrane of the RBs also facilitates the uptake of many needed host cell metabolites including ATP.

Chlamydia trachomatis can be grown in large quantities in embryonated chicken eggs. It can also be propagated in cultures of susceptible host cells, especially if the cells have been irradiated or treated with growth-inhibiting chemicals.

F. RESISTANCE TO PHYSICAL AND CHEMICAL AGENTS

Extracellular EBs are more readily inactivated than most bacteria. At room temperature or above most strains lose infectivity within 8 hours to several days depending on the strain and the nature of the suspending medium. Many strains are inactivated within 5 to 10 minutes at 56°C; however, they are resistant to freezing and remain viable indefinitely at −70°C.

Exposure to high and low pH, 0.1% formalin, 0.5% phenol and other disinfectants is rapidly lethal. *Chlamydia trachomatis* remains infective for 2 to 3 days at room temperature in unchlorinated tap water. *Chlamydia trachomatis* is susceptible to sulfonamides and to broad-spectrum antibiotics, including tetracycline and erythromycin. Systemic tetracycline is not advised for pregnant women and infants because of tooth staining.

G. EXPERIMENTAL MODELS

Man, certain subhuman primates and rodents are the natural hosts of *Chl. trachomatis.** A variety of monkeys, rodents, and apes are used as experimental animal models.

TRACHOMA

Classical ocular trachoma is a worldwide disease of poverty and poor hygienic conditions; it is spread principally by human contact and to a lesser

*Trachoma has been reported to occur in the apes of Java.

extent by fomites and insect vectors. It predominates in arid and semiarid regions, evidently bcause flies in arid areas are attracted to moisture of the eyes and thus contribute to the chain of transmission. Trachoma proceeds in four stages beginning as a mild conjunctivitis and ending in a healed stage with or without permanent injury. Since acquired immunity is weak, long periods of latent infection with recurrence occur. Recurrence, repeated exposure, reinfection and superinfection with other bacteria are common events that lead to severe ocular damage and often blindness.

Although the mechanisms of the pathogenesis of trachoma are not clear, heavily parasitized host cells rupture and release their load of bacteria (Sect. E). Host cell death may result from "energy parasitism" by the chlamydiae, which presumably usurp and utilize substances that the cell needs for survival or alternatively, from mechanical effects, osmostic effects or parasite-induced membrane lysis. The organisms live in vivo only *within epithelial cells of the conjunctivae;* consequently these cells are the primary sites of injury. However, two characteristics of those strains of organisms that cause trachoma make eventual blindness a probability. First, the growth of the organism in the epithelial cells of the conjunctiva leads to the formation of hard persistent lesions on the inner aspects of the eyelid. Even mild infections can lead to some scratching of the cornea which predisposes to both repeated and secondary infections that often result in impaired vision. Second, perhaps because of the scratching of the cornea, the infecting organisms are able to infiltrate the cornea and cause excessive granulation tissue, vascularization and total blindness. *Host immune responses may also contribute to the disease process.* For example, pannus formation in the cornea may represent an allergic response to chlamydial Ags. Animal studies indicate that the monokine, interleukin I, contributes to scarring. Although it is probable that the toxic type-specific proteins of the cell surface contribute to pathogenesis, this remains to be proved.

Acquired immunity to Chl. trachomatis is low and of short duration as indicated by the high frequency of latent disease and relapses, together with the observation that vaccine-induced immunity is likewise low, strain-specific and of short duration.

It has long been suspected that CMI may contribute to immunity to trachoma despite a lack of evidence to indicate that activated macrophages destroy *Chl. trachomatis.* However, it has recently been found that the addition of interferon λ to host cells prior to infecting them with *Chl. psittaci* enables the cells to inhibit intracellular growth of the organisms. This suggests the possibility that T cell lymphokines of CMI might act similarly against *Chl. trachomatis.* Work with *Chl. psittaci* also indicates that immune T cells can effect specific killing in vitro of parasitized L cells.

A role for Abs against surface proteins of Chl. trachomatis in acquired immunity is more certain since such Abs neutralize mouse toxicity and totally block infectivity of the organisms for cultured host cells. These effects may involve neutralization of adhesins and the activities of the organisms that obstruct phagosome-lysosome fusion. Such Abs might also serve in other unknown ways. Antibodies produced at the infection site, especially secretory IgA Abs, probably contribute importantly to acquired immunity.

Definitive diagnosis of trachoma rests on culture and identification of *Chl. trachomatis.* Scrapings of epithelial cells from conjunctivae of patients with trachoma usually contain typical cytoplasmic inclusions rich in infecting organisms; they can be demonstrated by the direct fluorescent Ab technique or by growth, either in the yolk sac of embryonated eggs or in cell cultures. Culture should preferably be conducted with McCoy cells treated with cyclohexamide. Serologic methods are commonly used to identify chlamydiae that grow in embryonated eggs or mice. Strains of *Chl. trachomatis* can be serotyped by use of the direct microimmunofluorescence technique (micro-IF technique) using monoclonal Abs. Low levels of type specific Abs against cell wall Ags of EBs detected by the indirect micro IF technique may be present in the sera and local exudates of infected patients; high levels of such Abs constitute a provisional diagnosis of chlamydial infection.

Antibiotic therapy is moderately successful; trachoma usually responds to local and systemic treatment with sulfonamides, tetracyclines or erythromycin. Systemic treatment has but limited effects against chlamydiae in ocular lesions but is valuable for preventing secondary infections. *Antibiotic therapy* may overcome the disease but nevertheless may allow the persistence of some chlamydiae within cells, thus leading to a *carrier state* that can last for years. Carriers are also frequent among untreated persons with latent disease who have developed enough immunity to suppress growth of the organisms, but not enough to eliminate carriage.

Man is regarded to be the principal if not the sole reservoir of classical human trachoma.

The best method for preventing trachoma is by good personal hygiene. At present, immunization procedures remain in the experimental stages. Current vaccines afford but limited strain-specific protection of short duration. Indeed, if administered during infection, they may sometimes exacerbate the disease.

INCLUSION CONJUNCTIVITIS-UROGENITAL TRACT COMPLEX

The inclusion conjunctivitis-urogenital tract complex of infections are caused by variants of *Chl. trachomatis* collectively called the "TRIC agent" because they cause a spectrum of diseases *between classical trachoma (TR) and classical inclusion conjunctivitis (IC).* This group of strain variants of *Chl. trachomatis* also cause a spectrum of sexually-transmitted infections including cervicitis, salpingitis, nongonococcal urethritis, epididymitis, prostatitis, proctitis and postpartum fever; secondarily they contribute to premature delivery and ectopic pregnancy. Diagnostic procedures for symptomatic TRIC agent infections are similar to those used for trachoma. Improved techniques for detecting asymptomatic TRIC agent carriers are sorely needed.

Urogenital tract infections are largely the result of sexual promiscuity; such infections in pregnant women often lead to inclusion conjunctivitis in their offspring, together with associated infections, including neonatal pneumonia, mucopurulent rhinitis, vulvovaginitis and ear infection. Urogenital tract infections with these agents are frequent; in developed countries 20 to 25% of adults are infected.

Although *neonatal inclusion conjunctivitis* is usually mild and self-limiting, it can lead to pannus of the cornea and conjunctival scarring; local treatment with tetracycline or erythromycin is effective. Prophylactic instillation of antibiotic-containing eye drops at birth is effective.

Neonatal pneumonia which develops at about 1 to 6 months of age presents as an atypical pneumonia with a pertussis-like cough but without appreciable fever or toxicity; although the disease is rarely fatal, it is debilitating and therapy with erythromycin or sulfonamides is advisable. Although the nature of the low levels of immunity generated in TRIC agent infections is obscure, such immunity probably rests on the forces of CMI and various Abs, including secretory IgA Abs; attempts to produce protective vaccines are planned.

LYMPHOGRANULOMA VENEREUM

Lymphogranuloma venereum (LGV) or "tropical bubo" is a long-recognized worldwide sexually-transmitted disease caused by a variant of *Chl. trachomatis.* The disease was described as a distinct entity during the 18th century; LGV is characterized by enlarged inguinal lymph nodes (bubos), and is largely confined to the tropics. The LGV agent is distinct from other variants of *Chl. trachomatis* in a number of ways, including lethal toxicity for chick embryos and mice injected intracerebrally, the possession of distinct serotype Ags (Table 32–1), and its predisposition to attack lymphoid cells, thus enabling it to spread through draining lymph nodes.

The disease is most frequent in male homosexuals and prostitutes and is commonly accompanied by other sexual diseases. The usual primary infections are urethritis, vaginitis, colitis and proctitis; latency and relapses are common and some asymptomatic infections occur. The disease may remain unnoticed during its early stages. After an incubation period of 1 to 2 weeks, constitutional symptoms arise and a *painless herpes-like lesion appears at the site of infection,* usually in the genital tract. This lesion is self-limiting and soon heals; however, enlarged tender lymph nodes become apparent 1 week to 2 months later. These so-called *"venereal buboes"* represent a DH type *allergic granulomatous response,* which frequently results in scarring with consequent obstruction of lymphatic vessels and sometimes the formation of sinus tracts. The scarring, in turn, leads to various symptoms, depending on the site of draining lymph nodes; for example, obstruction of the genital tract or rectum may occur. In women destruction of the vulva and urethra sometimes eventuates.

Laboratory diagnosis is based on successful culture of lesion specimens on chick embryos and serologic tests. Titers of Abs by the CF test constitute presumptive evidence of LGV if they exceed 1:64 and signify exposure to infection if they exceed 1:16. The Frei test for DH is conducted by intracutaneous injection of an Ag preparation from infected chick embryos. Although the strongest Frei reactions occur in patients with LGV, false positives occur in individuals infected with other chlamydiae that carry group-specific Ags.

Acquired immunity to LGV probably results from the combined forces of CMI and humoral Abs; Abs have been shown to prevent infection of cultured L cells by blocking adherence of EBs to host cells.

Therapeutic drugs include sulfadiazine and tetracycline with penicillin as the alternate drug of choice.

CHLAMYDIA PSITTACI

Many of the characteristics of the *Chl. psittaci* group of organisms are shared with the bacteria of the *Chl. trachomatis* group discussed above.

PSITTACOSIS

Psittacosis (ornithosis) or parrot fever was so-named because the first cases, described by Ritter in 1880, were contracted from psittacine birds. Some past outbreaks in the USA are: (1929–30) 169 cases with 39 deaths; (1956) 568 cases and (1974) a large midwest outbreak due to infected turkeys. Cases continue to occur in the USA, especially among workers in the poultry industry and others exposed to infected birds. The current annual case rate in the USA is about 120 to 140. Even with laws designed to prevent importation of infected birds, the disease will undoubtedly continue to occur in the USA because it is impossible to eliminate domestic avian carriers. Laboratory work with *Chl. psittaci* is especially hazardous and should only be conducted by experienced technicians working in specially equipped laboratories.*

Organisms of the Chl. psittaci group are natural pathogens of a wide range of birds and various mammals including most domestic animals and certain wild animals; they only *infect man as an accidental host.* The infectivity of strains of *Chl. psittaci* varies greatly; those adapted to birds have high infectivity for man, whereas those adapted to lower mammals have low infectivity for man; infections in domestic animals and birds have caused large financial losses. Although *Chl. psittaci* can establish long-term carriage in man as well as birds, the latter are the usual source of human infection. The disease is commonly contracted from birds by inhalation of droplets of nasal discharges or dusts from dried feces. Rare local outbreaks have resulted from the sputum of humans with either active or latent infection; for example, the carrier-induced Louisiana outbreak in 1942–43 involved 19 cases and 8 deaths.

Human infections range from those that are subclinical to those that are fatal. Most clinical cases present as an atypical pneumonia after an incubation period of 1 to 3 weeks. The infection may remain mild and resolve or spread systemically and either resolve or prove fatal. Systemic spread displays a wide spectrum of symptoms, many of which are not unlike those of typhus and typhoid fevers. A few of these symptoms are high fever, chills, chest pains, cough, vomiting, rose spots, enlargement of spleen and liver and cardiac and CNS symptoms. Death usually results from a combination of pulmonary insufficiency, toxic shock and circulatory collapse.

The wide range of symptoms stems from the marked toxicity of the organisms and their predilection for cells of the RES, properties which enable it to spread readily and cause marked injury in many organs including lung, kidney, liver, heart, spleen and the CNS. The organisms are abundant in Kupffer cells and splenic macrophages.

Since *Chl. psittaci* readily infects and destroys macrophages, this activity can account for much of the inflammation and injury produced in psittacosis, including damage to small vessels, thrombophlebitis and focal necrosis in various organs, such as the liver, spleen and lungs. The intensive pulmonary inflammation with accompanying edema and proliferation of alveolar epithelium is probably also due to extensive destruction of macrophages. Some of the CNS damage may be due to free circulating toxins.

---

*Ritter's description of parrot fever in 1880 as a distinct disease entity resulted from a study of 7 persons exposed to infected birds in his brother's home; 3 of them died of the disease which he named "pneumotyphus."

Acquired immunity to psittacosis is low as evidenced by the frequency of latent disease and relapse. As in the case of *Chl. trachomatis,* acquired immunity probably rests on both humoral Abs and CMI. The observation that treatment of cultured cells with interferon enables them to inhibit intracellular growth of *Chl. psittaci* suggests a role for CMI. Support for the role of humoral Abs are the observations that specific Abs block not only the adherence of EBs to cultured host cells, but also intracellular growth of the organisms.

Early definitive diagnosis of psittacosis is important because antibiotic therapy is usually effective. However, diagnosis is clinically difficult because the disease resembles many other infectious diseases. The organisms are usually present in the sputum and blood by the 5th to the 7th day of illness. Laboratory diagnosis rests on successfully growing the organisms in tissue culture on egg embryos and the death of mice injected intracerebrally. Complement fixation tests on serum are usually positive by 2 weeks and titers rise to 1:64 to 1:256 by the 3rd week; a single specimen titer of 1:64 or above suggests current or recent infection. Infected cells present typical inclusion bodies that do not stain with iodine; they are called Levinthal-Code-Lille bodies (LCL bodies).

Early treatment of psittacosis with tetracycline or erythromycin is usually effective, tetracycline being the drug of choice. Treatment should be continued for at least 2 weeks; recovery is usually protracted and treatment sometimes fails.

Reservoirs of human psittacosis are principally birds, with the rare exception of man-to-man transmission. *Measures to control disease transmission* include restrictions on the importation of birds, avoidance of contact with infected birds, eradication of infection, either latent or active, in domestic poultry with tetracycline, disposal of infected flocks and sanitary practices on the part of poultry raisers and processors. Unfortunately, the above measures have been *only partially effective.* Effective vaccines for use by persons in high risk groups have not been developed.

REFERENCES

Benes, S., McCormack, W.: Chlamydial Pneumonias: Seminars Infect. Dis. *5*:251, 1983.

Caldwell, H.D., and Schachter, J.: Antigenic analysis of the major outer membrane protein of *Chlamydia spp.* Infect. Immun. *35*:1024, 1982.

Darougar, S.: The humoral immune response to chlamydial infection in humans. Rev. Infect. Dis. *7*:726, 1985.

De La Maza, L.M. et al.: Interferon-induced inhibition of *Chlamydia trachomatis:* Dissociation from antiviral and antiproliferative effects. Infect. Immun. *47*:719, 1985.

Eissenberg, L.G., et al.: *Chlamydia psittaci* elementary body envelopes: ingestion and inhibition of phagolysosome fusion. Infect. Immun. *40*:741, 1983.

Grayston, J.T., et al.: Importance of reinfection in the pathogenesis of trachoma. Rev. Infect. Dis. *7*:717, 1985.

Harris, R.L. and Williams, T.W., Jr.: "Contribution to the question of pneumotyphus," A discussion of the original article by J. Ritter in 1880. Rev. Infect. Dis. *7*:119, 1985.

Johnson, A.P.: Pathogenesis and immunology of chlamydial infections of the genital tract. Rev. Infect. Dis. *7*:741, 1985.

Kampmeier, R.H.: Infection with *Chlamydia trachomatis* (TRIC Agent). Sex. Transm. Dis. *11*:169, 1984.

Kunimoto, D., and Brunham, R.C.: Human immune response and *Chlamydia trachomatis* infection. Rev. Infect. Dis. *7*:665, 1985.

Lipkin, E.S. et al.: Comparison of monoclonal antibody staining and culture in diagnosing cervical chlamydial infection. J. Clin. Microbiol. *23*:114, 1986.

Mardh, P.A., et al.: *Chlamydia Infections.* New York, Elsevier Biomedical Press, 1982.

Martin, D.H., et al.: Prematurity and perinatal mortality in pregnancies complicated by maternal *Chlamydia trachomatis* infections. JAMA *247*:1585, 1982.
Moulder, J.W.: Comparative biology of intracellular parasitism. Microbiol. Rev. *49*:298, 1985.
Sarov, I., et al.: Specific serum IgA antibodies in the diagnosis of active viral and chlamydial infections. *New Horizons in Microbiology,* (eds.) Sanna, A. and Morace, G., New York, Elsevier Science Pub. 1984.
Schachter, J.: Chlamydiae (Psittacosis-Lymphogranuloma Venereum-Trachoma Group). *Manual of Clinical Microbiology,* 4th Ed., E.H. Lennette, Ed. Wash. DC, Am. Soc. Microbiol., 1985.
Stephens, R.S., et al.: Molecular cloning and expression of *Chlamydia trachomatis* major outer membrane protein antigens in *Escherichia coli.* Infect. Immun. *47*:713, 1985.
Williams, D.M. and Schachter, J.: Role of cell-mediated immunity in chlamydial infection: Implications for ocular immunity. Rev. Infect. Dis. *7*:754, 1985.

33

LEGIONELLA

During an American Legion Convention held in Philadelphia in 1976 an outbreak of a "new" and often fatal pneumonic disease occurred that shocked the nation. The disease was named *Legionnaire's disease,* and the causative agent *Legionella pneumophila.* Subsequently it was discovered that a mild systemic nonpneumonic disease, named *Pontiac fever,* is also caused by *L. pneumophila.* It is presumed that these two forms of disease reflect strain differences among the infecting organisms. Human infections can also be produced by other species of the genus *Legionella.*

Although the classification of the legionellae is still undergoing constant change, it is presently structured to comprise the family *Legionellaceae,* the genus *Legionella* and twenty-three species; however, it has been proposed that two additional genera be created to accommodate all species except *L. pneumophila,* the cause of Legionnaire's disease. The legionellae are aquatic saprophytes that are ubiquitous in fresh water and soil. There is probably an element of opportunism in all severe infections produced by these organisms. Their discovery was delayed for many years because they stain poorly and require special culture media. To date only five species have been reported to infect man, *L. pneumophila* being the prototype species; the other four species are: *L. bozemanii, L. dumoffii, L. micdadei* and *L. longbeachae.* Legionella infections occur worldwide; the estimated number of symptomatic cases in the USA per annum is 25,000.

The legionellae are aerobic Gram-negative, nonsporing, pleomorphic, coccobacilli, which in infected tissues range in size from 0.5 to 0.7 × 1.0 to 2.0 μm; however, filamentous forms as long as 100 μm occur in cultures. Most species are motile by means of 1 or 2 flagella which may be polar, subpolar or lateral. The legionellae stain poorly and contain lipid-rich inclusion granules. They are nutritionally fastidious but can be isolated on special media containing a source of ferric iron and L-cysteine as an essential amino acid for primary culture (Sect. K). The organisms do not ferment sugars but use amino acids as a source of energy. They grow best at 35° to 42°C in a humid atmosphere containing 2 to 5% CO_2 but fail to grow below 25°C. Growth is favored, both in nature and in vitro, by the presence of algae and amoebae in which they grow. They also grow within cultured macrophages and fibroblasts.

The legionellae have not been thoroughly studied and speciation is extremely difficult even in the most advanced laboratories. Although morphologic, biochemical, cultural and serologic methods are helpful, the only reliable tool for accurate speciation is the highly sophisticated procedure of DNA hybridization. Much remains to be learned about the various *Legionella* species and the nature and management of the diseases they produce.

This chapter is focused principally on the strains of *L. pneumophila* that cause Legionnaire's disease.

LEGIONELLA PNEUMOPHILA

A. MEDICAL PERSPECTIVES

Recognition of the various legionelloses has stemmed from the discovery of *L. pneumophila* as the cause of Legionnaire's disease in 1976. As judged from retrospective studies, earlier outbreaks and sporadic cases occurred; for example, an outbreak of Legionnaire's disease took place in 1965 and an outbreak of Pontiac fever occurred in 1968. Outbreaks and sporadic cases of the two diseases continue to appear and knowledge about the organisms and the diseases they produce is gradually accumulating. Small, local, clustered epidemics of Legionnaire's disease commonly peak in August and September; cases predominate in upper middle-aged males who smoke. Sources of infection are principally large buildings, such as hotels, factories and hospitals that have air conditioning and water purification systems in which water condensers, coolant columns and humidifiers are used. Cases have recently been reported that resulted from contaminated shower heads in shower baths; contaminated drinking water has not been implicated. Inhaled aerosols that circulate through ventilation ducts are the most common cause of Legionnaire's disease. However, individual cases have occurred sporadically as the result of breathing dusty air in construction areas. During outbreaks about 1 to 2% of those exposed develop overt Legionnaire's disease; they apparently represent predisposed individuals (Sect. H). Mortality rates have ranged as high as 60% in some outbreaks and average about 15%. During epidemics, patients range from those with mild respiratory symptoms to those with life-threatening pneumonia. The number of asymptomatic infections that occur during outbreaks is unknown but may be substantial and occur in all age groups as judged by retrospective serologic studies; the incidence of sporadic cases in the population at large may be as high as 25%.

Although physicians are well aware of *L. pneumophila* as a cause of pneumonia, many cases of Legionnaire's disease go undiagnosed and unreported. The future outlook for severe cases is better than in past years because of the awareness of physicians and the prospects for improving diagnostic, therapeutic and preventive measures (Sect. N). Nevertheless serious outbreaks continue to occur such as the 1985 outbreak at Stafford, England that claimed 39 lives.

B. PHYSICAL AND CHEMICAL STRUCTURE

Under the electron microscope *Legionella* cells show double envelopes, each comprising a triple-layered unit membrane. Although *L. pneumophila* has been divided into 10 serogroups (serotypes) on the basis of the direct fluorescent Ab test (DFA test) using whole cells, the nature of the Ags involved is not known. Two serogroups each are recognized for *L. bozemanni* and *L. longbeachae.* Most cases of Legionnaire's disease are due to *L. pneumophila* strains that belong to serogroup 1. A pooled antiserum specific for *L. pneumophila* has been prepared which has a specificity near 99% but a sensitivity of only 60 to 80%. Consequently it only serves to identify *L. pneumophila* when positive.

An identifying characteristic of *L. pneumophila* is that the cells are uniquely rich (80 to 90%) in certain branched-chain fatty acids some 14 to 20 carbon atoms in length.

C. GENETICS

Little is known about the genetics of legionellae; however, *Legionella* species can be clearly distinguished by DNA hybridization.

D. EXTRACELLULAR PRODUCTS

A small heat-stable peptide with cytotoxic activity has been described that decreases hexose monophosphate and oxygen consumption of phagocytes during phagocytosis. *Legionella pneumophila* produces β-lactamase and other enzymes. The only *Legionella* sp. that does not produce β-lactamase is *L. micdadei.*

E. CULTURE

Legionella pneumophila grows well on selective and selective-differential media used for differentiating the organism from other species of legionellae (Sect. K); among the legionellae, *L. pneumophila* is uniquely capable of hydrolysing hippurate.

F. RESISTANCE TO PHYSICAL AND CHEMICAL AGENTS

Legionellae can survive for over a year in tap water. Even though they require a humid atmosphere for growth, they apparently can resist drying because breathing of contaminated dust has been noted to cause infection. Various disinfectants are effective against *L. pneumophila,* including calcium hypochlorite, isopropanol and didecylmethylammonium chloride. These and other agents are potentially useful for disinfecting air conditioning equipment.

G. EXPERIMENTAL MODELS

Guinea pigs and mice can be infected with *L. pneumophila* and can be protected by immunization with a partially purified Ag.

H. INFECTIONS IN MAN

Although Legionnaire's disease can occur at any age, the attack rate in exposed individuals increases with age, being greatest in the elderly. Since upper middle-aged groups have the greatest exposure in hotels, etc., the median age in outbreaks is 55 to 60 years. Many of the predisposing conditions are associated with age and sex differences as depicted below. The attack rate is higher in males than in females and higher in smokers than nonsmokers.

The rates of attack and death are highest in individuals whose antimicrobial defenses are naturally low or have been compromised in some way; for example, individuals whose local pulmonary defenses have been compromised by emphysema, bronchitis, existing infection due to other agents or smoking. Systemic immunodeficiencies constitute another category of predisposing conditions; they may be either genetic or acquired as the result of aging, immunosuppressive drugs, malignancy, existing infections, alcoholism and malnutrition. Severe Legionnaire's disease begins in the lung, but soon progresses to become a multisystemic disease. The incubation period has a wide range but averages about 5 days. The onset of disease is heralded by symptoms resembling influenza, including fever, malaise, myalgia, and usually headache and chills. In severe cases, the disease progresses rapidly to a state of *toxic pneumonia,* often with tachypnea, dyspnea, pleuritic pain, rales, and a dry cough that later becomes productive. Other symptoms include hemoptysis, bradycardia, vomiting, diarrhea, lethargy, confusion and delerium. Individuals with Legionnaire's disease are more likely to have early diarrhea and nonfocal encephalopathy than the non-*Legionella* pneumonias. Impaired renal function with hematuria and proteinuria often occurs and because of systemic spread, organisms can sometimes be isolated from the blood and internal organs. Abnormalities in liver function may also be present. In cases that later prove fatal, infection may spread from the lung to adjacent structures and often through lymphatics to the blood and distant sites. Death may occur at about a week from shock and/or respiratory failure.

In nonfatal cases of Legionnaire's disease the fever drops by 1 to 2 weeks and recovery is gradual.

I. MECHANISMS OF PATHOGENESIS

The lung commonly presents as a patchy interstitial pneumonia with focal coagulative necrosis; however extensive consolidation can occur. The neutrophil-rich exudate contains both free bacteria and bacteria within macrophages. It is presumed that after engulfment the organisms survive and grow because they inhibit the oxidative burst associated with phagocytosis and antimicrobial activities of these phagocytes.

J. MECHANISMS OF IMMUNITY

Essentially nothing is known about acquired immunity to the legionellae except that guinea pigs can be immunized with a partially-purified Ag. Whether activated macrophages may be able to effect phagocytic destruction of the organisms with or without help of Ab and C and whether Ab and C may exert bactericidal effects is not known. As with tubercle bacilli, resistance to legionellae may be influenced by the degree of exposure and bacillary load. A better understanding of host-parasite relationships is sorely needed, especially as such relationships pertain to pathogenesis and immunity.

K. LABORATORY DIAGNOSIS

The value of isolating and identifying legionellae from clinical specimens is 3-fold, diagnostic, epidemiologic and taxonomic.

Methods used to diagnose *Legionella* infections are complex and technically exacting; hence laboratories lacking the expertise or equipment needed should send specimens to suitable reference laboratories.

Although most cases of severe *Legionella* pneumonia are due to *L. pneumophila* (serogroup 1), there is no need for species identification from the clinical standpoint because the disease is similar irrespective of the causative species and because the therapeutic drug of choice is uniformly erythromycin.

Since early antibiotic therapy is crucial in all cases of severe *Legionella* pneumonia, the latter part of the present discussion will be focused on genus rather than species identification of isolates.

Legionella pneumonia should be suspected in any patient with a rapidly progressing severe pneumonia that cannot be shown to result from some other agent. Early diagnosis rests primarily on direct microscopic examination and culture of transtracheal aspirates, sputum, and if needed, open lung biopsies; early culturing of blood and various cavity fluids occasionally yields positive results, especially in cases that later prove fatal.

Examination of sputum is of limited value because the organisms are often sparse or absent. When specimens are examined microscopically by direct nonspecific staining methods, the legionellae stain well with the Gimenez stain used for rickettsiae and the Dieterle silver impregnation stain used for spirochetes but fail to stain with the routine Gram-stain using either safranin or 0.01% basic fuchsin for counterstaining; however, if counterstaining time is lengthened, the organisms tend to stain lightly Gram-negative. With the silver impregnation stain the organisms appear intensely brown to black with a pale yellow background.

Differentiation of the legionellae by antigenic analysis has not been fully explored and the antisera presently available do not permit uniform clear-cut differentiation between species, let alone strains within species. Work with monoclonal Abs should eventually permit such differentiation. The virtual certainty that not all *Legionella* species with pathogenetic potential have been discovered detracts from the value of specificity determinations using existing serologic tests. The lack of sensitivity of serologic tests is often a greater drawback than the lack of specificity. Despite their shortcomings serologic tests can often provide strong presumptive evidence of value in diagnosis and species identification.

The genus-specific direct fluorescent Ab test (DFA test) often aids in early diagnosis of pneumonia due to legionellae because it is over 99% specific and can be completed within 2 hours; when positive, it constitutes an almost certain diagnosis but when negative does not rule out such an infection. Since the objective is to detect causative organisms belonging to any species of the genus *Legionella,* a polyvalent pool(s) of strong serogroup antisera covering as many of the known pathogenic species as possible is used. Whereas false-positive tests are almost invariably due to technical errors, false-negative tests, as measured by positive culture tests of the same specimens, are high (range 20 to 50%). False negative tests are high because of the high concentration of organisms (10^4 cells per ml) required for microscopic visualization and because the pool(s) of antisera currently used lack the breadth of specificity and strength needed to detect all of the strains of legionellae that are capable of producing pneumonia.

Culture is conducted using several different media, including a routine medium, such as blood agar, that supports the growth of other pathogens but not *legionellae*; the purpose of a routine medium is to rule out other pathogens. Other media used include: (a) special growth medium that assures the primary isolation of legionellae as well as other organisms, namely, buffered charcoal yeast extract medium containing α-ketoglutaric acid (BCYEa medium); (b) a special L-cysteine deficient medium for confirming the L-cysteine requirement of legionellae-like colonies, namely, either (BCYE-Lcys medium) or tryptic soy 5% sheep blood agar (BAP medium); (c) a selective medium designed to selectively suppress the growth of contaminating organisms, namely BCYEa medium supplemented with cefamandole, polymyxin B and anisomycin (BMPAa medium) or BEYEa medium supplemented with glycine, vancomycin, polymyxin B and antisomycin (MWY medium); and (d) a selective-differential medium designed to aid in the recognition of colonies belonging to different species of *Legionella,* namely, MWY medium to which either bromcresol purple or bromthymol blue is added. Since the above media are not strongly selective, it is often advantageous to decontaminate sputum or other externally contaminated specimens with 10 volumes of KCl-HCl solution (pH 2.2) followed by plating 10 minutes later on selective and nonselective media. Decontamination can also be effected by the more time-consuming procedure if injecting the specimen i.p. into guinea pigs followed by later recovery from the peritoneal fluid. In atmospheres containing 5% CO_2 legionellae colonies appear after 2 to 5 days. They vary in morphology and color and when examined under the Wood's long-wave UV lamp emit light of different colors. Several newly discovered Legionella sp. display an electric blue-light, whereas other species exhibit a yellow-green light. Colonies are picked and subjected to direct staining procedures as well as sero-grouping, fatty acid analysis, tests for enzyme activities and DNA homology. Taken together, identification of various species of *Legionella* is complex and exacting and should only be done by highly competent and experienced technicians.

Other approaches for diagnosing *Legionnaire's disease* are to determine serum Ab levels, preferably by the indirect immunofluorescent method using heat-killed Ag to avoid cross reactions with *Mycoplasma pneumoniae,* and to examine the urine for *L. pneumophila* Ag with a specific antiserum.

Since Abs develop slowly and significant rises in the titer of serum Abs may not appear before 3 weeks to 2 months, the diagnostic value of serum Ab determinants is perforce retrospective. Although serum Ab tests in serogroup 1 infections have a specificity of 95 to 99%, their sensitivity is only 60 to 80%. A rise in titer from <32 to 128 signifies Legionnaire's disease. However, the titer on a single serum specimen must be >256 to provide strong presumptive evidence of Legionnaire's disease.

The test recently introduced for detecting serogroup Ag 1 in urine has a specificity of 99% and a sensitivity of 80%. Since tests for *Legionella* Ags in urine promise to become highly useful for early diagnosis, it is hoped that pooled antisera for this purpose will soon be developed, including genus specific and species specific as well as group specific antisera. Many of the above media are available in dehydrated commercial form, but quality control is often lacking.

L. THERAPY

Strong presumptive evidence of Legionnaire's disease or, indeed, pneumonia due to any of the legionellae justifies the initiation of therapy.

Erythromycin is the most effective antimicrobic for treating severe pneumonia due to any of the legionellae and if administered early in the disease often effects a cure; tetracyclines are less effective and β-lactamase degradable antimicrobics are of no value, except in *L. micdadei* infections. Rifampin is effective in experimental systems, although the development of rifampin-resistant organisms is a potential problem. Combined erythromycin-rifampin treatment is recommended for patients who fail to respond to erythromycin alone. Prolonged therapy for about 3 weeks is advisable.

Antimicrobial sensitivity testing is not done because there is no correlation between in vitro susceptibility and therapeutic efficacy.

From the therapeutic standpoint, a presumptive diagnosis of pneumonia due to legionellae is established by (a) failure to implicate any other causative agent and (b) the finding of typical bacilli on direct nonspecific staining and DFA testing; diagnosis is further strengthened by the appearance of typical colonies on BCYEa medium that fail to grow when subcultured on BCYEa-Lcys medium.

M. RESERVOIRS OF INFECTION

Whereas the vast majority of cases of Legionnaire's disease are contracted by breathing contaminated air from ventilation ducts in large buildings, a few are contracted in other ways; for example, workers who clean or repair air conditioning systems or breathe exhausts from auxiliary cooling towers on buildings are at high risk. Various sources of infection that account for sporadic cases include bath showers and dusty air at construction sites.

N. PREVENTION OF INFECTION

Routine cleaning and disinfection of air conditioning and water systems in large buildings promises to be useful for preventing outbreaks of Legionnaire's disease, although appropriate routines remain to be developed. Cleaning must be frequent enough to meet the threat of recolonization by these ubiquitous organisms; the organisms have been reported to grow within various amoebae that inhabit the water in air conditioning systems. Although a vaccine for human use has not been developed, its potential prophylactic use would probably be limited to high risk groups, such as individuals who work on air conditioning systems.

Pontiac fever, which was first described as a disease entity during an outbreak of the disease in Pontiac Michigan in 1968, is a nonpneumonic self-limiting, nonfatal form of legionellosis, caused in large part at least, by strains of *L. pneumophila.* Pontiac fever or its counterpart is also produced by other species of *Legionella.*

The disease is characterized by moderate leukocytosis, fever, chills, myalgia and headache; a dry cough develops in about 50% of cases but is a minor symptom. Symptoms usually appear after a short incubation period of some 20 to 48 hours and subside within 2 to 5 days.

Since the attack rate in exposed individuals during epidemics is about 90% and appears to occur equally in healthy and immunodeficient individuals with a peak incidence in the 30 to 40 year old group, variation in host resistance evidently plays little or no part as a determinant of infection. This is in sharp contrast to the severe legionelloses which are clearly opportunistic infections. Antimicrobial therapy is not indicated in Pontiac fever.

OTHER *LEGIONELLA* SPECIES

Although the number of infections due to species other than *L. pneumophila* are comparatively few, the spectrum of diseases, the properties of the organisms, and the methods of culture and speciation are largely in common with *L. pneumophila.* Other species can cause encephalopathy and valvular endocarditis, independent of pneumonia; in common with *L. pneumophila* erythromycin is the therapeutic drug of choice.

It is notable that all reported cases of infection due to *L. micdadei* have occurred in patients known to be immunosuppressed.

REFERENCES

Broome, C.V.: Pneumonia due to Legionella species: *Seminars in Infectious Disease.* Weinstein, L., and Fields, B.N. (eds), New York, Thieme-Stratton Inc., 1983.

Edelstein, P.H.: Legionella. *Manual of Clinical Microbiology,* 4th Ed., Ed. Lennette, E.H., Wash., D.C., Am. Soc. Microbiol., 1985.
Edelstein, P.H.: Environmental aspects of Legionella. Am. Soc. Microbiol. News, *51*:460, 1985.
McDade, J.E., et al.: Legionnaire's disease: Isolation of a bacterium and demonstration of its role in other respiratory disease. N. Engl. J. Med. *297*:1197, 1977.
Medoff, G., et al.: Morphogenesis and pathogenicity of *Histoplasma capsulatum.* Infect. Immun. *55*:1355, 1987.
Nolte, F.S., et al.: Electrophoretic and serological characterization of the lipopolysaccharides of *Legionella pneumophila.* Infect. Immun. *52*:676, 1986.

34

ACTINOMYCES AND *NOCARDIA*

Actinomycetous diseases are caused by "actinomycetes," a term used to designate certain bacteria in the order *Actinomycetales* that are characterized by long filaments about 1 μm or less in diameter, composed of cells arranged in chains because they fail to separate after binary fission. They are true bacteria in all respects despite the hyphal-like appearance of the bacterial filaments they present. The actinomycetes belong to a number of genera of different families; three of the medically most important genera are of the family *Actinomycetaceae,* namely, *Actinomyces, Arachnia* and *Nocardia.* Other genera belonging to the actinomycetes important to this discussion are *Bifidobacterium, Dermatophilus, Streptomyces,* and *Actinomadura.*

ACTINOMYCOSIS

The old terms, actinomycetes, and actinomycosis, with endings suggestive of fungi, continue to be used since the bacterial filaments they present resemble the hyphae and mycelia of fungi. Identification of the various genera of actinomycetes is based principally on cell wall composition.

Actinomycosis can be caused by any one of three closely-related species in decreasing order of frequency as follows: *Actinomyces israelii, Arachnia propionica,* and *Actinomyces naeslundii;* all three species are *members of the normal flora* of the mucosae of the human mouth and GI tract. Although the term, *actinomycosis,* is sometimes reserved for infections due solely to *Actinomyces israelii,* infections caused by any one of the above three organisms are essentially identical. Two other species of the normal flora of the mouth rarely cause actinomycosis but often contribute to dental caries; they are *Actinomyces viscosus* and *Actinomyces odontolyticus; Actinomyces* strains range from strict anaerobes to microaerophiles and aerobes that are facultatively anaerobic in the presence of CO_2.

ACTINOMYCES ISRAELII

Since *A. israelii, A. naeslundii* and *Arachnia propionica* are similar with respect to their morphology, physiology and pathogenicity, this discussion will be limited principally to infections due to *A. israelii* as the prototype of the group.

A. MEDICAL PERSPECTIVES

Bovine actinomycosis was first described in 1877 in cattle and was given the aptly descriptive name of "lumpy jaw"; *Actinomyces bovis* was found to be the causative agent. The next year Israel discovered that a similar disease in man results from infection with another species, *Actinomyces israelii.* This organism is one of the normal flora of the oropharynx and GI tract of man, its sole natural host. Infection usually follows a break in mucosal epithelium, particularly as the result of trauma that allows these bacteria to become established in deeper tissues where low O_2 tensions exist. Local conditions that promote anaerobiasis in tissue, such as necrosis, favor infection.

It is probable that actinomycosis will continue to occur with the same low frequency as it has in the past.

B. PHYSICAL AND CHEMICAL STRUCTURE

Actinomyces israelii is pleomorphic; it may occur either as slender, tree-like branching filaments 0.5 to 1.0 μm in diameter, as elongated diphtheroid-like rods, or as bacillary or coccoid forms. The organism is Gram-positive, nonacid-fast, nonsporing, noncapsulated and nonmotile.

When growing in pus, *A. israelii* often forms small club-shaped aggregates of radiating filaments surrounded by eosinophilic material of host origin. The yellowish sulfur color of these clusters of organisms originally prompted the name, "sulfur granules." The ray-like appearance of the filaments is most easily seen at the periphery of the granules (Fig. 34–1).

C. EXTRACELLULAR PRODUCTS

Although actinomycetes produce many extracellular products, there is little evidence available concerning the nature of these substances or the roles they may play in pathogenesis and immunity to actinomycosis.

D. CULTURE

Other anaerobic bacteria are almost invariably present in tissues infected with *A. israelii.* Consequently, the isolation of *A. israelii* is often difficult. Clinical specimens from draining lesions usually contain sulfur granules which should be separated from surrounding material and crushed before being added to culture media. Since strains of the organisms are either strict anaerobes or microaerophiles that are facultatively anaerobic, a good culture medium is thioglycollate broth in which growth occurs in the form of small "cotton balls." Cultures on solid media, such as brain heart infusion agar (with or without added blood), must be incubated under anaerobic conditions. Optimal growth occurs at 30° to 37°C. Compared with many bacteria, growth is slow and colonies do not appear before 3 to 10 days of incubation.

E. EXPERIMENTAL MODELS

There is no good animal model for human actinomycosis. Hamsters and suckling mice can be infected and have been used for experimental studies.

F. INFECTIONS IN MAN

The course of actinomycosis in man depends in part on the site of primary infection. This suppurative-granulomatous disease occurs chiefly in one of three patterns; *cervicofacial, thoracic,* or *abdominal.* Although hematogenous spread may occur, the disease usually remains limited to the general area of initial infection, where it erodes all tissues in its path and burrows to form draining sinuses. Lesions consist of suppurative areas surrounded by fibrosing granulation tissue.

Cervicofacial actinomycosis, the most common type of the actinomycoses, follows tooth extraction, periodontal abscess, dental caries, jaw fracture, or other conditions involving trauma; bronchiectasis, bronchitis and emphysema are also predisposing causes. Chronic lesions occur that often involve bone; interconnecting sinus tracts develop which open and discharge pus and necrotic material through the skin. The characteristic large, firm, indurated lesions due to fibrosis may be mistaken for malignant tumors; indeed, unnecessary

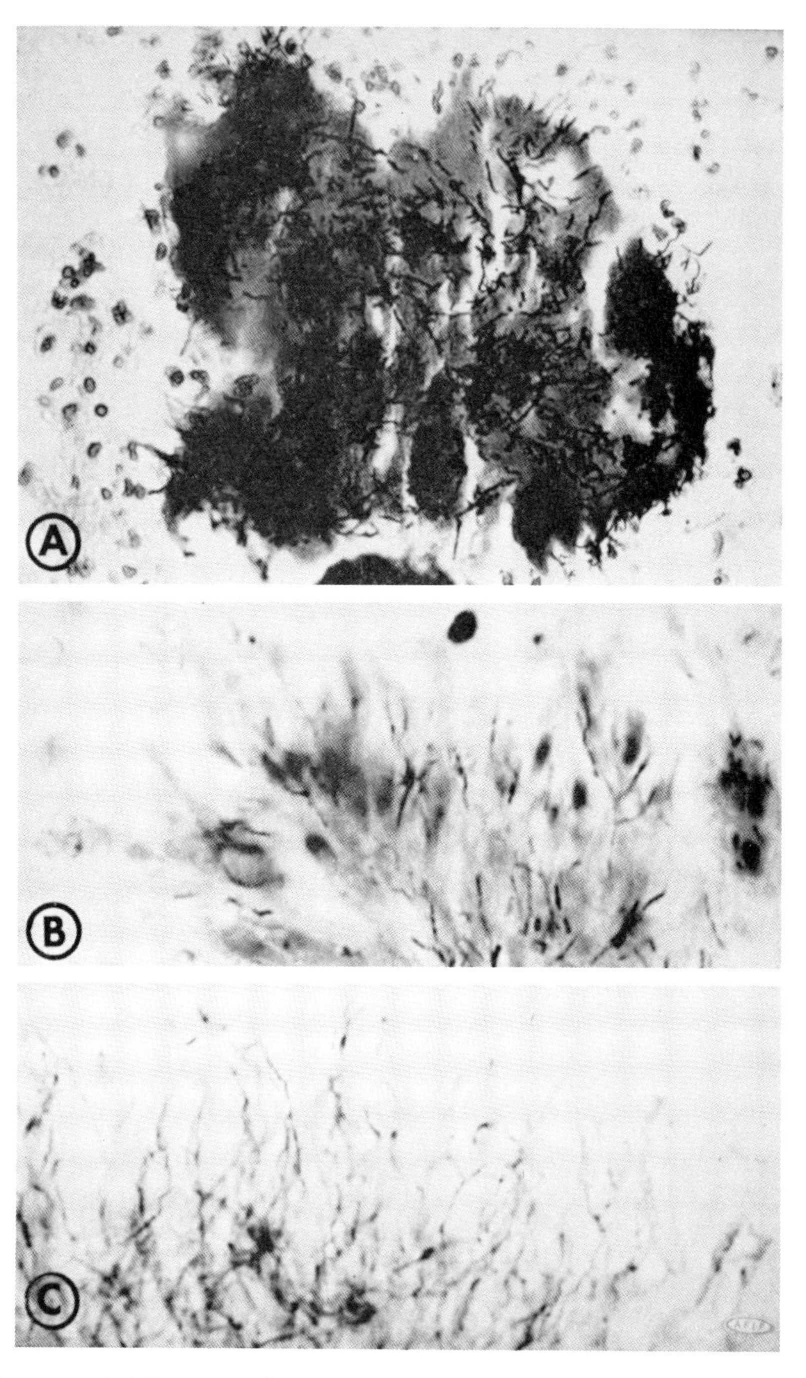

Fig. 34–1. *A.* A Gram stain of an actinomycotic granule obtained in a fragment of tissue removed for improving drainage of a thoracic wall sinus tract. Under direct microscopic observation some of the filaments were branched. ×1350. *B.* The periphery of an actinomycotic granule stained by the Brown and Brenn Gram method. Observe the delicate Gram-positive filaments within some of the "clubs." Branching can be seen. ×1350. *C.* A granule from the case shown in "B" but stained by the Gomori methenamine-silver method that reveals the filaments distinctly. ×1350. (From Emmons, C.W., et al. *Medical Mycology,* 3rd Ed., Philadelphia, Lea & Febiger, 1977.)

surgical procedures have been performed to remove a supposed cancer when the lesion could have been treated successfully with antibacterial agents.

Thoracic actinomycosis is caused by aspiration of oral secretions or occasionally by extension of cervicofacial lesions; it is a chronic lung disease, characterized by consolidation and by extension of the primary lesion through the mediastinum, ribs, pleura and chest wall. Signs of systemic infection may be mild or lacking and sputum cultures are often negative. Multiple draining sinus tracts are common. There may be fever, cough and production of bloody, purulent sputum. Hematogenous dissemination occurs in some 10% of cases.

Abdominal actinomycosis is a rare disease resulting from constantly swallowed organisms that colonize the oral cavity or organisms colonizing the gut; apparently they invade through any sort of perforation or break in the intestinal tract, including rupture of the appendix and surgical incisions. Extension occurs from the primary site of infection and may involve any adjacent organ. The indurated lesion has the consistency of a tumor. Actinomycosis of the urogenital tract is rare and can be mistaken for malignancy or tuberculosis; it sometimes results from intrauterine contraceptive devices.

G. MECHANISMS OF PATHOGENESIS

Once established, the ability to persist and produce a chronic infection is the principal pathogenetic attribute of *A. israelii;* however, the factors or mechanisms that enable the organism to produce a persistent infection and injury are not known. In any event, it is reasonably certain that lowered host defenses, either local or systemic, are predisposing.

Although other anaerobic bacteria are often present in the lesions, their possible contributions to the persistence of *A. israelii* or the overall disease process are unknown.

H. MECHANISMS OF IMMUNITY

Little is known about the mechanisms of resistance and acquired immunity to actinomycosis. However, immunosuppressants in general use are not frequently predisposing. It is probable that subtle cellular immune deficiencies are among the factors that predispose to infection with the various causative agents, including *A. israelii.*

I. LABORATORY DIAGNOSIS

The diagnosis of primary actinomycosis is often difficult because: (1) secondary invaders are often present in actinomycotic lesions, (2) actinomycotic agents may sometimes serve as secondary invaders in other diseases that mimic actinomycosis, such as tuberculosis, and in necrosing tumors, (3) constitutional symptoms are absent or variable and nonspecific, such as weight loss, cough, chest pain and fever, (4) x-ray findings are nonspecific, (5) routine sputum cultures are seldom positive, (6) "sulfur granules" characterized by peripheral club-shaped masses of filaments are often sparse and difficult to find, (7) because the actinomycotic agents are among the normal flora, the sole presence of the agent alone is meaningless in the absence of other evidence. Pus or other clinical materials from lesions commonly contain Gram-positive nonacid-fast bacterial filaments, or "sulfur granules," however sparse they may be. Wet mounts examined with reduced light often reveal tangled bacterial filaments or filament fragments. *Sulfur granules,* when present, are characteristic; they should be crushed between slides before examination and culture. The finding of sulfur granules with a radiating fringe of *eosinophilic clubs* surrounded by lipid-laden macrophages is virtually certain evidence of actinomycosis.

Direct culture of clinical specimens from patients not receiving antibiotics (Sect. D) often yields colonies after 3 to 10 days of incubation. Speciation is based on biochemical, cultural and cell-wall characteristics.

Biopsy is sometimes necessary in order to provide material for culture and for serial sectioning and staining to detect "sulfur granules" in tissue. Serologic diagnosis based on immunoelectrophoresis, and a known monospecific Ag-Ab reference system holds promise but practical methods remain to be developed.

I. THERAPY

Long-term (6 to 8 months) high-dose therapy with penicillin G or one of its penicillin alternatives, often in conjunction with surgical drainage of lesions, is the treatment of choice. Tetracycline, parenteral cephalosporins, lincomycin, erythromycin and clindamycin alone or in combinations are second-choice drugs for treating patients allergic to penicillin. Improvement is usually not seen before a month of therapy has passed. In the absence of a definitive bacteriologic diagnosis, trial chemotherapy is sometimes necessary.

K. RESERVOIRS OF INFECTION

Man is the only known reservoir of the three principal agents that cause human actinomycosis, including *A. israelii*; being members of the normal flora infection is presumed to be invariably endogenous.

L. PREVENTION OF DISEASE

Since actinomycosis results from endogenous organisms, prevention of infection depends on preventing tissue trauma that would allow endogenous organisms to invade and/or correcting predisposing conditions that compromise host defense, such as dental caries and periodontal disease.

NOCARDIOSIS

NOCARDIA ASTEROIDES

A. MEDICAL PERSPECTIVES

Nocardiosis, like actinomycosis, was first recognized in cattle. The causative organism was isolated by Nocard (1888) from cattle with a lymphatic disease called farcy, and subsequently was named *Nocardia farcinica.* Soon thereafter, it was discovered that human nocardiosis usually results from infection with *N. asteroides,* the prototype species and center of this discussion.

Mycetoma, a clinically based disease, can be caused by various species of fungi or by either of three *Nocardia* species, *N. asteroides, N. caviae* and *N. brasiliensis,* the latter being the most common cause of bacterial mycetoma. Although nocardiae comprise many species of soil saprophytes, only a few are known to cause disease in man and animals; nocardiae are strictly exogenous and although they may occasionally colonize mucosae of the respiratory tract as nonpathogens, they are not considered to represent part of the normal flora.

Because of the worldwide distribution of nocardiae in soil and water it appears safe to predict that the incidence of nocardiosis will not decrease; in fact, it will probably increase due to the increasing general use of immunosuppressive agents.

B. PHYSICAL AND CHEMICAL STRUCTURE

Nocardia sp. form branching filaments less than 1 μm in diameter; the filaments of some species fragment to bacillary and coccoid forms. The nocardiae are Gram-positive and may present a beaded appearance; their property of acid-fastness emphasizes their relationship to the mycobacteria. However, only certain species of nocardiae are acid-fast, and these only weakly so; acid-fastness tends to be lost on subculture.

C. EXTRACELLULAR PRODUCTS

No extracellular products that contribute to the pathogenesis of nocardiosis or immunity to the disease are known.

D. CULTURE

A major difference between the various actinomycetes is that Nocardia sp. are strict aerobes, whereas *Actinomyces* sp. are either strict anaerobes or are aerobes or microaerophiles that are facultatively anaerobic. Nocardiae grow well on Sabouraud's agar, blood agar, glucose agar, Lowenstein-Jensen medium, and other media; growth is stimulated by 10% CO_2. Colonies, which are dry, adherent and sometimes orange in color, do not become visible before 2 days to 1 week of incubation at the optimal temperature range of 30° to 37°C. Various species of *Nocardia* can grow over temperature ranges from 10° to 50°C. The basis of speciation is largely biochemical.

E. EXPERIMENTAL MODELS

The use of virulent log-phase, filamentous organisms in mice has provided a good model for immunologic studies. Guinea pigs and rabbits can also be infected experimentally. Cattle, dogs, and other domestic animals are subject to natural infections with nocardiae and thus are potential experimental models.

F. INFECTIONS IN MAN

Nocardiosis is a combined suppurative-granulomatous disease in which the suppurative aspects dominate. Most cases of nocardiosis occur as the result of a known immunodeficiency either systemic (50%) or local (15%) and *N. asteroides* is responsible for over 85% of all cases. The list of predisposing background diseases and conditions is long and varied; most are concerned with depressed CMI or phagocyte and globulin disorders. Indeed, severe nocardiosis is commonly, if not always, an opportunistic infection.

Human nocardiosis frequently begins as an inhalation-induced pulmonary infection; it may range from mild and transient to severe and chronic, the latter being fatal in some 40% of cases.

Chronic lung lesions may be solitary and expanding or diffuse and extensive; cavities form on occasion. Chronic nocardiosis bears some resemblance to pulmonary tuberculosis and actinomycosis; indeed, nocardiosis may sometimes coexist with tuberculosis. As in actinomycosis, sinus tracts frequently form and discharge pus to the exterior through the chest wall, or into abdominal viscera. Hematogenous spread often occurs to the brain and sometimes to other organs with a resulting fatality rate of some 80%. A history of undiagnosed pneumonia followed by brain abscess is highly suggestive of nocardiosis. Small brain abscesses tend to coalesce to form a large abscess. Infections of kidney, liver, spleen, and adrenals are also secondary to chronic lung involvement; however, it is of interest that bone seldom becomes infected.

Cutaneous and subcutaneous lesions of nocardiosis, which are due to *N. brasiliensis* in over 50% of cases, result from the introduction of organisms directly through a break in the skin, often by the agency of a thorn or splinter. If the infection remains limited a pustule results, but if deep tissues become infected a mycetoma may develop (Sect. on Actinomycetoma); the symptoms include local abscess, fever and tender draining lymph nodes.

G. MECHANISMS OF PATHOGENESIS

Little is known about the manner by which nocardiae cause disease; however, their similarities to species of the genera *Actinomyces* and *Mycobacterium* suggest that at least some pathogenetic mechanisms are common to all of these organisms.

H. MECHANISMS OF IMMUNITY

Although the final answers regarding mechanisms of immunity to nocardiosis are not at hand, recent animal experiments makes it reasonably certain that total immunity rests with the combined forces of T cell-activated macrophages and specific Abs. The Abs have been shown to act as opsonins to promote phagocytosis by macrophages. Studies with *N. asteroides in nonactivated mouse macrophages* have shown that virulence rests with the log-phase filamentous-form of the organism. Whereas the relatively avirulent stationary-phase coccoid-form organisms are readily phagocytized and inhibited and killed, filamentous organisms resist phagocytosis and if engulfed, grow within and kill these macrophages. They evidently avoid killing within nonactivated macrophages by blocking lysosome-phagosome fusion. However, as in tuberculosis, *activated macrophages* resist the toxicity of ingested filamentous organisms and prevent their reproduction and/or kill them.

I. LABORATORY DIAGNOSIS

Direct examination of pus, either in a wet mount under reduced light or in a Gram-stained preparation, usually reveals an abundance of characteristic bacterial filaments. Clinical specimens should be crushed between two glass slides to break up clusters or masses of filaments and then spread on the slides. Acid-fast staining, involving a brief period (5 to 10 seconds) of decolorization with acid-alcohol or careful destaining with 0.5% aqueous sulfuric acid instead of acid-alcohol, is of differential value. Periods of exposure to acid-alcohol longer than 5 to 10 seconds will completely decolorize weakly acid-fast nocardiae such as *N. asteroides* and *N. brasiliensis.*

Tissue sections may or may not contain readily detectable colonies of radiating filaments or scattered, branching filaments. Because the filaments are quite narrow, a routine hematoxylin-eosin (H-E) stain will usually not suffice, and either methenamine-silver or Gram staining may be needed to visualize the filaments. In tissue sections from mycetoma patients, however, sulfur granules of nocardiae can be visualized by H-E staining.

Cultures should be prepared as described in Section D, and examined for colony formation at intervals ranging from 2 days to 2 weeks. Biochemical tests in media incubated at 27°C help to differentiate nocardiae from certain species of *Streptomyces.* If cultures are incubated at 45 to 50°C, *N. asteroides* will grow, whereas most contaminating bacteria will be inhibited.

J. THERAPY

Although antimicrobics are effective in primary nocardiosis, about half of all past cases have been fatal, chiefly because the disease was not diagnosed until after dissemination to the brain or other vital organs had occurred. Surgical drainage of abscesses and other lesions is an essential adjunct to antimicrobial therapy. Because of late metastasis and the high frequency of relapses after treatment, large doses of sulfonamides should be continued for months and extended for at least 6 weeks after the disease has apparently cleared. Other anti-nocardial agents include ampicillin, aminoglycosides, cephalosporins, cycloserine, minocycline and trimethoprim-sulfamethoxazole.

Since many patients are immunodeficient, secondary infections with other opportunists are frequent. Nocardiae are present in soil worldwide and man is continually exposed. The infrequency of clinical nocardiosis suggests that most human beings possess or readily develop a high degree of resistance to these organisms. In vitro testing has heretofore been of limited value for choosing anti-nocardial drugs or synergistic combinations of drugs because in vitro and in vivo activities do not necessarily correlate. However, recent techniques hold much promise for drug selection.

K. RESERVOIRS OF INFECTION

The only known primary reservoir of nocardiae is soil.

L. PREVENTION OF DISEASE

The prevalence of primary pulmonary nocardiosis indicates that inhalation is a major mechanism of infection. Because the nocardiae are widespread in nature and heavy exposure is frequent, it is reasonable to assume that subclinical pulmonary infections sometimes take place and result in the development of protective immunity analogous to the responses seen in histoplasmosis, coccidioidomycosis, and a number of other diseases that follow inhalation of soil organisms. However, definitive proof for this is lacking. In any event, nocardiae would appear to have much lower pathogenetic potential than the causative organisms of histoplasmosis and coccidioidomycosis (Chap. 37). It is probable that development of disease depends more on the general immunologic status of the host than on exposure to the nocardiae; consequently, no measures are available for prevention of nocardiosis.

OTHER ACTINOMYCETOUS DISEASE

DERMATOPHILOSIS

Dermatophilosis (streptothricosis), a pustular exudative staphylococcal-like disease of skin due to *Dermatophilus congolensis,* is rare in man but frequent in cattle, sheep, horses and other animals. Dermatophilosis responds to antibiotics used to treat staphylococcosis and to topical treatment with metallic compounds.

ACTINOMYCETOMA

Actinomycetoma (Madura foot) is a subcutaneous infection due principally to one of several saprophytic species of actinomycetes present in tropical and subtropic soils; these organisms are members of the genera: *Actinomadura, Nocardia* (Sect. A) and *Streptomyces.* Mycetoma can also be caused by various filamentous fungi and rarely by *A. israelii.* Infection usually involves injury of hands or feet, particularly the feet as the result of going barefoot. Other injuries leading to infection include the bites of snakes and insects and, piercing by thorns. *Actinomycetoma* is a slowly progressing, suppurative-granulomatous infection characterized by lesion enlargement at the infection site, draining sinuses and ultimate local destruction of muscle and bone.

Laboratory diagnosis rests on direct examination of exudates for actinomycotic granules and organisms, together with culture on routine media. Speciation is determined by studies on morphology, physiology and cell wall composition.

Therapy of actinomycetoma involves long-term treatment with appropriate antimicrobics directed against the infecting species, together with surgery.

In infections due to *Actinomadura* and *Streptomyces* sp., streptomycin is the antibiotic of choice.

FARMER'S LUNG

Farmer's lung is an allergy-based lung infection which is usually due to inhalation of either *Aspergillus* sp. (Chap. 40), *Thermoactinomyces vulgaris* or *Micropolyspora faeni.* The causative organisms are present in spoiling grains or forage. The allergies involved are of the atopic and Arthus types.

Treatment consists of administering anti-allergic drugs and avoiding further exposure.

REFERENCES

Brennan, M.J., et al.: A 160-Kilodalton epithelial cell surface glycoprotein recognized by plant lectins that inhibit the adherence of *Actinomyces naeslundii.* Infect. Immun. *52*:840, 1986.

Emmons, C.W., et al.: *Medical Mycology,* 3rd Ed. Philadelphia, Lea & Febiger, 1977.

Lerner, P.I.: Pneumonia caused by the pathogenic actinomycetes *(Actinomyces, Arachnia, Nocardia). Seminars Infectious Disease.* Weinstein, L. and Fields, B.N. (eds.), New York, Thieme-Stratton, Inc. 1983, p. 110.

Sandberg, A.L., et al.: Type 2 fimbrial lectin-mediated phagocytosis of oral *Actinomyces spp.* by polymorphonuclear leukocytes. Infect. Immun. *54*:472, 1986.

Smego, R.A., and Gallis, H.A.: The clinical spectrum of *Nocardia brasiliensis* infection in the United States. Rev. Infect. Dis. *6*:164, 1984.

35

MISCELLANEOUS PATHOGENIC BACTERIA

Spirillum minus

Streptobacillus moniliformis

Bartonella bacilliformis

Erysipelothrix rhusiopathiae

Cardiobacterium hominis

A variety of organisms that cause infections rarely, or about which little is known, will be considered briefly in this chapter.

SPIRILLUM MINUS

Spirillum minus previously was considered to be a member of the family *Spirillaceae* and the order *Pseudomonadales;* however, the taxonomic classification of this bacterium is now considered uncertain in the current edition of *Bergey's Manual of Systematic Bacteriology* (9th Ed.). *Spirillum minus* is Gram-negative, has polar flagella (bipolar tufts), and is aerobic. The cells are short and thick (0.5 × 3.0 μm), and usually have 2 or 3 spirals per cell.

Spirillum minus is the cause of the human disease, *Soduku rat-bite fever.* The organism is widespread in rats and other rodents, and is usually transmitted to humans by the bite of infected rodents or the bites of cats and other animals that ingest rodents. The disease is rare in the USA and occurs most often in the Far East. After an incubation period of about 2 weeks, an abrupt febrile illness occurs, with an erythematous or purplish rash adjacent to the original wound. The fever may recur over a period of weeks to months, and polyarthritis is common. A chancre-like lesion may develop at the wound site during the disease, and regional lymphadenitis is common. Mortality has been reported to range from about 5 to 10%. Either penicillin or streptomycin has been used successfully to treat Soduku rat-bite fever.

The bacterium has not been cultured in vitro; however, it can be recovered from lesion exudates, lymph node aspirates, or blood by inoculation of these materials into spirillum-free laboratory mice or guinea pigs.

For diagnostic purposes, identification of the organism is usually made on the basis of typical morphologic appearance in wet mounts examined by dark-field or phase microscopy. Treatment with Giemsa or Wright's stain, or silver impregnation of the bacterium is also useful in aiding visualization of the bacterium in clinical specimens. If direct mounts are negative, animal inoculation should be performed.

STREPTOBACILLUS MONILIFORMIS

Streptobacillus moniliformis is a Gram-negative, facultative anaerobe indigenous to the nasopharynx of rodents, and causes a form of rat-bite fever differing from that due to *Spirillum minus.* The disease can be transmitted by the bites of rats *or* by ingestion of food that has been contaminated by rats. Rat-bite fever following ingestion of contaminated milk is called *Haverhill fever* and has occurred in epidemics. *Streptobacillus*-induced rat-bite fever resembles Soduku, described above, in that it presents as a persistent febrile disease with polyarthritis; however, lymphadenitis is not observed, and the rash in streptobacillus-induced disease is petechial, rubelliform, or morbilliform, rather than erythematous. Reported complications of the disease include subacute bacterial endocarditis, brain abscesses, and pneumonia. There are reports of successful treatment with either penicillin, streptomycin, or tetracyclines.

Streptobacillus moniliformis is notable for its pleomorphism. Rods, 0.1–0.7 × 1–5 μm long, with rounded or pointed ends occur singly or form long, wavy chains or filaments 10 to 150 μm long. Single rods may show central swelling, and chains or filaments may have a series of swellings resulting in a "string of beads" appearance.

An important characteristic of the organism is its ability to grow in either the bacterial phase with cell walls or in the wall-less L-form phase, in which the colonies are indistinguishable from *Mycoplasma* colonies. In fact, much of the original work on L-forms was done with naturally occurring L-forms of streptobacilli.

Streptobacillus moniliformis is fastidious on culture; it requires rich media containing either blood, serum, or ascitic fluid, and growth is enhanced by an increased CO_2 atmosphere. A rise in titer of specific serum agglutinins during infection is useful in diagnosis.

BARTONELLA BACILLIFORMIS

The genus *Bartonella* (family *Bartonellaceae*) belongs to the order *Rickettsiales* and is related to the rickettsiae (Chap. 31) and chlamydiae (Chap. 32).

Infections with *Bartonella bacilliformis* are transmitted by the sandfly, *Phlebotomus,* and occur only in the Andes mountains of South America, where both the vectors and long-term human carriers are found. Infections may remain latent or may become clinically apparent. Two forms of clinical disease occur, often in succession: *Oroya fever,* which is a severe, febrile, hemolytic anemia, and *verruga peruana* characterized by Peruvian warts. Oroya fever is characterized by the growth of organisms on or within erythrocytes and within endothelial cells. As many as 90% of the red cells may be infected, and the mortality rate in untreated patients is about 40%. Patients who survive the febrile disease often develop verruga peruana, a benign, generalized, granulomatous skin disease in which purplish papules and deep nodules occur, often in crops, during a period of several weeks to a year. Even if the patient is not treated, the mortality rate is less than 5%. Differences in host response undoubtedly contribute to the different disease processes.

In 1885, Carrion proved that Oroya fever and verruga peruana are caused by the same organism. He volunteered to be inoculated with material from verrucous lesions, and died 39 days later of Oroya fever! In his honor, bartonellosis is often referred to as Carrion's disease. At present, the disease is readily controlled by antibiotic therapy. In addition, insecticide control of sandflies has greatly decreased the incidence of bartonellosis.

Bartonella bacilliformis is a Gram-negative, motile, obligately aerobic, nonsporulating, pleomorphic bacterium that usually assumes the form of rods about 0.5 × 2.0 μm. It can be cultured in vitro in semisolid media containing either human, horse, or rabbit blood. Growth becomes apparent after about 10 days incubation at 28° to 37°C. The organism can also be grown in the chorioallantoic fluid and, like the rickettsiae, in the yolk sac of the chicken embryo.

ERYSIPELOTHRIX RHUSIOPATHIAE

Erysipelothrix rhusiophathiae is in the family *Corynebacteriaceae* and is, therefore, related to *Corynebacterium* and *Listeria* species. The organism is primarily a pathogen of swine and other animal species, and may live as a saprophyte in decaying organic matter. It causes erysipelas in swine and other animals and erysipeloid skin infections in humans. The human disease develops in skin abrasions after contact with tissues of infected animals or contaminated animal products, and is most often seen in abbatoir workers, butchers, veterinarians, fish-handlers, and housewives. The lesions are painful, edematous, and erythematous. Septicemia and endocarditis are possible complications of erysipeloid skin infections. Penicillin therapy is used to treat diseases caused by the bacterium.

Erysipelothrix rhusiopathiae is a Gram-positive, facultatively anaerobic, nonsporulating, nonmotile rod that tends to form long, nonbranching filaments. It can grow on a variety of rich media at an optimal growth temperature of about 37°C. It may be necessary to use a hand lens to see the colonies, which grow to a diameter of only 0.7 to 1.0 mm in 48 hours. Mice can be readily infected with the organism, and the mouse protection test with specific antiserum, along with cultural and biochemical characteristics, aids in identification. Also, the production of H_2S in the butt portion of triple sugar iron agar can be used to differentiate this organism from other morphologically similar Gram-positive bacteria.

CARDIOBACTERIUM HOMINIS

Cardiobacterium hominis is an etiologic agent of endocarditis in humans and is part of the normal flora of the human nasopharynx. The bacterium is a Gram-negative, pleomorphic organism consisting of filamentous forms, bacilli with swollen ends resembling "teardrops," and clusters resembling a rosette. It forms small nonhemolytic colonies (1 mm diameter) on blood agar after incubation for 48 hours at 35° to 37°C in a candle jar (increased CO_2 atmosphere). Some isolates of the bacterium produce colonies that pit the agar. Infections caused by this organism can be effectively treated with various antibiotics.

REFERENCES

Anderson, L.C., et al.: Rat-bite fever in animal research laboratory personnel. Lab. Anim. Sci. *33*:292, 1983.

Chandler, D.S., and Craven, J.A.: Persistence and distribution of *Erysipelothrix rhusiopathiae* and bacterial indicator organisms on land used for disposal of piggery effluent. J. Appl. Bacteriol. *48*:367, 1980.

Dijkmans, B.A. et al.: Brain abscess due to *Streptobacillus moniliformis* and *Actinobacterium meyerii*. Infection *12*:262, 1984.

Fliegelman, R.M., et al.: *Erysipelothrix rhusiopathiae* endocarditis: report of a case and review of the literature. J. Am. Osteopath. Assoc. *85*:94, 1985.

Knobloch, J., et al.: Antibodies to *Bartonella bacilliformis* as determined by fluorescence antibody test, indirect hemagglutination and ELISA. Trop. Med. Parasitol. *36*:183, 1985.

Normann, B., and Kihlstrom, E.: *Erysipelothrix rhusiopathiae* septicaemia. Scand. J. Infect. Dis. *17*:123, 1985.

Ognibene, F.P., et al.: *Erysipelothrix rhusiopathiae* bacteremia presenting as septic shock. Am. J. Med. *78*:861, 1985.

Rogosa, M.: *Streptobacillus moniliformis* and *Spirillum minus*. In *Manual of Clinical Microbiology*, 4th Ed., Chapter 35, Lennette, E.H., (ed.) Wash., D.C., Amer. Soc. Microbiol., 1985.

Savage, N.L., et al.: Clinical, microbiological, and histological manifestations of *Streptobacillus moniliformis*-induced arthritis in mice. Infect. Immun. *34*:605, 1981.
Shanson, D.C., et al.: *Streptobacillus moniliformis* isolated from blood in four cases of Haverhill fever. Lancet *2*:92, 1983.
Simon M.W., and Wilson, H.D.: *Streptobacillus moniliformis* endocarditis. A case report. Clin. Pediat. *25*:110, 1986.
Walker, T.S., and Winkler, H.H.: *Bartonella bacilliformis:* colonial types and erythrocyte adherence. Infect. Immun. *31*:480, 1981.
Weaver, R.E.: *Erysipelothrix.* In *Manual of Clinical Microbiology,* 4th Ed., Chapter 20, Lennette, E.H., (ed.) Wash., D.C., Am. Soc. Microbiol., 1985.
Wormser, G.P., and Bottone, E.J.: *Cardiobacterium hominis:* review of microbiologic and clinical features. Rev. Infect. Dis. *5*:680, 1983.

36

AN INTRODUCTION TO MEDICAL MYCOLOGY

Mycology is the study of fungi, a diverse group that includes molds, yeasts, mushrooms, and related organisms. Although over 100,000 species of fungi are recognized, fewer than 50 species are responsible for the majority of the fungal infections (mycoses) of man. The majority of fungi exist as soil saprophytes and only a few as parasites of animals and man. Most fungal species that infect man do so by acting opportunistically as facultative pathogens. A few of the dermatophytes are readily transmissible pathogens for man as their natural host. Man can also serve as an accidental host for some infecting fungi that exist in animals as their natural hosts.

The fungi that infect man are the subjects of study in medical mycology. Other fungi may be harmful or even lethal by virtue of forming hallucinogens or toxins that are ingested, such as aflatoxin and mushroom poisons. However, this discussion will be limited principally to the fungi that produce clinical infections in man. The mycoses develop slowly and are characterized by their chronicity and tendency to relapse. The incidence of infections and deaths due to the fungi has been grossly underestimated. Moreover, the list of fungal species known to be capable of producing disease in immunocompromised persons is mounting rapidly because of the increasing use of immunosuppressants.

A. CHARACTERISTICS OF FUNGI

The fungi are eucaryotic protists that differ from the simple bacteria and other procaryotic protists in many ways. Some of the important differences between fungi and bacteria are presented in Table 36–1.

The fungi have several *distinguishing characteristics.* They possess *unique, rigid cell walls* that differ considerably from cell walls of bacteria. The basic units of fungal cell walls are certain high MW polysaccharides, such as chitin and mannan, which are present in a thatched arrangement in the wall; they lack the teichoic and muramic acids present in the cell walls of bacteria. The cytoplasmic membranes of fungi contain *sterols,* a property that distinguishes them from virtually all bacteria except the mycoplasmas. All fungi reproduce

Table 36–1. Some Important Differences between Fungi and Bacteria

Property	*Fungi*	*Bacteria*
Cell structure	Eucaryotic	Procaryotic
Diameter of representative species	$>5\ \mu m$	$<2\ \mu m$
Cell wall composition	Contain chitin, mannan, other polysaccharides	Contain murein
Cytoplasmic membranes	Sterols present	Lack sterols (except for mycoplasmas)
Cytoplasmic contents	Include mitochondria, endoplasmic reticulum; cytoplasmic streaming	Lack mitochondria, endoplasmic reticulum; no cytoplasmic streaming
Nucleus	True nucleus with nuclear membrane: chromosomes, in pairs	Nuclear body equivalent to a single chromosome, without nuclear membrane
Mode of reproduction	Either sexual or asexual, with spore formation	Binary fission

asexually (anamorph state) and most can also reproduce sexually (telemorph state). However, in most medically important fungi asexual reproduction is so efficient that the sexual forms seldom, if ever, occur.

Molds are the filamentous forms of a heterogeneous group of fungi. From either a single conidium (asexual spore) or a sexual spore extension of the cell wall and an increase in cytoplasmic content leads to the formation of a long tubular structure, called a *hypha.* Hyphae can branch and extend at their tips, as diagrammed in Figure 36–1, to present an intertwined filamentous cottony mass known as a *mycelium*; hyphae can also form spores.

Hyphae may or may not present cross walls or *septa* which define the boundaries of individual mononucleated cells within each hypha. The septa are usually incomplete or porous and allow the passage of cytoplasm between cells. A *Nonoseptate* hypha represents a communicating hollow tube that is *coenocytic,* i.e. the hypha is bounded by a single continuous cell wall containing cytoplasm carrying multiple nuclei.

Some hyphae grow into the medium as "submerged hyphae," whereas others grow upward as "aerial hyphae" and send out stalk-like structures called condiophores or sporangiophores that bear asexual spore-like structures called conidia; the former as chains of exogenous conidia and the latter as endogenous conidia borne in a sac-like structure called the sporangium. Additional types of asexual conidia exist (Sect. B). Since conidia are spore-like in the sense that they are more resistant to physical and chemical agents than hyphae, they have been designated by such terms as condiospores and sporangiospores. In any event, when asexual conidia are referred to as spores, they should be clearly distinguished from sexual spores, including the *ascospores* of the *Ascomycetes* and the *basidiospores* of the *Basidiomycetes* (Sect. B).

Most yeasts are oval-shaped, single-celled, fungi that reproduce by budding at the smaller end of the cell. Spherical yeasts may bud at either single sites or simultaneously at multiple sites. Bud-derived daughter cells usually pinch off from the mother cell before they reach full size. However, under special environmental conditions the daughter cell may fail to separate but instead swells and elongates to form a hyphal-like structure called a *pseudohypha.* When this budding is repeated by succeeding daughter cells, the resulting chain of swollen, elongated cells with constrictions at their joining points form a *pseudomycelium* resembling a mycelium composed of true hyphae. The ability to convert from yeast-type growth to mold-type growth or vice versa is called

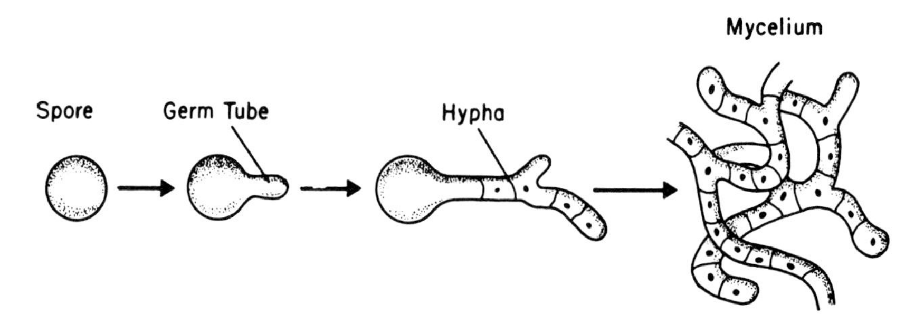

Fig. 36–1. Development of fungal mycelium.

dimorphism. Many of the pathogenic fungi are dimorphic, existing in nature as saprophytes in the mold form and in the human host as parasites in the yeast form; when cultured, some of them grow as yeasts at 37°C and as molds at 25°C.

Dimorphism is well illustrated by the opportunistic pathogen, *C. albicans* which can be induced to form several growth structures simultaneously, including true hyphae, pseudohyphae, chlamydospores, blastospores and germ tubes (Chap. 40). *Germ tubes* are narrow elongated buds that can mature to form true hyphae. They can be stimulated to form with great rapidity by incubating *C. albicans* in human serum for a few hours at 37°C.

Soil is especially suited for supporting mycelial growth and is the natural habitat of many fungi. Because of their rigid cell walls, fungi must acquire nutrients in soluble form, either by absorption or by pinocytosis. It is of interest that some fungi growing as saprophytes in nature possess means for trapping amoebae or worms, which then die and serve as a source of nutrients for the fungi. Others, notably some of those of medical importance, can lead an exclusive parasitic existence and derive their nourishment from the host. In either case, the fungi secrete a large variety of extracellular enzymes that aid in the degradation of substances of high molecular weight to small molecules that can be transported into the fungal cell as nutrients. Fungi are pathogenic for plants as well as animals and are of paramount importance in the recycling of organic matter, especially organic matter of plant origin.

B. CLASSES OF FUNGI

The fungi comprise four classes, the *Phycomycetes, Ascomycetes, Basidiomycetes* and the *Deuteromycetes (Fungi Imperfecti).* The lower fungi make up the class called *Phycomycetes.* The root of this word, phyco, means seaweed or algae, and the phycomycetes are sometimes called the algal fungi; some of them are aquatic and some are terrestrial. A familiar example of a terrestrial phycomycete is black bread mold, a member of the genus *Rhizopus.* Phycomycetes differ from higher fungi in that they form endogenous *asexual spores* (sporangiospores) carried *in sac-like structures called, sporangia,* and that they have nonseptate hyphae.

The three classes comprising the higher fungi, are characterized by *exogenous asexual spores* called *conidiospores,* which are formed on conidiophores and by the presence of septate hyphae with pores that permit the passage of nuclei and cytoplasm from one part of the mycelium to another.

The class *Ascomycetes* includes yeasts, morels, truffles, and many of the common molds. The word *"ascus"* means a bag or *sac-like structure,* and the ascomycetes were so named because they bear their *sexual spores in an ascus.* The ascomycetes include many fungi beneficial to man, such as the yeasts of the genus *Saccharomyces* that leaven bread and ferment alcoholic beverages.

The class *Basidiomycetes* consists of higher fungi that produce exogenous *sexual spores on a basidium or base;* this class includes the mushrooms.

Most of the fungi of medical importance belong to the class *Deuteromycetes (Fungi Imperfecti). These fungi cannot be classified on the basis of their mode of sexual reproduction, because their sexual stages are lacking or unknown.* They are probably members of other classes but are included in the so-called "taxonomic dump-heap" *Deuteromycetes,* because they cannot be properly identified. Classification of the multitude of fungal species is based primarily on their sexual-cycles which are often hidden and difficult to reveal. Consequently reclassification is a continuing process as newly discovered sexual cycles accumulate and demand attention.

C. CHARACTERISTICS THAT AID IN IDENTIFICATION OF FUNGI

The *asexual spores* of the higher fungi *aid in the identification* of some species (Fig. 36–2). These spores may be small, single-celled microconidia, or large, single or multi-celled macroconidia. Some fungi have both microconidia and macroconidia; other species have one or the other. The shapes and arrangement of the conidia are characteristic for a given species of fungus.

Other asexual reproductive structures that are useful in the identification of medically important fungi *include blastospores, arthrospores, and chlamydospores* (Fig. 36–2). The *blastospores* of yeasts are formed by budding. *Arthrospores* are generated by mycelial segmentation; the segments disarticulate to yield asexual spores, each of which is capable of giving rise to a hypha and new mycelium. *Chlamydospores* are thick-walled resting spores, formed by rounding-up and thickening of a segment of a hypha. Arthrospores and other conidia promote the aerial dissemination of fungi, because they break away from the mycelium and are dispersed in air.

Non-morphologic characteristics of fungi are also exploited in common lab-

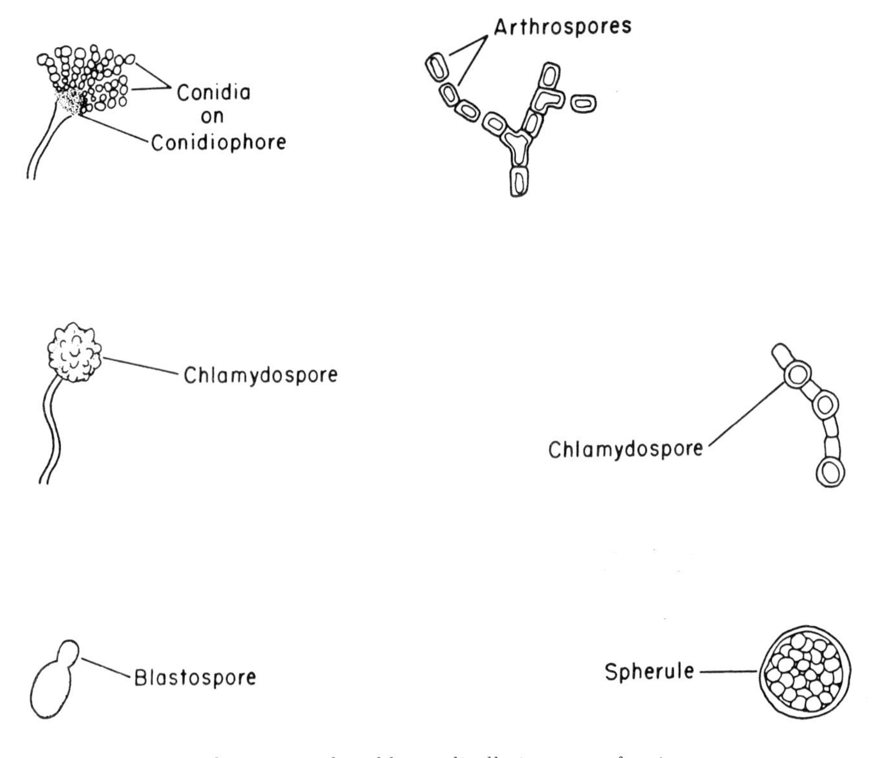

Fig. 36–2. Asexual spores produced by medically important fungi.

Table 36–2. Common Laboratory Procedures Used to Identify Fungi of Medical Importance

Procedure	*Advantage*
Wet mount of tissue or mucus-containing specimens in 10% KOH	Strong alkali degrades tissue and mucus and permits visualization of fungi
Wet mount of portions teased from fungal colonies and mounted in lactophenol blue	Permits observation of fungal morphology and presence of spores
Fixed slides stained with periodic-acid-Schiff (PAS) or methenamine silver stains	Both PAS and silver stain fungal cell walls to give good contrast with background in tissue sections or other clinical materials
Sabouraud's glucose agar for culture; incubation at room temperature (RT) for up to 6 weeks, or at 37°C for days	Low pH of the medium and RT incubation favor growth of fungi over bacteria Antibiotics may also be added to discourage bacterial growth
Slide cultures, with inoculated blocks of Sabouraud's glucose agar (about 1 cm square and 2 or 3 mm deep) on a glass slide covered with a coverslip and incubated in a moist chamber at RT; when spores form, the coverslip is carefully removed and examined in a lactophenol-blue wet mount	Permits observation of relatively undisturbed fungal growth

oratory tests and aid in identification (Table 36–2). Fungal cell walls are resistant to strong alkali (10% KOH or NaOH) which will degrade tissue and mucus, consequently, fungi can be observed more readily in many specimens following alkali treatment than without such treatment. The cell wall can also be stained with periodic-acid-Schiff (PAS) reagent or methenamine silver stains. Fungi are not affected by penicillin because they lack murein mucocomplex in their cell walls; as a consequence, penicillin and certain other antibiotics can be used to suppress bacterial growth and in so doing facilitate the isolation of fungi from clinical specimens. Unlike most bacterial pathogens fungi generally prefer a low pH and the majority of them will grow readily at room temperature; they are either strict aerobes or facultative anaerobes.

D. GENERAL CHARACTERISTICS OF MYCOSES

The fungal diseases have been placed in four major categories; the *superficial mycoses,* the *intermediate (subcutaneous) mycoses,* the *deep (systemic) mycoses* and the *opportunistic mycoses.*

Two properties of fungi appear to be of paramount importance in the *pathogenesis* of the diseases they cause. First, the tissue tropism of a fungus determines the site(s) of infection produced. Tissue tropism denotes the predilection of a parasite for a particular tissue or organ of the host. Although many pathogenic bacteria exhibit tissue tropism (e.g., gonorrheal infection of epithelial cells of genital mucous membranes and Shigella infection of the gut epithelium), the basis for such selective localization of fungi has not been established. The tissue tropism of many fungi is marked. Examples are *Histoplasma capsulatum,* which lives within macrophages of the RES; *Cryptococcus neofor-*

mans, which thrives in the CNS; *Paracoccidioides brasiliensis,* which lives selectively in mucous membranes and lymph nodes; and *Blastomyces dermatitidis,* which elects to reside in skin. The dermatophytes responsible for superficial mycoses thrive on keratin; indeed, some are so selective that they prefer the keratin of hair, some the keratin of nails, and others the keratin of skin. Among fungal diseases communicability is limited to dermatophyte infections.

In addition to tissue tropism, a second property of fungi that is important in *pathogenesis* is the ability of many of them to incite sensitivity, especially DH. Often, the sensitivity of the host to fungal Ags accounts in large part for the pathologic effects produced. Unlike bacteria many fungi incite a combined pyogenic granulomatous response. Moreover, the common pathogenetic mechanisms of bacteria, e.g., toxin production or capsule formation, are not characteristic of pathogenic fungi even though some nonpathogenic fungi produce toxins which, if ingested in the manner of staphylococcal enterotoxin or botulinum toxin, may be toxic. However, there is some evidence suggesting an endotoxin-like activity for some species of pathogenic fungi. Capsule formation is not characteristic of most pathogenic fungi; it is prominent in only one pathogenic species, *Cryptococcus neoformans.* Fungal cell wall polysaccharides are resistant to tissue enzymes and are cleared very slowly from the body.

Acquired immunity to fungal diseases is principally CMI, although collaboration between CMI and humoral immunity may sometimes occur. It is commonly observed that DH to Ag(s) of a fungus is associated with a favorable prognosis or immunity to reinfection. On the other hand, humoral Abs usually reflect the extent of infection and the gravity of prognosis rather than the strength of immunity.

Other compelling evidence that fungal infections are resisted and controlled in large part by cellular immune mechanisms comes from clinical experiences with immunologically-deficient patients. It is well established that defective or depressed CMI, either genetic or acquired, predisposes to serious fungal infections, whereas defective or depressed humoral immunity usually does not. Often the invading fungi in such instances are *opportunists* that are ubiquitous in the environment. Such opportunistic infections with species of *Candida, Aspergillus,* and many other genera present a major problem in the management of patients whose CMI has been suppressed by drugs or other therapy used to prevent rejection of organ transplants or to treat tumors. Cancer patients are at particular risk of infection because the late general suppression of CMI produced by the malignant disease itself is exaggerated by treatment with anticancer drugs, many of which have strong immunosuppressive activity.

A role for protective forces other than specific CMI in resisting and controlling fungal infections is indicated by both clinical and experimental data. Neutrophils have been implicated as playing a part in protection. For example, in candidiasis, the intact neutrophil appears to be important in preventing systemic spread of the organisms. In addition, it is well established that diabetics and patients with certain endocrine and other disorders are highly susceptible to certain opportunistic fungi, even though their CMI appears to be essentially normal. It is possible that this increase in susceptibility is related to neutrophil dysfunction. Lymphokines and natural killer cells have been reported to be active against certain fungi.

Immunosuppression can be induced by certain fungi thus enhancing their invasiveness and pathogenicity.

Clearly, much remains to be learned about the pathogenesis and both innate and acquired immunity to fungi. The current increase in opportunistic fungal infections, together with the difficulties in treating mycoses, emphasizes the need for a better understanding of pathogenesis and immunity in fungal infections.

Diagnosis of mycoses is often difficult and many clinically evident cases go undiagnosed or are misdiagnosed.

E. DRUG TREATMENT OF MYCOSES

The mycoses are often more difficult to treat than bacterial infections because only a few effective antifungal agents are available (Chaps. 6, 37, 38, 39 and 40). Treatment of new emerging mycoses is especially difficult.

REFERENCES

Ajello, L.: Emerging Mycoses of Present and Future Concern. In *New Horizons in Microbiology.* A. Sanna and G. Morace (eds.), Amsterdam, Elsevier Science Publishers, 1984.

Black, C.M., et al.: Modulation of lysosomal protease-esterase and lysozyme in Kupffer cells and peritoneal macrophages infected with *Nocardia asteroides.* Infect. Immun. *54*:917, 1986.

Drauhet, E., et al.: Evolution of antifungal agents: past, present and future. Rev. Infect. Dis. *9*:1:54, 1987.

Emmons, C.W., et al.: *Medical Mycology,* 3rd Ed. Philadelphia, Lea & Febiger, 1977.

Frontling, R.A., et al.: An overview of macrophage-fungal interactions. Mycopathologica *93*:77, 1986.

McGinnis, M.R.: *Laboratory Handbook of Medical Mycology.* New York, Academic Press, 1980.

Rippon, J.W.: *Medical Mycology,* 2nd Ed. Philadelphia, W.B. Saunders Co., 1982.

Young, L.S.: Current needs in chemotherapy for bacterial and fungal infections. Rev. Infect. Dis. *3*:5380, 1985.

37

DEEP-SEATED MYCOSES

Coccidioides immitis

Histoplasma capsulatum

Blastomyces dermatitidis

Cryptococcus neoformans

Paracoccidioides brasiliensis

The etiologically based *deep-seated mycoses* are often referred to as *systemic mycoses*. They comprise a group of similar diseases, each of which may range in severity from inapparent to lethal infections. The various deep-seated mycoses and their respective etiologic agents will be considered separately despite their many similarities.

COCCIDIOIDES IMMITIS

Coccidioides immitis is a soil fungus that characteristically produces lung lesions resembling tuberculosis. Prior to 1930 the disease was not identified before autopsy. However, the initiation of a long series of subsequent investigations by Charles E. Smith and colleagues at Stanford University soon led to detailed information about the organism and the infections it produces in the San Joaquin Valley area of California. Such infections are indigenous in animals as well as in man.

A. MEDICAL PERSPECTIVES

The etiologic agent of coccidioidomycosis, *C. immitis,* exists in large numbers in the soils of endemic areas of the Southwestern USA, Mexico, Central America, and South America where weather conditions favor growth and sporulation. Coccidioidomycosis is seen so often in the San Joaquin Valley of California that it has gained the common name of "valley fever". The deserts of the Southwestern USA offer the unique climate essential for the growth and dissemination of *C. immitis;* a wet, rainy period allows the germination and reproduction of spores and an ensuing hot dry period permits air-and dust-borne spread of the highly-infectious arthrospores. The infection rate is high following dust storms and among airfield and construction workers. As high as 90% of the population in endemic areas give positive skin tests to Ags of the fungus, indicating that most people who are regularly exposed to *C. immitis* become infected. However, only about 40% of infected persons become overtly ill, and fewer than 1% develop severe disease. Although the wearing of masks to reduce high exposure is useful, it is reasonable to expect that the future incidence of disease in endemic areas will not decline, but instead may rise due to the increasing use of immunosuppressants.

B. PHYSICAL AND CHEMICAL STRUCTURE

Being a typical dimorphic fungus, *C. immitis* varies in structure between a mold-form found in soil and a spherule-form found in infected tissues (Fig. 37–1). Since the mold-form grows most readily in vitro, the Ags used for skin testing and other immunologic studies have been obtained exclusively from this form; however, spherule Ags are currently being evaluated. A variety of polysaccharides and proteins are present in coccidioidin, the filtrate of mycelial cultures that is commonly used as an Ag preparation. This mixture elicits positive DH reactions in sensitized individuals, and positive precipitin and CF reactions with serum from patients with extensive infection. The components responsible for these various activities are poorly characterized.

As shown in Figure 37–1A, barrel-shaped arthrospores are formed in the mycelium. Within tissues the infecting arthrospore (asexual spore) germinates to give rise to a sac-like structure, the spherule, which increases in size as the number of contained endospores increases. The original thick wall of each

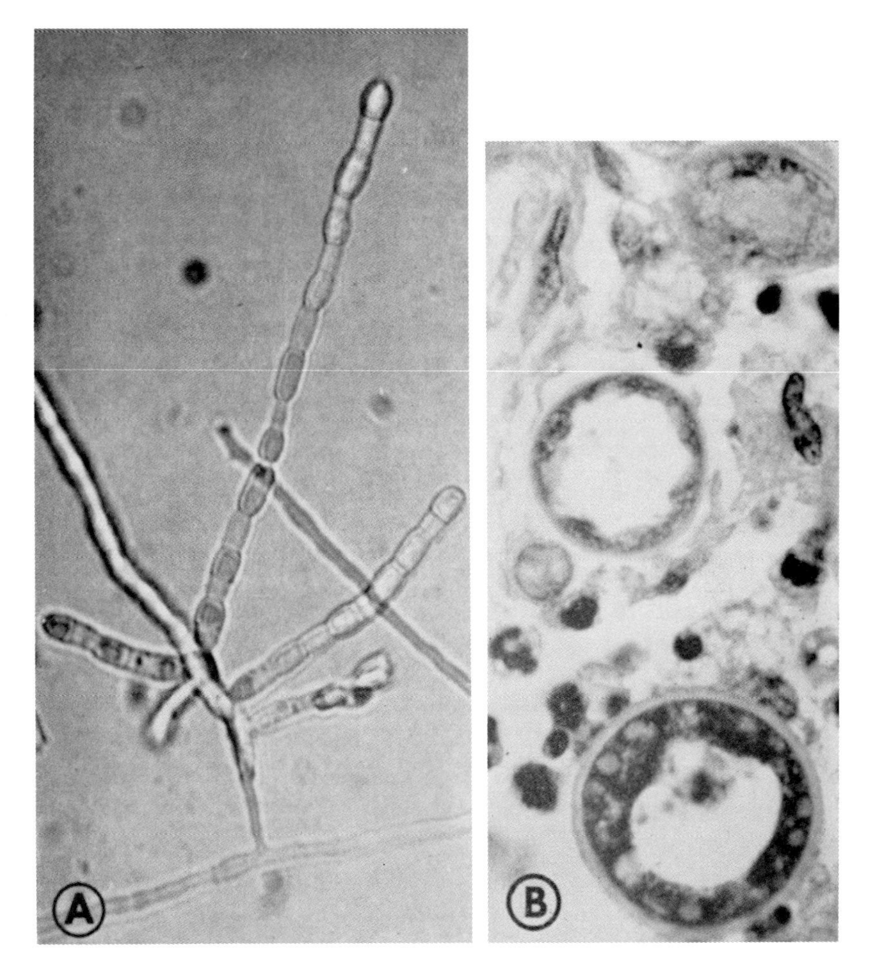

Fig. 37–1. *Coccidioides immitis.* A, Mold form with arthrospores. B, Tissue form in the lung, showing an intact sporangium with well-defined endospores at the lower right and in the center a ruptured sporangium, releasing small endospores. (From Emmons, C. W., et al.: Medical Mycology, 3rd Ed. Philadelphia, Lea & Febiger, 1977.)

spherule becomes thinner as the diameter increases from approximately 15 μm to about 75 μm. At this point the spherule bursts and releases hundreds of endospores, each of which can form a new spherule (Fig. 37–1B).

C. EXTRACELLULAR PRODUCTS

Few data are available concerning extracellular fungal products of possible importance in the pathogenesis or immunology of coccidioidomycosis. There is no evidence that either endotoxins or exotoxins are produced.

D. CULTURE

Sabouraud's glucose agar supports the growth of *C. immitis* at room temperature. Specimens, such as sputum, exudate from lesions, and spinal fluid, contain the spherule form of the fungus; however, when this form is plated onto Sabouraud's medium, growth of the mycelial form occurs. Extreme caution is necessary when dealing with cultures of *C. immitis* because the *arthrospores readily become airborne by the slightest air current. Special techniques must be used for handling spore-bearing materials to prevent infection.* Petri dishes should not be used for culture work. Many laboratory workers have become infected and some have developed severe disease.

The colonies form a white cottony mycelium at first, which soon turns buff or grayish brown. At this stage of the growth cycle the appearance of the mycelium has been described as "moth-eaten." Sporulation begins within 3 days to 2 weeks, depending on cultural conditions.

The mycelial form (mold-form) of *C. immitis* is readily isolated from contaminated soil. The hyphae are relatively wide, branching and septate.

The spherule form has been cultivated in vitro on special media.

E. EXPERIMENTAL MODELS

Mice and guinea pigs can be infected with *C. immitis* by either inhalation or injection and have been used extensively in the study of immunologic responses.

Virulence apparently depends on the ability of arthroconidia and endospores to resist phagocytosis and intracellular killing by nonactivated macrophages and PMNs; only activated macrophages can ingest and kill these fungal forms.

F. INFECTIONS IN MAN

Infection usually results from the inhalation of airborne arthrospores. About 60% of infected individuals remain asymptomatic but become skin-sensitive to coccidioidin by about 3 weeks after infection. *Disease production evidently rests on the degree of exposure and the susceptibility of the individual.*

The acute and most common form of the disease resembles influenza. The initial symptoms of pulmonary infection consist of cough, fever and malaise. One to 2 weeks later, about 3 to 5% of patients develop sequelae attributable to a hypersensitivity reaction, the most frequent being skin lesions (erythema nodosum or erythema multiforme) or joint symptoms (desert rheumatism). Most of the acute cases resolve in a few weeks. A chronic, progressive, granulomatous disease occurs in fewer than 1% of those infected, especially immunosuppressed individuals; this is often fatal unless intensive treatment is given and may persist for years. The granulomatous lesions resemble the tubercles of tuberculosis and cavitation sometimes occurs. The infection may disseminate to many parts of the body, including bones, viscera and the CNS; however, the GI tract is remarkably resistant to infection.

G. MECHANISMS OF PATHOGENESIS

The fact that primary infection with *C. immitis* usually causes self-limiting or asymptomatic disease and that no known toxin exists suggests that the principal mechanism of pathogenesis of this fungus is related to its ability to induce hypersensitivity reactions. It is probable that the early transient allergic lesions in the skin and joints of some 20% of patients represent immune complex hypersensitivity reactions. The arthroconidia are antiphagocytic; moreover, arthroconidia and endospores engulfed by nonactivated macrophages prevent phagosome-lysosome fusion and killing of organisms.

H. MECHANISMS OF IMMUNITY

Cellular immunity normally develops after exposure to *C. immitis* and its level is probably correlated with the degree of DH in the skin. As in tuberculosis, progressive disseminated disease leads to anergy. Also CMI may be depressed due to an increase in prostaglandin production. The development of anergy signals a poor prognosis. Evidently, the chronic form of disseminated disease and reactivation disease rests on a depression of CMI, often induced by anti-cancer drugs. It is also of interest that the incidence of dissemination varies greatly, depending on genetic infuences. Whereas about 1 in 400 Caucasians with diagnosed coccidioidomycosis develops disseminated disease, the incidence is about 14 times higher in blacks and 100 times higher in Filipinos! The reason for these differences are unknown, but are obviously related to different capabilities for developing protective immunologic responses to the fungus. Other high-susceptibility groups include American Indians, age-extreme groups and women in late pregnancy. Susceptibility in pregnancy may be accounted for by the growth stimulating effects of estradiol and progesterone on *C. immitis* and/or the immunosuppression associated with pregnancy.

I. LABORATORY DIAGNOSIS

Although the diagnosis of coccidioidomycois is usually made on the basis of clinical findings, skin tests, serology and culture are helpful in chronic disseminated forms of disease.

Sputum or other clinical specimens can be examined directly by means of a KOH wet-mount slide preparation, which reveals characteristic spherules. Culture of such material on Sabouraud's glucose agar yields typical colonies, usually within several days at room temperature or at 37°C. Isolates are identified by animal inoculation and immunodiffusion tests for specific Ags present in culture broth.

Serologic tests on spaced serum specimens are highly valuable for diagnosis and prognosis. Whereas IgM precipitins against a heat-stable component of coccidioidin appear about 2 weeks following onset in 90% of cases and usually recede in a few months, IgG Abs detected by the CF test against heat-labile Ags appear in 80% of the patients at about 3 weeks and recede soon after symptoms subside. Exclusive of coccidioidal meningitis, a CF titer of 1:32 or higher usually reflects disseminated disease; a continuing rise in titer above 1:32 signifies progressing infection and a poor prognosis. In most cases of meningitis the CF test on spinal fluid is positive.

Skin tests with coccidioidin become positive in 1 to 4 weeks after infection and persist indefinitely except in the face of progressive disseminating disease when anergy develops. Positive skin tests in healthy individuals indicate immunity to reinfection disease.

J. THERAPY

The therapeutic drug, amphotericin B, is effective for treating severe primary and systemic coccidioidomycosis. Newer alternative drugs are miconazole and ketoconazole, although they are less well established. Another therapeutic measure is surgery.

K. RESERVOIRS OF INFECTION

Except for accidental laboratory infections, soil is the only source of infection.

L. PREVENTION OF DISEASE

Only under rare circumstances is coccidioidomycosis transmitted by any means except inhalation of airborne spores. Therefore, effective control depends principally on prevention of dust dissemination by such measures as installing lawns or paving, oiling or watering soil. The efficacy of this kind of control was proven during World War II; it was observed that a number of workers involved in building U.S. Air Force bases in endemic areas became ill with coccidioidomycosis following exposure to contaminated dust. Measures were taken to decrease dust inhalation, with a consequent dramatic decrease in the case rate. An obvious rule is that workers who must be heavily exposed should wear masks. A new and promising vaccine for protecting persons at high risk is currently being tested. Such high-risk persons should avoid areas of endemic disease.

HISTOPLASMA CAPSULATUM

A. MEDICAL PERSPECTIVES

Samuel Darling, who discovered *H. capsulatum* in about 1900, so-named the organism because he mistook it for a capsulated animal parasite. Since it is now known to be a *noncapsulated* saprophytic soil fungus, its present name is obviously inappropriate. Histoplasmosis, caused by *H. capsulatum,* shares many characteristics with coccidioidomycosis. Both diseases extend over a spectrum from acute self-limited to chronic, granulomatous infections, and both are endemic in man and some animals residing in certain geographic locations. However, histoplasmosis is the more wide-spread of the two diseases and occurs in many parts of the world, notably in the USA along the Mississippi River, along other large rivers draining into the Mississippi, and in many areas along the East Coast. The organisms grow best in nitrogen-rich soils fertilized by bird droppings (Sect. J).

For many years, only the severe disseminated form of histoplasmosis was recognized. Largely as a result of massive screening tests for tuberculosis in the USA, it was realized that the rate of infection with *H. capsulatum* is high in endemic areas. Roentgenographic examination revealed that many individuals in these regions harbored healed, calcified lesions resembling those of primary tuberculosis, even though a surprising number failed to react to the tuberculin skin test. However, these subjects were found to react to skin tests with *histoplasmin,* an antigenic extract of *H. capsulatum.* Thus, infections with *H. capsulatum,* like those with *C. immitis,* result in clinical disease in a small minority of persons infected and severe disseminated disease is a rare event.

In the USA it is estimated that over a half million people become infected each year and that over 25,000 clinical cases occur.

B. PHYSICAL AND CHEMICAL STRUCTURE

Histoplasma capsulatum is a dimorphic fungus, occurring in the mold-form in soil and in vivo as a typical oval yeast (about 2 to 4 μm in length) within macrophages (Fig. 37–2A). The yeast cells multiply by forming single buds

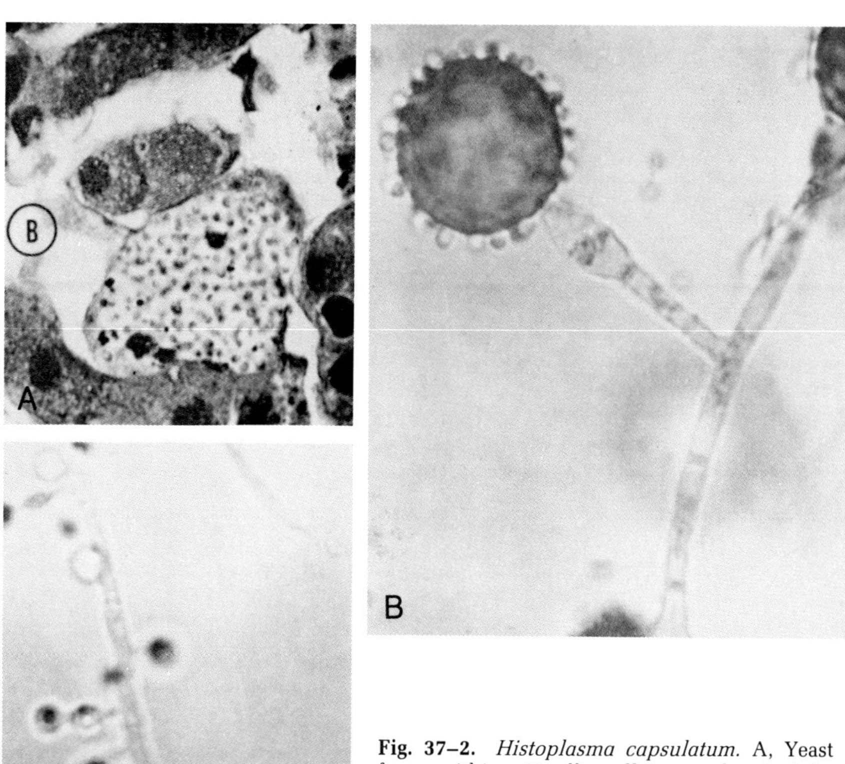

Fig. 37–2. *Histoplasma capsulatum.* A, Yeast forms within a Kupffer cell (macrophage) of the liver. B, Moldform hyphae with microconidia. C, Macroconidia. (From Emmons, C. W., et al. Medical Mycology, 3rd Ed. Philadelphia, Lea & Febiger, 1977.)

and propagate until the macrophage is packed with organisms. The few extracellular yeast-form organisms found in vivo probably represent organisms released by the disruption of infected macrophages.

The mold-form, found in soil, produces microconidia (2 to 5 μm in diameter), which are infectious (Fig. 37–2B). The mold-form also produces characteristic macroconidia, which are large (8 to 15 μm in diameter), round or pear-shaped structures with thick walls and many surface projections (Fig. 37–2C). Although these macroconidia are usually erroneously referred to as *tuberculate chlamydospores,* it is more accurate to call them spiny or warty spores.

Distinctive Ags have been prepared from both the yeast- and mold-forms of *H. capsulatum,* and generally a battery of Ags is used for serologic testing.

C. EXTRACELLULAR PRODUCTS

Although *H. capsulatum* produces many extracellular products, none of them is known to be directly related to the pathogenesis of histoplasmosis or protective immunity.

D. CULTURE

Histoplasma capsulatum, a thermally dimorphic fungus, can be cultured to yield either the yeast- or mold-form. On blood agar at 37°C, the yeast form develops. Grossly, the colonies are small, smooth, moist and either white or cream-colored.

On Sabouraud's glucose agar at room temperature up to 35°C, the mycelial form predominates; colonies are white and cottony, deepening to a buff or brown color with age. The organism grows slowly and colony development may require as long as 3 to 12 weeks.

E. EXPERIMENTAL MODELS

Many laboratory animals are susceptible to histoplasmosis. Although a single spore or as few as 10 yeast cells may infect a mouse, a dose of at least 10^6 yeast cells is required to produce a fatal infection.

F. INFECTIONS IN MAN

Being airborne, the spores of *H. capsulatum* most often reach and infect the respiratory tract where they are promptly engulfed by macrophages in which they grow. The yeast-laden nonactivated macrophages occasionally migrate to reach the circulation and cause disseminated disease.

Whereas, the initial lung infection is usually inapparent, clinical disease develops in some 5% of those infected. In such individuals the infection may occur either as an *acute disease,* as a *chronic cavitary disease* or as a severe blood-borne *disseminated disease.*

When symptoms are present, they may include fever, cough and chest pain or alternatively, more severe manifestations and complications of respiratory infection, such as dyspnea and pleurisy with effusion. The initial infection is pyogenic but soon converts to a granulomatous response. Localized pulmonary histoplasmosis resembles tuberculosis in its histopathology. A unique form of the disease, "African histoplasmosis," is due to the variant organism, *H. capsulatum var. Duboissi.* A rare but serious lesion is fibrosing mediastinitis.

In the occasional patient who develops severe disseminated histoplasmosis, infection may spread to virtually any or all portions of the RES. Lesions are often present in the spleen, liver, lungs, lymph nodes, adrenals, and other organs.

Acute pulmonary histoplasmosis can exhibit mild flu-like symptoms that soon subside, or alternatively, present painful, tuberculosis-like symptoms including fever, dyspnea, night sweats, etc. Lesions are usually multiple and late healing with calcification is the rule; this latter type of infection probably results from unusually heavy exposure.

Chronic pulmonary histoplasmosis is essentially always superimposed on chronic obstructive lung disease with associated emphysema; it may be the outcome of the continued progression of the initial infection or the result of reactivation of an old quiescent lesion; cavitation is frequent.

Disseminated histoplasmosis is commonly a generalized disease of the RES that presents lesions in almost every organ of the body. If the immune response is adequate, the lesions regress and become calcified; alternately, if the CMI response of the patient is markedly depressed, the disease may be rapidly fatal.

Primary cutaneous histoplasmosis is a rare infection that results from entrance of organisms through a traumatic break in the skin. The infection may remain localized or spread through draining lymph nodes to occasionally cause chronic disseminated disease.

G. MECHANISMS OF PATHOGENESIS

The propensity of *H. capsulatum* to propagate within nonactivated macrophages accounts in large measure for the virulence of this fungus. However, little is known about the ways in which the organism causes disease. Presumably injury due to DH plays a role in pathogenesis since caseation necrosis often occurs.

H. MECHANISMS OF IMMUNITY

Infection stimulates prolonged immunity, which evidently rests on CMI involving the ability of activated macrophages to destroy the organism. Although neutrophils can ingest and destroy *H. capsulatum,* they probably do not contribute significantly to defense.

Some evidence indicates that when the yeast burden is high, T suppressor cells increase and CMI is suppressed.

Susceptibility to *H. capsulatum* probably rests on ill-understood immune deficiencies either innate or acquired (Chap. 36).

I. LABORATORY DIAGNOSIS

Histoplasma capsulatum can be cultured from clinical specimens such as sputum. Biopsies of lymph nodes or other infected organs or bone marrow, spinal fluid and buffy coat may contain large numbers of the fungi within macrophages (Sect. B). Although the organisms can be seen in tissue sections stained with hematoxylin and eosin, special stains such as methenamine-silver or periodic-acid-Schiff (PAS) reagent improve their visualization. Definitive diagnosis rests on identification of *H. capsulatum* as the etiologic agent by conversion to the yeast form and the detection of specific Ag.

Histoplasmin skin tests are usually contraindicated in histoplasmosis, first because of extensive cross-reactivity with other fungal pathogens and, more importantly, because skin tests alter the levels of humoral Abs important in serologic diagnosis.

Complement fixation tests, using histoplasmin or heat-killed yeast cells, are useful for diagnosis and prognosis provided parallel tests with cross-reactive Ags of related fungi are done; specific CF Abs arise in 90% of patients 2 to 4 weeks following initial infection; after resolution they usually decline below a titer of 1:8 by about 9 months. In patients with an established diagnosis, rising titers of specific CF Abs that reach 1:32 or above and persist for a few weeks indicate progressive disease.

Precipitating Abs measured by immunodiffusion tests show that specific precipitins against two histoplasmin Ags, M and H, appear about 1 month after infection in 80% of patients; whereas the M-band precipitins persist for several years after infection; the H-band precipitins disappear soon after infection. Consequently, H-band precipitins are a valuable signal of active infection. Other tests for specific Abs include a fluorescent-Ab test and a counterimmunoelectrophoresis test. New tests for circulating Ag also promise to be of diagnostic value.

J. THERAPY

Treatment of progressive disease is carried out with amphotericin B, the drug of choice. Since old healed lesions sometimes reactivate, they are often removed surgically. Exogenous reinfection sometimes occurs especially in immunosuppressed persons.

K. RESERVOIRS OF INFECTION

Nitrogen-rich soil is the principal reservoir of *H. capsulatum,* especially soil that is enriched by the feces of birds or bats. Thus areas around chicken houses, under starling roosts, and in bat caves frequently contain large numbers of the fungal spores. Birds do not become infected, apparently because their high body temperature inhibits the organism. Bats, however, are susceptible to histoplasmosis and *H. capsulatum* can be recovered from bat guano. A number of cases of histoplasmosis have occurred among cave explorers as the result of inhaling large numbers of infectious conidiospores. For this reason, histoplasmosis has been called "spelunker's disease."

L. PREVENTION OF DISEASE

The prevention of histoplasmosis depends on avoiding or eliminating highly-contaminated soil foci. It is virtually impossible to eliminate *H. capsulatum* from soil once it becomes contaminated. Soil can be decontaminated to some extent by the use of formaldehyde or other chemicals and the dissemination of spores can be prevented by covering contaminated dusty areas with a layer of clay or shale several inches thick.

BLASTOMYCES DERMATITIDIS

A. MEDICAL PERSPECTIVES

Blastomycosis is caused by the saprophytic fungus, *Blastomyces dermatitidis,* which has been irregularly isolated from soil. The organism is evidently distributed in many areas of the world in a regional manner. Endemic disease in man and animals, particularly dogs, appears to be largely limited to local pockets within larger regional areas. Mini-epidemics of clustered cases have been reported in young children and teenagers who are presumed to have shared common airborne exposure; the majority being males.

Blastomycosis is seen most often in Canada and the USA, especially in the Northern Mississippi River Valley and the Ohio River Valley. The disease is noted for its chronicity and the suppurative and granulomatous lesions it presents. It is principally a pulmonary infection that often spreads widely to other parts of the body even though the primary lung infection may be asymptomatic. Future improvements in therapy may occur, but no change in the incidence of disease can be anticipated.

B. PHYSICAL AND CHEMICAL STRUCTURE

Blastomyces dermatitidis is a temperature-inducible dimorphic fungus that grows in tissues or cultures at 37°C as a thick-walled budding yeast, and in cultures at room temperature as a mold form. Lateral, round conidia (3 to 5 μm) are borne along the hyphae, and presumably are the infectious form. The yeast forms are spherical; they are 7 to 20 μm in diameter, with a wall almost

1 μm thick. They produce only one bud at a time, which helps to distinguish the organism from similar species. When the organism enters the sexual reproduction cycle, it is designated *Ajellomyces dermatitidis.*

C. EXTRACELLULAR PRODUCTS

A culture filtrate containing extracellular products, called *blastomycin,* analogous to coccidioidin and histoplasmin, is used as an Ag preparation for skin-testing and serologic tests.

D. CULTURE

Blastomyces dermatitidis can be cultured on blood agar at 37°C to yield yeast-form colonies that appear waxy and become wrinkled. On Sabouraud's glucose agar at room temperature, mold-form colonies arise. Growth is slow and full development of colonies requires 2 to 3 weeks or more; mold-form colonies, which are first white and cottony, turn brown with age. Certain identification requires conversion to the yeast form.

E. EXPERIMENTAL MODELS

Guinea pigs and mice have been used for studying *B. dermatitidis.* Intraperitoneal injection of the organism into mice gives rise to abscesses on the omentum. Variations in strain virulence for mice occur; such virulence correlates with the presence of a granulomagenic cell wall component.

F. INFECTIONS IN MAN

Blastomycosis cases tend to occur in small-cluster outbreaks among associated individuals sharing the same environment, those engaged in such activities as outdoor agricultural or construction work, especially work involving high exposure to dust and wood or wood products. Being male, middle-aged or black predisposes to infection, evidently because of male sex hormones and genetic factors.

Infection is almost always initiated in the lung but the outcome varies to include (1) cases of mild pulmonary infection that soon heal with or without developing symptoms and with or without early spread to distant sites, especially the skin and (2) cases in which the primary pulmonary lesions progress to severe disease that may or may not subsequently heal with or without calcification. Primary lung infections rarely remain limited to the lung but instead usually spread to distant sites and often cause progressive generalized disease. In consequence of the varied course of primary pulmonary infection, most cases of blastomycosis present either generalized disease, disease confined to the skin or alternatively to the lung and the skin.

Severe pulmonary blastomycosis resembles tuberculosis and sometimes progresses to cavitation; however, caseation is rare.

Disseminated blastomycosis is a progressive systemic disease that follows hematogenous spread from unresolving pulmonary lesions; in descending order of frequency, spread involves the skin, bones, genitourinary tract, CNS, spleen, kidney, adrenals, lymph nodes, heart and other viscera.

Cutaneous blastomycosis, secondary to pulmonary infection, usually begins as one or more subcutaneous lesions that progress to ulcerate through the overlying skin and then extend laterally over months to years. The lesions, which commonly first appear on exposed skin surfaces, present advancing granulomatous borders and healing centers.

Primary cutaneous blastomycosis is a rare disease that results from entrance of organisms through breaks in the skin. Although infection extends from skin lesions through lymphatics to local lymph nodes, systemic spread seldom, if ever, occurs and spontaneous resolution is frequent.

G. MECHANISMS OF PATHOGENESIS

Little is known about the mechanisms of pathogenesis of blastomycosis. It is of interest that the cutaneous form evidently occurs subsequent to a primary respiratory tract infection, which is usually inapparent. The skin lesions contain few organisms but huge numbers of lymphocytes and plasma cells are present, suggesting that an intense chronic immunologic response is a characteristic of this disease. Although neutrophils and macrophages dominate in the initial lesions, typical granulomas soon develop that express resistance to the organisms. In contrast, neutrophils are alleged to stimulate growth of the organisms.

H. MECHANISMS OF IMMUNITY

Although definitive information on immunity to blastomycosis is meager, it is highly probable that CMI and activated macrophages play the major role. Individuals with immunodeficiencies, either genetic or acquired, show increased susceptibility to the disease. The mechanisms responsible for the genetic susceptibility of blacks have not been defined, but the susceptibility of males over females evidently relates to sex hormones as indicated by animal studies.

I. LABORATORY DIAGNOSIS

A direct mount in 10% KOH may reveal the characteristic yeast form in pus or sputum. Culture on media containing antibiotics is useful in diagnosis, but serologic tests and skin tests are of limited value. The organism grows slowly and differentiation from *H. capsulatum* is important. Mouse inoculation is also of diagnostic value. Biopsy is sometimes necessary for establishing a diagnosis.

Serologic tests for specific Abs against yeast-phase Ag A using the immunodiffusion technique is of some diagnostic value but more importantly serves as an indicator of the effects of therapy; the Abs disappear a few months after successful treatment. How early anti-A Abs appear after infection is not known. The specificity of the "K" Ag remains to be determined.

Skin tests with blastomycin are usually of no diagnostic value because the test is often falsely negative or falsely positive due to high cross-reactivity with other pathogenic fungi, such as *H. capsulatum*. Occasionally, however, strongly positive monospecific reactions can be demonstrated that are diagnostically significant.

J. THERAPY

Chemotherapy with amphotericin B is the treatment of choice, especially in severe pulmonary and systemic disseminated disease. The drug 2-hydroxystilbamidine is also useful for noncavitary lung disease or disease involving only lung and/or skin.

Treatment with either drug must be monitored for side effects; relapses after treatment are frequent. The therapeutic value of ketoconazole or miconazole is still controversial. Corrective surgery is sometimes carried out during chemotherapy. All patients with extrapulmonary disease, relapsing or chronic pulmonary disease should be treated with amphotericin B.

K. RESERVOIRS OF INFECTION

Although there is some question as to the reservoirs of *B. dermatitidis,* it is probably a soil saprophyte (Sect. A). The organism appears to be distributed in a regional manner that leads to numerous areas of endemic disease in man and certain animals as accidental hosts, especially domestic hunting dogs. As with *H. capsulatum,* bird droppings favor growth in soil.

L. PREVENTION OF DISEASE

Blastomycosis is not transmitted person to person and since definitive reservoirs other than soil have not been found, there are no known effective means to control transmission except to avoid locales where point outbreaks occur.

CRYPTOCOCCUS NEOFORMANS

The genus *Cryptococcus* comprises nineteen species, only one of which is a quasi-opportunistic pathogen, namely *Cryptococcus neoformans.* This organism is a capsulated yeast-form fungus; it is a world-wide soil saprophyte which, like *H. capsulatum* and *B. dermatitidis,* is most abundant in soils fertilized by bird droppings. Primary pulmonary infection occurs in man and animals as accidental hosts, infection being the result of inhalation of contaminated dust. The incidence and severity of overt pulmonary disease is usually minimal but the organism can metastasize to produce serious disease in virtually any part of the body, especially the CNS.

A. MEDICAL PERSPECTIVES

Primary pulmonary infection with *C. neoformans* is usually inapparent or mild; however, pulmonary infections often metastasize to the CNS with resulting subacute or chronic meningitis. In untreated patients the meningitis is uniformly fatal after a period ranging from a few months to as long as 20 years. It was formerly called "torula meningitis" because the old name for this fungus was *Torula histolytica.* The success of new treatment regimens has greatly improved the future outlook for patients who need treatment.

B. PHYSICAL AND CHEMICAL STRUCTURE

Cryptococcus neoformans occurs in the form of a budding yeast, 4 to 20 μm in diameter, with a strain-specific polysaccharide capsule (Fig. 37–3). The cell wall is thin as compared with other yeasts of comparable size and tends to collapse to yield crescent forms.

The variable-sized capsules of *C. neoformans* are often enormous and may measure up to 10 to 20 μm in thickness. Their weakly-antigenic capsular polysaccharides have been classified into four antigenic types: A,B,C and D which tend to be regionally distributed. Recently, it has been shown that some strains of *C. neoformans* have a sexual phase that inolves two mating types.

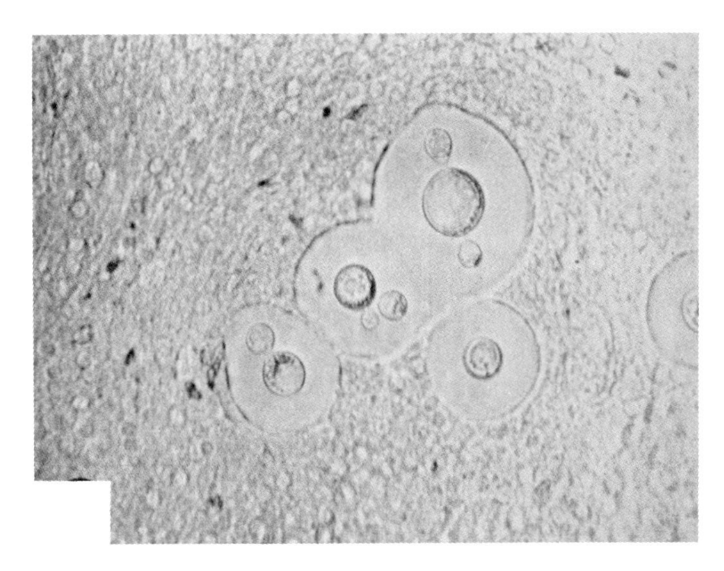

Fig. 37–3. *Cryptococcus neoformans.* Note the thin-walled, budding yeasts surrounded by large clear areas of polysaccharide capsules. (From Emmons, C. W., et al.: Medical Mycology, 3rd Ed. Philadelphia, Lea & Febiger, 1977.)

C. EXTRACELLULAR PRODUCTS

The only products known to be important in virulence and pathogenicity are the antiphagocytic capsular polysaccharide, and diphenol oxidase (Sect. G).

D. CULTURE

Cryptococcus neoformans is readily cultured on Sabouraud's glucose agar at room temperature. Pathogenic-strain organisms are killed at 41°C but will grow at 37°C which helps to distinguish pathogenic from nonpathogenic strains. Capsulated strains form mucoid colonies in 2 to 3 days that are cream-colored at first but later darken to a brownish shade.

E. EXPERIMENTAL MODELS

Mice and rabbits are susceptible to *Cryptococcus neoformans.*

F. INFECTIONS IN MAN

Like many other mycoses, severe cryptococcosis occurs principally in immunocompromised individuals and presents various clinical pictures. The common mode of infection is via inhalation; most primary pulmonary infections may be inapparent or mild. On x-ray, lesions may be found to be either large and solitary, or diffuse and widespread. Occasionally, workmen who inhale large numbers of organisms develop a severe influenza-like pneumonitis, but more often the lung disease is discovered by chance because the symptoms are so mild. It is of interest that severe granulomatous cryptococcal lung lesions do not calcify during healing; consequently, it is probable that many cases of pulmonary cryptococcosis heal without being diagnosed.

From the lung, but rarely from other primary sites of infection, the organisms may *disseminate* to involve virtually any organ, especially in immunodepressed persons. *Visceral forms* often mimic tuberculosis or cancer. In about 10% of those infected, bones and joints are involved. *Cutaneous* and *mucosal* lesions can vary from superficial ulcers to nodules to granulomas or to carcinoma-like lesions.

Persistent primary pulmonary disease that does not disseminate evidently occurs principally in persons who are not systemically immunosuppressed, but instead have preexisting pulmonary disorders as a background. Diagnostic biopsy for these and other lesions should be covered by preoperative chemotherapy to reduce the risk of meningitis following surgery.

Cryptococcal meningitis, the most dangerous form of disseminated disease, resembles tuberculosis or other chronic forms of meningitis. The onset is commonly insidious, with intermittent headaches; fever may be present or absent. Vertigo or other CNS symptoms may be observed, depending on the location and extent of the lesion. At autopsy, granulomatous lesions of the meninges or mucoid masses of organisms are found. In untreated subjects the disease usually progresses with increasing CNS deterioration and ends in death after a few weeks or months, especially if the patient has received steroids. Alternatively, the course of this eventually fatal disease may extend for years, marked by remissions and exacerbations. Cryptococcal meningitis is sometimes mistakenly diagnosed as a psychotic disorder!

G. MECHANISMS OF PATHOGENESIS

The pathogenesis of cryptococcal infections is poorly understood. Clearly, the antigenic polysaccharide of the capsule is an important virulence factor; it is antiphagocytic, and is produced in such large quantities that it may induce a state of immunologic paralysis, thereby allowing the infection to progress. Despite its capsule the organism is often engulfed and destroyed by neutrophils and macrophages. A second but puzzling factor of virulence concerns the production of diphenol oxidase.

Cryptococcus neoformans seems to have a predilection for the CNS, but the basis for this is not known. Continued growth of the organism leads to the accumulation of masses of highly mucoid material that have a gelatinous or myxomatous appearance. In the CNS especially, these accumulations can lead to damage from pressure effects.

H. MECHANISMS OF IMMUNITY

The study of immune mechanisms in cryptococcosis is hampered by lack of suitable Ag preparations. Although it is suspected that inapparent primary pulmonary infections occur frequently without subsequent disease (analogous to many mycoses), proof is lacking because no reliable skin-test Ag is available.

The weakness of the inflammatory response to the organism is a striking characteristic of cryptococcosis. For example, the discharge from the skin lesions that grossly resemble abscesses is not pus but, instead, is composed almost entirely of a mass of yeast-form cells. Similarly, tissue lesions contain few inflammatory cells but huge numbers of organisms. Whereas PMNs and non-activated macrophages are ineffective, *C. neoformans*-activated macrophages can ingest and kill them. Immunodeficiencies in CMI, either innate or acquired, are evidently important determinants of susceptibility to cryptococcosis. Some notable predisposing conditions are: the leukemias, sarcoid, SLE and AIDS.

I. LABORATORY DIAGNOSIS

The usual specimens include sputum, biopsied tissue, bone marrow, lesion aspirates and spinal fluid. Capsules are evident as clear halos in tissue sections and India ink preparations. Blood cultures are negative except in the most severe cases. Speciation of isolates is established by morphologic studies, substrate utilization, growth at 37°C and the unique ability to synthesize diphenol oxidase. The organisms are sparse in the sputum of patients with pulmonary disease and lung biopsy may be necessary.

The titration of polysaccharide Ag in serum and spinal fluid is a highly sensitive and valuable diagnostic and prognostic tool. It entails the agglutination of latex particles coated with specific Abs produced in rabbits. A positive test for specific Ag signifies cryptococcosis and a constant or rising titer indicates a grave prognosis; as resolution proceeds the Ag titer drops and disappears. Levels of specific Abs vary and their determination is of no value. Interfering reactions due to nonspecific proteins can be eliminated by boiling the specimen. Enzyme immunoassay is a more sensitive test but is more difficult to perform.

J. THERAPY

The usual therapy is combined treatment with amphotericin B and flucytosine.

K. RESERVOIRS OF INFECTION

The only known reservoir of human cryptococcosis is soil, especially soil rich in bird droppings.

L. PREVENTION OF DISEASE

The only apparent means of preventing cryptococcosis is to avoid inhalation of dusts from soils rich in bird droppings; this applies especially to persons whose CMI is known to be depressed, either naturally or as the result of receiving immunosuppressants.

PARACOCCIDIOIDES BRASILIENSIS

Paracoccidioides brasiliensis is alleged to be a soil saprophyte that appears to be regionally distributed in various parts of Central and South America. The organism produces opportunistic infections limited to man in local areas of endemic disease which are most prevalent in Brazil.

Primary infection virtually always occurs in the lung; although infection is initially asymptomatic, the DH skin reaction to *paracoccidioidin* becomes positive. The initial infection may resolve or alternatively, after a latent period of a few weeks to several decades, symptomatic paracoccidioidomycosis (formerly called South American blastomycosis) may develop which varies in its course depending on age.

Juvenile type disease in children and young adults often involves wide dissemination to skin, bone, lymph nodes and various internal organs; it is sometimes fatal within weeks to months.

Adult type disease usually develops after a long latent period of years to decades. It may remain localized in the lung or spread widely to various sites and organs, particularly the skin and mucosae of the nose, mouth, and GI tract where granulomatous ulcers develop. Cavitary lesions may develop in the lung but calcification is rare.

Paracoccidioidomycosis is most frequent in highly-exposed outdoor workers and malnutrition appears to be a predisposing factor.

Paracoccidioides brasiliensis resembles *B. dermatitidis* but is distinguished from the latter by its unique characteristic of multiple budding as contrasted with the usual single budding of yeast-forms.

Diagnosis rests on serologic tests, microscopic examination of specimens for multiple-budding yeast forms, and culture at 25°C with or without antibacterial agents. Culture on enriched media at 35°C to promote yeast-form growth and multiple budding should also be done.

The immunodiffusion test for specific Abs is highly valuable for diagnosis, prognosis, and monitoring treatment. The test is always positive in active disease and disappears with regression. Skin tests show cross-reactivity involving Ags of other fungi. The immunodiffusion test is also valuable for identifying the fungus.

The therapeutic drug of choice for prolonged treatment is ketoconazole; amphotericin B is also highly effective. Sulfa drugs, including sulfamethoxypyridazine, are usually suitable for treating outpatients.

REFERENCES

Bhattacharjee, A.K., et al.: Capsular polysaccharides of *Cryptococcus neoformans.* Rev. Infect. Dis. *6*:619, 1984.

Blackstock, R., et al.: Induction of a macrophage-suppressive lymphokine by soluble cryptoccal antigens and its association with models of immunologic tolerance. Infect. Immun. *55*:29, 1987.

Cole, G.T., et al.: An immunoreactive, water-soluble conidial wall fraction of *Coccidioides immitis.* Infect. Immun. *55*:657, 1987.

Drutz, D.J. and Huppert, M.: Coccidioidomycosis: Factors affecting the host-parasite interaction. J. Infect. Dis. *147*:372, 1983.

Eissenberg, L.G. and Goldman, W.E.: *Histoplasma capsulatum* fails to trigger release of superoxide from macrophages. Invest. Immun. *55*:29, 1987.

Emmons, C.W., et al.: *Medical Mycology,* 3rd Ed. Philadelphia, Lea & Febiger, 1977.

Kaufman, L.: Antigen Detection: Its role in the diagnosis of mycotic disease and the identification of fungi. In *New Horizons in Microbiology,* Sanna, A. and Morace, G. (eds.), Amsterdam, Elsevier Science Publications, 1984, pg. 193.

Pennington, J.E., and Meyer, R.D.: Fungal Pneumonia. In *Seminars in Infectious Disease.* Weinstein, L. Fields, B.N., (eds.), New York, Thieme-Stratton, Inc., 1983, p. 209.

38

INTERMEDIATE MYCOSES

Sporotrichosis

Chromomycosis

Rhinosporidiosis

Rhinoentomophthoromycosis

Lobomycosis

Subcutaneous Phycomycosis

Phaeohyphomycosis

The categorization of fungal diseases into deep, intermediate and superficial mycoses is arbitrary and is based on the usual but not the only sites of infection. The group, *"opportunistic mycoses"*, based on immunodeficiency is also arbitrary since immunodeficiencies, to one degree or another, play an important role in susceptibility to almost all fungal infections. Likewise, the groups *chromoblastomycosis* and *mycetoma* have a clinical rather than an etiologic basis.

Although *intermediate mycoses* (subcutaneous mycoses) usually result from breaks in the skin, infection can occur by other routes.

Since *mycetoma* is a clinical syndrome caused by a number of fungal and bacterial agents, especially members of the *Actinomycetaceae,* it has been included in Chapter 34.

SPOROTRICHOSIS

Sporotrichosis is a worldwide disease that shows some tendency toward local regional endemicity. It is essentially always caused by *Sporothrix schenckii;* an occasional case may be due to a closely related organism, *Ceratocystis stenoceras.*

A. MEDICAL PERSPECTIVES

Sporothrix schenckii occurs in the mold form in soil and on vegetation with a widespread distribution over most of the world. It is probable that infections with this fungus occur fairly often via either the skin or the respiratory route; in common with many fungal infections, clinical disease is infrequent. Skin testing for DH is not routinely done; however, in one experimental study of various populations, positive skin-test reactions to Ags of sporothrix were obtained in 10% of hospital patients tested, in 20% of all gardeners and nursery workers and in 60% of long-term nursery workers. These results suggest that continued exposure to the organism eventually leads to subclinical infection and the development of DH. Because of extensive exposure involving many people, it is likely that sporotrichosis will continue to occur with about the same incidence as it has in the past.

B. PHYSICAL AND CHEMICAL STRUCTURE

Sporothrix schenckii is a dimorphic fungus that occurs in tissues, or in cultures at 37°C, as elongated, banana-shaped or cigar-shaped, budding yeast forms about 2 × 6 μm (Fig. 38–1). The mold form develops at 25°C and has slender hyphae, about 1 to 2 μm in diameter, with flower-like collections of small conidiospores, each of a size approximately 3 × 6 μm, borne on sterigma (Fig. 38–2). Each hair-like sterigma is so slender as to be barely visible under light microscopy. The hair-like appearance of the sterigmata gave rise to the genus name, *Sporotrichum* (trichum = hair).

Although the cells contain Ags that are active in agglutination, precipitation and CF tests, they have not been well characterized.

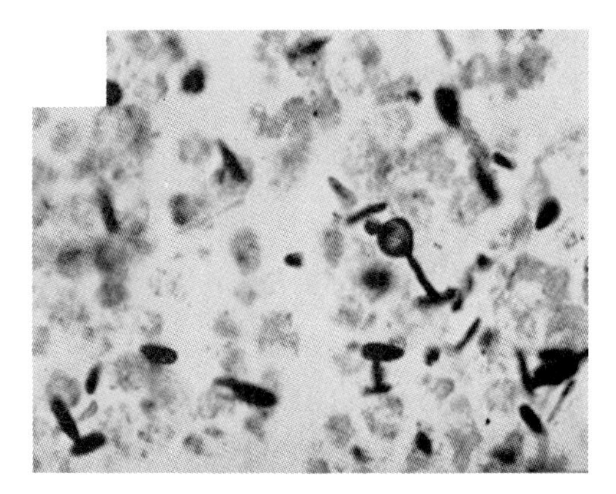

Fig. 38–1. *Sporothrix schenckii.* Elongated yeast form in tissues. (From Emmons, C. W., et al.: Medical Mycology, 3rd Ed. Philadelphia, Lea & Febiger, 1977.)

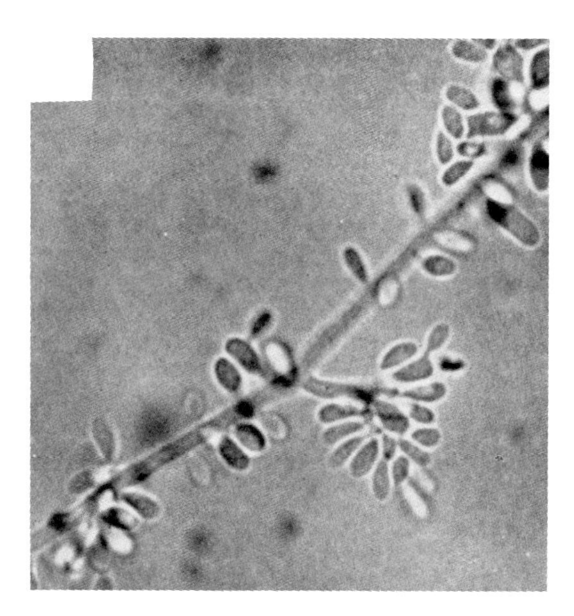

Fig. 38–2. *Sporothrix schenckii.* Mold form in culture. Note the slender hyphae with conidia. (From Emmons, C. W., et al.: Medical Mycology, 3rd Ed. Philadelphia, Lea & Febiger, 1977.)

C. EXTRACELLULAR PRODUCTS

Extracts from culture filtrates have been used as Ag preparations in tests for cutaneous DH to *S. schenckii.* There are no extracellular products that are known to contribute to pathogenesis.

D. CULTURE

At room temperature on Sabouraud's agar, containing added antibacterial antibiotics, *S. schenckii* usually grows in the conidia-producing mycelial form as blackish to brownish, wrinkled mold colonies that appear in about 3 to 5 days. On blood agar at 35° to 37°C, typical creamy yeast-form colonies arise composed of multiple-budding cells that are usually fusiform in shape; most strains of *S. schenckii* will not grow at 38°C.

Fungi characterized by brown-to olive-to black pigments are sometimes grouped together and referred to as "dematiaceous fungi;" the genera include *Sporothrix, Cladosporium, Alternaria* and *Wangiella.*

E. EXPERIMENTAL MODELS

Rodents can be infected by intraperitoneal injection of *S. schenckii.*

F. INFECTIONS IN MAN

Primary infection with *S. schenckii* usually results from soiled thorns, splinters or other objects that penetrate the skin. On rare occasion inhaled organisms produce primary, progressive, disseminating lung disease, especially in individuals with depressed CMI. The organisms are ubiquitous in decaying organic matter and on many living plants. Sporotrichosis is not limited to man, but occurs frequently in certain domestic animals, especially horses; it can be transmitted directly from man to man. Rarely, transmission can occur through hair follicles of normal skin by direct contact with an infected person; however, by far the most frequent means of infection is by accidental cutaneous introduction of organisms present on vegetation or dead organic materials.

Lymphocutaneous sporotrichosis is the most common form of disease (Fig. 38–3). About 3 weeks, or occasionally months, after injection a painless lesion resembling the chancre of syphilis forms at the site of inoculation (the chancriform lesion). Within 1 to 2 weeks, similar lesions begin to appear along the draining lymph vessels, resulting in a characteristic chain of ulcers. Lymphocutaneous lesions often heal after a few weeks but may persist for many months or even years, during which time the patient does not feel ill. As infected tissues become necrotic, pus is discharged to the exterior and granulomatous lesions may develop; spontaneous healing rarely occurs.

Cutaneous sporotrichosis is limited to the skin and does not involve the lymphatics, perhaps because of difference between strains of the fungus, or because of different host responses; cutaneous lesions usually heal.

Disseminated sporotrichosis is rare; spread can occur from even small skin lesions to involve various tissues and organs, including the CNS, bones, joints, muscle, genitourinary tract and lungs. The lesions may resemble the pyogenic ulcers of the primary cutaneous form, or they may be granulomatous. *Mucocutaneous sporotrichosis* is a rare form of disseminated disease and, in common with all forms of disseminated disease, carries a grave prognosis.

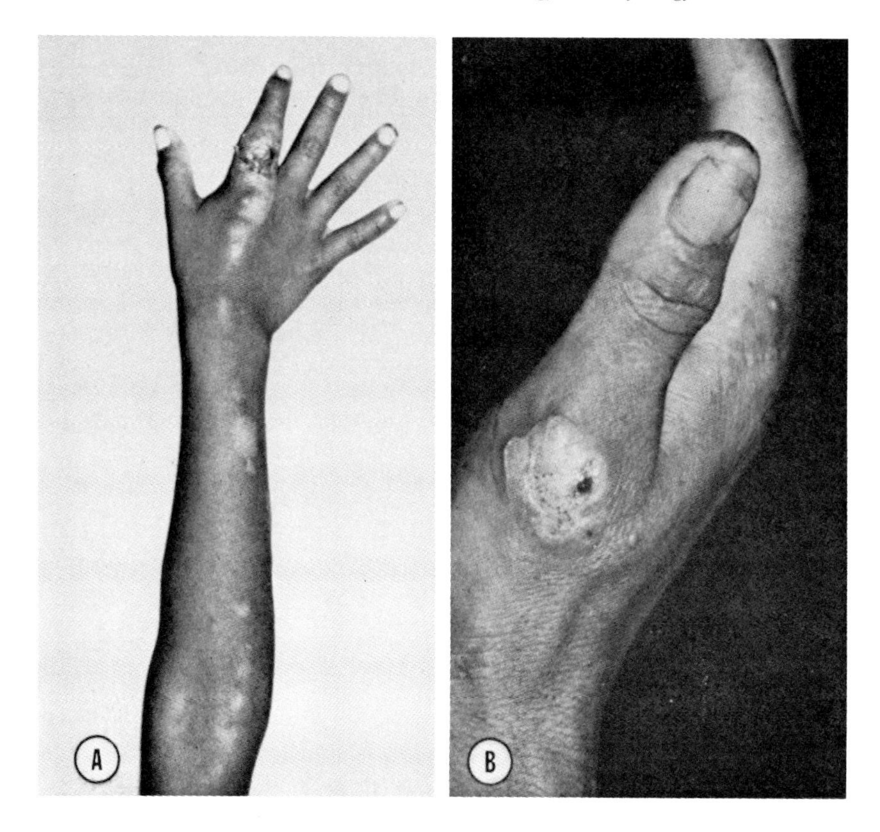

Fig. 38–3. Chancriform lesions of primary cutaneous sporotrichosis, extending along lymphatic channels. (From Emmons, C. W. et al.: Medical Mycology, 3rd Ed. Philadelphia, Lea & Febiger, 1977.)

G. MECHANISMS OF PATHOGENESIS

The means by which *S. schenckii* causes disease are not known. However, the predominance of lesions in low-temperature areas of the body may be because most strains of the organism will not grow at 38°C, as indicated by animal studies.

H. MECHANISMS OF IMMUNITY

It is generally held that CMI is of major importance in resistance to sporotrichosis. Alcoholism and malnutrition appear to be predisposing, perhaps through their immunodepressing effects. Experimental studies of small samples of human populations have shown that cutaneous DH develops in parallel with the extent or duration of exposure to the fungus. Normal individuals often possess specific Abs without any history of sporotrichosis; however, cross-reactions between Ags of *S. schenckii* and other fungi are common and may sometimes interfere with tests for specific Abs.

I. LABORATORY DIAGNOSIS

Definitive diagnosis is based on the clinical picture and specimen culture (Sect. D). The yeast-form of *S. schenckii* in tissue is sparse and difficult to see and is not often observable in KOH mounts. Special staining techniques such as PAS reagent, or fluorescent-Ab microscopy is helpful for locating the organisms by direct examination. Tissue sections sometimes reveal quasi-specific "asteroid bodies" composed of clusters of yeast cells with peripheral eosinophilic rays consisting of Ag-Ab complexes. Identification of cultured isolates from pus is made by morphologic study of asexual spores and cultivation of the yeast form at 35° to 37°C. Animal inoculation can also be used to induce the yeast phase. Skin tests for DH are of little or no value and tests for agglutinating Ab are only meaningful in late disease with titers above 1:40. A specific Ag has been demonstrated in serum that awaits diagnostic evaluation.

J. THERAPY

Amphotericin B is apparently effective for treating the occasional patient with disseminated sporotrichosis. Localized cutaneous lesions respond well to the age-old treatment with orally-administrated potassium iodide. Other antifungal agents available include griseofulvin, 5-flucytosine and dihydroxystilbamidine.

K. RESERVOIRS OF INFECTION

The reservoirs of *S. schenckii* are widespread in nature and include both dead organic materials and living plants. On rare occasion infected animals and man can serve as sources of human infection.

L. PREVENTION OF DISEASE

There are no generally accepted practical methods for preventing disease. Protective clothing or gloves can help to prevent inoculation by thorns or splinters. Arrest of an epidemic in South African gold mines caused by sporothrix growth in mine-timbers is a special example of transmission control; the timbers were effectively treated with fungicides.

CHROMOMYCOSIS

Chromomycosis (chromoblastomycosis) is a worldwide clinical syndrome that results from the traumatic subcutaneous implantation of one of the pigment-producing dematiaceous fungi, including *Fonsecaea campactum, Fonsecaea pedrosoi, Phialophora verrucosa* and *Cladosporium carrioni.* Some species can be readily identified in pus as pigmented yeast-form cells. Dark melanin-like pigment is also apparent in the mycelial colonies that form slowly on Sabouraud's medium at 20°C. Various species of dematiaceous fungi, including those mentioned above, can produce diseases other than chromomycosis; they will not be discussed in this chapter.

Chromomycosis is most common in warm climates and occurs most frequently on the feet and legs of laborers whose skin is exposed to soil; secondary bacterial infection is a frequent complication. The lesions first appear as small painless nodules that grow and expand to become itching cauliflower-like granulomatous ulcers that may persist for months to years. The disease does not involve lymphatic spread and systemic dissemination, which is rare, is virtually limited to the CNS.

Diagnosis rests on culture and identification of isolates by cultural and morphologic studies, particularly on hyphae and conidia. *Therapeutic* measures, including surgical procedures and treatment with flucytosine or potassium iodide, have been of limited benefit.

RHINOSPORIDIOSIS

Rhinosporidiosis is a serious chronic disease characterized by polypoid lesions in the nasal mucosa and often in the skin and various mucosae of the mouth, throat, genitalia and rectum; it is caused by *Rhinosporidium seeberi.* The organism is presumed to have a reservoir in nature but has not been cultivated from any source even though unique sporangia and sporangiospores are present in lesions. The disease occurs in cattle, horses and various other animals as well as in man. Many cases of human disease, chiefly in children and young adults, have been recorded in India, Argentina and Southeast Asia. *Treatment* includes topical antifungal agents, surgery and ethylstilbamidine.

RHINOENTOMOPHTHOROMYCOSIS

Rhinoentomophthoromycosis is a rare widespread disease of the tropics that presents lesions in the nasal mucosa. The disease can be acute or chronic in man and is due to the fungus, *Entomophthora coronatus,* a soil saprophyte and insect parasite. Horses also serve as accidental hosts. *Diagnosis* is based on morphologic studies on cultured isolates and hyphae in tissue sections. *Treatment* includes potassium iodide, amphotericin B and surgery.

LOBOMYCOSIS

Lobomycosis is a chronic subcutaneous disease in man, caused by a fungus currently designed as *Loboa loboi.* The organism also produces disease in the bottle-nose dolphin. Human infection is rare and, to date, has only been reported in the Amazon river region. The initial lesions are painless granulomatous nodules that arise on the extremities and facial area; they progress to become ulcerative and expand slowly over many years. Tissue sections reveal sparse "asteroid bodies" and large yeast-form organisms that exhibit multiple budding. The organism has not been cultured. *Therapeutic measures* include treatment with sulfa drugs and surgery.

SUBCUTANEOUS PHYCOMYCOSIS

Subcutaneous phycomycosis is caused by *Basidiobolus haptosporus.* It is a rare disease that presents granulomatous, subcutaneous lesions characterized by the associated development of massive local edema. The disease recedes spontaneously after a few months. It is most prevalent among children in tropical areas of Africa, India and Southeast Asia. The organism is naturally present in the intestinal tract of frogs, lizards and other related animals. *Diagnosis* is based on the morphology of cultured isolates and *treatment* with iodides is effective.

PHAEOHYPHOMYCOSIS

Phaeohyphomycosis is a clinically-based disease that occurs in two forms, namely a cutaneous form characterized by an encapsulated, subcutaneous cyst on one of the extremities and a systemic form characterized by brain abscess. The lesions contain darkly-pigmented septate hyphae. Whereas the most frequent etiologic agent of the systemic form of disease with brain abscess is *Cladosporium bantianum,* the cutaneous form can be caused by various species belonging to several genera including *Exophiala, Wangiella* and *Phialophora.*

Identification of the etiologic agents is based principally on physiologic and morphologic studies on cultured isolates.

Treatment by surgical removal is effective in the case of subcutaneous cysts but not in the case of brain abscess, which is usually lethal.

REFERENCES

Aram, H.: Sporotrichasis. A historical approach. Int. J. Dermatol. *25*:203, 1986.

Emmons, C.W., et al.: *Medical Mycology,* 3rd Ed., Philadelphia, Lea & Febiger, 1977.

Joklik, W.K., et al. (ed.): *Zinsser Microbiology,* 18th Ed. Stamford, Conn., Appleton-Century-Crofts, 1984.

Vollum, D.I.: Chromomycosis: Review. Br. J. Dermatol. *96*:454, 1977.

Waldorf, A.R.: Host-parasite relationship in opportunistic mycoses. CRC Crit. Rev. Microbiol. *13*:133, 1986.

Yamamoto, H., et al.: Chromoblastomycoses of the maxillary sinus. Arch. Otorhinolaryngol. *242*:129, 1985.

39

DERMATOPHYTES

The superficial mycoses are clinically-based infections of the cornified non-living keratin layer of the skin and the keratin of hair and nails. The keratinophilic fungi (dermatophytes) that cause these infections (dermatophytoses) owe their unique ability to utilize keratin in large measure to their capacity to digest keratin. Few other organisms produce keratinases. The dermatophytes do not invade tissues below the epidermis evidently because of antifungal factors normally present in serum, specific Abs and in some instances because the infecting fungal species cannot tolerate temperatures above 35°C (Sect. H).

The dermatophytes that cause human infections comprise forty species of fungi that belong to three genera of the *Fungi Imperfecti,* namely *Microsporum* (17 sp), *Epidermophyton* (2 sp), and *Trichophyton* (21 sp). Whereas a few species are soil saprophytes (geophilic group) and some have animals as their natural hosts (zoophilic group), most have man as their natural host (anthropophilic group). Although species of the latter group have evolved to approach near-perfect parasitism and produce mild infections that tend to be chronic, other species, especially of the zoophilic group, tend to produce more severe acute infections, apparently because they induce a more intense nonimmunologic inflammatory response than the anthropophiles. This strong inflammatory response, together with the DH response accounts for the relative severity of zoophilic infections.

Most dermatophytes are strikingly similar to each other in many respects, including surface Ags; currently their identification rests principally on the morphology of conidia and colonial properties.

An occasional dermatophyte species may be able to cause several clinically different infections or alternatively, a given clinical entity may be caused by any one of several species. Since infection is in the epidermis and Ag exposure is of the contact type the DH response tends to be strong and the Ab response weak.

Since dermatophytes have little if any capacity to invade deep tissues, immunosuppression is not an important determinant of susceptibility.

A. MEDICAL PERSPECTIVES

The dermatophytes occur in virtually all parts of the world, however, environmental conditions sometimes favor a higher incidence of some species in certain geographic locations, as compared to other locations. For example, *Trichophyton schoenleinni* is largely limited to the Mediterranean countries, whereas *T. rubrum* is more widespread, and is common in the USA.

It is probable that every normal person becomes infected with at least one of the dermatophytes at some time during life, often in childhood. Most infections are trivial and self-limiting. However, part of the population is plagued by long-lasting infections with the anthropophilic dermatophytes, many of which are resistant to treatment. Even though they are seldom dangerous or life-threatening, the dermatophytoses are responsible for a tremendous amount of discomfort and annoyance for many people.

Although past regional and global changes have occurred in the distribution of various species of dermatophytes and the incidence of various dermatophytoses, it is difficult to predict future changes. Circumstances responsible for changes include wars, alterations in social, economic and hygienic conditions, immigration patterns, and improved diagnostic, therapeutic and preventive measures. Hopefully, circumstances that lessen the problem of dermatophytoses will improve in the future; advances in diagnosis, therapy and prevention are reasonably certain. A few examples of past changes in the USA are: favus, an old scourge in local areas, has decreased to near extinction; cases of tinea capitis due to *M. audouinii* have been reduced to near zero by use of the Wood's lamp and improved therapy terminating with oral griseofulvin; the most common cause of tinea pedis shifted after World War II from *T. mentagrophytes* to *T. rubrum.*

B. PHYSICAL AND CHEMICAL STRUCTURE

Most of the dermatophytes are molds that form conidia characteristic of the species; therefore, conidia are important aids in speciation. Hyphae grow in superficial layers of dead keratinized cells where they form conidia. The cell-wall material, typical of fungi, is present in the hyphae and conidia, but capsular substances are lacking.

Microsporum. Members of the genus *Microsporum* have macroconidia as their predominant spore form. These are large (20 to 125 μm long), multicellular conidia formed on the ends of hyphae. Figure 39–1 shows the spindle-shaped macroconidia of *M. gypseum* and *M. canis,* two of the common dermatophytes.

Colonies of *Microsporum* sp. are usually tan to brown in color, and become cottony after 2 to 4 weeks of culture. They grow well on Sabouraud's glucose agar at room temperature.

Trichophyton. Microconidia are the prominent spore forms of members of the genus *Trichophyton,* although macroconidia may also be present. The species most common in the USA are *T. rubrum, T. tonsurans* and *T. mentagrophytes;* spores of the latter species are shown in Figure 39–2. Coiled hyphae are formed by *T. mentagrophytes;* the macroconidia of this species are elongated

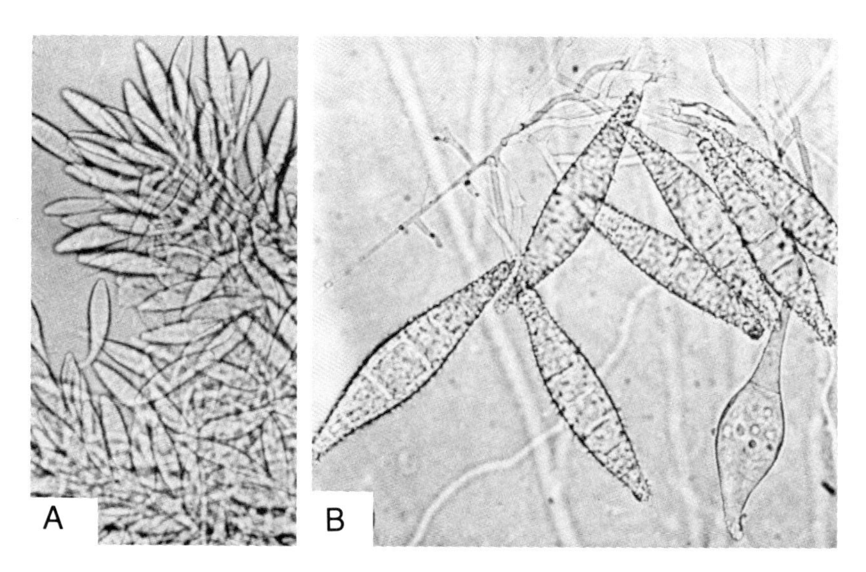

Fig. 39–1. Macroconidia of *Microsporum* sp. A, *M. gypseum.* B, *M. canis.* (From Emmons, C. W., et al.: Medical Mycology, 3rd Ed. Philadelphia, Lea & Febiger, 1977.)

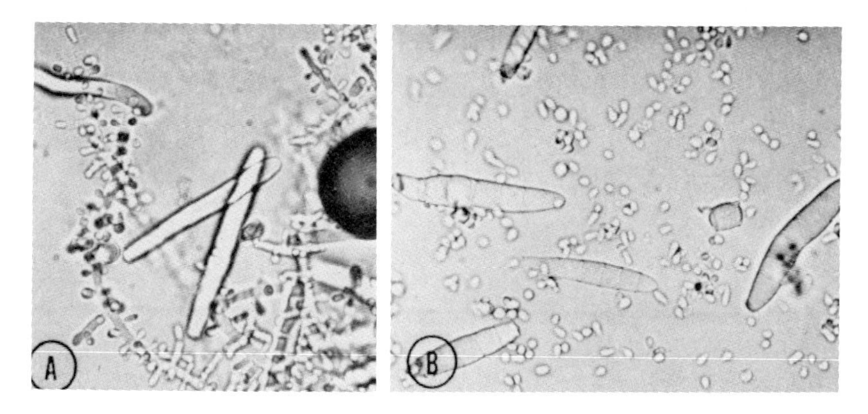

Fig. 39–2. *Trichophyton mentagrophytes.* A, Microconidia and B, macroconidia. (From Emmons, C. W., et al.: Medical Mycology, 3rd Ed. Philadelphia, Lea & Febiger, 1977.)

(8 to 50 μm), and grape-like clusters of spherical microconidia occur along the sides of the hyphae.

The colonial forms of *Trichophyton* sp. vary considerably. Colonies may be smooth and powdery, and may range in color from white, pink, red, or purple to yellow or brown.

Epidermophyton. In the USA, *E. floccosum* is often seen. It bears clusters of oval or club-shaped macroconidia about 8 to 15 μm long (Fig. 39–3), but no microconidia. Its colonies are greenish-yellow and powdery in appearance.

C. EXTRACELLULAR PRODUCTS

Little is known about extracellular products of the dermatophytes that may be important in pathogenesis and immunogenicity. All of these fungi utilize keratin and some species are known to produce extracellular keratinases and proteolytic enzymes that allow them to degrade host components to provide nutrients. The products of some species may induce intense inflammation.

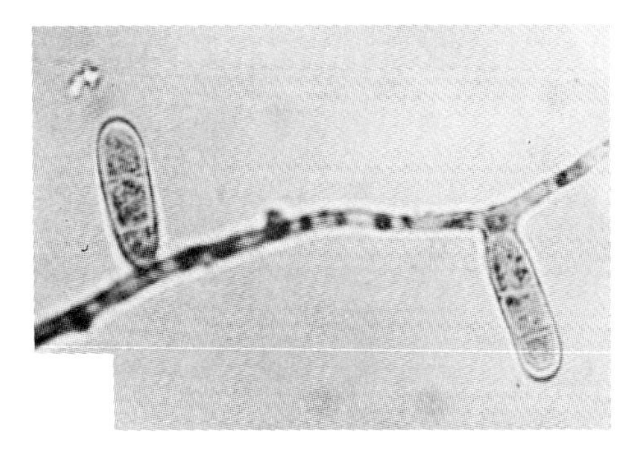

Fig. 39–3. Macroconidia of *Epidermophyton floccosum.* (From Emmons, C. W., et al.: Medical Mycology, 3rd Ed. Philadelphia: Lea & Febiger, 1977.)

D. CULTURE

The dermatophytes are slow-growing, compared with most bacteria; some species are inhibited at 37°C. When grown on Sabouraud's glucose agar at room temperature, the colonies of some species are apparent after 1 week, but other species may require several weeks to form visible colonies. Addition of special nutrients, such as vitamins, to the medium enhances the growth of some species. For example, thiamin supplementation stimulates the growth of *T. tonsurans.*

E. EXPERIMENTAL MODELS

Innate and acquired immunity to the dermatophytoses is so strong that it has proven impossible to infect most human volunteers with the dermatophytes. Experimental animal models are also of little value for this reason.

F. INFECTIONS IN MAN

The dermatophytes cause *tinea* and related diseases. The word *tinea* comes from a Latin root meaning a gnawing worm, and a common name for some forms of tinea is ringworm. This circular form of tinea results because the fungal infection spreads in an expanding circular manner resulting in a ringshaped, scaly, itching area on the skin. Other species grow in or on hair or nails. Still other dermatophytoses are recognized by such common names as athlete's foot, jock itch and barber's itch.

Tinea is further classified by the area in which it is found:

Tinea capitis, a disease of the scalp and hair, is most frequently caused by members of the genus *Microsporum,* although some *Trichophyton* species sometimes serve as etiologic agents. In the USA, the species that cause tinea capitis in order of frequency are *M. tonsurans, M. canis,* and *M. audouinii.*

Tinea corporis (tinea circinata) or ringworm of the smooth skin is caused by many species of the genera *Microsporum* and *Trichophyton; M. canis* and *T. mentagrophytes* being the most common offenders.

Tinea barbae or barber's itch occurs on bearded areas of the face and neck; it is usually caused by either *T. mentagrophytes,* or other species of *Trichophyton* or *M. canis.*

Tinea pedis (athlete's foot) which commonly involves itching and sloughing of epidermis and splitting of the toe-web skin has been found in well over half of the individuals in all populations that have been studied. Its incidence is higher in adults than in children, and higher in men than in women; the disease is exacerbated by local hot, moist conditions. It is frequent among athletes who use common shower stalls. The most frequent etiologic agents of tinea pedis are *T. rubrum, Epidermophyton floccosum* and *T. mentagrophytes.*

Tinea cruris, which designates infections of the groin, perineum and perianal regions is most often caused by *E. floccosum.*

Tinea unguium (onychomycosis), or tinea of the nails, is caused by the same fungi that cause athlete's foot; it is especially prone to occur in 40- to 50-year-old women and can result in distortion and loss of nails.

Tinea versicolor is a tinea of smooth skin characterized by the changing color of the lesion; hence the term versicolor. The etiologic agent is *Pityrosporum orbiculare.*

It is of singular interest that *Pityrosporum orbiculare* prefers only keratin of the skin and not of the hair and nails; thus, this infection is limited to the outermost layer of the skin.

Tinea nigra usually is a tropical disease of the hands and feet caused by the soil fungus, *Cladosporium werneckii.* The black color of the lesions is due to the pigmented septate hyphae produced.

Tinea can sometimes result in pustular abscesses of hair folicles, called *kerions.* The *kerion* type of infection occurs most often in *tinea barbae.*

Favus is a severe chronic type of ringworm, characterized by dermal crusts resembling a honeycomb. It can be caused by a number of different dermatophytes, but *T. schoenleinii* is the most frequent agent. Favus is endemic in the Mediterranean area, parts of Southern Asia, Africa and the Orient. Patients known to have favus are not admitted to the USA; however, some cases occur in this country, apparently as a result of previous introduction of the fungus by immigrants from endemic areas.

G. MECHANISMS OF PATHOGENESIS

All dermatophytes are keratinophilic, but different genera vary as to the kind of keratin they prefer. This amazing preference evidently determines the site of infection of the various dermatophytes.

Members of the genus *Microsporum* seem to prefer only the keratin of skin and hair, and consequently they rarely invade nails. They grow both inside (endothrix) and outside (exothrix) the hair shaft, forming a sheath of spores on the surface of each hair. Infected hairs tend to break off just above the scalp, leaving a stubble.

Epidermophyton sp. invade skin and nails, but do not attack hair. Thus, they are often implicated as the cause of tinea pedis, tinea cruris, and tinea unguium (onychomycosis), but not tinea capitis.

Trichophyton sp. are not fastidious, and will live on keratin of hair, skin or nails. Species such as *T. tonsurans* grow principally within hair shafts, causing the hair to break off below the scalp surface. This results in bald areas with "black dots" where hairs have been lost. Other species, such as *T. mentagrophytes,* are ectothrix fungi. *Trichophyton mentagrophytes* often causes *kerion,* probably because of strong inflammatory responses to the fungi within hair follicles.

In general, the symptoms of dermatophytoses are more pronounced when man is the unnatural rather than the natural host. Thus, *T. mentagrophytes* and *M. canis,* which have animals as natural hosts, tend to incite much more vigorous nonimmunologic inflammatory responses in man than *M. audouinii* or other anthropophilic fungi, which have man as their natural hosts. The combined immunologic-nonimmunologic inflammatory responses of the host are often the principal factors in the pathogenesis of the disease.

The patterns of growth of the dermatophytes account for the usual clinical picture. In ringworm, the fungi grow in a widening circle or ring that appears brownish to red at the periphery as a result of the inflammatory response at this site. The center of the ring is scaly; at this site thickened, infected areas of the keratinized layer of the skin tend to scale off. Hypersensitivity to fungal Ags probably leads to the vesicle formation that is often noted.

The sensitivity usually elicited by dermatophytoses can result in *dermato-*

phytids or *"id" reactions.* These contact-type dermatitis reactions occur in regions distant from the site of infection with the fungus. The id reaction is often manifested by vesicular lesions on the hands or elsewhere, which itch and may become secondarily infected and painful.

Common Ags shared by the dermatophytes are capable of inciting a DH reaction in the skin of sensitized patients tested with trichophytin, an antigenic preparation made from cultures of various dermatophytes. Apparently the "id" reaction is analogous to the trichophytin skin test in that the sensitive patient gives a sensitivity reaction to soluble Ags often at a site far removed from the focus of infection. Dermatophytid reactions may be more severe and troublesome than actual dermatophytosis. Thus, the pathogenesis of disease may rest principally on allergic or possibly other inflammatory responses of the host to the infecting fungus.

Although dermatophytes often persist in infected superficial tissues for many years, they rarely if ever invade deeply, presumably in some instances because they cannot tolerate 37°C. Either the pathogenetic capacities they exhibit on body surfaces are totally ineffective in deep body tissues or the organisms are unable to persist in deep tissues.

H. MECHANISMS OF IMMUNITY

The fact that the skin is the site of exposure to Ags of infecting dermatophytes raises unresolved questions with respect to potential immune responses, namely that the skin as a route of Ag entrance into the body involves the participation of Langerhan's cells in the immune response; these cells incite the generation of suppressor and contrasuppressor T cells and the development of DH and CMI (Myrvik & Weiser: *Fundamentals of Immunology,* pp. 145–146).

Delayed hypersensitivity reactions may be a distinct advantage in some superficial mycoses because the reactions cause sloughing of infected skin, thus ridding the body of most of the parasites.

The human host is able to limit dermatophyte infections to superficial areas, at least in part, as a result of undefined antifungal factors naturally present in plasma. The dermatophytes cannot live in the presence of human plasma or serum. It has been shown that, whereas high concentrations of specific Ab added to media completely inhibit growth of *T. mentagrophytes,* lower concentrations of Ab cause structural changes in the hyphae and conidia. There is no evidence that C contributes to these effects.

Despite host resistance the dermatophytes often establish long-lasting states of parasitism. They are able to withstand the hypersensitivity responses of the host and, to a large extent, are protected in their exterior environment from both humoral and cellular immune mechanisms. They may often, as in favus, persist for the host's lifetime.

Hormonal influences can affect the course of fungal infections profoundly. In tinea capitis caused by *M. audouinii,* for example, children commonly remain infected until they reach the age of puberty, at which time the infection becomes self-limiting. It has been suggested that the increase in long-chain fatty acids in the skin after puberty may be related to clinical improvement. There is experimental evidence that fatty acids and certain oils used on hair inhibit the penetration of some dermatophytes into hairs.

I. LABORATORY DIAGNOSIS

Dermatophytoses must be distinguished from other diseases, particularly psoriasis and contact dermatitis. The dermatophytes may be recovered from skin scrapings, nails, or hair. Direct examination of KOH mounts (Chap. 36) often reveals hyphae or conidia. Ectothrix or endothrix infections of hairs can be distinguished by direct microscopic examination for hyphae and arthrospores. If microscopic examination is negative, culture should be done on carefully collected specimens.

Although the growth of dermatophytes may be optimal at a pH of 6.8 to 7.0, they tolerate a more acid environment and usually grow well on Sabouraud's glucose agar at pH 5.6 to 6.0. Characteristic colonies appear commonly after about 1 to 3 weeks of incubation at room temperature.

Slide cultures are especially useful for identifying dermatophytes; identification is based on the production of characteristic macroconidia or microconidia, as well as on colonial appearance, and occasionally on certain nutritional requirements.

The fluorescence of *M. audouinii* and other *Microsporum* species is helpful in distinguishing infections with these fungi from *tinea* caused by *Trichophyton.* Hairs that are infected with *Microsporum* usually display a green fluorescence under ultraviolet light, whereas those infected with *Trichophyton* do not fluoresce; an exception is *T. schoenleinii* which fluoresces. The light source used is the Wood's lamp. Monoclonal Abs promise to soon provide a more efficient method for the speciation of dermatophytes.

J. THERAPY

The dermatophytoses are often self-limiting; alternatively they may persist for many years. Patients with severe or disfiguring lesions require prolonged oral treatment with griseofulvin or the alternate drug ketoconazole. Miconazole and clotrimazole are the most effective agents for topical treatment. Since tinea pedis is made worse by shoes that confine perspiring feet, powders or other means of keeping the feet dry are sometimes helpful. Antifungal agents, such as undecylenic acid, are often incorporated into powders. Ointments containing tolnaftate or fatty acid salts may be effective for treating tinea of the glabrous skin.

Table 39–1. Reservoirs of Some Common Dermatophytes

Genus and Species	*Reservoir*
Microsporum audouinii	Man
M. canis	Dogs, cats
M. gypseum	Soil
M. fluvum	Soil
Trichophyton schoenleinii	Man
T. mentagrophytes	Man, domestic animals, and perhaps soil
T. rubrum	Man
Epidermophyton floccosum	Man

K. RESERVOIRS OF INFECTION

Depending on the species, dermatophytes that infect man may be normal inhabitants of soil, animal species, or man only. The reservoirs of some of the common dermatophytes are listed in Table 39–1.

L. CONTROL OF DISEASE TRANSMISSION

The dermatophytoses are best prevented by avoiding exposure to causative fungi. In infections with *M. audouinii,* the organisms are transmitted from one child to another, especially by contact with caps or other heavily contaminated items. Infections with *M. canis,* on the other hand, are usually contracted from infected animals; owner's pets may be responsible, but stray cat and dogs fondled by children are more often the source.

The efficacy of disinfectant foot baths and similar measures intended to prevent the spread of tinea pedis is open to question. Although it is certain that skin scales may be infectious, experimental attempts to transfer tinea have often failed. Such experiments emphasize that marked differences in natural immunity must exist that determine the establishment of disease. Needless to say, avoidance of the source of dermatophytes is worthwhile.

Routine sterilization or disinfection of instruments in barber shops and hairdressers' salons has helped prevent the dissemination of certain dermatophytes.

REFERENCES

Ajello, L.: Emergency mycosis of present and future concern. In *New Horizons in Microbiology.* Sanna, A. and Morace, G. (eds.), New York, Elsevier Science Publishers, 1984, p. 303.

Deshmukh, S.K., et al.: Degradation of human hair by some dermatophytes and other keratinophilic fungi. Mykosen. *28*:463, 1985.

Emmons, C.W., et al.: *Medical Mycology,* 3rd Ed. Philadelphia, Lea & Febiger, 1977.

Pettit, J.H.: Treatment of superficial fungal infections. Trop. Doct. *16*:105, 1986.

Svejgaard, E.: Oral ketoconazole as an alternative to griseofulvin in recalcitrant dermatophyte infections and onychomycosis: Acta Derm. Venereol. *65*:143, 1985.

Van Cutsem, J., et al.: Activity of orally, topically, and parenterally administered intraconazole in the treatment of superficial and deep mycoses: Animal Models. Rev. Infect. Dis. *1*:515, 1987.

40

OPPORTUNISTIC MYCOSES

Candidiasis

Aspergillosis

Mucormycosis

Geotrichosis

Other Opportunistic Fungi

The common practice of grouping certain mycoses as being "opportunistic" is somewhat arbitrary because essentially all mycotic agents exhibit opportunism to one degree or another. However, the so-called "opportunistic fungi" are presumed to seldom, if ever, cause life-threatening infections unless host defenses have been severely breached in some way. The incidence of such infections is increasing rapidly because of the increasing use of immunosuppressive agents in cancer therapy and organ transplantation. As stated in Chapter 36, there are doubtless many more potentially pathogenic opportunistic fungi in nature than are presently recognized; moreover, their discovery is being hastened by the increasing use of immunosuppressants.

CANDIDIASIS

Candidiasis is caused principally by three species of the genus *Candida,* namely, *C. krusei, C. tropicalis* and *C. albicans,* with the latter being the most frequent. These three fungi are among the normal flora of the mucous membranes and skin of man. Candidae are also among the normal flora of many species of animals. Many candidae occur in nature as saprophytes, a few of which infect man on rare occasion. Man becomes colonized by candidae of the normal flora at birth; candidiasis can develop at any age and may be local or disseminated.

A. MEDICAL PERSPECTIVES

Candidiasis is the most common of the systemic mycoses and in the USA causes about 25% of all deaths due to the fungi. The disease develops in a high percentage of patients with leukemia (30%), AIDS, and severe combined immunodeficiency (100%).

Candidiasis presents a wide spectrum of clinical states, ranging from acute, self-limiting infections of the mucous membranes and skin to chronic and sometimes fatal systemic disease. Infection is favored by abridgment of one or more of the many factors of host defense, both general and local, including minute breaks in the mucosae or the skin. The incidence of candidiasis and the death rate will probably continue to increase, because predisposing circumstances are expected to increase in the future (Chap. 36). Effective antifungal agents are available for treatment but benefits are only temporary unless the underlying host defect can be corrected.

B. PHYSICAL AND CHEMICAL STRUCTURE

Candida albicans occurs among the normal microbial flora in man, principally as an ovoid to spherical budding yeast, approximately 4 to 5 μm in diameter. Although it is a dimorphic fungus wih the ability to form true hyphae and true mycelia, it also often forms *pseudomycelia* composed of pseudohyphae. Either of these two filamentous forms is referred to as the mycelial or mold (M) form; the budding yeast form is called the Y form (Chap. 36). Hyphal forms are produced in vivo only when tissues are invaded.

Groups of blastospores are formed along mycelia under certain growth conditions in vivo (Fig. 40–1); in vitro culture reveals that additionally round, thick-walled chlamydospores occur at the ends of true hyphae or between hyphal cells (Fig. 40–2). Chlamydospores represent resting, resistant spores

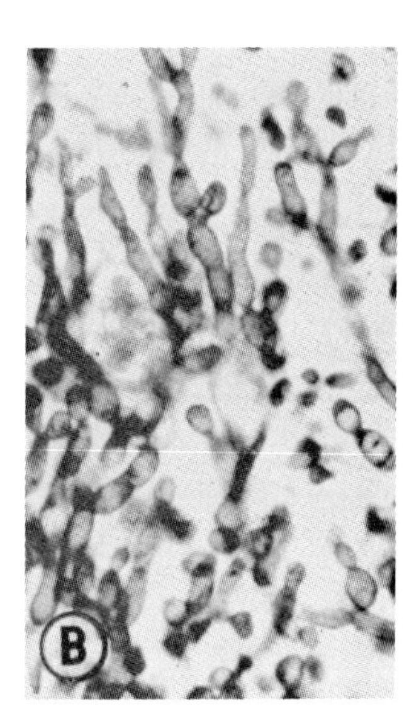

Fig. 40–1. *Candida albicans.* A section from the tongue of a patient who died of acute leukemia complicated by candidiasis. Both blastospores and hyphae are frequently seen in tissues during acute candidiasis. (From Emmons, C. W., et al.: Medical Mycology, 3rd Ed. Philadelphia, Lea & Febiger, 1977.)

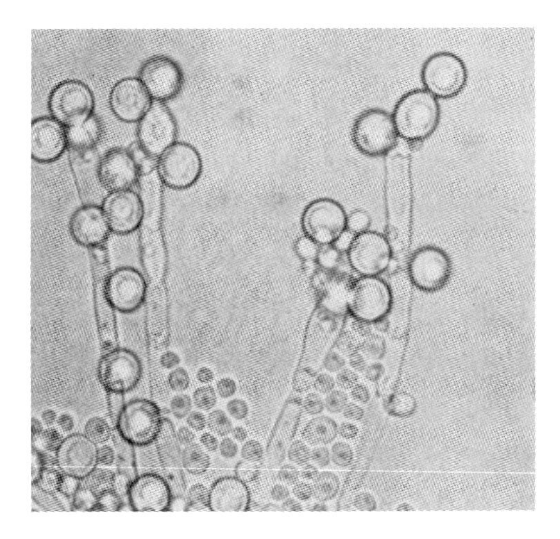

Fig. 40–2. *Candida albicans.* Organisms in culture, showing yeast blastospores (smaller spheres) and chlamydospores (larger spheres). (From Emmons, C. W., et al.: Medical Mycology, 3rd Ed. Philadelphia, Lea & Febiger, 1977.)

and consequently are formed in old cultures or on relatively poor media, such as cornmeal agar, at a temperature of about 21°C.

The cell walls of *C. albicans* contain typical fungal constituents and, in addition, possess unidentified components that are toxic for mice (Sect. G).

C. EXTRACELLULAR PRODUCTS

Candida albicans is not known to produce extracellular products of importance in the pathogenesis of disease. Culture filtrates are used as antigenic preparations in testing for cutaneous DH and for in vitro immunologic assays.

D. CULTURE

Growth is rapid on Sabouraud's medium, blood agar, trypticase soy, and many other rich media. Creamy yeast colonies are formed after overnight incubation at 21° or 37°C; the optimum growth temperature is around 30°C. After a few days' growth on Sabouraud's agar, colonies contain largely Y forms on the surface; however, both true mycelium and pseudomycelium grow into the agar. The characteristic clusters of blastospores are formed along the hyphae, and chlamydospores are often present (Fig. 40–2). When incubated in serum at 37°C, the organisms are stimulated to rapidly form *germ tubes* within 1 to 4 hours (germ tube test).

E. EXPERIMENTAL MODELS

Laboratory rodents, such as guinea pigs, rabbits and mice, can be infected with *C. albicans* and are often used in experimental studies. Like man, the mouse and other rodents are resistant to infection, and large numbers of yeast-form cells must be injected in order to establish an infection. Most often, the yeast cells are inoculated intravenously or intraperitoneally; in either case, the result is widespread abscesses in many tissues, especially the kidney, and death occurs within a week with most dosages used. Another model involves injection of yeast cells into the mouse thigh to produce a lesion that becomes self-limiting by about 4 to 6 weeks.

Attempts to produce lesions in the skin and mucous membranes of normal human subjects and animals fail so long as integuments are intact. However, if mucous membranes are abraded before inoculation, infection occurs briefly. Similarly, if skin is abraded and macerated, infection can be initiated, but lasts only so long as the tissue is kept macerated.

It has been possible to colonize, and occasionally to produce infection, by feeding *C. albicans* to germ-free animals that lack normal bacterial flora. Increased susceptibility to infection can also be induced in mice by neonatal thymectomy or x-irradiation. Animal models have been of some benefit for evaluating the therapeutic usefulness of antifungal drugs.

F. INFECTIONS IN MAN

Complex schemes for classifying different forms of human candidiasis have been presented. Perhaps the simplest approach is to consider these infections as a spectrum with a few important prototypes (Table 40–1).

By far the most common *Candida* infections are the acute pseudomembranous lesions of mucous membranes that occur as the result of relatively minor aberrations in host defense. Oral thrush and widespread mucosal and skin

Table 40–1. Manifestations of Candidiasis in Man

Disease Process	*Common Predisposing Factors*	*Probable Outcome*
Lesions of mucous membranes:		
Oral (thrush)	Infancy, old-age, diabetes, antibiotic therapy, steroid therapy	Usually self-limiting; may be chronic over many months
Vulvovaginal	Pregnancy, diabetes	Self-limiting when predisposing condition is cleared
Gastrointestinal, perianal	Antibiotic therapy	Self-limiting after use of antibiotics is discontinued
Perlèche-lesions at corners of the mouth	Wearing dentures	May be self-limiting when dentures are properly fitted
Intertrigo-lesions within folds of tissue such as groin	Obesity, perspiration	Clears with aid of medicated powders to keep affected areas dry
Paronychia-lesions around nails	Maceration	Usually self-limiting
Candida granuloma—production of horny, granulomatous lesions	Immune deficiency	May respond to antifungal agents, including amphotericin B, at least temporarily
Chronic mucocutaneous candidisis—fungus grows in superficial layers, does not invade past epithelium	Immune deficiency, usually cellular or combined	Recurs after treatment; few instances of long-term success with immunologic reconstitution have been reported (Sect. J)
Disseminated	Malignant disease; other serious diseases that compromise host defenses	Often fatal. May respond to amphotericin B and/or flucytosine (Sect. J)

lesions in newborn infants are good examples. The infant's mucous membranes are contaminated during passage through the birth canal. Candidiasis develops in some 4% of the newborn but when the normal bacterial flora develops a short time later, the yeast infection usually subsides. However, premature babies, who are immunologically less mature than full-term babies, may experience more severe and persisting infections. Other conditions that predispose to infections *of mucosae and skin* include hypothyroidism, diabetes or other endocrine abnormalities, cancer and treatment with immunosuppressants or broad-spectrum antibiotics that disturb the normal microbial flora. Vulvovaginal and gastrointestinal manifestations are also common in these patients. Cutaneous infections occur most often as a direct result of mechanical forces that cause abrasion, or from continued excess moisture, or both. For example, *perléche* is seen in patients who have deep folds around the corners of the mouth where saliva collects to provide a moist environment for fungal growth. *Intertrigo* also occurs in moist folds of skin between obese buttocks and groins, and beneath pendulous breasts; these conditions are intensified by perspiration during hot weather. Intertrigo may occur between the fingers of workers who keep their hands in water all day and regularly have macerated skin.

Paronychia and other lesions around the nails are not likely to be caused by candidae unless there is maceration of some predisposing abnormality of the nail bed.

Certain rare forms of candidiasis occur as a result of serious immunodeficiencies. *Candida* induced granulomas develop into horny, disfiguring, granulomatous lesions of the skin. More frequent, but still rare, is *chronic mucocutaneous candidiasis* (CMC), in which the fungi are found only in the superficial layers of skin and mucous membranes. *Disseminated candidiasis* is usually a terminal infection complicating other severe disorders.

Outstanding features of CMC are its *chronic nature,* the ease with which the infection is often cleared by vigorous therapy only to return soon after cessation of therapy, and its association with *cellular immunodeficiencies* or *severe combined immunodeficiencies;* indeed CMC invariably develops in patients with severe combined immunodeficiencies who survive for any appreciable length of time. It is also common among patients with deficiencies involving CMI. On the other hand CMC does not present a problem in patients with primary deficiencies in humoral immunity.

Disseminated candidiasis is often a terminal event in patients with other overwhelming disorders, such as malignant disease. During dissemination the infection may spread to virtually any or many organs especially the kidneys, CNS, heart and eyes.

Use of contaminated equipment by "mainliner" drug addicts has resulted in persistent or fatal endocarditis. In addition, certain intravenous catheters and shunts used for medical therapy can introduce organisms into the blood and establish infections that usually become self-limiting when the offending catheter is removed.

G. MECHANISMS OF PATHOGENESIS

Despite the fact that candidiasis is the most common opportunistic fungal disease, almost nothing is known about the mechanisms of pathogenesis. It has been shown that the cell walls of *C. albicans* contain endotoxin-like materials which when injected are lethal for mice. However, there is no evidence that this kind of toxicity occurs during the course of human infection. Adherence of *C. albicans* to mucosal epithelium is mediated through a specific ligand-receptor interaction involving a fungal glycoprotein. Adherence contributes to pathogenesis as exemplified by the in vivo transformation of yeast-form cells to the more adherent and more pathogenic hyphal-form cells. It is of singular interest that antifungal drugs inhibit the transformation of yeast cells to hyphal-type cells.

H. MECHANISMS OF IMMUNITY

Most immunologic studies on candidiasis have been conducted on patients with CMC, which usually does not involve invasion of the bloodstream or deep tissues. Most of these patients are deficient in CMI due to one or more T cell-dependent immunologic defects as indicated by: a failure of T cells to respond to specific Ag with lymphokine production, T cell unresponsiveness to mitogens, failure of patients to react to common skin tests for DH and failure of the in vitro activation of T cells by specific Ag. Other alleged defects include failure of Ag processing by macrophages, an increase in T suppressor cells, and the production of blocking Abs. It is alleged that among CMC patients, young persons between 10 to 30 years of age carry the best prognosis because they have the best chance of regaining competence in CMI following chemotherapy. Claims that transfer factor may be beneficial in candidiasis remain to be confirmed. The most compelling evidence for the concept that CMI plays the major role in immunity to candidiasis is that the disease is much more frequent in patients who exhibit defects in CMI than in patients with defective humoral immunity. This is supported by the observation that selective suppression of CMI in animals increases their susceptibility to candidiasis.

Even though cellular immunity is undoubtedly a major factor in resistance, total immunity to candidiasis probably depends on a combination of defense forces some of which are nonspecific. Indeed, the majority of patients with candidiasis have no evidence of immunodeficiency and moreover, many patients with such deficiencies do not become infected. A variety of nonspecific humoral factors have been shown to inhibit or kill *C. albicans.* Human serum contains a factor that clumps the yeast, and transferrin and other iron-binding substances inhibit the organism by competing for available iron. The microbicidal peroxidase-H_2O_2-halide ion system found in neutrophils and certain body fluids can kill *C. albicans.* It is of singular interest that patients with chronic granulomatous disease are highly susceptible to candidiasis. This may be related to an observation in rabbits, namely that their neutrophils contain six cystine-rich antimicrobial peptides, three of which can bind to and kill *C. albicans* in vitro.

In support of a role for humoral immunity, resistance can be effected by immune serum produced by immunizing animals with killed organisms. It is probable that Abs effect protection by aggregating the organisms and preventing their adherence to mucosal cells.

Assuming that CMI is of major importance in resistance to *Candida sp.* the principal question is: what are the operative mechanisms? Most bacterial infections that serve as models of antimicrobial cellular immunity involve intracellular pathogens that are destroyed by T cell-activated macrophages (Chap. 8). Candidae are not intracellular pathogens, and moreover, in experimental candidiasis the cellular response is principally, granulocytic; mature macrophages do not kill candidae to a significant extent, although precursor macrophages of mice can bind to and kill *C. albicans.* However, it is possible that CMI is effected by certain lymphokines that may be antifungal in vivo, as they are in vitro; indeed Ab-dependent cellular cytotoxicity may also be an effector mechanism.

Specific immunosuppression has been reported in severe cases of CMC; it is presumed to result from a cell wall glycoprotein produced by *C. albicans.*

Much remains to be learned about immunity to *C. albicans,* nevertheless, some new tools are now at hand for unraveling the mysterious tangle of interacting factors concerned in host resistance. The possibility that the Langerhan's cell of the skin may be an important determinant in immunity to CMC deserves exploration.

I. LABORATORY DIAGNOSIS

Exposed lesions can usually be easily diagnosed by clinical appearance, together with finding typical budding yeast cells and pseudohyphae and/or true hyphae in lesion scrapings treated with KOH.

Deep lesions are more difficult to diagnose because of the problem of obtaining specimens free of skin and mucosal contaminants; resort to aspiration or biopsy may be necessary. Blood culture may sometimes be falsely positive because of contaminating organisms introduced by intravenous catheters.

Candidae are readily seen in stained pus or exudates from patients with candidiasis. Within tissue sections, special stains, such as PAS or methenamine silver, make the fungi more readily visible. The presence of pseudohyphae and/or true hyphae in tissue is pathognomonic of invasive infection with *C. albicans.* The organisms are easily cultured, and colonies become apparent within a day or two (Sect. D).

The major problem in laboratory diagnosis lies in determining whether demonstrable *C. albicans* is the etiologic agent of the infection. The organism is commonly present in normal flora, and the small numbers initially present in a specimen may propagate rapidly before the sample reaches the laboratory, for example, a sputum sample. Therefore, prompt examination and culture is necessary; large numbers of organisms, including hyphal forms, should be initially present in fresh specimens if candidiasis exists. Tissue invasion leads to the formation of pseudohyphae and/or true hyphae.

Although *serologic tests for specific Abs* are not diagnostically reliable, various immunologic techniques using monoclonal Abs for detecting specific *Candida* Ags in sera and techniques for detecting circulating *Candida* metabolites are highly promising. The immunologic techniques include immunodiffusion, enzyme immunoassay and radioimmunoassay.

A useful test for distinguishing *C. albicans* from similar organisms is the *germ-tube test. Candida albicans* in yeast form, inoculated into undiluted human serum at 37°C, will produce germ tubes within 4 hours, and usually by 1 hour. Other yeasts, and most other species of *Candida,* fail the test, except certain strains of *C. stellatoidea.* Sugar fermentations also aid in the identification of *Candida sp.*

J. THERAPY

Therapy consists of correcting predisposing conditions and treatment with antifungal agents. *Disseminated disease* may be treated with amphotericin B and/or flucytosine, provided the organism is not resistant to flucytosine; ketoconazole may also be useful.

Cutaneous candidiasis responds to topical application of gentian violet or antibiotics, including ketoconazole, nystatin and miconazole. Agents effective against CMC include amphotericin B, flucytosine, miconazole and ketoconazole.

K. RESERVOIRS OF INFECTION

Man is the principal reservoir of opportunistic *Candida sp.* which are part of the normal flora of most mucous membranes and skin.

L. PREVENTION OF DISEASE

At present, prevention of candidiasis depends on correcting or eliminating predisposing conditions.

ASPERGILLOSIS

Aspergillosis comprises a group of worldwide human diseases caused by about a dozen fungal species of the genus *Aspergillus, Aspergillus fumigatus* being the most common etiologic agent. Since *Aspergillus sp.* are many and are widespread in soil and decaying vegetable matter many persons are heavily exposed to inhalation of spores. Despite heavy exposure, infection is rare because normal individuals are highly resistant to aspergillosis; infection occurs principally in individuals who have predisposing abnormalities or diseases (cystic fibrosis, bronchiectasis, chronic bronchitis, tuberculosis, sarcoid, etc.) or who are systemically immunosuppressed. Several clinical types of aspergillosis occur, allergy being a frequent component because the organisms are highly allergenic. Animals, birds and insects can become infected as well as man.

Aspergillus sp. are not dimorphic; they are characterized by chains of conidia borne on conidiophores. Speciation is based principally on morphology; some 150 species are recognized.

Extracellular products of *Aspergillus sp.* include metabolites that are both toxic and oncogenic.

Culture of Aspergillus sp. can be carried out on most any routine medium; growth is rapid and aerial mycelium and conidia form in abundance.

Aspergillosis predominates in adult males and tends to be concentrated in excessive-exposure groups, such as malt workers and allergy-prone agricultural workers who handle moldy grain and forage (farmer's lung). However, as stated above, underlying lung abnormalities and immunocompromised states attending malignancy, immunosuppressants, neutropenia, granulocytopenia, etc. are strongly predisposing. Allergy-based aspergillosis is a true infection caused by organisms growing on mucosae.

Asthmatic individuals are especially prone to develop the *atopictype of allergic aspergillosis* following heavy inhalation of spores; it is characterized by eosinophilia and bronchoconstriction.

The Arthus type of allergic aspergillosis, with alveolitis as the predominant lesion, is attended by dyspnea, fever, leukocytosis, cough and rales. *Pulmonary aspergillosis* can also be caused by mixed types of allergy including DH.

Invasive aspergillosis occurs almost exclusively in severely immunosuppressed individuals in whom it is always fatal unless treated early. The infection usually progresses rapidly in the lung from which it spreads widely to many organs in the body. The hyphae invade vessel walls with resulting hemorrhage and thrombosis. On rare occasion invasive pulmonary aspergillosis can occur in an apparently normal person; prognosis is good in such cases.

Colonization aspergillosis is a form of aspergillosis in which organisms heavily colonize various cavities and channels that have external openings, such as tuberculous cavities, ears, nose, eyes and nasal sinuses. The organisms often grow so lavishly that they form "fungus balls." Colonization may be asymptomatic or symptomatic. Allergic reactions and trauma may sometimes be predisposing to *colonization aspergillosis.* Aspergillosis of the skin and nails has also been reported.

The pathogenesis of invasive aspergillosis rests on the capacity of the hyphae to invade blood vessels, presumably because of the toxic metabolites it produces. Immunologic tissue injury undoubtedly contributes to pathogenesis in those forms of aspergillosis that have an allergic component.

Although natural immunity depends largely on the capacity of neutrophils and macrophages to engulf and destroy hyphae and conidia, it is not known whether the host responds with acquired immunity and if so, what its nature might be.

Laboratory diagnosis depends on microscopic examination of KOH preparations of sputum and of biopsied tissue, culture of clinical specimens and serology (Fig. 40–3). Typical septate hyphae and conidia should be present in sections of tissue from patients with invasive aspergillosis; the hyphae are usually seen in parallel array invading blood vessels. However, in invasive disease, culture of sputum may be necessary because sputum is usually negative on microscopic examination.

Immunodiffusion tests for specific precipitins against *A. fumigatus* Ags show positive findings in 80 to 100% of cases involving allergy; immediate and Arthus type skin tests are always positive and eosinophilia is frequently found. Serologic tests in patients with invasive aspergillosis are unreliable because they often yield false-negative or false-positive reactions.

Treatment of invasive aspergillosis is carried out with amphotericin B as the drug of choice; flucytosine is another recommended drug. Pulmonary lavage has been proposed as a method for administering drugs. *Colonization aspergillosis* based on structural abnormalities of the lung is not benefitted by drugs. *Aspergillosis based on allergy* is usually treated with a combination of antifungal drugs and corticosteroids. Superficial lesions may be treated wth nystatin.

MUCORMYCOSIS

Mucormycosis is caused by a number of saprophytic species of opportunistic soil fungi of the order *Mucorales.* They belong to the genera *Rhizopus, Absidia, Mucor,* and *Cunninghamella.* Species of *Mucorales* are ubiquitious in soil, air and water, including hospital environments. *Mucormycosis* occurs in individuals whose defenses are compromised in some way. Individuals with ketoacidosis are especially vulnerable, such as diabetics and uremia patients; other predisposing causes are malignancy, severe burns, natural immunodeficiencies and artificial immunosuppression.

The organisms produce *broad nonseptate hyphae,* aerial mycelia, sporangia, sporangiospores and sexual zygospores; they grow rapidly on routine media to produce profuse cottony colonies.

The hyphae usually invade through the upper or lower respiratory tracts or

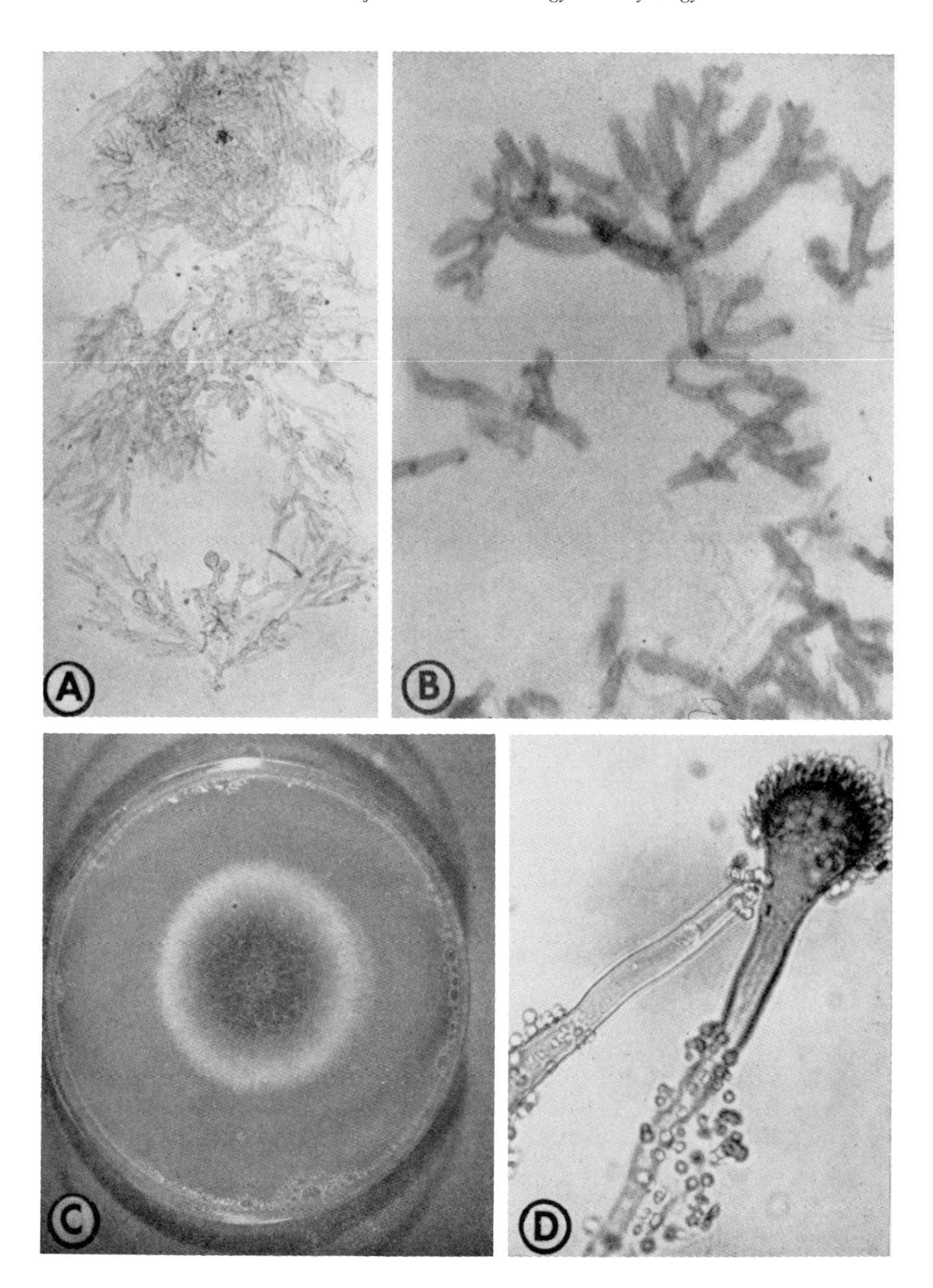

Fig. 40–3. *Aspergillus fumigatus. A,* Hyphae from pulmonary tissue digested with NaOH × 212. *B,* Same as *A.* × 690. *C,* Colony on Sabouraud's agar. *D,* Conidiophore. × 690. (From Emmons, C. W. et al: *Medical Mycology,* 3rd Ed. Philadelphia, Lea & Febiger, 1977.)

through burn wounds of the skin. Lesion progression is usually rapid, spores do not form in vivo.

Primary pulmonary mucormycosis following spore inhalation may initially be focal or diffuse. However, the growing hyphae soon invade blood vessels and produce highly destructive pulmonary lesions; death usually results within days to a few weeks at most.

Rhinocerebral mucormycosis begins in the nasal mucosa but soon progresses to involve the sinuses, eyes and CNS; suppuration, local necrosis, hemorrhage, cranial nerve palsy, and thrombosis are prominent features. Death usually occurs within a week.

Laboratory diagnosis rests on microscopic examination and culture of sputum, exudates and biopsied tissue; blood culture is always negative. Speciation is based on morphology and is often difficult. Tests for Abs that react with a pool of specific Ags of 11 *Mucor sp.* (Zs-Ag) by immunodiffusion is positive in about 70% of cases.

Early treatment is carried out with amphotericin B supplemented, when indicated, by surgical debridement.

GEOTRICHOSIS

Geotrichosis is a rare disease due to *Geotrichum candidum,* a saprophytic soil fungus that exists as a parasite among the normal flora of the mouth, GI tract, and genitourinary tract. Sites of infection include the lung, mouth, and GI and genitourinary tracts. *Diagnosis* rests on culture and *therapeutic drugs* consist of amphotericin B, iodides and nystatin.

OTHER OPPORTUNISTIC FUNGI

Additional opportunistic fungi are constantly being discovered, principally as the result of the increasing use of immunosuppressants. The number of known opportunistic fungi has now reached some 300 species among a total of about 100,000 species. The emerging opportunistic fungal infections have been assigned to two umbrella groups, the *hyalohyphomycosis group,* involving 14 known genera and the *phaeohyphomycosis group,* involving 26 known genera.

Unfortunately, infections due to this wide range of "new" opportunistic fungi, which are often life-threatening, present continuing and difficult diagnostic and therapeutic problems.

REFERENCES

Ajello, L.: Emerging mycoses of present and future concern. In *New Horizons in Microbiology.* Sanna, A. and Morace, G. (eds.), Amsterdam, Elsevier Science Publishers, 1984, p. 303.

Carrow, E.W., and Domer, J.E.: Immunoregulation in experimental murine candidiasis: Specific suppression induced by *Candida albicans* cell wall glycoprotein. Infect. Immun. *49*:172, 1985.

de Repentigny, L. and Reiss, E.: Current trends in immunodiagnosis of candidiasis and aspergillosis. Rev. Infect. Dis. *6*:301, 1984.

Domer, J.E.: Contrasts between candida and other pathogenic fungi in their interactions with the immune system of the host. In *New Horizons in Microbiology.* Sanna, A. and Morace G. (eds.), Amsterdam, Elsevier Sciences Publishers, 1984, p. 179.

Epstein, J.B., et al.: Oral candidiasis pathogenesis and host defense. Rev. Infect. Dis. *6*:96, 1984.

Horn, R., et al.: Fungemia in a Cancer Hospital: Changing frequency, earlier onset and results of therapy. Rev. Infect. Dis. *7*:646, 1985.

Lehrer, R.I., et al.: Correlation of binding of rabbit granulocyte peptides to *Candida albicans* with candidacidal activity. Infect. Immun. *49*:207, 1985.

Levine, H.B., et al.: Pre-clinical evaluation of antifungal drugs for deep mycoses. In *New Horizons in Microbiology.* Sanna, A. and Morace, G. (eds.), Amsterdam, Elsevier Science Publishers, 1984, p. 317.

McGinnis, M.R.: Taxonomy and nomenclature of some medically important yeasts. In *New Horizons in Microbiology.* Sanna, A. and Morace, G. (eds.), Amsterdam, Elsevier Science Publishers, 1984, p. 311.

Shibl, A.M.: Effect of antibiotics on adherence of microorganisms to epithelial cell surfaces. Rev. Infect. Dis. *7*:51, 1985.

Weitzman, I.: Saprophytic molds as agents of cutaneous and subcutaneous infection in the immunocompromised host. Arch. Dermatol. *122*:1161, 1986.

Glossary

This glossary is limited mostly to terms and abbreviations not readily found in dictionaries, or for which varying definitions may be given. These are, for the most part, much-used terms, not often included in the index. Special attention has been given to abbreviations.

Abscess: A local collection of pus that represents the liquefied remains of leukocytes (principally neutrophils) and disintegrated tissue.

Accessory cells: Cells that aid in immune responses to thymus-dependent Ags effected by lymphocytes, namely, macrophages and dendritic cells.

Accidental host: See Unnatural host.

Acid-fast: Not decolorized by acid treatment after staining; i.e., the mycobacteria retain certain stains even when treated with acid alcohol.

Acidosis: Abnormal increase in concentration of hydrogen ions in the blood and compensatory mechanisms that alter blood buffers.

Acquired immune deficiency syndrome (AIDS): A disease caused by a sexually transmitted virus which severely damages the immune system. It is characterized by a tendency to develop Kaposi's sarcoma and fatal secondary infections.

Acquired immunity: Immunity acquired during the lifetime of an individual, naturally or artificially; can be either active or passive in nature. Usually specific but can be nonspecific; an example of nonspecific immunity is immunity due to interferon.

Actinomycetes: Genera of bacteria of the order *Actinomycetales*, characterized by long filament-like chains of cells; all are facultative pathogens.

Active immunity: Immunity acquired as a result of an immune response mounted by the host; may be nonspecific or specific.

Acute phase serum: Serum derived from a patient with an acute inflammatory disease; such serum contains "acute phase substances" consisting principally of normal serum components present at elevated levels. They include the antibacterial agents lysozyme and beta-lysin.

Adenyl cyclase: An enzyme bound to the cytoplasmic membranes of most animal cells; catalyzes the formation of cyclic adenosine monophosphate from ATP.

Adhesins: Microbial surface components that bind to corresponding surface receptors on host cells, thus permitting microbes to adhere to and colonize on host cells; often serve as virulence factors.

Adjuvant: A substance that increases or diversifies the immune response to an antigen (Ag). Example: Freund's adjuvants (a) complete water in-oil emulsion of mineral oil, plant waxes, and killed tubercle bacilli; (b) incomplete-Freund's (complete emulsion minus the bacilli).

Adoptive immunity: Active immunity developed in a recipient as a result of having received immunocompetent donor cells.

Aerobe: A microbe requiring O_2 for growth.
Agglutination: The clumping together of particulates, caused by either electrostatic forces of attraction or by the binding action of ligands; e.g., antibodies (Abs).
Agglutinin: An agent that agglutinates particulates, including nonspecific ligands and Abs acting as specific ligands.
Aggressin: A microbial product or component that opposes host defense mechanisms, e.g., capsules that oppose phagocytosis.
AIDS: See Acquired immune deficiency syndrome.
Allergen: An Ag or hapten-carrier complex that engenders allergy.
Allergic granuloma: A granuloma resulting from an allergic response to the inciting agent (see Granuloma); a cluster of epithelioid cells surrounded by a mantle of infiltrating lymphocytes and macrophages; giant cells are usually also present.
Allergy: A state of specific hypersensitivity, most often used to designate hypersensitivities due to IgE Ab.
Alloantibody: An Ab produced against a foreign alloantigen.
Alloantigen: An Ag present in some, but not all, individuals of a species.
Allograft (homograft): An organ or tissue transplanted from one individual to a genetically dissimilar individual of the same species.
Alpha hemolysis: Partial hemolysis best seen on blood agar; often generates a greenish color.
Amyloidosis: A pathologic state characterized by widespread intercellular deposits of an abnormal body protein called amyloid; can lead to tissue and organ damage.
Anaerobe (strict anaerobe, obligate anerobe): A microbe that can grow only in the absence or virtual absence of O_2.
Anamnestic response (recall, secondary or memory phenomenon): An accelerated, intensified immune response to an Ag that occurs in an animal previously exposed to the same Ag or related Ags; may involve humoral and/or CMI responses; depends on memory cells.
Anergy: Unresponsiveness to an allergen; may be due to specific desensitization or nonspecific unresponsiveness.
Angina: Sore throat; this term also has other meanings.
Anthrax regions: Local regions where special soil and climatic conditions enable *B. anthracis* to persist in soil because they provide opportunity for periodic repetition of the cycle "vegetative cell growth—followed by sporulation"; such regions are permanent reservoirs of infection.
Anthropophilic: Having a preference for humans rather than animals.
Antibiosis: Inhibition exerted by one organism against another organism(s).
Antibody(ies) (Ab(s): An immunoglobulin molecule specific for an Ag or hapten, produced by B cells and evoked principally by Ags or haptens that conjugate with proteins to form hapten-carrier complexes.
Antigen(s) (Ag(s)): A substance that can evoke either a humoral and/or a CMI response; a substance that can react specifically with Abs in vitro.
Antigen-combining sites: The sites on Ab molecules that bind specifically to Ag-determinant sites.
Antigen masking: The coverup of microbial outer cell wall and membrane Ags by nonantigenic capsules or adsorbed host components.

Antigen mimicry: The circumstance in which a microbe produces an Ag that mimics and is cross-reactive with an Ag of the host.

Antigenic competition: That which occurs when an individual is exposed simultaneously to 2 or more Ags of different strengths; the "strong Ags" commonly suppress the immune response to the "weak Ags" (see Fundamentals of Immunology, Myrvik and Weiser, 2nd Ed. Philadelphia, Lea & Febiger, 1984, page 186).

Antigenic determinants (epitopes, antigen determinants): The local molecular configuration(s) on the Ag molecule that incite the formation of specifically reacting Abs (See Valence and specificity of Ags and Abs).

Antimicrobial cell-mediated immunity: Specific acquired immunity effected against microbes by special T effector cells (T_E cells) that function by producing lymphokines that act by: (1) maintaining a population of T_E memory cells, (2) mobilizing and activating macrophages, and (3) initiating and promoting interactions among the various cells concerned in the formation of the allergic granuloma.

Antimicrobial defenses: See Defenses.

Antiseptics: Disinfectants used topically on external surface wounds.

Antiserum: A serum that contains Abs; the term is usually used to designate a serum containing Abs induced by injection of an Ag.

Anuria: A state in which urine volume is severely or totally suppressed.

Apparent infection: Infection associated with clinically evident disease.

Arthropods: Invertebrates with jointed limbs, including mites and ticks.

Aschoff body: The allergic-type myocardial granuloma that characterizes rheumatic fever.

Ascites: Excess serous fluid in the peritoneal cavity.

Ascospores: Sexual spores of fungi borne in a sac called an ascus.

ASO test: A test for Abs against streptolysin O; used as an indicator of recent streptococcal infection in the diagnosis of rheumatic fever.

Asymptomatic infections: Infections without clinical symptoms; if infecting microbes are shed, such infections also represent a state of carriage.

Atelectasis: Collapse of alveoli of the functioning lung; usually due to blockage of bronchioles by mucus and exudates or pressure effects.

ATP (adenosine triphosphate): A chemical compound that is the major storage form of energy in cells; ATP possesses two high-energy phosphate bonds that release energy on hydrolysis.

Attenuated pathogen: A mutant strain derived from a pathogen that is avirulent or of low virulence. Attenuation may be achieved by genetic selection or by physical or chemical treatment.

Atypical mycobacteria ("unclassified" or "anonymous" mycobacteria): Facultatively pathogenic mycobacteria in humans; comprises species other than the recognized pathogenic species *M. tuberculosis, M. bovis*, and *M. leprae*. They are not transmitted from man to man, and most species are probably saprophytes that act as opportunists.

Autoclave: An apparatus for sterilizing various articles and materials such as foods and media; uses superheated steam under pressure.

Autoimmune diseases: Diseases caused by immune responses to patient's own tissue Ags, such as autoimmune hemolytic anemia due to RBC Abs.

Auxotyping: Typing strains of microbes within a species based on their nutritional needs for growth.

Avirulent microbe: A nonpathogenic mutant strain organism derived from a pathogenic species (see Virulence).

Bacillus: The name for the genus *Bacillus*; also used loosely to designate rod-shaped bacteria.

Bacteremia: Presence of bacteria in the blood stream; associated with nonseptic as well as septic states.

Bacterial antigens: Antigens of bacteria often designated by terms to indicate various structures and functions, e.g., the H Ags of flagella and the virulence Ags (Vi Ags) of cell envelopes.

Bacteriocins: Protein molecules, produced by bacteria, that kill other strains of the same or closely related species.

Bacteriophages: Special viruses that parasitize bacteria.

Bacteriostatic agent: A nonlethal agent that arrests the growth of bacteria.

Bacteriuria: Presence of bacteria in the urine; may occur in subclinical as well as clinical states.

Basophils: Granulocytes of blood characterized by a few large basophilic granules in the cytoplasm and a coarsely-lobed nucleus; they carry heavy loads of mediators concerned in inflammation and allergic reactions.

B cells: Lymphocytes of the B cell class concerned primarily in Ab formation; the plasma cell represents a mature Ab-forming cell.

BCG (bacillus of Calmette and Guerin): An attenuated strain of *Mycobacterium bovis* used as vaccine for preventing tuberculosis.

Beta hemolysis: Complete lysis of RBCs; colonies of beta-hemolytic bacteria on blood agar are surrounded by clear areas of complete hemolysis.

B-F group organisms: Bacteria of the *Bacterioides* and *Fusobacterium* genera.

Biopsy: The act of removing tissue from a living animal or person.

Biotypes: Phenotypic variants of the prototype strain of a microbe that best represents the species.

Blocking antibody: An Ab that can block a specific activity of another Ab or an immune cell. Examples are the incomplete Ab of brucellosis that blocks agglutination by brucella agglutinins and the incomplete Ab of Rh disease that blocks the "saline agglutinin" (See Incomplete Abs).

Blood-brain barrier: The walls of CNS capillaries, structured as "barriers" to limit passage of blood components, including all classes of Abs, blood cells, and foreign particulates; the barrier often contributes importantly to antimicrobial defense by sheltering the vitally important structure, the CNS, from infection.

Booster immunization: The practice of administering secondary doses of immunogens to reawaken waning immunity.

Bright's disease: A form of acute hemorrhagic glomerulonephritis that constitutes a sequela of infection with nephritogenic strains of *Strep. pyogenes* (principally type 12) and occasionally type C streptococci; nephritogenic strains produce a "nephritic-strain"- associated protein (NSAP), which probably represents the nephritogenic agent.

Brill's disease: A mild form of typhus fever which represents reactivation in patients with long-latent infection due to *R. prowazekii*.

Broad spectrum antibiotics: Antibiotics that act against a broad range of microbes, including both Gram-negative and Gram-positive species.
Bronchiectasis: A pulmonary disease characterized by dilated bronchi which may sometimes contain purulent secretions; may lack productive cough except in the morning.
Bronchopneumonia: Pneumonia characterized by patchy areas of inflammation.
Buboes: Enlarged lymph nodes, usually in the groin or axilla, commonly caused by infections by such agents as *Y. pestis, H. ducreyi* and *Chl. trachomatis*; buboes are often painful and exhibit variable areas of necrosis, suppuration, and granuloma formation.

cAMP (cyclic adenosine monophosphate): A mediator of the actions of many hormones; within cells the level of cAMP regulates cellular function.
Capsid: The protein coat of a virus.
Capsular swelling test (Quellung test): Swelling of the bacterial capsule as the result of a specific reaction with Ab; used in serologic identification of capsulated bacteria.
Capsulated (encapsulated) microbes: Microbes with a detectable coat of material on the microbial surface.
Carbuncle: A multifocal suppurative infection in deep tissue that usually results from spread of an initial infection in overlying skin; commonly due to *Staph. aureus.*
Carriage: The state of serving as a carrier of a pathogen.
Carrier: A host that harbors and releases pathogens but shows no apparent disease. A carrier may shed pathogens during convalescence (convalescent carrier) for long periods (chronic carrier) or periodically (intermittent carrier).
Caseation necrosis: Necrosis of tissue that yields residual material of cheese-like consistency, the prototype being the caseous tubercle of tuberculosis; when caseous material liquefies, a cavity results.
Cell activation: Conversion of a dormant cell to an activated cell capable of effecting its potential physiologic function(s).
Cell-mediated immunity (CMI), cellular immunity: Specific acquired immunity; the effector arm rests on cells rather than humoral Abs; CMI can be passively transferred with "immune cells" but not with "immune serum."
Cellulitis: Diffuse nonsuppurative inflammation of cellular tissues; especially frequent in subcutaneous tissues as a result of infection.
Cervicofacial: Pertaining to both the neck and the face.
CF (complement fixation): Binding of the components of C with other substances, especially Ag-Ab complexes.
CFU (colony forming unit): One or more cells that replicate and give rise to a single colony (see Colony).
Chancre: The small, firm, painless primary lesion of syphilis characterized by late ulceration; usually presents as a solitary lesion on the genitalia or lips.
Chancroid (soft chancre): The chancre-like genital lesion due to *H. ducreyi.*
Chemotaxin: An agent that can cause chemotaxis.
Chemotaxis: Attraction or repulsion of a motile cell by a chemical.
Child: A person between age 2 years and puberty.

Childbed fever: See Puerperal fever.
Chocolate blood agar: Heated blood agar; contains the growth-promoting factors X (hemin) and V (NAD) needed by *H. influenzae.*
Cholecystectomy: Surgical removal of the gallbladder and its duct.
Clean-voided urine specimen: Urine collected in midstream after careful cleansing of external urogenital areas to minimize contamination by normal flora.
Clone: A population of identical cells; usually derived by asexual reproduction of a single progenitor cell.
CMC (chronic mucocutaneous candidiasis): Infection due to *C. albicans* which usually occurs in immunodeficient individuals.
CMI: see Cell-mediated immunity.
CNS: Central nervous system; composed of the brain and spinal cord.
Coccobacillus: An oval-shaped bacterium that resembles both a sphere and a rod.
Coccoid: Resembling cocci; rounded or spherical cells.
Coccus: A spherical microbe.
Cold agglutinins: Immunoglobulins in blood plasma or serum that cause the agglutination of RBCs at 0° to 4°C but not at 37°C; observed in primary atypical pneumonia and certain other disorders.
Cold antibodies: Antibodies that act best at temperatures between 4°C and 25°C.
Cold autoantibodies: Autoantibodies that act best at 4°C to 25°C, e.g., the cold autohemagglutinins of congenital and tertiary syphilis and of listeriosis.
Colicins: Bacteriocins produced by *Escherichia coli.*
Colitis: Inflammation of the colon.
Colonization: Growth of organisms on external integuments; may or may not lead to invasion of underlying tissues.
Colony: A cluster of growing cells derived from one or a few cells; if the latter are identical, it is called a "colony-forming unit" (CFU).
Colostrum ("first milk"): The viscous mammary secretion produced at or near the time of parturition; it is rich in minerals and contains secretory IgA Abs that afford neonates some protection against intestinal infections.
Commensalism: See parasitism.
Complement (C): A multifactorial system comprising some 9 interacting serum components that are primarily concerned in antimicrobial immunity, both specific and nonspecific; activation of the C system results in the cascading interaction of C components and can be initiated in the classical manner by Ag-Ab reactions (classical pathway) or nonspecifically by nonantibody agents, such as microbial polysaccharides of *Strep. pneumoniae* (alternate pathway). Activation leads to the fixation and "consumption" of C components. Antimicrobial activities effected by C include lysis and opsonization.
Compound microscope: A light microscope having two or more lenses.
Conidiospores (Conidia): Asexual spores of fungi borne on branched stalk-like structures called conidiophores.
Conjugation: Exchange of chromosomal material by sexual recombination between male and female bacterial cells.
Consumption coagulopathy: Reduction in one or more of the blood-clotting factors due to extensive intravascular blood clotting.

Contact inhibition: Inhibition of growth of normal cells in monolayer cultures that occurs when cells contact each other in the monolayer.

Coombs antiglobulin test: An agglutinin test in which antiglobulin Abs are used to detect serum globulins (principally incomplete Abs) bound to blood cells.

Coproantibodies: Secretory IgA Abs shed into the intestinal tract; increased amounts are shed during infection.

Cord factor: A component of *M. tuberculosis* associated with the capacity of the organism to grow in aggregates resembling skein-like cords.

Cornification: Conversion of a tissue into a hard or horny mass, e.g., replacement of an area of consolidation in pneumococcal pneumonia with dense scar tissue.

Corticosteroids: Steroids with certain chemical or biological properties characteristic of the hormones secreted by the adrenal cortex.

Coryneforms: Species of *Corynebacterium* that are morphologically similar to *C. diphtheriae*.

Counterimmunoelectrophoresis: Immunoelectrophoresis performed with two electrical currents applied sequentially in perpendicular directions to give greater resolution to the precipitin lines formed.

C-reactive protein: An abnormal beta-globulin that appears in the serum of various patients with acute inflammatory diseases. Combines with the C-substance of the pneumococus to form a precipitate.

Crisis in pneumoccal pneumonia: Sudden improvement of symptoms beginning about the 7th to the 8th day of illness; usually attended by the appearance of free Ab in the blood; slow recovery without crisis is termed "recovery by lysis."

Cross-reacting Ag: An Ag that cross-reacts with Abs elicited by another Ag because the two Ags carry identical or near-identical Ag determinants (epitopes).

Croup: A severe throat infection accompanied by noisy respiration and hoarse coughing; occurs commonly in children and may present a pseudomembrane.

CSF: Cerebrospinal fluid.

C-substance: The specific Forssman-like carbohydrate in the cell wall of *Strep. pneumoniae*.

Culture media: Media used for growing microbes; include special purpose media as follows: (1) selective media that favor the isolation of desired organisms by suppressing the growth of other organisms, (2) differential media that permit the differentiation of colonies of the desired organism from the colonies of other organisms, (3) selective-differential media that act both selectively and differentially, and (4) transport media designed to preserve delicate pathogens during transport to the laboratory.

Cystitis: Inflammation of the urinary bladder, usually due to infection.

Cytophilic antibodies: Antibodies which, in the free state, have an affinity for cells that depends on attraction forces independent of those that bind Ag to Ab. Examples are binding of IgE Abs to mast cells and the binding of certain IgM and IgG Abs to macrophages.

Cytoplasmic membrane (protoplasmic membrane, plasma membrane): The membrane enclosing the cytoplasm of a cell.

Darkfield illumination: A procedure based on the principle of light reflected from the object being observed.

Debilitated person: A person who is physically weakened.

Débridement: The procedure of removing foreign material and devitalized tissue from an area of living tissue.

Decubitus ulcers (bedsores): Ulcers of the skin that result from ischemia at local pressure points in dependent parts of the body.

Defenses: Systemic and local body components that defend the body against harmful agents. Antimicrobial defenses consist of *external defenses*, which include surface integuments and associated organs (eyes, ears, etc.); and *internal defenses*, both specific and nonspecific, which operate in underlying tissues.

Defervescence: Disappearance of fever.

Delayed hypersensitivity (delayed sensitivity) (DH): A specific T-cell-dependent sensitive state characterized by a delay of many hours in onset time and course of reaction following challenge with Ag. It can be transferred with cells but not with serum.

Dendritic cells: Various dendrite-bearing cells that serve as important accessory cells in immune responses; they occur principally in reticular tissues (see Fundamentals of Immunology, Myrvik and Weiser, 2nd Ed. Philadelphia, Lea & Febiger, 1984, page 147).

Dermatophytes: A group of keratinophilic fungi that invade superficial keratinized areas of the body (hair, skin, nails).

Desensitization (hyposensitization): Abolition or diminution of sensitive state by administration of Ag, usually in small repeated doses.

Desquamation: Excessive sloughing of the surface layer(s) of an integument such as the epidermis of the skin.

Differential medium: A medium that permits visual differentiation of colonies of different kinds of organisms (see Culture media).

Dimorphic fungi: Fungi that can undergo morphologic conversion and grow in either yeast or mold form depending on environmental conditions; many fungal pathogens are dimorphic.

Diphtheritic membrane: A pseudomembrane in the throats of patients with diphtheria; it consists of necrotic epithelium embedded in a fibrinous exudate; mechanical removal leaves a bleeding surface.

Diphtheroid: A bacterium that is morphologically similar to *C. diphtheriae*.

Disinfectants: Chemicals that can kill pathogens and potential pathogens at any site.

DNA: Deoxyribonucleic acid.

DNA homology: The study of DNA relatedness in which the DNAs from different microbes are matched for composition; DNA homology is one of the best indicators of the relatedness of respective microbes.

DPT triple vaccine: A vaccine containing three immunogens: diphtheria toxoid, killed *Bordetella pertussis*, and tetanus toxoid.

Dysentery: Inflammation of the intestine, characterized by abdominal pain, tenesmus, and diarrhea with blood and mucus in the feces.

Edema: A state characterized by an excess of intercellular tissue fluid, often resulting from lymphatic blockade or increased venous pressure; can be either local or generalized.

ELISA (enzyme-linked immunosorbent assay): A serologic test in which enzyme-linked Abs are used.

Emaciation: A state in which an individual is underweight because of extreme loss of flesh, usually from illness or starvation.

Emphysema (pulmonary): A disorder characterized by dilated alveolar spaces, usually resulting from atrophy and rupture of alveolar walls.

Empyema: Pus in a body cavity, often the pleural cavity.

Endemic: The continuing prevalence of a disease in a population at a relatively low level.

Endemic typhus (murine typhus): A mild form of typhus due to *R. typhi* transmitted from the rat, a natural host, to man, an unnatural host, by the rat flea (see Epidemic typhus).

Endocytosis: A broad term designating the engulfment of fluid droplets (pinocytosis) and/or particulates (phagocytosis); the term phagocytosis is commonly used to designate the engulfment of large particulates by "professional phagocytes," namely neutrophils and macrophages; the chamber formed by invagination of the cytoplasmic membrane during endocytosis has been variously termed, cytoplasmic vesicle, endocytic vesicle, endocytic vacuole, and phagosome. Certain ingested microbes may interfere with normal phagosome-lysosome fusion (see Phagolysosome).

Endogenous pyrogens: Substance produced by granulocytes, macrophages, and other cells that act on the hypothalamus to cause fever.

Endospore: A highly resistant resting form of bacteria produced within the vegetative cell.

Endotoxin (LPS): A toxic moiety of complexed protein, lipid, and polysaccharide present in the cell walls of Gram-negative bacteria.

ENL (erythema nodosum leprosum): Immune complex lesions that occur in some patients with leprosy, particularly in drug-treated patients with lepromatous and intermediate forms of leprosy, caused by Ag-Ab-C complexes deposited in vessels of the skin.

Enrichment medium: A medium that favors the growth of one organism over others in a mixture.

Enteric bacteria: Bacteria that naturally inhabit or often invade the GI tract, including nonpathogens and pathogens.

Enterocolitis: Inflammation involving both the small and large intestine.

Enteropathic *E. coli*: Strains of *E. coli* that can cause intestinal disease.

Enterotoxin: A toxin that produces pathologic changes in the alimentary tract, e.g., cholera toxin.

Eosinophil: A blood granulocyte characterized by a few coarse eosinophilic granules in the cytoplasm and low phagocytic activity; it plays important roles in allergic reactions and immunity to animal parasites (see Granulocytes).

Epidemic: A disease outbreak in a human population in which an increasing number of cases arise with time, usually due to an infectious agent; commonly subsides within months or years at most.

Epidemic typhus: A severe and often fatal form of typhus due to *R. prowazekii* which is transmitted from man to man, a natural host, by human body and head lice. The only other known natural host is the flying squirrel.

Episome: A fragment of DNA (carrying genetic information) that can exist either free in the cytoplasm or integrated into the chromosome.
Epithelioid cell: A nonmotile transformed macrophage that characterizes the allergic granuloma; epithelioid cells typically occur in the center of the primary granuloma as closely adherent collections of pale-staining, elongated, epithelial-like cells.
Epithet: The species name or second part of a Latin binomial.
Epitopes (Antigenic determinants): Local chemical configurations on the Ag molecule that incite Abs which react specifically with their respective epitopes; also used to designate configurational sites on nonantigenic ligands which bind to corresponding receptors (paratopes) on molecules or cells.
Epizootic: An epidemic in an animal population.
Erysipelas: A cellulitis due to *Strep. pyogenes*; usually arises around the nares and mouth and spreads over the face; tends to recur repeatedly in the same individual over many years and, if untreated, may spread to produce septicemia and meningitis.
Erythema multiforma: An acute inflammatory skin disease in which red papules, macules, or tubercles appear, usually on the head, neck, and extremities.
Erythema nodosum: A painful self-limiting allergic lesion in the dermis, particularly over bony prominences of the extremities, characterized by an infiltrate rich in lymphocytes and occurring in many diseases, including rheumatic fever.
Erythema nodosum leprosum: See ENL.
Erythematous rash: A red skin eruption occurring in patches of varying size and shape.
Erythrogenic toxin: A toxin produced by group A streptococci that causes erythema of the skin (scarlet fever).
Eschar: A dry mass of necrotic tissue that separates from living tissue, as in certain burn wounds and some infections.
Eucaryotic cells (eucaryotes): Cells characterized by membrane-bound nuclei.
Eumycetes: True eucaryotic fungi.
Exogenous fungal spores: Spores formed on the exterior of hyphae.
Exotoxin: A toxin released from the cell that produces it.
External defenses: See Defenses.
Exudate: An excess of tissue fluid made up of extravasates from damaged vessels in areas of inflammation, usually rich in blood proteins and cells.

FA test: Fluorescent Ab test.
Facultative aerobe: An organism that grows best anaerobically but can adapt to an aerobic environment.
Facultative anaerobe: An organism that grows best aerobically but can adapt to an anaerobic environment.
Facultative intracellular pathogen: A pathogen that can survive and grow intracellularly as well as extracellularly.
Facultative parasite: A saprophyte that can assume a parasitic existence.
Facultative pathogen (opportunistic pathogen): A nonpathogenic microbe (parasite or saprophyte) that, under abnormal circumstances, can invade and

produce disease. An example is *Candida albicans* infection in immunosuppressed individuals.

Facultative saprophyte: A parasite that can assume a saprophytic existence.

"Farmer's lung": An allergy-related lung infection caused by inhalation of various fungal spores present in spoiling grains and forage.

Faucial area (throat): The space between the oral cavity and the pharynx.

Febrile illness: An illness characterized by fever.

Fermentation: The metabolic process in which the final electron acceptor is an organic compound.

Fertility factor (F factor): A piece of DNA in male (F+) cells that may pass through the male sex pilus to female (F−) cells during mating.

Fibrillae: Unusually fine fimbriae.

Fibrosis: The replacement of normal tissues with fibrous scar tissue.

Fimbriae (pili): Hair-like surface structures on bacteria that serve various functions (see Pili).

Fistula: A pathologic sinus that often transports pus from an abscess through external integuments.

Flagella: Whiplike surface structures on bacteria that effect locomotion; they may be positioned: (1) around the cell (peritrichous), (2) as a polar tuft of flagella (lophotrichous), (3) as a single polar flagellum (monotrichous), and (4) as a single flagellum or tuft at each end of the cell (amphotrichous).

Flora: See Normal microbial flora.

Fluorescent Ab: Antibody coupled with a fluorescent dye.

Fomites: Inanimate objects or materials that transport pathogens to infect new hosts.

Food poisoning: A loosely used term to designate illness due to eating food contaminated with pathogenic microbes and/or their products, i.e., illness may be due to contaminating salmonellae or, alternatively, to the contaminating toxins of *Cl. botulinum, B. cereus*, or *Staph. aureus*.

Formalin: A saturated aqueous solution of gaseous formaldehyde at room temperature (about 40% formaldehyde).

Forssman Ag: A widely distributed heterophile Ag. For example, when sheep RBCs that contain Forssman Ag are injected into the rabbit (an animal lacking Forssman Ag), so-called "Forssman Ab" is formed.

Frei reaction: Specific DH skin test reaction to the causative agent of lymphogranuloma venereum, a variant of *Chl. trachomatis.*

Freund's adjuvants: See Adjuvant.

FTA test: Serologic test for treponemal Abs in which a slide preparation of *T. pallidum* is exposed to the patient's serum, washed, and finally exposed to a fluorescein-labeled antiglobulin serum.

FTA-Abs test (fluorescent treponemal antibody absorption test): A test for syphilis in which the patient's serum is absorbed with Ag of the nonpathogenic Reiter strain treponeme before applying the FTA test.

Fulminating disease: A disease that develops suddenly and worsens rapidly.

Fungi: Eucaryotic protists that differ from bacteria principally on the basis of cell-wall composition and modes of reproduction.

Furuncle: Localized abscess developing from an infected hair follicle.

Gamma globulins (immunoglobulins): Serum proteins with antibody activity comprising 5 classes, namely IgA, IgE, IgG, IgM, and IgD.
Genetic transformation: Genetic change(s) induced in recipient bacteria by the uptake and incorporation of short pieces of free DNA from donor bacteria.
Genome: The complete set of genetic factors possessed by a cell.
Genotype: the genetic constitution of an organism.
Genus abbreviations: *Borr.-Borrelia, Br.-Brucella, Chl.-Chlamydia, Cl.-Clostridium, Fr.-Francisella, Ps.-Pseudomonas, Sal.-Salmonella, Sh.-Shigella, Staph.-Staphylococcus, Strep.-Streptococcus.*
Germ tubes: Elongated buds that develop when dimorphic fungi convert from the yeast form to the mold form of growth and mature to form true hyphae.
Germicide: A chemical agent that can destroy microbes irrespective of their potential pathogenicity.
Ghon complex: A healed lesion comprising the residuum of a solitary lesion (Ghon tubercle) of primary tuberculosis and infected draining lymph nodes; usually develops in childhood.
GI tract: Gastrointestinal tract.
Giant cells of Langhans: Large multinucleated nonmotile cells formed in allergic granulomas by the fusion of macrophages.
Giemsa stain: A stain commonly used for hemopoietic tissue and blood and for visualizing certain microbes that parasitize blood cells.
Glabrous skin: Smooth, hairless skin.
Globi: Collections of cell debris and "foamy macrophages" filled with *M. leprae* and bacterial lipid; they characterize the lesions of lepromatous leprosy.
Glycocalyx: A layer of material present on the surface of many bacterial species, including medically important bacteria, which enables them to persist in unfavorable environments; it can often be demonstrated as a capsule and often serves as a virulence factor.
Gnotobiotic animals: Animals that are maintained in a germ-free state; they are usually obtained by caesarian section.
Graft versus host disease (GVHD): A disease caused by the destruction of host cells (principally lymphocytes) by donor-type "immune cells" that arise from the donor allograft.
Gram-negative sepsis: A severe toxic, febrile state resulting from infection by enteric Gram-negative bacteria from parenteral administration of endotoxins of such bacteria.
Granulation tissue: Early tissue formed in the repair of a wound; it is rich in ingrowing capillaries, loops of which convey a granular appearance to the wound surface.
Granulocytes: Blood leukocytes characterized by granules in the cytoplasm and multilobed nuclei; they include polymorphonuclear neutrophils, basophils, and eosinophils (see respective cells).
Granuloma: Local nodular lesion composed of collections of macrophages, macrophage-derived cells, and lymphocytes; all granulomas induced by infectious organisms are essentially allergy-based (see Allergic granuloma).
Growth cycle of bacteria: The cycle that occurs when a stationary-phase inoculum of bacteria is placed in a fresh batch of medium; a growth-death

cycle comprising 4 phases, namely, the (1) lag phase, (2) log phase, (3) stationary phase and (4) death phase (see Chap. 2, p.22)

Gumma: The granulomatous lesion of tertiary syphilis.

Halophilic (salt-loving) bacteria: Bacteria that grow best in high concentrations of salt above 2 to 6%.

Hapten: A nonantigenic substance that cannot evoke Abs because it lacks a carrier molecule but nevertheless can react specifically with Abs.

Hemagglutinin: A substance that agglutinates RBCs; often specific Abs. Examples are the anti-A and anti-B Abs of the ABO blood group system, and the hemagglutinating enzyme of influenza virus.

Hematopoietic tissue (hemopoietic tissue): Blood cell-forming tissue.

Hematuria: The passage of urine containing RBCs.

Hemolysin: A substance that can lyse RBCs.

Herd resistance (herd immunity): The overall resistance possessed by a human or animal population.

Herpes-like lesion: Vesicular lesion resembling the lesions caused by herpes viruses, e.g., the cold sore of herpes simplex virus infection.

Herxheimer reaction (Jarish-Herxheimer reaction): A violent and occasionally fatal reaction of probable allergic nature that sometimes arises in syphilis patients hours after the initiation of penicillin therapy.

Heterophile Ags: Antigens common to different species.

Heterotrophic microbes (heterotrophs): Microbes requiring organic material for growth.

Histamine sensitizing factor: A factor produced by *H. pertussis* that enhances the host's sensitivity to histamine.

Hold-fast pili: Special pili that enable bacteria to adhere to and colonize on mucous membranes.

Hosts: Species or individual plants and animals that harbor parasites (see Parasitism); hosts are classified as natural hosts, unnatural hosts (accidental or incidental hosts), intermittent hosts, alternate hosts, intermediate hosts, transport hosts, and terminal hosts. A given host may fit one or more of the above categories.

HSF: see Histamine sensitizing factor.

Humoral immunity: Immunity initated by the specific Ab response of B cells to Ag, which in some instances requires the helper activity of T cells.

Hutchinson's triad: Stigmata that characterize late congenital syphilis; they are: (1) notched permanent incisors, (2) saddle nose, and (3) interstitial keratitis leading to corneal opacity and often blindness.

Hybrid cells: Cells produced by the fusion of two cells of different types; hybrid cells occasionally form spontaneously in mixed-cell cultures, although they do not perpetuate themselves indefinitely in culture. Cell hybridization can be speeded by special techniques.

Hybridoma: A clone of hybrid cells representing the progeny of a hybrid cell formed by fusing two cells of different types, one being a normal cell and the other a malignant cell. The malignant cell conveys to the hybrid cell the ability to perpetuate itself indefinitely in culture.

Hybridoma techniques: Special techniques for producing various hybridomas, the most notable being the fusion of a plasmablast with a myeloma tumor cell to yield a hybridoma that produces monoclonal Abs.

Hypha (pl. hyphae): Vegetative tubular fungal structures which may or may not possess cross walls (septa) for separating individual cells; the hyphae of some fungi lack septa for separating individual cells. Such multinuclear nonseptate hyphae are called *coenocytic hyphae.*

Hypovolemic shock: Shock caused by reduced blood volume.

Iatrogenic infection: Infection transmitted to patients by physicians.

IC (inclusion conjunctivitis): An eye disease caused by *Chlamydia.*

Icterus (jaundice): A yellowish staining of integuments and deep tissues with blood-borne bile pigments.

"Id": A sterile skin lesion of allergic nature distant from the site of infection, resulting from systemic transport of microbial Ag from the site of infection to the reaction site.

Immediate sensitivity (immediate hypersensitivity): A state of specific sensitivity mediated by Abs and characterized by a short onset time and course of reaction after contact with Ag; e.g., the wheal and flare skin reaction of a hay-fever patient in response to a pollen scratch test and anaphylactic reactions to penicillin.

Immune: First used in the biologic sense to designate resistance to injury resulting from infectious agents. Recently the meaning of this term has been extended to include the ability to react specifically with an Ag irrespective of whether or not the Ag is injurious. For example, an animal which, because it possesses Abs, can react specifically with a bland substance such as egg albumin is said to be "immune."

Immune clearance: Removal of Ag (soluble or particulate) from the circulation or tissues as the result of binding to specific Abs or Ag-primed lymphocytes.

Immune complex disease: Disease resulting from the depositon of circulating soluble Ag-Ab complexes in tissues, especially along basement membranes and in vessel walls. The deposited complexes usually contain C. Examples are poststreptococcal glomerulonephritis, systemic lupus erythematosus, and vascular lesions of the Arthus reaction.

Immune deficiencies: Deficiencies in the immune system involving components concerned in both innate and acquired immunity; can be either primary (genetic or congenital) or secondary to some other abnormality. Deficiencies concerned with specific immune responses may involve either the B cell system and/or the T cell system. Defects that suppress both innate and acquired immunity include defects in the C system and accessory cells (see Fundamentals of Immunology. Myrvik and Weiser, 2nd Ed. Philadelphia, Lea & Febiger, 1984, page 214).

Immune response(s): Specific response to an Ag involving either B-cells and/or T-cells; includes both humoral and CMI responses.

Immunization: The act or process of rendering an individual resistant or immune to a harmful agent. Often used loosely to designate a specific response to an Ag, irrespective of whether or not protection results.

Immunogen: A substance that engenders an immune response, either cellular or humoral, irrespective of whether or not the host is protected.

Immunologic tolerance (antigenic paralysis): A state of immunologic unresponsiveness to an Ag as the result of some unique prior experience with the Ag, such as excessive exposure.

Immunosuppressive agent: Any agent, physical or chemical, that can suppress specific immune responses. Examples are x-irradiation, azathioprine, and antilymphocyte globulin.

Imperfect fungi: Fungi for which no known means of sexual reproduction exists.

Impetigo contagiosa: A cutaneous infection with a "skin strain" of *Strep. pyogenes*; frequently seen in epidemic form among underprivileged infants and children inhabiting tropical climates.

Inapparent infection (subclinical infection, latent infection, silent infection): An infection lacking clinical symptoms.

Inclusion bodies: Intracellular masses of abnormal material, often foreign, e.g., viruses or chlamydiae within host cells.

Incomplete Abs (nonprecipitating Abs, univalent Abs, nonagglutinating Abs, and blocking Abs): Antibodies that can react with specific Ag-determinants but fail to bring about precipitation or agglutination. They are usually IgG Abs and are detected by such tests as the Coombs' antiglobulin test or by adding albumin to the system.

Induration: Abnormal hardening of a tissue; usually due to extensive infiltration with cells and fluids or the formation of fibrous tissue.

Infant: A person under 2 years of age.

Infant botulism: A fatal disease in young infants due to intestinal growth of *Cl. botulinum*; causes 10% of all "crib deaths" and is the leading cause of infant mortality in developed countries; sIgA antitoxic Abs in some breast milks afford some protection.

Infection: The presence of living microbes in parenteral tissues.

Infection immunity: Immunity that accompanies infection but drops sharply after infection ceases; usually, if not always, due to CMI.

Inflammation: A vascular response to injury; involves local changes in blood pressure in small vessels and vascular permeability, etc.

INH (isonicotinic acid hydrazide): Drug used for therapy of tuberculosis.

Innate immunity: Immunity that is not acquired but is inherent to the species; immunity that does not rest on previous experience with a harmful agent or related agents. Innate immunity is in contrast to immunity acquired during the life of the individual as a consequence of experience with the agent or a related agent.

Integration of DNA: The covalent bonding of DNA from a donor organism with the chromosome of the recipient organism.

Intercurrent infection: See Secondary infection.

Interferons (IFNs): Biogically active proteins produced by various cells under the stimulus of infecting viruses and other agents; IFNs are of 3 types: alpha, beta, and gamma. Whereas gamma IFNs have immunoregulatory activity, alpha and beta IFNs render host cells resistant to viruses by inducing them to produce a viral inhibitory protein.

Interleukins: Leukocyte products that promote leukocyte interactions. Interleukin 1 is a monokine called lymphocyte-activating factor (LAF) which causes a special subset of lymphocytes to produce the lymphokine IL2, a T cell growth factor (TCGF); IL2 is also called lymphocyte growth factor.

Intermittent host: An individual host who is parasitized intermittently by the same species of parasite.

Internal defenses: See Defenses.

Intracellular parasite: An organism that normally grows inside a cell; can be obligately or facultatively intracellular.

In utero: Within the uterus.

In vitro: Outside of the living body.

In vivo: Within the living body.

Ischemia (local anemia): Lack of blood circulation in a local area of tissue.

K antigens: Bacterial capsule and envelope polysaccharides and proteins, particularly of Gram-negative bacteria; often incite protective opsonins. (*C. diphtheriae* also carries an Ag called K Ag.)

Karyotype: The chromosomal characteristics of an individual or a cell line.

Keratinized cells: Cells containing keratins, the water-insoluble proteins that give hair, skin, and nails their horny, tough properties.

Keratinophilic: Having an affinity for keratinized cells, e.g., certain fungi.

Keratitis: Inflammation of the cornea; often caused by *Chl. trachomatis* (see Trachoma).

Koch phenomenon: A phenomenon in experimental tuberculosis described by Koch which demonstrated that acquired immunity to the disease limits the spread of bacilli from a site of superinfection.

Koch's postulates: Criteria for establishing the etiologic agent of an infectious disease, namely, it must: (1) be present in all cases of the disease, (2) be grown in pure culture, and (3) reproduce the disease upon inoculation into new hosts.

Kupffer cells: Fixed macrophages lining the sinusoids of the liver.

Labile: Susceptible to destructive agents or unfavorable environments.

Lactoferrin: An antibacterial iron chelator in neutrophils that restricts the iron needs of ingested organisms.

Latent infection: See Inapparent infection.

Latent period: The interval between penetration of phage into a bacterial cell and cell lysis.

Lepromatous leprosy (LL): The most severe and progressive form of leprosy; LL patients are Mitsuda-negative and lack specific acquired immunity.

Leukocytes: White blood cells, including granulocytes, lymphocytes, and macrophages.

Leukopenia: Abnormally low number of leukocytes in the blood.

Limbs of the immune response (afferent, central, and efferent): The *afferent limb* relates to recognition and processing Ag by macrophages, and recognition of Ag by lymphocytes; the *central limb* relates to the responses of lymphocytes to Ag, including blastogenesis, proliferation, and differentiation leading to the production of Abs and/or "immune cells" (lymphocytes and macrophages); the *efferent limb* relates to the various activities of Abs and immune cells.

L-forms of bacteria: Bacteria that lack intact cell walls; can occur as penicillin-resistant natural variants.

Ligand: A chemical agent that can bind molecules, cells, or other particulates together.
Lipopolysaccharide (LPS): A complex present in Gram-negative bacteria called the O Ag; it possesses endotoxic activity and can serve as a polyclonal B-cell activator; in severe infections it causes fever, shock, and hemorrhage.
Littoral cells: Fixed macrophages that line sinuses within lymph nodes and spleen.
Lobar pneumonia: Pneumonia characterized by large areas of inflammation involving entire lobes or major portions of lobes.
Lockjaw: A symptom of tetanus involving spasmodic contraction of masseter muscles; contraction of lumbar muscles may also occur.
LPF (Lymphocyte-promoting factor): The factor of *Bordetella pertussis* that causes the lymphocytosis of pertussis.
Ludwig's angina: A spreading streptococcal infection in the floor of the mouth with associated abscess formation in the neck.
Lumbar puncture: Puncture of the spinal canal, usually for the removal of cerebrospinal fluid or introduction of medications.
Lyme disease: A newly recognized worldwide disease due to *Borrelia burgdorferi*.
Lymphadenopathy: Any pathologic disorder that affects a lymph gland(s) such as infection and tumor metastases; usually involves enlargement of draining lymph nodes.
Lymphocytes: Principal cells of the immune system comprising two major classes, B cells concerned solely in Ab production and T cells that contribute to both the humoral and CMI responses (see B cells and T cells).
Lymphocyte-promoting factor: A factor produced by *Bord. pertussis* which elicits polyclonal T cell activation and lymphocytosis.
Lymphokines: Biologically active mediators formed by lymphocytes exposed to various stimuli, including foreign Ags and mitogens.
Lymphoreticular system: The system comprising lymphoid tissue and associated components, including reticular cells and reticular fibrils.
Lyophilization: The process of vacuum-drying materials from the frozen state; used for preserving microbes in a dormant state.
Lysins: Agents that can dissolve cells.
Lysogenic conversion of bacteria: A change in a bacterium that results from carriage of a prophage, e.g., the conversion of a nontoxin-producing bacterium to a toxin producer by the action of a prophage.
Lysogeny: See Prophage.
Lysosomes: Small hydrolase-rich cytoplasmic vesicles that fuse with phagosomes to form phagolysosomes where the killing and digestion of engulfed microbes often occurs.
Lysozyme (muramidase): An enzyme that degrades murein, a component of the cell walls of many bacteria; most effective against Gram-positive bacteria.

Maceration: Softening and disintegration of tissue resulting from excessive exposure to a liquid or other agent; e.g., the skin of the hands often become macerated following long exposure to moist conditions.

Macroconidium (pl. macroconidia): The larger of two types of conidia produced by fungi that bear both large and small conidia.

Macrophages: Large mononuclear phagocytes, including immature blood monocytes, wandering macrophages in tissues (histiocytes), and fixed macrophages, such as Kupffer cells of the liver and littoral cells lining sinusoids of the spleen and lymph nodes. Macrophages can transform into epithelioid cells or fuse to form giant cells; they play major roles in innate and acquired immunity.

Macrophage fusion factor: A lymphokine that causes macrophages to fuse to form giant cells.

Macular rash: A skin eruption characterized by flat, patchy lesions (macules).

Malaise: A general feeling of mild discomfort and weakness; often occurs in the early stages of infection.

Mantoux test: A test for tuberculin hypersensitivity conducted by injecting the tuberculin preparation, PPD, intracutaneously (see Tuberculin test).

Mast cells: Specialized tissue cells rich in vasoactive amines; they resemble basophils in form and function and participate in anaphylactic-type reactions.

Medium: A confluent substance; usually a liquid in which particulates such as bacteria are suspended for the purpose of culture (see Culture media).

Memory cells: Dormant long-lived antigen-primed lymphocytes that can rapidly mount a heightened specific immune response.

Menadione: 2-methyl-1, 4-naphthoquinone; a water-insoluble substance with vitamin-K activity.

Meninges: The membrane that envelopes the brain and spinal cord.

Meningocele: A protrusion of the meninges through a defect in the skull or vertebral column, resulting in a cyst filled with CSF.

Mesophiles: Microbes that grow best at temperatures between 20°C and 45°C.

Metachromatic granules: Granules in the cytoplasm of certain bacteria that do not stain true to the color of the dye used, e.g., the red granules in *C. diphtheriae* stained with methylene blue.

Metastatic lesions of infection: Lesions distant from the local site of infection; they arise from blood-borne organisms and are usually small and multiple.

Microaerophile: A microbe that grows best at O_2 tensions below one atmosphere.

Microbes: Organisms of microscopic or submicroscopic size; microbes include principally viruses, bacteria, fungi, and other unicellular plants and animals.

Microbial colonization: Attachment to and growth of microbes on surfaces, e.g., colonization of the teeth by members of the normal flora.

Microconidium (pl. microconidia): The smaller of two types of conidia in fungi that bear both large (macroconidia) and small conidia.

MIF (migration inhibition factor): A lymphokine that inhibits the normal migration of macrophages in culture.

Miliary lesions: Small metastatic lesions that resemble millet seeds in size and appearance, e.g., the lesions of miliary tuberculosis.

Minimum lethal dose (MLD): The smallest dose of an infecting organism that will kill 100% of the animals injected; the dose that will kill 50% of the animals injected constitutes the LD/50.

Mitsuda test: A test conducted by injecting heat-killed *M. leprae* into the skin. A positive reaction is characterized by the development of an allergic granuloma in 3 to 4 weeks. Most normal persons and patients with tuberculoid leprosy have positive reactions, whereas patients with lepromatous leprosy do not.

Molds: A heterogeneous group of fungi that exhibit filamentous growth; a few are dimorphic; i.e., they can exhibit either mold or yeast type growth.

Monoclonal antibody preparation: A preparation containing Abs directed specifically against a single antigenic determinant; commonly derived from a hybridoma.

Monocytes: Immature macrophages in circulating blood.

Monokines: Biologically active mediators produced by macrophages.

Morbidity: The ratio of ill to well individuals in a population; a term generally used with respect to disease due to a single infectious agent.

Morbilliform rash: A skin rash that resembles the fine, rose-red maculopapular rash of measles.

Mortality (fatality rate): The ratio of the number of deaths to the total number of cases of a particular disease in a given population.

M protein: A surface antiphagocytic protein of *Strep. pyogenes* which serves as a virulence factor and protective immunogen; includes many types. The M proteins of *Strep. pneumoniae* are related but distinct.

Muramyl dipeptide: A cell wall component of bacteria which can activate macrophages.

Mutagens: Agents that enhance the rate of natural mutation.

Mutation: A heritable change in the genome of an organism which does not result from the incorporation of genetic material from another organism.

Mutualism: See Parasitism.

Mycelium (pl. Mycelia): The mass of hyphae, often cotton-like, that comprises a colony of mold.

Mycetoma: A chronic infection caused by various fungi and actinomycetes.

Myxomatous: Mucinous, mucoid, resembles the tumor myxoma.

NADase: An enzyme that degrades nicotinamide-adenine dinucleotide.

Natural antibodies: Antibodies formed naturally with or without known exposure of the individual to respective Ags.

Natural host: A host in which an organism naturally perpetuates itself.

Necrosis: The local death of cells or tissues.

Neonate: An infant under 4 weeks of age.

Neutrophil: See Polymorphonuclear neutrophil and Granulocytes.

Nonimmune clearance: Phagocytic clearance of foreign particulates from the blood in the absence of known agents of specific immunity, such as Abs.

Normal microbial flora (microflora): Microbes that normally perpetuate themselves on and in external integuments of essentially every healthy individual of a species; they cause no harm to the healthy host (see Parasitism).

Nosocomial infections: Hospital acquired infections.

Obligate intracellular pathogens: Highly adapted pathogens that can live and grow only within host cells, e.g., *Chlamydia trachomatis*.

Obligate parasites: Parasites that cannot live as free-living forms; includes both pathogens and nonpathogens.
Ophthalmia neonatorum: Acute purulent gonococcal conjunctivitis in the newborn; usually acquired during passage through the birth canal.
Opportunistic pathogen: See Facultative pathogen.
Opsonins: Substances that combine with particulates and, in so doing, promote phagocytosis, e.g., Abs and C.
Osteitis: Inflammation of bone.
Osteomyelitis: Inflammation of bone marrow, adjacent bone, and cartilage.

Pandemic: An epidemic of worldwide or nearly worldwide extent.
PAP: Primary atypical pneumonia caused by *Mycoplasma pneumoniae*.
Papule: A small, firm, elevated lesion in the skin.
Parasite: An organism that lives in or on another organism (see Parasitism).
Parasitism: Interrelations between hosts and parasites including: (1) a state in which the parasite is benefited and the host is neither harmed nor benefited (commensalism); (2) a state in which both parasite and host are benefited (mutualism or "perfect parasitism"); and (3) a state in which the parasite is benefited and the host is harmed (pathogenism). Symbiosis merely means living together, but is sometimes used incorrectly in place of mutualism.
Parenteral (outside the intestine): Parenteral tissues are tissues underlying the surface integuments of the body. For example, materials are injected "parenterally," i.e., subcutaneously, intramuscularly, etc.
Paronychia: Inflammation at the edge of the nailbed, usually due to infection.
Paroxysm: A spasm or convulsion that heralds a worsening of symptoms.
Parrot fever: Psittacosis due to *Chl. psittaci*, a natural pathogen of birds, especially parrots.
Parturition: The act of giving birth.
PAS: Periodic-acid-schiff stain, which imparts a bright-red color to fungal cell walls and other carbohydrate-rich substances; this abbreviation is also used for para-aminosalicylic acid.
Passive hemagglutination: Hemagglutination produced by Abs reacting with Ags adsorbed to RBCs.
Passive immunity: Immunity resulting from the passive acquisition of either "immune serum" or "immune cells."
Pasteurization: Elimination of vegetative pathogens from a material, e.g., a fluid food, such as milk, by a limited period of heating below the boiling point (usually 65°C for 30 min); subsequent preservation is needed.
Pathogen: A parasite that is capable of producing illness in a significant number of healthy individuals lacking acquired immunity.
Pathogenesis: The sequence of events that leads to disease, including modes of injury and development of lesions.
Pathogenism: see Parasitism.
Pathognomonic: Pertains to a symptom or lesion which is so unique that it serves as a diagnostic hallmark of a disease.
Perfect parasitism (mutualism): see Parasitism.
Perinatal period: The period at or near the time of birth.
Periodontal: Surrounding a tooth, as the periodontal membrane; a periodontal abscess involves the side of the tooth root.

Periostitis: Inflammation of the periosteum of bone.

Periplasmic enzymes: Enzymes loosely attached to the outer surface of the cytoplasmic membrane.

Perleche: An inflammatory condition at the angles of the mouth, often caused by infection with *Candida albicans*.

Petechial rash: Tiny, rounded, hemorrhagic spots in the skin or mucosae.

Peyer's patches: Grossly visible patches in the mucosa of the ileum consisting of specialized lymphoid tissues; play an important role in the IgA Ab response to intestinal Ags (see Fundamentals of Immunology, Myrvik and Weiser, 2nd Ed. Philadelphia, Lea & Febiger, 1984, pages 33 to 35).

Pfeiffer's bacillus: *Hemophilus influenzae*.

Phage(s) (bacteriophage(s)): Viruses that parasitize bacteria.

Phage typing: A procedure for differentiating closely related bacteria on the basis of their susceptibility to different phages.

Phagocytes: Cells possessing the specialized function of engulfing particulates; also called "professional phagocytes" to differentiate them from cells that rarely phagocytize particulates (see Endocytosis).

Phagocytosis: The process by which phagocytes ingest particulates and enclose them within phagosomes; the particles may be engulfed by the action of pseudopods or by simple invagination of the cytoplasmic membrane followed by internalization of the phagosome (see Endocytosis).

Phagolysosome: A vesicle resulting from the fusion of a phagosome with a lysosome(s); the site where microbes are often killed and digested.

Phagosome: The cytoplasmic ingestion vesicle of phagocytosis bounded by cytoplasmic membrane (see Endocytosis).

Phenol coefficient: The disinfection capacity of a substance compared with that of phenol as measured under standard conditions.

Phenotype: The outward expression of the genotype.

Phlebitis: Inflammation of veins, usually due to infection.

Phthisis (consumption): A wasting disease; a term commonly used to designate far-advanced tuberculosis.

Phylogenetic: Relates to the evolutionary development of a plant or animal.

Pili or fimbriae: Hairlike appendages of bacteria; hold-fast pili enable bacteria to adhere to surfaces such as mucosae, whereas sex pili of male bacteria enable them to attach to and conjugate with female bacteria.

Pinocytosis (cell drinking): The ingestion of fluid droplets and small particulates by cells; involves ingestion mechanisms similar to those of phagocytosis (see Endocytosis).

Plasma: The fluid portion of unclotted blood.

Plasma cell: Mature Ab-forming B cell; presents an eccentric nucleus.

Plasmids: Circular intracytoplasmic structures in bacteria, consisting of double-stranded DNA that can replicate independent of chromosomal DNA.

Plating: A common procedure used to identify microbes and obtain pure cultures; it consists of distributing specimens on semisolid media in culture plates and incubating the plates until colonies develop for further study. A pure culture can usually be obtained by picking a typical colony and replating it several times.

Pneumococcal shock: A toxemia that sometimes develops early in pneumococcal lobar pneumonia; it may lead to a sudden drop in fever ("pseudocrisis") and the development of life-threatening shock; toxemia is probably due to pneumococcal neuraminidase.

Pneumonic plague: A highly fatal form of epidemic plague involving person-to-person respiratory droplet transmission.

Polymorphonuclear neutrophil (PMN, poly, neutrophil): A highly phagocytic granulocyte of blood characterized by many fine neutrophilic granules in the cytoplasm and high motility (see Granulocytes).

Polyvalent antiserum: An antiserum containing Abs specific for more than one Ag or microbe; often used for prophylaxis or therapy.

Pontiac fever: A mild self-limiting disease caused by variants of *Legionella pneumophilia* and other *Legionella sp.*

Portals of infection: Portals through which infecting microbes breach external integuments to invade underlying tissues.

Pott's disease: Tuberculosis of the spine.

Precipitin: An Ab that can unite with Ag to form a precipitate.

Procaryotae: Members of a kingdom of organisms characterized by nuclei composed of naked DNA.

Prodromal: Pertaining to early symptoms of a disease.

Prophage: A phage chromosome which integrates with the bacterial chromosome and then replicates along with the bacterial chromosome as a silent partner (lysogenization); the prophage renders the bacterial cell immune to superinfection by extracellular phage. On rare occasions (1×10^5), prophage can mature to form lytic phage with resulting host cell lysis.

Protein A: A *Staph. aureus* protein which binds to the Fc segment of human IgG, subclasses 1, 2, and 4, as well as to certain globulins of animals.

Protoplast: A cell that lacks a cell wall.

PRP vaccine: Polyribosylribitol phosphate Ag of type b *H. influenzae.*

Pruritis: Itching.

Pseudomembrane: A "false membrane"; a membrane-like material consisting of fibrin and necrotic material (see Diphtheritic membrane).

Pseudomembranous colitis: An antibiotic therapy-associated disease due to growth of *Cl. difficile* in the intestine.

Psychrophile: A cold-loving organism, e.g., a species of microbe that grows best at temperatures below 20°C.

Puerperal fever (childbed fever): A postpartum infection of the uterus which frequently leads to septicemia and death.

Puerperium: The period immediately after childbirth.

Pure culture techniques: Various techniques and combinations of techniques used for obtaining pure cultures of microbes; they include animal inoculation, filtration, selective physical and chemical destruction, enrichment culture, and the classical methods of plate culture (see Plating).

Purpura: Splotches of hemorrhage in external integuments.

Purulent: Pertains to exudates rich in dead and dying neutrophils (pus).

Pus: The semifluid residue of necrotic cells and tissues liquefied by the enzymes of the numerous dying neutrophils present.

Putrefaction: Decomposition of proteinaceous material with resulting malodorous products.

Pyelonephritis: Inflammation of the kidney including the renal pelvis; usually results from infections of the kidney.

Pyocins: Bacteriocins produced by *Pseudomonas aeruginosa.*
Pyogenic microbe: A microbe that induces the formation of pus.
Pyrexia: Fever.
Pyrogens: Substances that induce fever; endogenous fever-inducing substances liberated from body cells, including neutrophils and macrophages. Exogeneous pyrogens include substances liberated from bacteria that cause fever apparently by injuring cells, with resulting liberation of endogenous pyrogen.

Quellung reaction: Swelling of the bacterial capsule as the result of combining with specific Ab (see Capsular swelling test).
Quinsy: Severe pharyngitis (usually due to *Strep. pyogenes*); characterized by abscess formation and purulent exudates which coalesce to cover the faucial tonsils and peritonsillar areas; frequently constitutes a mixed infection including *Bacteroides* sp. and anaerobic streptococci as well as *Strep. pyogenes.*

R factor: Plasmid that codes for resistance to antibiotics.
RBCs: Red blood cells (erythrocytes).
Reactivation disease: Disease that reappears due to the renewed activity of residual organisms of preexisting disease; sometimes erroneously termed "endogenous reinfection." An example is reactivation tuberculosis due to residual bacilli.
Reagins (reaginic Abs, Wasserman Abs): Anti-tissue IgG and IgM Abs that arise in various diseases, especially syphilis; they react not only with animal lipids but with treponemal Ags as well. The term reagin is sometimes used inappropriately to designate IgE Abs of atopic allergy.
Recirculating lymphocytes: Lymphocytes that recirculate continuously from blood to lymph and back to blood again; they include virgin cells and memory cells largely of the T cell class.
Recrudescence: Reawakening of a disease after a period of improvement or remission.
Reinfection: An infection that occurs after recovery from a previous infection with the same agent. Reinfection is due to exogenous organisms.
Relapse: The recurrence of symptoms of a disease after a period of remission.
Replication: A duplication process requiring a template.
Replicon: A segment of a chromosome that can replicate independently of the chromosome in which it is located.
Reservoir of infection: A source or storehouse of an infectious agent.
Resolving power of the microscope: Power to distinguish between neighboring objects or points.
Reticular cell: The cell that forms reticulin, the component of reticular fibrils. It has a pale-staining nucleus and a fine cytoplasmic membrane that is difficult to visualize.
Reticular tissue: Tissue made up of reticular cells and associated reticular fibrils; found in lymphoid structures, e.g., lymph nodes, bone marrow, and spleen.
Reticuloendothelial system (RES): An ill-defined term, initially used to des-

ignate large mononuclear phagocytes associated with endothelium and reticular tissues, namely, cells of the macrophage group, both mature and immature and both circulating and fixed. They include blood monocytes and fixed macrophages, such as the Kupffer cells of the liver and the littoral cells lining the sinuses of the spleen and lymph nodes. Additional cells with little or no phagocytic activity, which appear to represent part of the RES because they contribute importantly as accessory cells in specific immune responses, comprise the dendritic cells of reticular tissues (see Fundamentals of Immunology, Myrvik and Weiser, 2nd Ed. Philadelphia, Lea & Febiger, 1984, page 147).

RF: Rheumatic fever.

Ribosomes: Cytoplasmic ribonucleoprotein particles which direct cell synthesis of protein in a cell.

Ricewater stool: Cholera stool, composed largely of water and mucus.

Ringworm: A circular lesion caused by a keratinophilic fungi (dermatophytes); the lesion is scaly and itchy and spreads peripherally.

Rose spots: Small red areas in the skin of typhoid fever patients.

Rubelliform rash: Resembling the pale-pink rash of German measles.

Sabouraud's medium: A medium designed especially for growing fungi.

Salpingitis: Inflammation of the fallopian tubes.

Saprophyte: A microbe that normally exists on dead organic material.

Satellite phenomenon: Enhanced growth of *H. influenzae* colonies in proximity to *Staph. aureus* growth.

Scalded skin syndrome (Ritter's disease): A toxic epidermal necrolytic disease due to strains of *Staph. aureus* which produce an epidermolytic exotoxin called "*exfoliatin.*" Ritter's disease occurs most often in young children and immunocompromised adults and is characterized by the formation of bullae and desquamation of epithelium.

Schick test: A test for antitoxic immunity to diphtheria conducted by the intracutaneous injection of a small dose of diphtheria toxin into one arm and a control dose of toxoid into the other arm as a test for possible allergy to the toxin. A "true negative" Schick test indicates immunity, whereas a "true positive" test indicates lack of immunity.

Schwartzman phenomenon: An endotoxin-induced nonimmunologic inflammatory reaction characterized by hemorrhage and necrosis; can be induced by the injection of sequential doses of endotoxin; occurs naturally under special circumstances and may be generalized as well as localized.

Scrofula: Tuberculosis of the cervical lymph nodes due to the drinking of milk contaminated with bovine tubercle bacilli.

Secondary infection (superinfection, intercurrent infection): Commonly used to designate an infection that is superimposed on an existing infection due to another agent. The term has also been used improperly in rare instances in which the superimposed exogenous infection is caused by the same microbial species (reinfection).

Secretory component (secretory piece, transport piece): A 6000 M.W. polypeptide which binds to dimeric IgA to form secretory IgA.

Secretory IgA: An IgA-secretory complex consisting of an IgA dimer of 2 monomers joined by the J chain peptide to which the secretory component becomes bound.

Selective medium: A medium that restricts the growth of unwanted, but not wanted, organisms in a mixture.

Sepsis: Illness due to the presence of pathogenic microbes and their toxins in the body, especially pyogenic microbes.

Septic shock: Shock (often fatal) that results from a state of acute sepsis.

Septicemia: Presence of microbes in the blood with associated sepsis.

Sequela (pl. sequelae): A morbid condition that follows as a consequence of a foregoing disease.

Serology: The study of Abs in serum.

Serotyping: Immunologic procedure using specific antisera to distinguish between closely related organisms (commonly strains within a species); it is based on antigenic differences.

Serum: The clear fluid remaining after blood clots; lacks the plasma elements consumed during clot formation.

Silent infection: see Inapparent infection.

Somatic: Pertaining to the body, in bacteria, somatic Ags are a part of the cell, e.g., the lipopolysaccharide Ags of the cell walls of Gram-negative bacteria.

Spheroplast: A bacterium with a partially deficient cell wall.

Splenomegaly: Enlargement of the spleen.

Sporangiospores: Asexual fungal spores borne in a sac (sporangium) carried on a stalk-like structure (sporangiophore).

Sporulation: The act of forming spores.

S→R mutation: Mutation of bacteria involving organisms that produce smooth (S) and rough (R) colonies.

Sterilant: Any microbicidal agent used for sterilization.

Sterilization: the act of freeing an object or material of all living organisms by removing or destroying them.

Stomatitis: Inflammation of the oral mucosa.

Streptolysin O: An oxygen labile β-hemolysin protein produced by *Strep. pyogenes.*

Streptolysin S: An oxygen-stable β-hemolysin peptide produced by *Strep. pyogenes.*

Subacute infection: An infection of intermediate duration as contrasted to acute infections of short duration and chronic infections of long duration.

Subclinical infection: see Inapparent infection.

Sulfur granules: Gold-colored granules composed of Gram-positive, non-acid-fast, club-shaped masses of radiating bacterial filaments, occurring in the lesions of actinomycosis due to *A. israelii.*

Summer complaint: A term used to designate epidemic shigellosis in infants and children; it is largely fly-transmitted and, if untreated, is often fatal.

Superinfection: See Secondary infection.

Suppuration (pyosis): The formation of pus.

Surface phagocytosis: Phagocytosis which demands that the particle be trapped against a surface by the phagocyte.

Symbiosis: The living together of 2 or more organisms (see Parasitism).

Syndrome: A group of symptoms and signs which, taken together, characterize a disease or lesion.

Synergism: Enhancement by cooperation; e.g., two or more chemicals acting together yield a combined effect that is greater than the algebraic sum of their individual effects.

TAB vaccine: A pool of heat-killed *Sal. typhi, Sal. paratyphoid-A* and *Sal. paratyphoid-B*; provides limited protection against these pathogens.
T cells: Thymus-derived lymphocytes comprising numerous subtypes including, T helper cells, T effector cells, T suppressor cells, and T contrasuppressor cells, etc.; T cells contribute to both humoral and CMI responses (see Lymphocytes).
Tenesmus: Painful but unsuccessful efforts to expel feces or urine.
Terminal host: An individual host that does not transmit a carried parasite to another host.
Therapeutic index: Ratio of the toxic to the effective dose of a drug.
Thermophiles: Microbes that grow best above 45°C.
Thrombocytopenic purpura: Hemorrhages in the skin, mucous membranes, and elsewhere (purpura) resulting from lack of platelets.
Thrombophlebitis: An inflammation of veins leading to the formation of blood clots (thrombi); often due to infection.
Thrombosis: The formation of thrombi (clots) within blood vessels.
Thrush: A fungal infection of the oral mucosa and URT.
Thymus-dependent Ags (TD Ags): Antigens that fail to evoke the humoral Ab response without the help of T cells.
Thymus-independent Ags (TI Ags): Antigens that can evoke the humoral Ab response without the help of T cells but never evoke the CMI response.
Tissue tropism: The property of a parasite that enables it to localize selectively in a tissue; often resulting from a specific affinity of the organism for a certain type of cell in a tissue; commonly due to adhesins.
Toxemia: A toxic state in which toxin is present in the blood.
Tox-gene: The structural gene of corynephage which directs the production of diphtheria toxin.
Toxic shock syndrome: A generalized staphylococcal toxemia characterized by fever, hypotension, skin rash, and desquamation of skin on palms and soles; most frequent in children and menstruating women who use vaginal tampons.
Toxigenicity: The ability of a microbe to produce a toxin.
Toxinogenicity: The ability of a microbe to engender a toxic state in its host.
Toxoid: A modified exotoxin that has been treated to destroy its toxicity but not its antigenicity.
TPI test (treponema immobilization test): A test for syphilis based on an Ab that acts with C to cause immobilization of *T. pallidum*.
Tracheostomy, tracheotomy: Surgical creation of an opening in the trachea.
Trachoma: An ocular disease due to *Chl. trachomatis*, a major cause of blindness in underdevelopd countries; worldwide cases exceed 500 million, including 6 million who are blind.
Transcription: The process of incorporating genetic information into messenger RNA. The incorporation of the DNA message into RNA is by a process that involves base pairing of complementary RNA with the DNA and polymerization of the bases so ordered.

Transduction: Transfer of genes from one bacterium to another by bacteriophages.
Transformation: The transfer of naked DNA from one bacterial cell to another.
Transient microflora: Microbes that propagate transiently on body surfaces.
Transport host: An individual host that transports a parasite from one host species to another.
Transposon: Segment of DNA that is able to replicate and insert one copy at a new site in the genome; it can migrate from one plasmid to another or to a bacterial chromosome or to a bacteriophage independent of the usual recombinant mechanisms.
TRIC agents: The chlamydiae that cause trachoma and inclusion conjunctivitis, including variants of *Chl. trachomatis.*
Tubercle: The granulomatous lesion of tuberculosis.
Tuberculin(s): Preparations derived from culture filtrates of tubercle bacilli; used in skin tests to detect infection with *M. tuberculosis;* old tuberculin (O.T.), a glycerin-containing heated culture filtrate; purified protein derivative (PPD), a preparation derived from O.T. The active components in tuberculins are proteins and protein fragments having molecular weights less than 10,000.
Tuberculin anergy: A state of tuberculin negativity in individuals who were formerly tuberculin-positive; may represent specific anergy as in far-advanced tuberculosis or nonspecific anergy as in individuals superinfected with the measles virus.
Tuberculin "converters": Individuals whose tuberculin test converts from negative to positive; early detection of "conversion" is valuable for determining the time of primary infection among high-risk groups, such as physicians and nurses.
Tuberculin shock: Delayed systemic shock following the injection of tuberculin into a tuberculous individual; shock begins in a few hours and peaks in 15 to 29 hr; fatal shock can be produced in tuberculous guinea pigs and in bygone years sometimes occurred in tuberculous patients on "tuberculin therapy."
Tuberculin test (Mantoux text): An intradermal test with PPD tuberculin. The test is useful as a screening measure for detecting tuberculous infection. A positive reaction indicates that the individual is, or in the very recent past has been infected with *M. tuberculosis.*
Tuberculoid leprosy: A form of leprosy characterized by a positive Mitsuda reaction and an appreciable level of acquired immunity; contrasts sharply with the more progressive form of leprosy, lepromatous leprosy.
Tyndallization (intermittent sterilization): Sterilization of media by heating near the boiling point after several successive intervals of incubation of 12 hours or more to allow spore germination.

Ulcer: A shallow, open-crater-shaped surface lesion in skin or mucosae due to superficial loss of tissue; often displays suppuration.
Undulant fever: Human brucellosis.
Unnatural host (incidental host, accidental host): A host species in which an organism does not naturally perpetuate itself.

URT: Upper respiratory tract.
UT: Urinary tract.

Vaccine: A suspension of organisms (commonly attenuated or killed) used for immunization; often used loosely to designate any material employed for active immunization.

Vaccine encephalopathy: A severe and often fatal complication seen occasionally in children given pertussis vaccine (1 death per 5 to 10 million vaccinated); may be allergic in nature.

Valence and specificity of antigens and antibodies: Since most Ag molecules, especially proteins, carry multiple epitopes of varying configurations, most Ags are both multispecific and multivalent. In contrast, all Ab molecules are monospecific and multivalent, although the valence of Ab molecules of different Ig classes varies.

Vasculitis (angiitis): Inflammation of blood vessels.

VDRL test (Venereal Disease Research Laboratory Test): A flocculation (precipitation) test for syphilis.

Vector: A living agent that transmits pathogens to infect new hosts, e.g., an arthropod that transmits rickettsiae between animal hosts. Conveyance may be mechanical (mechanical vector) or the microbe may multiply, transform, or undergo life cycle changes in the vector (biologic vector).

Vegetative microbe: The growing form of a microbe.

Vincent's disease (Vincent's angina, trench mouth, necrotizing ulcerative gingivitis): Infection of the oral mucosa characterized by ulceration and formation of a gray pseudomembrane; fusiform bacteria and spirochetes are found in abundance in the lesions.

Viridans streptococci: Streptococci that produce colonies on blood agar surrounded by zones of greenish color and associated partial hemolysis; (alpha hemolysis).

Virulence: The relative pathogenicity of different strains of pathogens within a species; avirulence designates lack of virulence.

Virulence factors: Factors that endow a microbe with the property of virulence. Only certain factors of virulence may be evident (key factors).

Walking pneumonia: Mycoplasmal pneumonia due to *M. pneumoniae*: characterized by relatively moderate symptoms of gradual onset, including headache and a persistent nonproductive cough.

Waterhouse-Friderichsen syndrome: A fatal syndrome characterized by bacteremia, seen in acute meningococcemia; it is manifested by massive skin hemorrhage, shock, and acute adrenal hemorrhage and insufficiency.

Weil-Felix test: A differential test for determining etiologic agents of rickettsial diseases using cross-reactive Ags from various strains of *Proteus vulgaris.*

White plague: A term formerly used to designate widespread tuberculosis of epidemic proportions.

WHO: World Health Organization.

Widal test: A specific agglutination test used in the diagnosis of typhoid fever, conducted with a suspension of *Sal. typhi.*

Wild-type strains: Strains of a microbe with the genetic constitution of the majority of strains of that microbe found in nature.
Woolsorter's disease: Pulmonary anthrax due to inhalation of dust from wool, hides, etc. contaminated with *B. anthracis*.

Yaws: A nonvenereal disease caused by *T. pertenue*, an organism closely related to *T. pallidum*.
Yeasts: Unicellular oval-shaped fungi characterized by their mode of asexual reproduction, namely, budding.

Zoonoses: Diseases of animals that are transmissible to man.

Index

Page numbers set in *italics* indicate illustrations; page numbers followed by "t" indicate tables.